MW01383104

Physics: Health and the Human Body

Physics: Health and the Human Body

Daniel R. Gustafson
Wayne State University

Wadsworth Publishing Company
Belmont, California

A Division of Wadsworth, Inc.

Physics Editor: Marshall Aronson
Production: Greg Hubit Bookworks
Designer: Joe di Chiarro
Copy Editor: Paul D. Monsour
Cover and chapter-opening illustrations: Mark Pechenik
Picture research: Research Reports, New York

Photo credits

p. 2, Golden State Warriors; p. 10, Robert Isaacs/Stanford University Medical Center; p. 20, Wide World Photos; p. 21 (top), Stanford University Medical Center; p. 21 (margin), Photo by the author; p. 28, Courtesy of Dover Publications; p. 29, United Press International; p. 46, U.S. Postal Service; p. 47, N.Y. Daily News Photo; p. 53, J. Maser/Peter Arnold Inc.; p. 85, Photo by the author; p. 97, Stanford University Medical Center; p. 115, Frederick B. Merk, Ph.D.; p. 118, L.V. Bergman & Assoc., Inc.; p. 141, Courtesy of Schwinn; p. 145, Wide World Photos; p. 153, Wide World Photos; p. 154, Wide World Photos; p. 156, Alain Dejean/Sygma, Jean Pierre Laffont/Sygma; p. 159, United Press International; p. 166, Mano Foto; p. 167, Photo by the author; p. 168, Prince Manufacturing, Inc.; p. 188, Loma Linda University, photo by Pat Hunter and Scott Roskelly; p. 206, New York University Medical Center; p. 211, Greg Ramsey; p. 213, Michael Weisbrot and Family; p. 219, Lester V. Bergman & Assoc., Inc.; p. 229, Ian G. MacIntyre/Smithsonian Institution; p. 232, New York University Medical Center; p. 236, Lovelace Medical Center; p. 269, H. Gritscher, Peter Arnold, Inc.; p. 280, George Resch, Fundamental Photographs; p. 290, Lester V. Bergman & Assoc., Inc.; p. 297, Steven Halpern; p. 320, John Marmaros/Woodfin Camp & Associates; p. 331, Francis Bitter National Magnet Laboratory; p. 337, Rapho/Photo Researchers, Inc.; p. 372, Fundamental Photographs; p. 407, Linda Hilsen; p. 418, Michael Weisbrot and Family; p. 455, Loma Linda University, photo by Dr. G. Gardiner; p. 456, Michael Weisbrot and Family; p. 472, Stanford Medical Center; p. 473, Loma Linda University, photo by Pat Hunter and Scott Roskelly; p. 492, © Joel Gordon 1979; p. 495, New York University Medical Center.

ISBN 0-534-00756-2

Printed in the United States of America

4 5 6 7 8 9 10—85

Library of Congress Cataloging in Publication Data

Gustafson, Daniel R.
Physics: Health and the Human Body.

Includes index.
1. Physics. 2. Medical physics. I. Title.
QC21.2.G87 530'.02'461 79-17708
ISBN 0-534-00756-2

Contents

Preface

The physics background that the allied health major needs is quite different from that needed by other students. The standard one-year physics course taken by pre-med students is both too long and too mathematical for most allied health professions. Conversely, the courses that introduce nonscience majors to the role of physics in society rarely contain the topics that the allied health major needs to study physiology, body mechanics, and the other subjects that use physics as a foundation.

This book was specifically written to meet the special needs of allied health students, nurses, and physical education majors. The topics in physics that these students will use in physiology have been covered in detail; in addition, most of the discussions clearly show the application of these topics to physiology. By studying these applications, the student should find it much easier to understand the important role of physics in the health professions. As always, bridges are easiest to build when people on both sides of the river cooperate.

The examples are taken mainly from the health professions, as are many of the discussion questions, problems, and experiments. Chapter 18 gives several examples of how physics is actually used in the health professions to reinforce the idea that physics is a useful subject to study. Incidentally, the students should glance at Chapter 18—without worrying at all about the details—before starting the course. This will give them a quick preview of how physics fits into their planned profession.

The student who has had algebra in either high school or college will have little problem with the math in this book, since it is confined to the manipulation of simple equations, powers, square roots, and the areas and volumes of simple shapes. The manipulation of equations is shown in detail the first time an operation is done so that the student who has not had algebra can pick up the basic skills with a little extra work. SI units are used for the most part, but the other units that are commonly used in the health professions are also discussed in detail.

Acknowledgments

I would like to thank everyone who helped me to write this book. In particular, I would like to express my thanks to the Occupational Therapy Department at Wayne State for suggesting such a course and to their students for helping me test much of this material; to William Gustafson, R.N., for his suggestions on the applications of physics in nursing; to Mike Snell and the Wadsworth staff; and to Mary Jo. For their helpful reviews and suggestions, I would also like to thank William T. Achor, Western Maryland College; Milo V. Anderson, Pacific Union College; W. Micque Brown, Edinboro State College; David L. Carleton, Southwest Missouri State University; Jerry H. Fullmer, Utah Technical College; Harold C. Glahe, Niagara University; Susan J. Grow, Vincennes University; Paul Hlayaty, Orange County Community College; Eugene D. Jacobson, Suffolk County Community College; Wilbur D. Kimbrough, Community College of Allegheney County; Clement Y. Lam, North Harris County College; Michael J. Matkovich, Oakton Community College; Sister Gonzaga Plantenberg, Saint Anselm's College; Charles W. Rogers, Southwestern Oklahoma State University; Richard M. Simpson, Grand View College; J. Robert Walker, Lehigh County Community College; Stanley J. Yarosewick, West Chester State College.

D. R. Gustafson

A Note to the Student

If you are curious about how physics relates to your future in the health professions, turn to Chapter 18. You won't understand all the details yet, but you will be able to see how physics is used in the real world of the health professions—the world you will be entering in a short time.

One last suggestion before we start on the subject matter. Do the problems that your instructor assigns. Physics can only be learned by doing physics. Learning physics by only reading and listening is as hard as learning to catch a ball by only reading and listening. The theory is important, but the practice is more important. The concepts of physics are like the tools of a surgeon: With practice, the tools can often be used to save lives, restore health, and improve people's lives. Without constant practice, the tools are but useless bits of metal with strange-sounding names.

CHAPTER ONE

Basic Science and the Health Professions

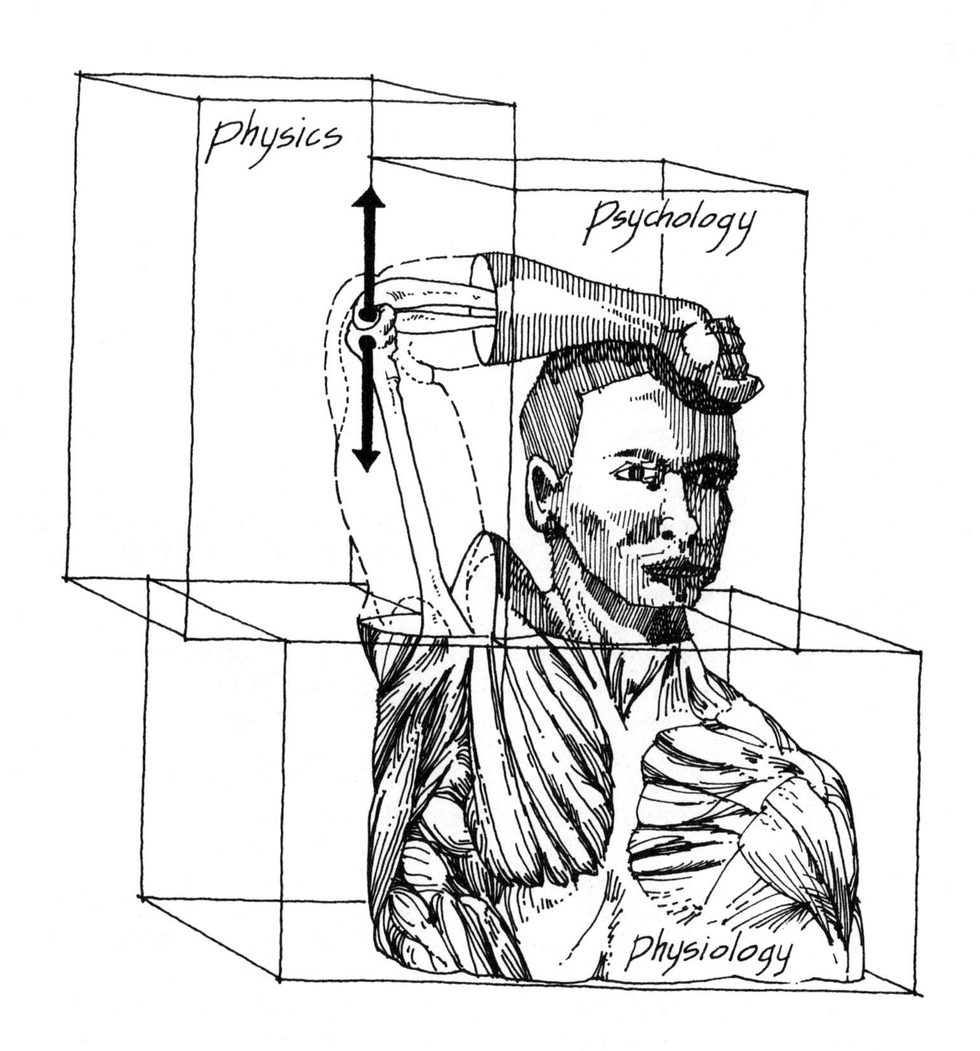

Understanding the laws of physics can help to coach players to move their bodies more effectively.

In helping people to improve their health and physical performance, the health professional uses information from three basic areas of science: physiology, psychology, and physics. Physiology is needed to understand how the parts of the body function, psychology to understand the role of the mind in controlling the body, and physical science to utilize the fundamental laws that all matter—including the matter from which the body is made—must obey. All three areas are essential to the health professional, since the majority of problems in health care require information from all three fields.

When one of the three areas is weak or missing, the results usually show it. As an example, consider the problem of how a basketball coach trains players to jump higher to get more rebounds. Physiology is absolutely essential to help the players develop the strength and endurance needed to play basketball. Motivating the players to train harder, to practice, and to get the utmost out of their bodies in spite of fatigue and discomfort requires a deep understanding of human psychology. But physiology and psychology without a knowledge of the physics of motion will not answer the question, What is the most effective way to make the body jump higher? For instance, how should the player move his or her arms in order to reach as high as possible at the top of the jump? Should the player hold his or her hands overhead before jumping? Or would it be better to raise the hands at the same time that the legs push upward? Or would it be best to raise the hands after the player leaves the floor? These questions can only be answered by applying the physical laws of motion to the specific case of the human body. (The answer to these questions is in the last chapter of the book, as are several other examples of how physics is used in the health professions.)

Physiology, psychology, and physical science are important in the healing arts, too. Consider the rehabilitation of a person weakened by injury or disease. The health professional must understand physiology to recover as much body function as possible; psychology is necessary to motivate the patient to exercise and to recover from the depression of the loss. Finally, teaching the patient to make the best use of the capabilities that remain usually comes down to the purely physical matters of limb leverage, muscle strength, and lifting angles.

In this book we will be concerned mainly with the basic laws of physics and how they relate to the human body and human activity. But I will also try to show how physics relates to physiology and psychology. The relationships among the three are as important as the separate parts, because the human body must always be treated as a whole. All three areas are essential for doing that.

Concepts

Physiological science
Psychological science
Physical science

CHAPTER TWO

Measurement

Educational Goals

The student's goals for Chapter 2 are to be able to:

1. Express measurements in SI units and to use the SI prefixes to form the names of larger and smaller units.
2. Convert non-SI metric units into SI units.
3. Use significant figures, error limits, and percentage errors to communicate the probable error of a measurement.
4. Properly round off the results of calculations.
5. Express large and small numbers using scientific notation.

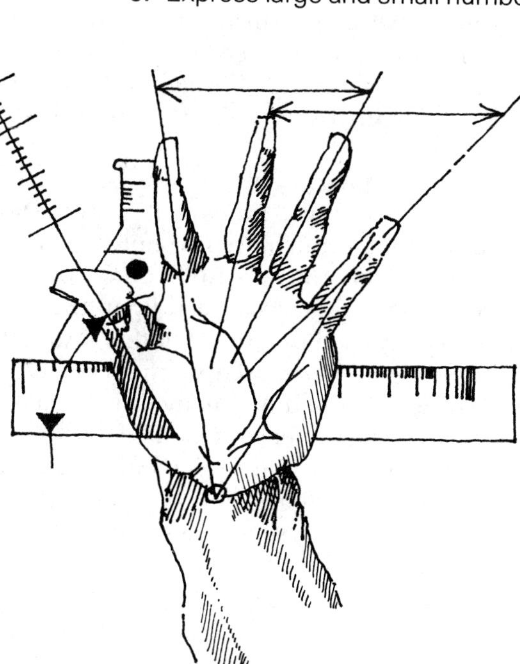

Measurement

The motions of the human body cannot be adequately understood with descriptive words alone. Words such as *high, heavy, fast,* and *slow* can convey only a small part of the information. We need specific facts, such as how high a person's blood pressure is and how fast a person can run to guide our efforts at improving the person's health or performance. That is, we need measurements to describe any motion of the body fully.

Over the centuries, people have used thousands of measurement systems. Length, for instance, has been measured in feet, yards, inches, miles, meters, rods, light-years, barleycorns, ems, bolts, fathoms, furlongs, hands, and cubits, just to mention a few. (Unfortunately, of the list above, all but barleycorns are still in common use.) Different professions often used different units to measure the same thing. When people from these different professions met to work on a mutual problem, they sometimes spent more time trying to straighten out their measurements than working on the problem. Obviously, the solution was to have one system of units used for all purposes throughout the world.

Only now are we close to this solution with the Système International d'Unités—SI for short. Although remnants of earlier systems remain, SI units are preferred for almost all purposes. The three fundamental SI units and their abbreviations are: the **meter** (m) for measuring length, the **second** (s) for measuring time, and the **kilogram** (kg) for measuring mass—which is the amount of matter in an object. (We will see later that the mass of an object is *not* the same thing as its weight. Weight refers to the pull of gravity on an object; like any other pull or push, weight is a force, not a mass.)

The word *kilogram* might seem to be an odd choice for the fundamental unit of mass. Frankly, it is. If we were designing a system of units from scratch, we certainly would not give a fundamental unit a name that means 1,000 times some other unit. However, in 1960, when scientists agreed upon the SI system, the kilogram was so widely used that they felt that it would be easier to live with the slight confusion caused by the term *kilogram* than to get the entire world to adopt a new name for the kilogram. This small but annoying defect may yet be remedied.

In addition to worldwide agreement on the size of the fundamental units, the SI system has two other important advantages. The first is that the SI system is a decimal system: All units are related to each other by multiples of 10. Thus, converting to a larger or smaller unit is simply a matter of moving a decimal point.

The names of units use a standard set of prefixes (see Table 2-1) to indicate size. Thus, a microsecond (μs) is one-millionth of a second, a kilometer (km) is 1,000 meters, a millimeter (mm) is one-thousandth of a meter, a milligram (mg) is one-thousandth of a gram, and so on.

The second advantage of the SI system is not so obvious, but it does even more to simplify calculations. In the SI system, the derived quantities—those other than the fundamental quantities of length, mass, and time—are always directly related to meters, kilograms, and seconds. Thus, we never need a conversion factor in a calculation. (Conversion factors are the multipliers used to convert from one system of measurement to another.)

As an example, consider the SI unit of volume. Volume is equal to *length* × *length* × *length* (length³). For a rectangular container, the volume is simply the product of its length, width, and height. That is,

$$V = lwh.$$

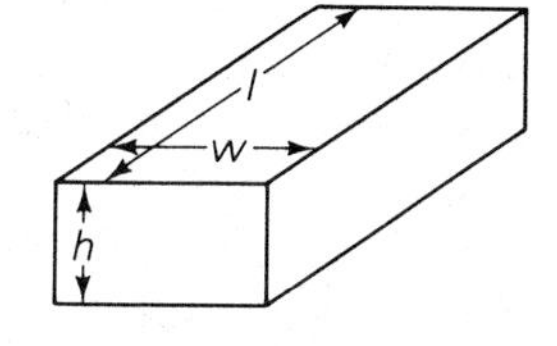

Figure 2-1 $V = lwh.$

In the SI system, we measure the three dimensions in meters. Thus, for a box 0.8 meter long by 0.5 meter wide by 0.03 meter high,

$$V = 0.8 \text{ m} \times 0.5 \text{ m} \times 0.03 \text{ m}$$
$$= 0.012 \text{ m}^3.$$

Table 2-1 Names and Abbreviations of SI Prefixes

Prefix	*Abbreviation*	*Meaning*	*Scientific Notation*
tera-	T	one trillion	(10^{12})
giga-	G	one billion	(10^{9})
mega-	M	one million	(10^{6})
kilo-	k	one thousand	(10^{3})
hecto-	h	one hundred	(10^{2})
deca-	da	ten	(10^{1})
deci-	d	one-tenth	(10^{-1})
centi-	c	one-hundredth	(10^{-2})
milli-	m	one-thousandth	(10^{-3})
micro-	μ	one-millionth	(10^{-6})
nano-	n	one-billionth	(10^{-9})
pico-	p	one-trillionth	(10^{-12})

Note: The preferred multiples are 1,000 and 1/1,000. This is why the prefixes *deca-* and *deci-* are rarely used. This is also why *centi-* and *hecto-* are only used in a few special cases, such as centimeter and hectare.

Only one volume unit eliminates the need for a conversion factor: the cubic meter (m^3). If any other volume unit was chosen—gallon, liter, fifth, or whatever—we could not calculate the volume directly from the measurements; we would have to use a conversion factor. But as long as the dimensions of the box are measured in the SI length unit, the meter, multiplying the length, width, and height automatically gives the volume in the SI volume unit, the cubic meter.

The same procedure was used to choose the other derived units of the SI system. For instance, speed is measured in meters per second (m/s), and the SI unit of area is the square meter (m^2). Thus, calculations are extremely easy in the SI system. Since conversion factors are never needed, the only things that we must do are (1) to make sure the quantities we start with are in SI units and (2) to use the correct formula. The answer will be in the correct SI unit. Remember that this always works automatically *only* for SI units. Non-SI units will usually require a conversion factor.

In doing a problem, a good procedure is to check that the quantities are in SI units; if they aren't, change them to SI units. For example, if we had a problem involving a 100-gram muscle, a bone 20 centimeters long, and a time of 2 minutes, we would express the 100 grams as 0.1 *kilogram,* the 20 centimeters as 0.2 *meter,* and the 2 minutes as 120 *seconds.*

For many physical quantities, the relationship to the fundamental units is quite complicated. Their complete names, in terms of the fundamental units, can also be complex. However, there is a simple solution to this problem: The complicated unit gets a simple name of its own. The SI unit of power is a good example:

$$\text{power} = \frac{\text{mass} \times \text{length}^2}{\text{time}^3},$$

so the SI unit of power is

$$\frac{\text{kg}\cdot\text{m}^2}{\text{s}^3}.$$

But the $kg \cdot m^2/s^3$ is given the simple name *watt* (W), sparing us from having to ask for a 100-kilogram-meter-squared-per-second-cubed light bulb. Most of the SI units have such simple names; a list of their names and abbreviations can be found in Appendix I.

Many often used non-SI units also have simple names. One example is the *liter* (l). The SI unit of volume, the cubic meter, is too large for most everyday purposes. The milli–cubic meter is a very convenient size, but not a very convenient name. Thus, the milli–cubic meter is called the *liter.* When you are doing calculations, you should keep in mind that the liter is *not* the SI unit of volume. If the volume is given in liters, you should convert it to cubic meters by dividing by 1,000.

The developers of the metric system tried to adjust the size of the units to make the mass of 1 liter of water equal to exactly 1 kilogram. They missed, but only by about one part out of a million. For all ordinary purposes, we can say that a liter of water (at 4°C) has a mass of 1 kilogram. Of course, this relationship is true only for water. A liter of another liquid will usually have a different mass. A liter of mercury, for instance, has a mass of more than 13 kilograms.

Another volume unit that you will often run into is the milliliter (ml). The milliliter is equal to 1 cubic centimeter (cm^3, or cc), thus in SI units the milliliter is one-millionth of a cubic meter.

Table 2-2 Common Non-SI Units

Unit Name	*Abbreviation*	*Value in SI Units*	*Physical Quantity*
liter	l	$0.001\ m^3$	volume
milliliter or cubic centimeter	ml, cm^3, cc	$0.000001\ m^3$	volume
are	a	$100\ m^2$	area
hectare	ha	$10{,}000\ m^2$	area
angstrom	Å	10^{-10} m	length
minute	min	60 s	time
hour	hr	3,600 s	time
year	yr	31.6×10^6 s	time
gram	g	0.001 kg	mass
tonne or metric ton	t	1,000 kg	mass

Dosages of liquid medications are usually measured in milliliters (or cm^3, or cc). Since a liter of water has a mass of 1 kilogram, a milliliter of water has a mass of 1 gram.

Accuracy in Measurement and Calculations

No measurement is ever totally accurate. The error might be very small, as it is when we use high-precision instruments, or it might be large, as it is when we make an eyeball estimate of a quantity. But there is always an error.

In many cases, the size of the error is just as important as the value of the measurement. For instance, a nurse would be concerned about any patient whose temperature was reported to be 39°C (about 102°F); however, that nurse would like to know whether the temperature was taken with a clinical thermometer that was accurate to a tenth of a degree or with a common household thermometer of questionable accuracy. Similarly, a track coach would be excited to hear of a young athlete who could run the 100-meter dash in 11 seconds. But the coach would lose a bit of enthusiasm if the time had been measured by a bystander counting "one thousand one, one thousand two, . . . , one thousand eleven" instead of with a stopwatch.

When a rough indication of the error is enough, measurements are written down in the following way: Starting on the left, as usual, write down the digits until you come to the first digit that you are not completely sure of. This last, doubtful digit is written down, but from that point on you just write zeros until you get to the decimal point. If the doubtful digit is already on the right-hand side of the decimal point, you just stop. For example, if you measured a sprinter running 100 meters in 12 seconds by counting "one thousand one, one thousand two, . . . , one thousand twelve," you would write the time as 12. seconds. This indicates that you are fairly sure that the time is more than 11 seconds, but you aren't sure whether the time was really 11, 12, or 13 seconds. That is, the 2 next to the decimal point is in doubt. If you measured the sprinter's time with a stopwatch that had tenth-second accuracy, the time would be written as 12.2 seconds to indicate that you are sure of the 12 seconds but not of the 2 one-tenths of a second. If you timed the sprinter with an electronic timer accurate to a millisecond, the time would be written as 12.197 seconds to indicate that you are sure of the 12 and 19 one-hundredths seconds but not of the 7 one-thousandths of a second. Basically, 12.197 seconds says that the time was between 12.196 and 12.198 seconds.

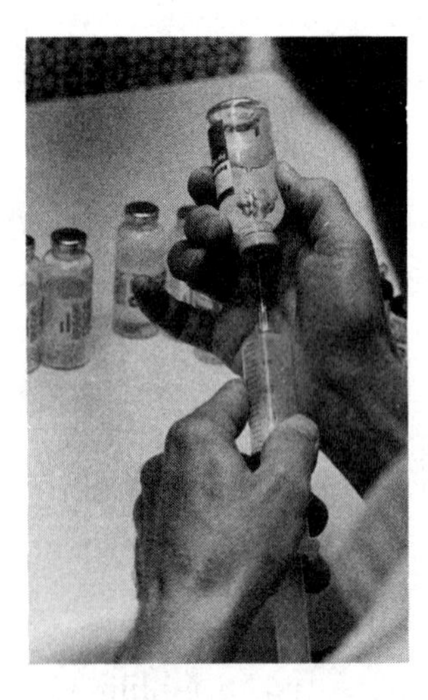
Accuracy is essential in the measurement of drugs.

The digits that are accurate and the last, doubtful digit are called significant digits (or significant figures) because they give significant information about the value of the measurement. Thus, 12. is said to be accurate to two significant figures, 12.2 is accurate to three significant figures, and 12.197 is accurate to five significant figures. The zeros that just locate the decimal point are not significant figures; thus, both 0.00017 and 23,000,000 have two significant figures.

EXAMPLE If you measured the length of a hypodermic needle with a common meter stick accurate to the nearest millimeter, how should the result be written: 0.02 meter, 0.023 meter, or 0.0232 meter?

Answer The length should be written as 0.023 meter, to indicate that the measurement is accurate to one-thousandth of a meter, that is, to the nearest millimeter. Note that 0.023 has two significant figures, not three or four. The zeros only locate the decimal point. We could alternately write the answer as 23 millimeters or 2.3 centimeters; in both cases, it is clear that there are just two significant figures.

Sometimes the amount of error must be known. For instance, some drugs must be given in very accurate doses. There are two common ways to describe the error in a measurement. The first is to attach plus and minus figures. For example, 101 ± 2 means the result is more than $101 - 2 = 99$ but less than $101 + 2 = 103$. Similarly, 375 ± 3 means the measurement is between 372 and 378.

When a measurement is made with a random sampling method, one can never be 100% sure that the measurement is within certain limits. For instance, a measurement such as 101 ± 2 based on a random sampling means that there is better than a 66% chance that the result is between 99 and 103.

The second common way of describing the error is to give the percentage error. For instance, saying that the measurement is 101 to within 2% means that the measurement is between $[101 - (101 \times 2\%)] = 99$ and $[101 + (101 \times 2\%)] = 103$. Saying the result is 470 to within 1% means the result is between 465 and 475.

Pocket calculators give answers to eight or ten significant figures nearly as fast as you can push the buttons. But when the numbers you start with are only accurate to two or three significant figures, it doesn't make much sense to give the answer to ten significant figures. For instance, suppose you had roughly 3 liters of soda pop and you divided it among seven children without any special care. How much soda pop does each child get? A calculator says that each child gets

$$\frac{3}{7}\ 1 = 0.428571429\ 1.$$

But we know perfectly well that no child got exactly 0.428571429 liter. What we do know is that each child's portion was probably between 0.3 liter and 0.5 liter; thus, it would make more sense to write the answer as 0.4 liter of soda pop per child.

The general rule for rounding off the answers to multiplication and division problems is to round off so that the answer has the same number of significant figures as the *least* accurate input number.

Before we see where this rule came from, you should note that there is a difference between numbers that come from measurements and numbers that come from counting. For instance, in the example of dividing about 3 liters of soda pop among 7 children, the 3 liters is a measurement that is accurate to one significant figure. The 7 is exact—it comes from counting the kids, and there are exactly 7. Thus, numbers that come from exact counts do not contribute errors.

Let's try a sample calculation to see the reason for the rounding rule. Suppose you measured a room to be 3.8 meters wide by 7.2 meters long. The width of 3.8 meters means that the room was between 3.7 and 3.9 meters wide, and the length of 7.2 meters means that the room was between 7.1 and 7.3 meters long. What is the area of the floor?

$$\begin{aligned} A &= lw \\ &= 7.2 \text{ m} \times 3.8 \text{ m} \\ &= 27.36 \text{ m}^2, \end{aligned}$$

which implies that you know the area is probably between 27.35 and 27.37 m². But you know no such thing. The area could be as small as

$$\begin{aligned} A_{\text{small}} &= 7.1 \text{ m} \times 3.7 \text{ m} \\ &= 26.27 \text{ m}^2 \end{aligned}$$

or as large as

$$\begin{aligned} A_{\text{large}} &= 7.3 \text{ m} \times 3.9 \text{ m} \\ &= 28.47 \text{ m}^2. \end{aligned}$$

Since all we know is that the area is between about 26 m² and 28 m², it would be appropriate to round off the answer to 27 m². That is, we round the answer off to two significant figures, the same number of significant figures that are in the measured length and width.

Scientific Number Notation

The ordinary way of writing numbers becomes quite awkward for very large and very small numbers. For instance, there are about 100,000,000,000,000,000,000 atoms in a sand grain, and each atom is approximately 0.0000000003 meter in diameter. Both these numbers are hard to write and hard to understand.

A better way of writing large and small numbers is with scientific number notation. The basic idea of scientific notation is simple: Instead of writing multiples of 10 as a 1 followed by a long string of zeros, we write the number as a power of 10. For example,

$$100 = 10 \times 10 = 10^2$$

$$1{,}000 = 10 \times 10 \times 10 = 10^3$$

$$1{,}000{,}000 = 10 \times 10 \times 10 \times 10 \times 10 \times 10 = 10^6$$

$$1{,}000{,}000{,}000 = 10 \times 10 \times 10 \times 10 \times 10 \times 10 \times 10 \times 10 \times 10 = 10^9.$$

Thus, the number of atoms in a sand grain, 1 followed by 20 zeros, is 10^{20} (the little superscript number to the right of the 10 is called the exponent).

What about numbers that are not multiples of 10—how do we write them in scientific notation? No problem. Any number can be broken down into the product of a small number and a power of 10. For instance, 2,300,000 can be written as

$$2.3 \times 1{,}000{,}000,$$

and the 1 million can be written as 10^6. Thus, in scientific notation,

$$2{,}300{,}000 = 2.3 \times 10^6.$$

If we wanted to, we could write 2,300,000 as 23×10^5 or 0.23×10^7, but the usual choice in scientific notation is to have one digit to the left of the decimal point. The following are some more examples of scientific notation:

$$3{,}500 = 3.5 \times 10^3$$

$$532{,}000{,}000{,}000 = 5.32 \times 10^{11}$$

$$91{,}000{,}000 = 9.1 \times 10^7$$

$$110{,}000 = 1.1 \times 10^5$$

Scientific notation has another advantage: The significant digits are separated from the zeros that locate the decimal point. In ordinary notation, whether a zero is significant or whether it just locates the decimal point is never clear. For example, if we carefully measured the length of a large building to the nearest meter and found it to be 200 meters long, the ordinary way of writing the result, 200 meters, doesn't tell us that both zeros are significant. We would tend to think that the 200 meters is just an estimate. In scientific notation there is no question. If we mean the length is 200 $\pm$ 1 meter, we write it as 2.00×10^2 meters. If we meant *about* 200 meters, we would write 2×10^2 meters.

Very small numbers can also be written in scientific notation. Recall that a negative exponent means division by that power, as in the examples below:

$$10^{-1} = \frac{1}{10} = 0.1$$

$$10^{-2} = \frac{1}{10 \times 10} = 0.01$$

$$10^{-3} = \frac{1}{10 \times 10 \times 10} = 0.001$$

$$10^{-6} = \frac{1}{10 \times 10 \times 10 \times 10 \times 10 \times 10} = 0.000001.$$

Again, any number can be broken down into the product of a small number and a power of 10. So 0.002 would be 2×0.001 or 2×10^{-3}, $0.0000536 = 5.36 \times 0.00001 = 5.36 \times 10^{-5}$, $0.0728 = 7.28 \times 10^{-2}$; the size of an atom, 0.0000000003 meter, is 3×10^{-10} meter.

We'll skip the details of how to do calculations with scientific notation, since we won't be doing very many. But we should note that multiplication and division are easy to do. Multiplication of powers of 10 is done simply by adding exponents. Thus,

$$10^8 \times 10^7 = 10^{8+7} = 10^{15}$$

$$10^3 \times 10^{-9} = 10^{3-9} = 10^{-6}$$

and so on. To multiply two numbers in scientific notation, you just regroup the numbers so that the powers of 10 are together, then multiply the ordinary numbers in the ordinary way and multiply the powers of 10 by adding the exponents. For example,

$$\begin{aligned}(3.4 \times 10^9) \times (1.8 \times 10^{-6}) &= (3.4 \times 1.8) \times (10^9 \times 10^{-6}) \\ &= (6.1) \times (10^{9-6}) \\ &= 6.1 \times 10^3\end{aligned}$$

(rounding off to two significant figures)

Division involves subtracting exponents:

$$\frac{10^4}{10^2} = 10^{4-2} = 10^2$$

$$\frac{10^3}{10^9} = 10^{3-9} = 10^{-6}$$

and so on. Thus, division in scientific notation is similar to multiplication; for instance,

$$\frac{9.7 \times 10^4}{3.2 \times 10^{-5}} = \left(\frac{9.7}{3.2}\right) \times \left(\frac{10^4}{10^{-5}}\right)$$

$$= 3.0 \times 10^{4-(-5)}$$

$$= 3.0 \times 10^9 \quad \text{(rounding off to two significant figures)}$$

Concepts

Measurement
Units (of measurement)
Système International d'Unités (SI)
Mass
Length
Time
Conversion factor
Accuracy
Error
Significant figures
Percentage error
Rounding answers
Scientific notation
Powers of 10
Exponent

Discussion Questions

1. Describe the main advantages of the SI metric system over other measurement systems such as the English system.
2. Since the derived units of the SI system were chosen to free us from conversion factors, the units are sometimes too large or too small for everyday measurements. Which do you think is more useful, getting rid of conversion factors or having a convenient size for the units? Explain why. Is there any way to have both?
3. Contrary to some people's opinion, the goal of science is not to measure everything as accurately as possible. We want to measure quantities as accurately as needed for the intended purpose. For instance, explain why it would be pointless for a physician, nurse, diet counselor, or athletic coach to measure a person's mass to the nearest milligram and height to the nearest micrometer.

4. Give some examples of measurements that must be made very accurately and other examples where rough estimates are good enough.

5. Explain why very large and very small numbers are easier to write and understand when expressed in scientific notation rather than in the ordinary way.

Problems

1. The advantages of a decimal system without conversion factors are most apparent when we try a problem in a system that has neither advantage: the English system. How many gallons of water are needed to fill a bathtub 2 yards long, 2 feet 5 inches wide, and 7 inches deep?

   Now see how much easier a problem is in the SI system. How many cubic meters of water are needed to fill a bathtub 2 meters long, 0.75 meter wide, and 0.2 meter deep?

2. Express the following quantities in the proper SI units:
   a. 27 centimeters
   b. 2,500 grams
   c. 1 day
   d. 5 minutes
   e. 50 milligrams
   f. 100 liters

3. Use prefixes with the units to express the following quantities more conveniently:
   a. 0.002 meter
   b. 50,000 grams
   c. 70,000,000 watts
   d. 0.000,000,01 second
   e. 0.01 liter

4. What is the mass of 3 milliliters of water?

5. Estimate the error in a measurement made as follows:
   a. with an ordinary meter stick
   b. with a clinical thermometer
   c. with a cheap outdoor thermometer
   d. with an ordinary bathroom scale
   e. with a beam-balance scale, such as is used in hospitals and doctor's offices
   f. by counting "one thousand one, one thousand two," etc. (for about a minute)
   g. with a coach's stopwatch
   h. by pacing off the distance (for distances under 20 meters)

6. How many significant figures do the following numbers have?
   a. 7.5
   b. 9,720,000

c. 9,720,000.1
d. 0.0009
e. 0.00374
f. 90.003
g. 81.030
h. 503.

7. Convert the following plus and minus errors to percentage errors:
a. 978 ± 9
b. 4.2 ± 0.6
c. 295.1 ± 0.7
d. 27 ± 8

8. Round off the results of the following calculations to the proper number of significant digits:
a. $1.71 \times 37 =$
b. $9{,}756 \times 0.027 \div 0.8 =$
c. $32{,}000 \div 969.77 =$
d. $75 \times 0.01385 \times 107 \div 69.305 =$
e. $7.35 \times 0.0983 \div 8.3 =$

9. Express the following numbers in scientific notation:
a. 83,000
b. 0.000027
c. 0.0034
d. 37,000,000,000
e. 3,291
f. 0.7

10. Carry out the following calculations using scientific number notation.
a. $(3.7 \times 10^5) \times (2.7 \times 10^{-3})$
b. $(8.2 \times 10^8) \div (6.1 \times 10^6)$
c. $(5.4 \times 10^5) \times (7.2 \times 10^{-8})$
d. $(8.8 \times 10^{-6}) \div (9.3 \times 10^{-2})$

Experiment

1. The ability to make a good eyeball estimate of a quantity is often just as important in the health sciences as the ability to make very accurate measurements. Try to develop your estimation skill by estimating the height and mass of a friend, the time length of a verse of a song, the temperature of warm tap water, the mass of this book, and so on. Next measure these things and calculate the percentage error in your estimates. With a little practice, you should be able to make estimates that are within 20% of the measured values.

CHAPTER THREE

Forces

Educational Goals

The student's goals for Chapter 3 are to be able to:

1. Recognize vector and nonvector quantities.
2. Explain the difference between mass and weight.
3. Calculate the weight of objects of known mass.
4. Use Newton's first law of motion to determine whether an object has a net force acting on it.
5. Identify all the outside forces acting on an object.
6. Add parallel forces.

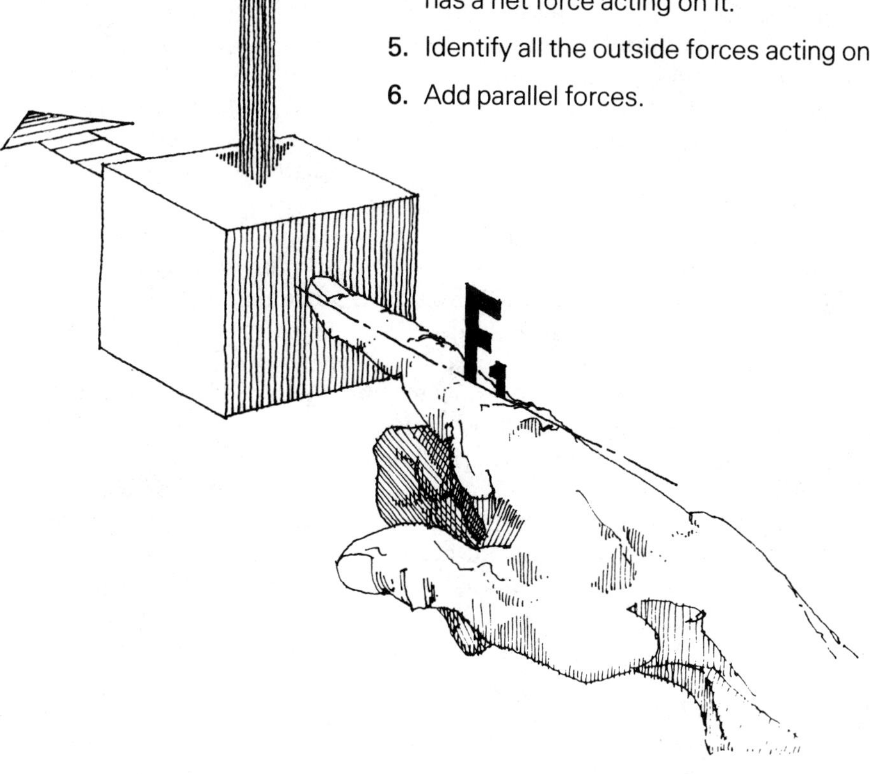

This gymnast must control the forces of his body with split-second precision.

Forces are important in every human activity. The muscles produce forces, which the bones and tendons transmit to other parts of the body and to other objects. Even the simplest motion involves several muscles, all producing forces that move the body with millimeter accuracy and split-second coordination. Of course, most healthy people take all this for granted—they have long since forgotten how hard it was learning how to walk, run, write, and talk. But if we watch a stroke victim struggling to write again, or a champion gymnast perfecting a turn, we quickly appreciate the complexity of the forces inside the body.

Helping people to control the forces within their bodies is a central part of the health professions. Because the body is so complex, this is not an easy task—particularly since no two people are exactly alike. Adapting a method of exercise or muscle control to individual peculiarities demands a sound knowledge of the mechanics of the human body, which must start with the fundamental laws of force, motion, and matter.

The physical quantity *force* can be defined in several ways, but the simple definition is a good one: A force is a push or a pull. Clearly, a push or a pull is neither a length, nor a mass, nor a time. Therefore, it must be measured in units of its own. The SI unit of force is the **newton** (N), honoring the discoverer of the laws of motion, Sir Isaac Newton (1642–1727). Like all derived units, the newton can be expressed as a combination of the fundamental units:

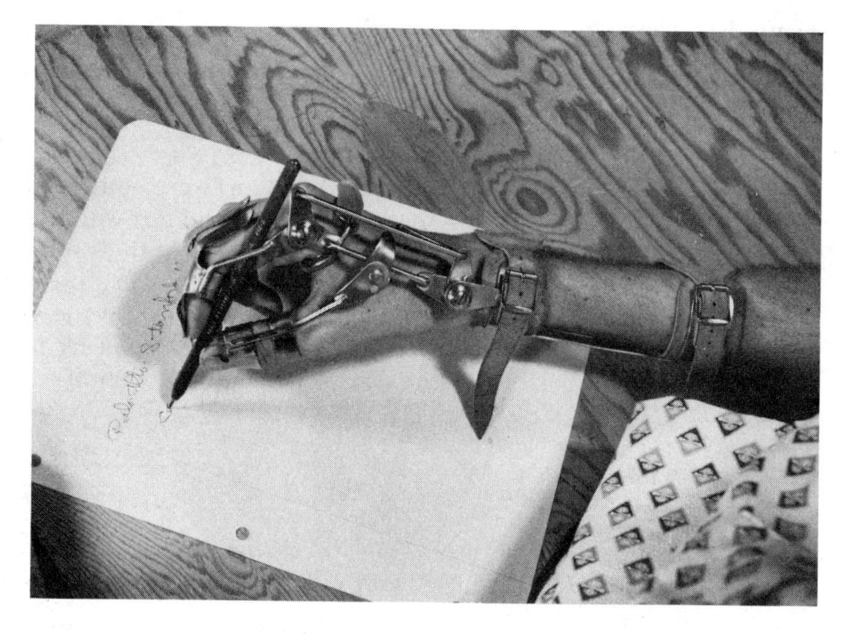

The designer of this hand brace clearly had a sound knowledge of the mechanics of the human body.

A six-pack of soda pop weighs about 23 newtons.

$$1 \text{ newton} = \frac{1 \text{ kilogram} \cdot \text{meter}}{\text{second}^2}.$$

We will understand this relationship better when we study motion, but as long as we use SI units, we need not remember the precise relationship.

The newton is about the right size for discussing the forces of human activity. For instance, you use about 1 newton of force to lift a small paper cup full of water and about 20 newtons to push moderately with your little finger. A very strong person can produce a force of about 3,000 newtons while pushing a stuck car or lifting a heavy weight.

Forces and Direction

To describe a force fully, we must give its *direction* as well as its size. For example, to throw a ball to someone, *both* the size and direction of the throwing force must be right.

Quantities that have both size and direction, such as forces, are called *vector* quantities. The properties of vectors are somewhat different from those of nonvector quantities, which are technically known as *scalar* quantities. The most important difference is that vectors do not add and subtract in the ordinary way. For instance,

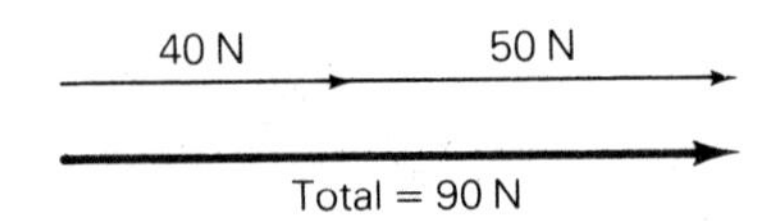

Figure 3-1 The vector sum of a 40-N force and a 50-N force is 90 N *only* when the two forces push in the *same* direction.

suppose a boy and a girl were both pushing on a chair, the boy with a 40-newton force and the girl with a 50-newton force. We would like to know what the sum of their forces is, but we can't unless we know in what directions they are pushing. The sum could be 90 newtons, but only if they push in the same direction (Figure 3-1). The sum could also be as small as 10 newtons if they push in opposite directions (Figure 3-2), or it could be anything between 10

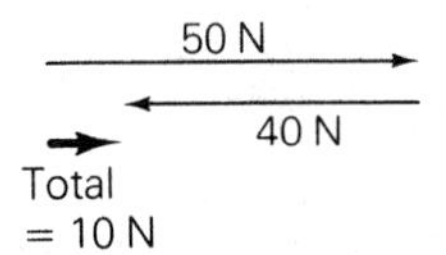

Figure 3-2 The vector sum of a 50-N force and a 40-N force is 10 N when the two forces push in opposite directions.

newtons and 90 newtons. Some examples of the vector addition of a 40-newton force and a 50-newton force are shown in Figure 3-3. We will look into the details of vector addition in the next chapter.

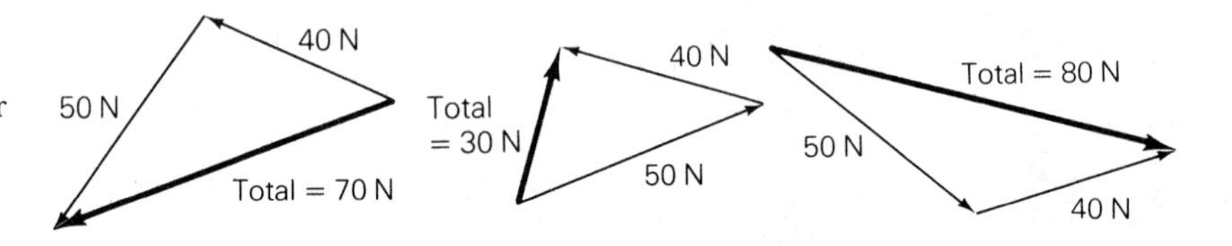

Figure 3-3 Three examples where the vector sum of a 50-N force and a 40-N force is between 10 N and 90 N.

We have to use vector addition to discuss forces in the body because the forces of the muscles and bones seldom act in the same direction. Consider the foot, shown at one instant while running in Figure 3-4. The Achilles tendon is pulling upward in one direction, the bones of the leg push in another direction, and the floor is pushing in yet a third. Since the three forces are all acting in different directions, we can't add them in the ordinary way; we must use vector addition.

To determine whether a particular physical quantity is a vector, all we have to do is ask the question, "Is the direction needed to describe this quantity fully?" If the answer is yes, the quantity is a vector. If no, the quantity is a nonvector (scalar quantity).

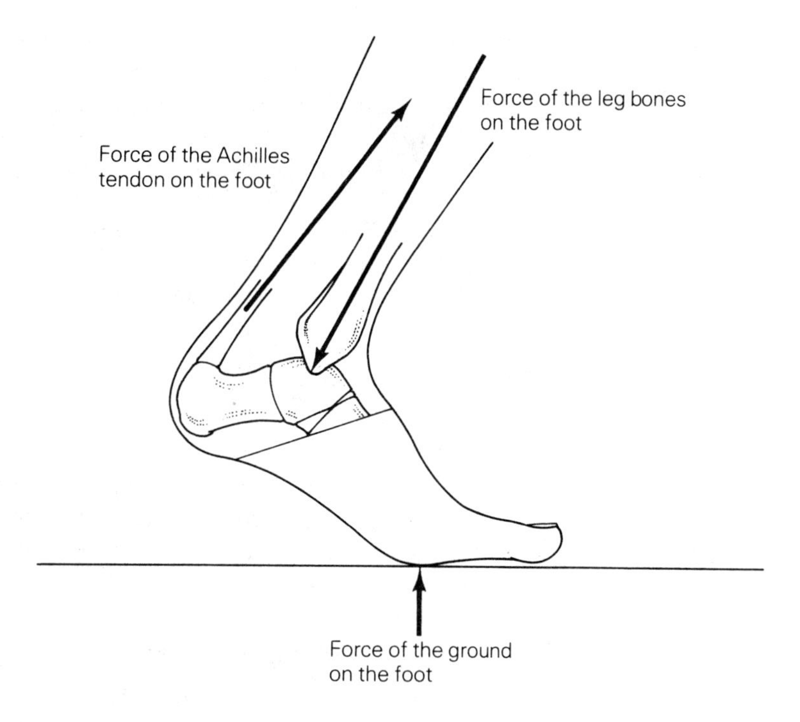

Figure 3-4 The forces on the foot while running.

Gravity and Weight

The weight of an object is defined as the *force* of gravity on that object. Thus, an object's weight is a completely different thing from its mass, which is the amount of material that the object contains. In fact, weight is a vector quantity (acting downward), while mass is a nonvector. Consequently, these quantities are measured in different units: Since weight is the force of gravity, it is measured in newtons, while mass is measured in kilograms.

Unfortunately, there is a great deal of confusion about weight and mass. The confusion arises because the words *mass* and *weight* are used interchangeably in everyday conversation. This isn't surprising; after all, our language is far older than the laws of motion. Naturally, before the laws of motion were understood, it was not obvious that weight and mass were two very different things.

Mass and weight are also confused because there is a direct relationship between them *on or near the surface of the earth.* We know that the more massive an object is, the more it will weigh, and vice versa. Two objects with the same mass will have the same weight provided that *they are in the same place on earth.* However, the weight of an object is *not* the same everywhere. It will vary slightly from place to place on the earth, it will be drastically smaller on the moon, and it will be zero in deep outer space. The mass of that object, by contrast, will be the same everywhere; the amount of material in the object won't change. Consider a bone with a mass of one kilogram. That bone will have a mass of one kilogram no matter where it is in the universe. Its weight, however, will vary considerably as it is moved from place to place. It would weigh 9.78 newtons in Egypt, 9.83 newtons at the North Pole, 1.63 newtons on the moon, 23.2 newtons on Jupiter, and zero in far outer space.

Although the weight of an object varies from place to place on the earth, the variation is quite small—less than 1%. Within this accuracy, the weight of an object on or near the surface of the earth can be found by multiplying the mass of the object by 9.8 m/s². This quantity is called the acceleration due to the earth's gravity, abbreviated in formulas as the letter g. That is,

$$g = 9.8\ \text{m/s}^2.$$

Thus, for any object near the surface of the earth.

$$\text{weight} = \text{mass} \times 9.8\ \text{m/s}^2;$$

or, expressed as a formula

$$W = mg.$$

We will see why the acceleration due to the earth's gravity has the size and units that it does when we study motion.

EXAMPLE What is the minimum force required to lift a 5.0-kilogram child?

Answer The force would have to be at least as large as the weight of the child, which is

$$\begin{aligned} W &= mg \\ &= 5.0\ \text{kg} \times 9.8\ \text{m/s}^2 \\ &= 49\ \text{kg}\cdot\text{m/s}^2 \\ &= 49\ \text{N}, \end{aligned}$$

since $1\ \text{kg}\cdot\text{m/s}^2 = 1$ newton. Of course, the force would have to be in the upward direction. Note that the answer automatically comes out in the SI unit of force, the newton, because we started with quantities expressed in SI units.

EXAMPLE Figure 3-5 shows a patient in Buck's traction. Suppose we wanted the rope to apply a force of approximately 20 newtons to the patient's foot. What mass should we hang on the other end of the rope?

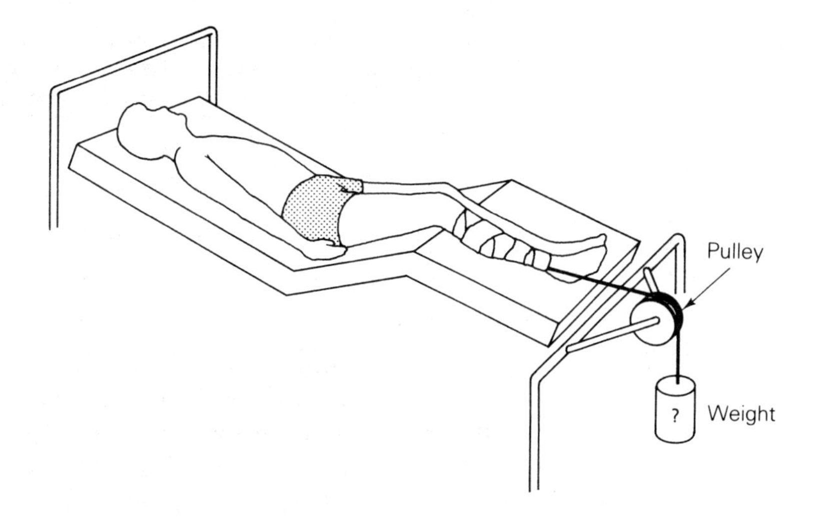

Figure 3-5 Buck's traction.

Answer As long as the pulley is reasonably free of friction, the rope will transmit the full weight of the suspended mass to the foot. That is, a friction-free pulley changes the direction but not the size of the force of the rope. Thus, the weight of the mass should be about 20 newtons. The mass that would weigh 20 newtons can be found by a slight rearrangement of the formula

$$W = mg.$$

Since we want to find m and not W, we divide both sides of the equation by g, which gives

$$\frac{W}{g} = \frac{mg}{g},$$

The two g's on the right side of the equation cancel,

$$\frac{W}{g} = \frac{m\not{g}}{\not{g}},$$

so we find that

$$\frac{W}{g} = m.$$

This can be turned around to give

$$m = \frac{W}{g}.$$

In our case,

$$m = \frac{20\ \text{N}}{9.8\ \text{m/s}^2}$$

$$m = 2.04\ \text{kg}.$$

We wanted a force of approximately 20 newtons; there is no need to produce exactly 20 newtons. Thus, we would hang a 2-kilogram mass on the end of the rope, which produces a force on the rope of

$$\begin{aligned} W &= mg \\ &= 2\ \text{kg} \times 9.8\ \text{m/s}^2 \\ &= 19.6\ \text{N}, \end{aligned}$$

which is close enough to 20 newtons for our purposes.

The force transmitted by a rope is referred to as the *tension* in the rope. For instance, if we pull on a rope with a force of 100 newtons, the tension in the rope is said to be 100 newtons. As long as no other outside forces act on a rope, the tension will be the same everywhere. That is, if we pull with a force of 50 newtons on one end of the rope, the rope will pull with a force of 50 newtons on whatever the other end is tied to. One outside force that we cannot eliminate is the weight of the rope, but modern, high-strength ropes are so light compared with the forces that they can transmit that the weight of the rope can be ignored in most practical problems.

Forces and Motion

Forces and motion are closely related. That is, we know that forces can cause objects to move. However, the exact relationship is not obvious, because we must consider *all* forces acting on an object, not just the force that seems to be causing the motion. For instance, consider a wagon being pulled along the sidewalk by a child. What forces are acting on the wagon? The pull of the child, of course; but that is only the most obvious. The pull of gravity downward, the force of the sidewalk that holds up the wagon, and friction with the sidewalk, which pushes backward and makes the wagon hard to pull, must be taken into account, too (Figure 3-6).

The frictional force is especially troublesome, and it stumped everyone until the time of Newton. We now know that forces are needed to start and to change motion; but before Newton's time,

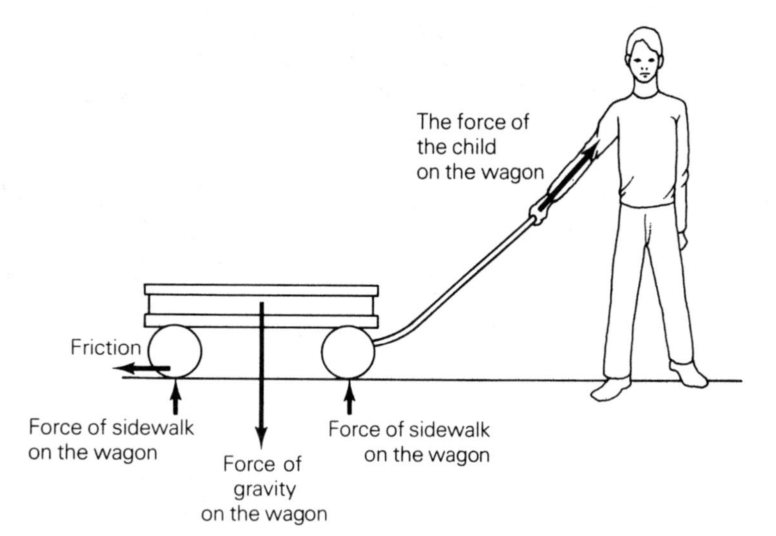

Figure 3-6 The forces on a wagon. Note that these are the only forces on the wagon. The force on the child and the sidewalk are not shown.

people thought that forces were also needed to keep motion going. After all, most motions stopped as soon as force was removed. For instance, a rock that was dragged down a dirt road by oxen stopped when the oxen stopped pulling. People didn't realize that the frictional force of the road had stopped the rock. There were a few counterexamples in which motion continued long after the force was removed: rocks thrown from a sling and arrows shot from a bow, for example. But such counterexamples were largely ignored.

Newton realized that *all* the forces acting on an object, including friction, must be taken into account. He realized that if friction didn't exist, motions would continue forever. For example, a rocket shot into deep space keeps on moving in a straight line at a constant speed after the engine is shut off. No force is needed to keep it going. In fact, just the opposite is true: A force would be needed to stop the rocket.

Newton's realization that the true nature of forces was to produce *changes* in motions was stated in his first law of motion:

Newton's First Law of Motion

An object at rest will remain at rest unless acted upon by an outside force. An object in motion will remain in uniform motion in a straight line unless acted upon by an outside force.

When the oxen stop pulling the plow, the force of friction quickly stops it.

The phrase "uniform motion" means motion at a constant speed. The distinction between outside forces and internal forces is necessary to distinguish between the outside applied forces, which can produce changes in the motion, and the internal forces, which just hold the object together and produce no changes in the motion. I have stated the first law nearly the way that Newton stated it around 1700. A more modern statement would be the following:

Newton's First Law of Motion

If there are no outside forces acting on an object, there will be no *changes* in the motion of the object.

It is clear that a stationary object will not move unless a force acts on it. What is not so obvious is that when an object is at rest, the *net* force acting on it is zero. By "net force" we mean the vector sum of all the outside forces acting on the object.

Let's see how this applies in a specific case. Consider the barbell held aloft by the weightlifter in the photograph. Since the barbell is stationary, the first law of motion tells us that the net force on it is zero. But doesn't the force of gravity act downward on the barbell? Of course, but the weightlifter's hands are also pushing upward. Since the barbell is not moving, we can be sure that the force of the weightlifter's hands is exactly equal in size but opposite in direction to the force of gravity, thus producing a net force of zero on the barbell.

The first law of motion also says that if we get rid of friction and other outside forces, a moving object will continue to move forever in a straight line. To eliminate friction altogether is very hard, but we can get fairly close with modern bearings and lubri-

The net force on the stationary barbell is zero.

cants. An example would be a modern bicycle with well-oiled ball bearings on a smooth, level road. When the bicyclist stops pedaling, thus stopping the force pushing forward, the bicycle doesn't stop. It coasts along at almost the same speed. This shows that a motion continues without change when the net force on the moving object is zero. Of course, the air and road friction slow the bicycle down eventually; but because they are so small, they take quite a long time to stop it. If the friction could somehow be eliminated entirely, the bicycle would go on forever.

An air hockey table reduces friction to a very low level by having the puck ride on jets of air from the table. Even at very low speeds the puck will glide across the table with almost no change in speed, thus demonstrating Newton's first law of motion. The air track found in many physics laboratories also uses air jets to reduce friction to a very low level.

An unfortunate example of motion continuing in a straight line occurs all too often on the highways. Suppose a person in a car has foolishly left the seat belt unfastened. Now suppose that the car hits a barrier and stops suddenly. What stops the *person*? Nothing

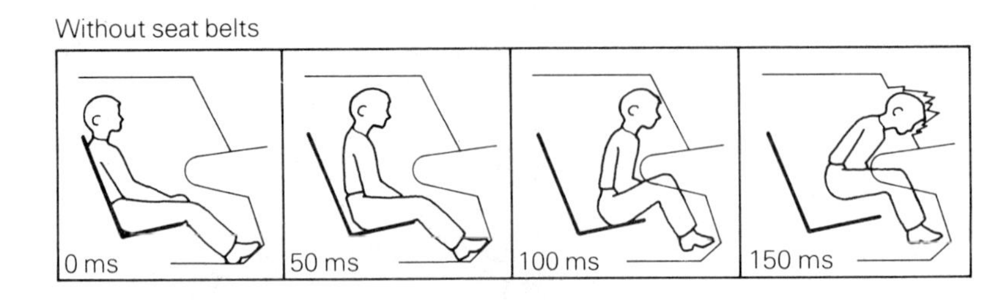

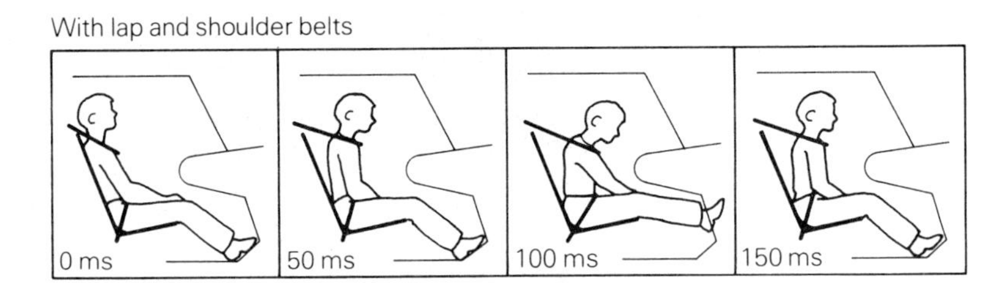

Figure 3-7 The positions of a passenger during a 50 km/hr barrier impact with and without seat belts. The times shown are milliseconds after impact. At 150 ms after impact with seat belts, the pull of the belts is starting to return the person to the seat.

does at first. There is *no* force, except the very small force of friction with the seat, to change his or her motion. The person therefore continues to move at the same speed at which the car was going until he or she hits the now stopped dashboard. The dashboard *is* capable of providing the force needed to bring the person to a stop, but this force is so large that the person is often killed or severely injured. A fastened seat belt would have slowed the person down as the car hit the barrier, preventing the person from striking the dashboard and minimizing the forces applied to the body; in this way seat belts reduce or even eliminate injuries. Figure 3-7 shows the positions of the body in a 50 km/hr collision with a barrier with and without a seat belt. The person wearing a seat belt clearly has a far better chance of surviving the accident.

Newton's first law also states that the *net force* on an object is *zero* when the object is moving at a steady speed in a straight line. If we think about a rocket in outer space with its engines off, we see that this must be true. Clearly, no force acts on the rocket whether the rocket is moving or not. However, when friction is present, we sometimes forget that the net force on an object moving steadily in a straight line is also zero. Consider a car going along the road at a steady 100 km/hr. Doesn't the engine produce a force on the car? Of course it does, but the force of air and road friction is exactly equal but opposite to the force produced by the engine. Thus, the *net force* on the car is zero. Similarly, the force of gravity pulling down is exactly counterbalanced by the force of the road pushing upward. Whenever an object is at rest or moving at a constant speed in a straight line, we can be absolutely sure that the *net* force acting on it is zero.

The Velocity Vector

Motion cannot be discussed in terms of numbers alone. Just as in the case of force, we must also know the direction. For instance, traveling at 90 km/hr with the traffic is a relatively safe thing to do; traveling at 90 km/hr against traffic is likely to result in a serious accident. In describing an object's motion, we must consider not only how fast it is going but also in what direction. In other words, it takes a vector to describe an object's motion fully. That vector, consisting of the object's speed and direction, is called the *velocity* of the object.

An important point about forces and motions is that a *change in direction* is a change in the motion, just as slowing down or speeding up is. Therefore, a *force* is also required to change the direction of motion, even when the speed is not changed. Consider a car that starts out from rest, accelerates to 50 km/hr, goes along a straight road at 50 km/hr, goes around a corner at a steady 50 km/hr, goes along another straight section at 50 km/hr, and then stops (Figure 3-8). There are three times when the net force on the car is not zero: while the car starts, as it goes around the corner, and while it is stopping. Before it starts, while it is moving steadily on the two straight sections, and after it stops, the net force on the car is zero.

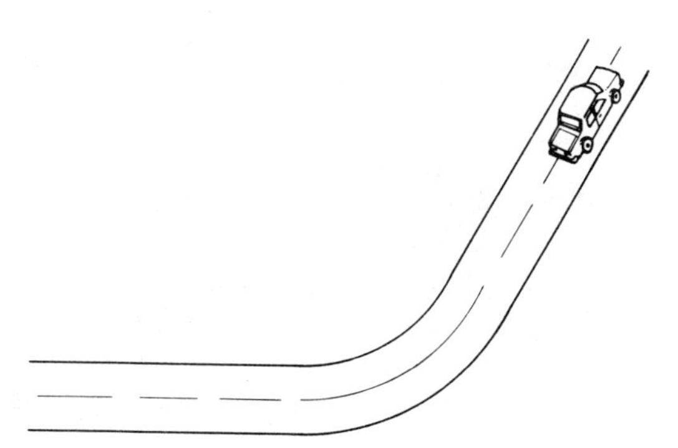

Figure 3-8

In Figure 3-9, an accident victim in an ambulance is receiving fluids intravenously (IV). However, keeping an IV running properly in a moving ambulance is not an easy task. Consider how the first law of motion applies to the victim's blood. If the ambulance

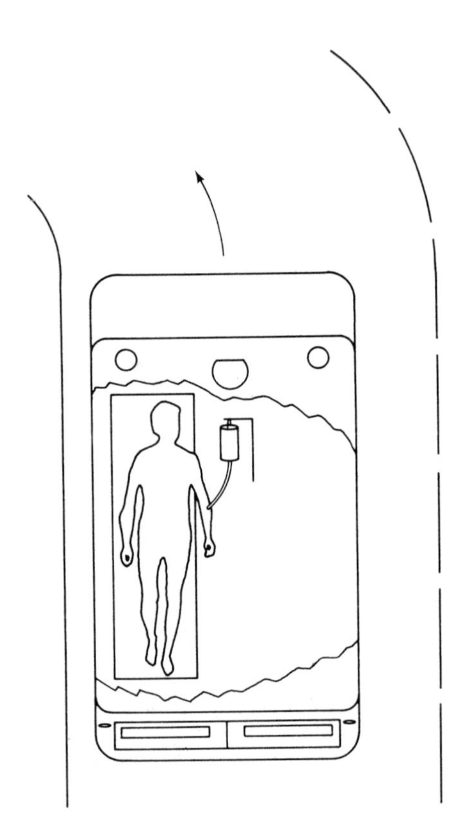

Figure 3-9

makes a left turn, the first law of motion tells us that a force, acting toward the left, must be applied to the ambulance, the victim, and the victim's blood to change their velocity toward the left. The force on the ambulance comes from friction with the road, the force on the victim comes from the straps on the stretcher, and normally the force on the blood would come from the walls of the victim's blood vessels. But there is a hole in the victim's vein where the IV needle enters. At that spot nothing provides a force to change the velocity of the blood toward the left. The first law of motion tells us that if no force acts on the blood, it will tend to continue its motion straight ahead—which means that the blood will tend to flow back up into the IV tube. Once the victim's blood gets into the IV tube, it will tend to coagulate and stop the flow. When you consider all the turns, bumps, and stops that an ambulance makes on an emergency run, you can see that keeping an IV running smoothly can be quite a chore. This problem is usually handled by adding a syringe to pump the fluid in whenever necessary.

Adding Parallel Forces

Whenever two vectors act in the same direction, they can be added like ordinary nonvector quantities. (If they act in exactly opposite directions, they would be subtracted.) This, and the fact that the net force on an object at rest is zero, is all that is needed to investigate the forces in a number of common situations.

EXAMPLE A 100-kilogram man is standing on one foot. The approximate masses of the various parts of his body are shown in Figure 3-10. What are the contact forces at the floor, both knees, and the neck? (The term *contact force* is given to the force that occurs at the point of contact between solid objects. It is, of course, an oversimplification to treat the various sections of the body as solid objects; nevertheless, the errors that this assumption causes are quite small, because most of the forces between the sections are transmitted by the bones, which are essentially solid.)

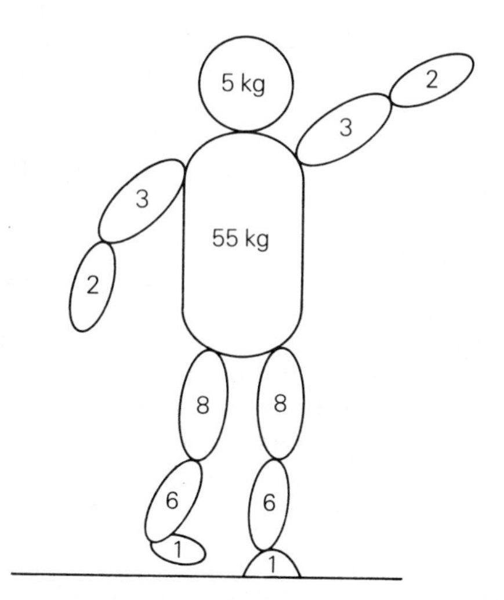

Figure 3-10

Answer The contact force of the foot on the floor is equal to the total force of gravity pulling down on the man. This is

$$\begin{aligned} W &= mg \\ &= 100 \text{ kg} \times 9.8 \text{ m/s}^2 \\ &= 980 \text{ N.} \end{aligned}$$

Next, look at the knee of the leg he is standing on. That knee is supporting all of his body except the lower leg and foot below that knee, which have a

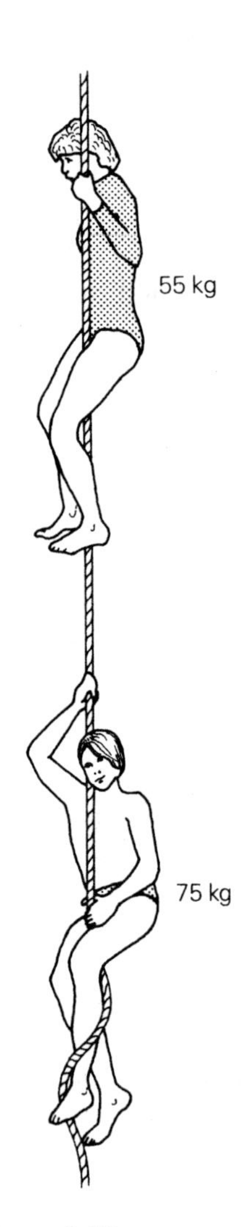

Figure 3-11

mass of 7 kilograms. Thus, the knee is supporting the remaining 93 kilograms. The force of gravity on 93 kilograms is

$$\begin{aligned} W &= mg \\ &= 93 \text{ kg} \times 9.8 \text{ m/s}^2 \\ &= 911 \text{ N}. \end{aligned}$$

Since the man is not moving, the net force on the part of his body above the knee must be zero. If gravity exerts a 911-newton force downward, a 911-newton force must be acting upward through the knee to balance the gravitational force. Thus, the contact force at the knee is 911 newtons.

Next, consider the contact force at the neck. Gravity pulls downward on the head with a force of

$$\begin{aligned} W &= mg \\ &= 5 \text{ kg} \times 9.8 \text{ m/s}^2 \\ &= 49 \text{ N}. \end{aligned}$$

But the net force on the head must be zero; therefore, the contact force at the neck must be 49 newtons upward.

What about the force at the knee of the leg held off the floor? The lower leg and foot have a mass of 7 kilograms; thus, gravity is pulling downward with a force of

$$\begin{aligned} W &= mg \\ &= 7 \text{ kg} \times 9.8 \text{ m/s}^2 \\ &= 69 \text{ N}. \end{aligned}$$

The knee joint must be pulling upward with a counterbalancing force of 69 newtons. Note that there is an important difference in the way that the two knee joints transmit the forces. The force on the knee joint of the leg he is standing on is one of *compression*—the force squeezes the bones closer together. The force on the other knee joint is one of *tension*—the force is pulling the bones apart at that joint.

EXAMPLE A 75-kilogram man and a 55-kilogram woman are climbing the same rope. The woman is above the man, and both are resting for the moment. What is the tension in the section of rope above them and the section between them (Figure 3-11)?

Answer First, let us calculate the weights. The man weighs:

$$\begin{aligned} W_m &= m_m g \\ &= 75 \text{ kg} \times 9.8 \text{ m/s}^2 \\ &= 735 \text{ N}. \end{aligned}$$

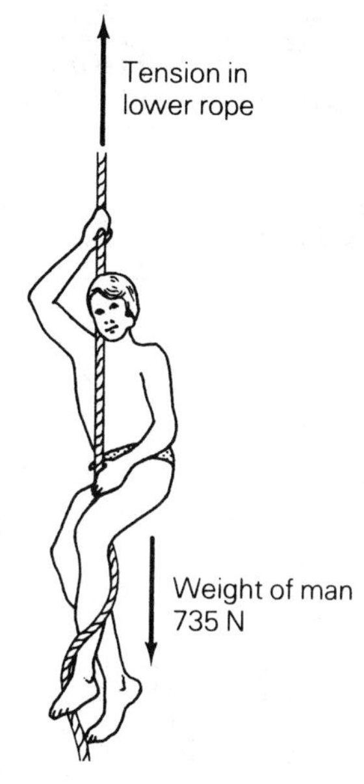

Figure 3-12 The forces on the man.

Figure 3-13 The forces on the woman.

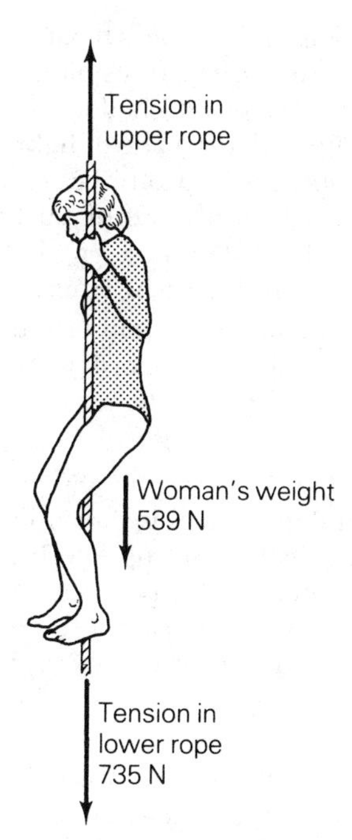

The woman weighs:

$$\begin{aligned} W_w &= m_w g \\ &= 55 \text{ kg} \times 9.8 \text{ m/s}^2 \\ &= 539 \text{ N}. \end{aligned}$$

Now look at the forces on the man (Figure 3-12). Since the man is not moving, the net force acting on him must be zero. Gravity is pulling down on him with a force of 735 newtons, so the rope must be pulling up with a force of 735 newtons. Thus, the tension in the section of the rope above him is 735 newtons. The rope transmits this force to the woman (Figure 3-13), so there are two downward forces on the woman: her weight of 539 newtons and 735 newtons from the rope below, for a total downward force of 1,274 newtons. She, too, is stationary, so the net force on her is also zero. Thus, the tension in the rope above her must be 1,274 newtons acting upward to counterbalance the downward forces on her exactly.

Pulleys

A pulley system is a very efficient and inexpensive way of multiplying the force that the body can produce. While they are not used as often now as they were in the days of the clipper ships, pulleys are still widely used in exercise apparatus, on sailboats, and for lifting heavy objects. They are an essential part of the traction apparatus used in the treatment of bone fractures and other injuries. Inside the body, the tendons transmit forces in exactly the same manner as ropes do. Watch the tendons on the top of your hand as you open and close your hand—the tendons will pass over the knuckles in much the same way that ropes move over pulleys (Figure 3-14).

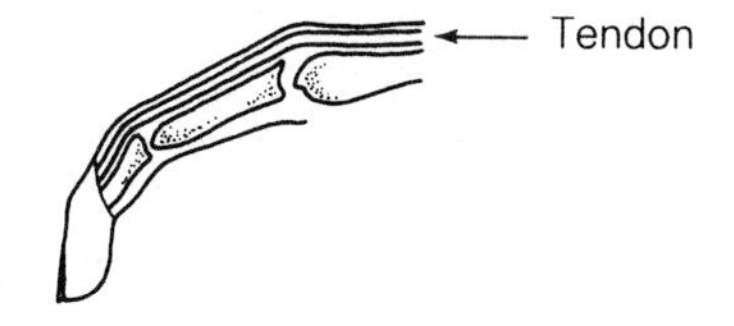

Figure 3-14 Bones and tendons of the index finger. Note how the tendons pass over the knuckles in the same way that ropes go over pulleys.

Pulley systems are easy to analyze when the main ropes of the system run parallel to each other. The only other condition is that the pulleys must be free of friction; this condition will be satisfied in most practical cases when the pulleys are oiled. As long as the ropes are parallel and the pulleys are friction-free, the systems can be analyzed using the concepts of parallel-force addition.

EXAMPLE A patient's leg is being supported by the traction arrangement shown in Figure 3-15. We would like to find the upward forces on the patient's foot.

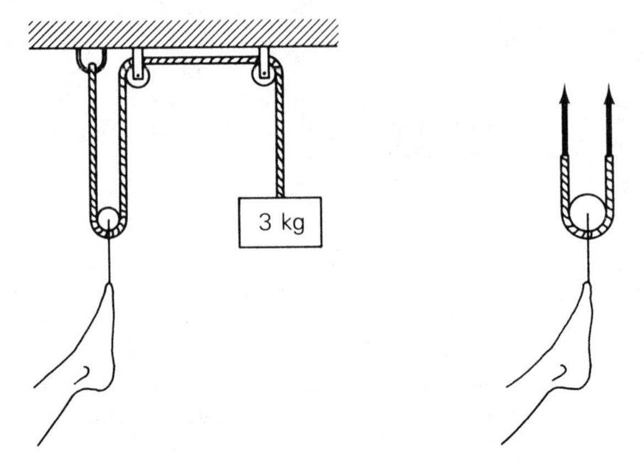

Figure 3-15 **Figure 3-16**

Answer The 3-kilogram mass has a weight of

$$W = mg$$
$$= 3 \text{ kg} \times 9.8 \text{ m/s}^2$$
$$= 29.4 \text{ N}.$$

Thus, the tension in the rope is 29.4 newtons. As long as the rope is flexible and the pulleys are reasonably friction-free, this tension will be transmitted along the entire length of the rope. Now look at the pulley attached to the patient's foot (Figure 3-16). (We will assume that the pulley is made of light plastic, so we can ignore its small weight.) The rope acts in two places on the pulley, and in both spots it is pulling upward with a force of 29.4 newtons. Thus, the rope is pulling upward on the pulley and on the patient's foot with a combined force of 2 × 29.4 newtons = 58.8 newtons.

We might note that this also means that the foot and the portion of the leg that is not supported by the bed must weigh 58.8 newtons, in order for the net force on the foot and the unsupported part of the leg to be zero.

EXAMPLE A pulley system of the type commonly used to multiply forces is shown in Figure 3-17. Such systems can be found wherever people use ropes to lift heavy objects. Assuming that the average person can pull on a rope with a force of 200 newtons without any danger of straining, how heavy a stone could a person lift with this system? (Two hundred newtons is around one-third the weight of the average person.)

Answer As long as the pulleys are reasonably friction-free, the tension in the rope will be 200 newtons at all points along the rope. The sections of

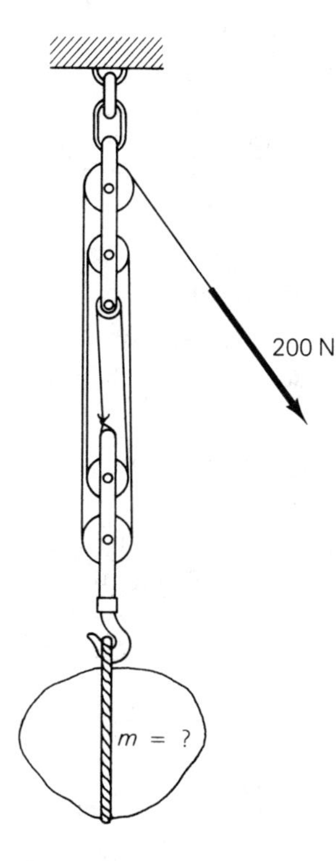

Figure 3-17

the rope leaving the movable lower pulley run parallel to each other, and there are five sections, each pulling upward with a force of 200 newtons. Thus, the total upward force on the movable pulley is 5 × 200 newtons = 1,000 newtons. Therefore, the average person could easily lift any stone weighing less than 1,000 newtons with the pulley system or, in terms of mass, any stone with a mass less than

$$m = \frac{W}{g} = \frac{1{,}000 \text{ N}}{9.8 \text{ m/s}^2} = 102 \text{ kg}.$$

The multiplication factor, or *mechanical advantage* as it is commonly called, is 1,000 newtons ÷ 200 newtons = 5. It isn't an accident that the mechanical advantage is equal to the number of sections of the rope leaving the movable pulley: This relationship is true for all pulley systems of this type.

The force of the person is multiplied very effectively by the pulley system; however, the person isn't getting something for nothing. The stone doesn't move nearly as far as the rope does. Suppose the stone is pulled up 1 meter. The five sections of the rope between the pulleys are each shortened by 1 meter, but the total length of the rope stays the same. Thus, the person must have pulled 5 meters of rope from the pulley system. So the mechanical advantage of 5 is gained at the expense of having to move the rope 5 times farther than the stone moves.

Concepts

Force
Vectors
Nonvector (scalar)
Gravity
Weight
Mass
Acceleration due to gravity (g)
Tension
Friction
Newton's first law of motion
Speed
Velocity
Parallel forces
Contact force
Pulley
Mechanical advantage

Discussion Questions

1. Why must the direction and size both be given to describe a vector fully?

2. Explain the difference between weight and mass. How are they related? How are they different?
3. Newton's first law of motion tells us that the *net* force is zero on a car traveling on a straight, level road at a steady speed. Does this mean that (a) there are no outside forces acting on the car or (b) the vector sum of all the outside forces on the car is zero?
4. At what point of the throw does a baseball pitcher release the ball?
5. Use Newton's first law of motion to explain why you are much more likely to have your car skid off at a turn in a slippery road than on a straight section.
6. Estimate the speed (in m/s) of a baby crawling, a person walking, and an athlete running as fast as possible for a short distance.
7. Give some examples of how you have used pulley systems or have seen them used. Were they used to multiply a force, to change the direction of a force, or both?

Problems

1. Calculate your weight in newtons. To find your mass in kilograms, divide your weight in pounds by 2.2.
2. Which of these quantities are vectors and which are nonvectors (scalars)?

mass of a person	location of a distant city
weight of a person	location of a nearby table
force of a magnet on a nail	velocity of a bullet
time of day	volume of a bottle
date of a certain battle	

3. A golf ball has a mass of 45 grams. What does it weigh in newtons?
4. Suppose the maximum force that a weightlifter can exert is 3,000 newtons. What is the largest mass that he could lift?
5. What force does the rope exert on the patient in the traction arrangement shown below?

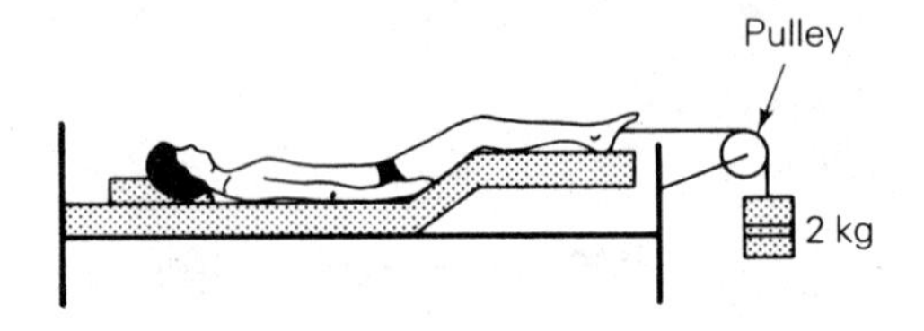

6. A rectangular water tank 1 meter wide, 2 meters long, and 3 meters high is full of water. What does the water weigh in newtons?

7. Find the contact forces at the indicated points of the snowperson.

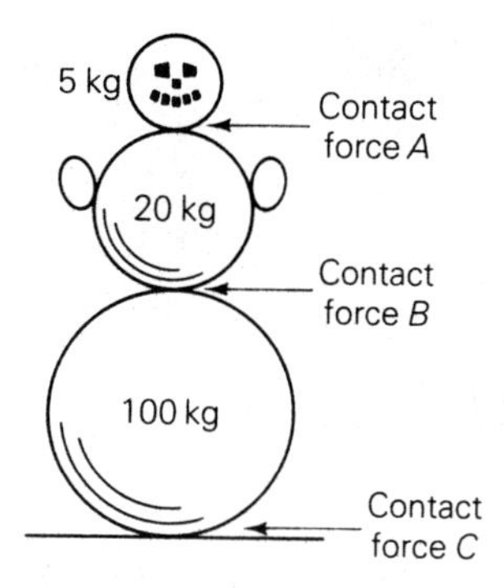

8. A small girl is whirling a ball on a string in a horizontal circle over her head. Is a net force acting on the ball? If the string breaks when the ball is in the position shown below, in what direction will the ball go?

9. A car is traveling at 73 km/hr. What is its speed in m/s, the SI velocity unit?

10. What is the mass of the heaviest object that a 100-kilogram man can lift with the pulley arrangement shown above (assuming that the man is not tied down to the floor, of course)? How many meters of rope must the man pull in order to lift the object 1 meter?

11. Find the tensions in ropes A, B, C, and D.

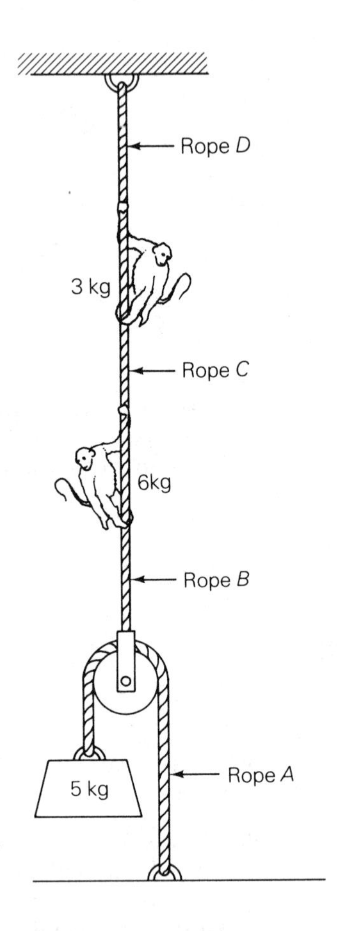

12. A 5-newton force and a 3-newton force are applied to the same object. Since no information is given about the directions of these two forces, the total force acting on the object cannot be found. However, we can find the greatest possible and the least possible values of the sum of these two forces. What are they?

Experiments

1. Check the speedometer in your car by measuring the time it takes to travel a known distance at a steady speed. Calculate your actual speed and compare it to the speedometer reading. The mileposts on expressways are generally located very accurately, so they can be used to mark off the known distance (1 mile = 1.61 kilometers).

2. You can get an idea of the size of the newton, the SI force unit, by picking up a full 1-liter bottle of milk, pop, or water. The full bottle will weigh a bit more than 10 newtons. (If you don't have a 1-liter bottle handy, use a full 1-quart bottle, which weighs almost the same.)

3. Put a sheet of paper on a table and place a full water glass on it. If you pull the paper slowly, the glass will slide along with the paper; but if you yank the paper very quickly, you can pull the paper out without moving the glass or spilling the water. Explain why.

4. Build a pulley system like the one shown in Figure 3-15 or Figure 3-17. You can probably find enough parts in your old Tinkertoy box or Erector set. Calculate the mechanical advantage of your pulley system; then measure the mechanical advantage by seeing how many blocks are needed on the movable pulley to balance one block on the end of the string. Use blocks that all weigh the same.

CHAPTER FOUR

The Forces between Objects

Educational Goals

The student's goals for Chapter 4 are to be able to:

1. Identify action-reaction force pairs.
2. Explain how reaction forces are important in activities such as walking, running, jumping, and swimming.
3. Add vectors.
4. Use vector addition to find the unknown forces on objects that are either stationary or moving at a constant velocity.

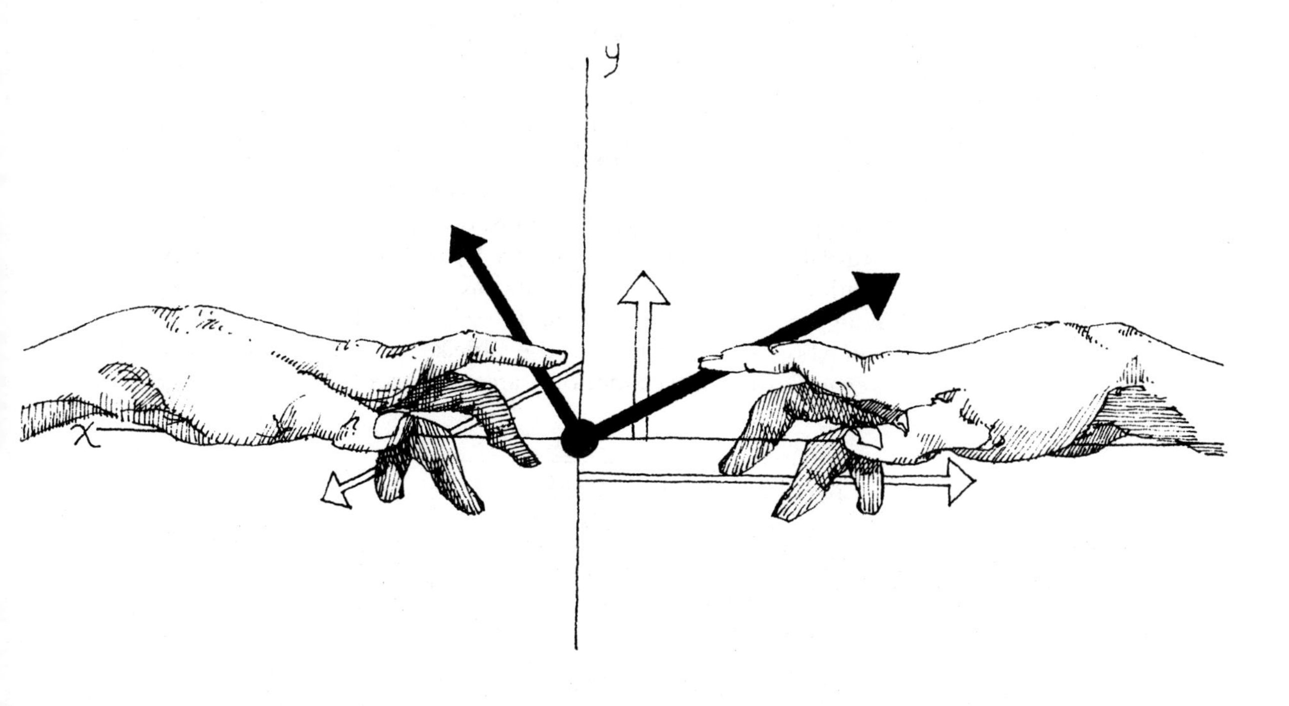

Whenever an outside force acts on an object, we know that there is another object somewhere producing that force. This second object might actually be touching the first object, or it might be some distance away—as it could be in the case of magnetic or gravitational forces. The point is that forces are something that can exist *only* between *pairs* of objects. Moreover, the *forces between objects always exist in pairs.* You cannot touch something without being touched by it in return. If you hit a nail with a hammer, the hammer exerts a force on the nail that drives the nail deeper, and *the nail exerts a reaction force on the hammer that stops the hammer.* When a car crashes into a tree, the car exerts a force on the tree that damages the tree, and *the tree exerts a reaction force on the car that stops the car,* wrecking it in the process. Even when the objects do not make direct contact, forces exist only in pairs. If you try to pick up a block of iron with a magnet, the magnet pulls on the iron—but the iron also pulls back on the magnet.

The force pairs that act between pairs of objects are called *action-reaction pairs.* It is important to note that the two forces always act on *different* objects. When a hammer strikes a nail, the reaction force acts on the hammer. When a car hits a tree, the reaction force acts on the car, and so on. It is important to remember that the action force and the reaction force act on different objects, because it is easy to make the mistake of thinking that both forces somehow act on the same object. This mistake leads to confusion and makes the effect of the forces impossible to understand.

What are the direction and size of the reaction force? Think of what happens when you kick a rock forward. The reaction of the rock pushes backward on your toes, squeezing them between your foot and the rock. This shows that the reaction force is opposite in direction to the action force (Figure 4-1). What about the size of the reaction force? We know that the harder we kick the rock, the more the rock will hurt our toes. That is, the larger the action force is, the larger the reaction force will be. In fact, as Newton realized, the relation between the action and reaction forces is quite simple: They are always exactly equal in size but exactly opposite in direction. This is true whether the objects are stationary or moving. Newton's third law of motion states this principle:

Newton's Third Law of Motion

When a body exerts a force on a second body, the second body exerts an equal but opposite force on the first body.

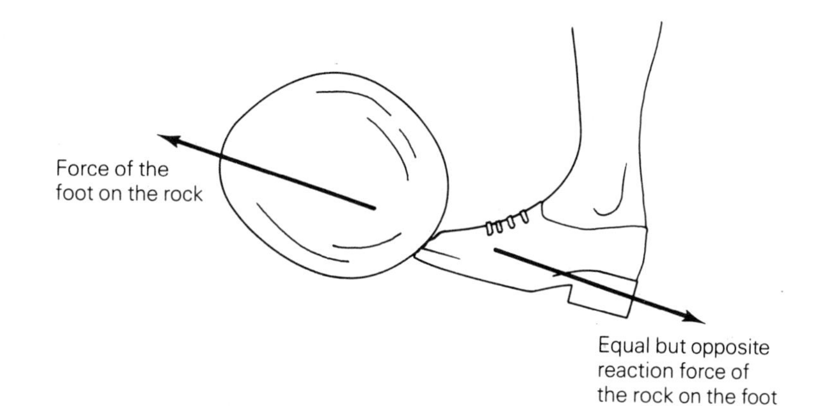

Figure 4-1 When you kick a rock, the rock exerts an equal but opposite reaction force on your foot.

The internal forces that hold objects together also occur in action-reaction pairs, and they are also equal in size but opposite in direction. Since both members of an internal force pair act on the same body, they produce a net force of zero. Thus, the vector sum of the internal forces is always zero, no matter how complex the forces are. As a result, internal forces cannot alter the motion of an object. Diving is a good example of this fact. Once a diver leaves the board, nothing that the diver can do will change the spot where he or she will hit the water. No matter how the diver twists and turns in the air, all the forces that the diver generates are internal forces; they cannot change the basic path of motion.

EXAMPLE Why can't you pull yourself off the floor by pulling very hard on your shoelaces (Figure 4-2)?

Answer When you pull up on your shoelaces, your shoelaces pull down on your hands with equal force in the opposite direction. Since both forces eventually act on your body, the net force from these two forces is exactly zero.

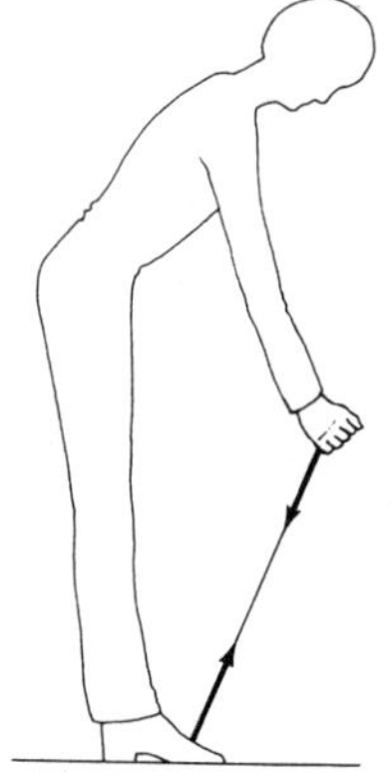

Figure 4-2

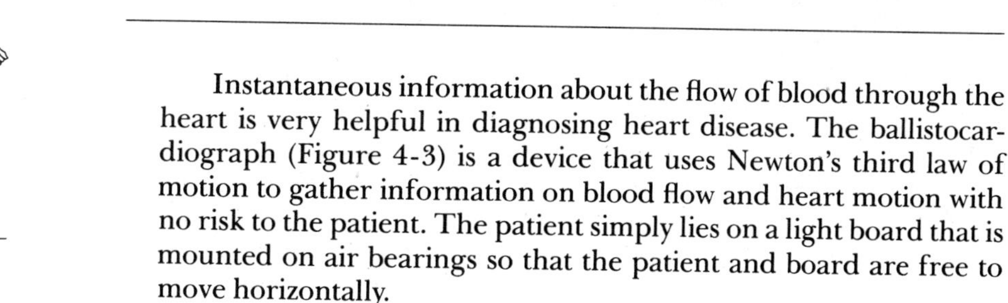

Instantaneous information about the flow of blood through the heart is very helpful in diagnosing heart disease. The ballistocardiograph (Figure 4-3) is a device that uses Newton's third law of motion to gather information on blood flow and heart motion with no risk to the patient. The patient simply lies on a light board that is mounted on air bearings so that the patient and board are free to move horizontally.

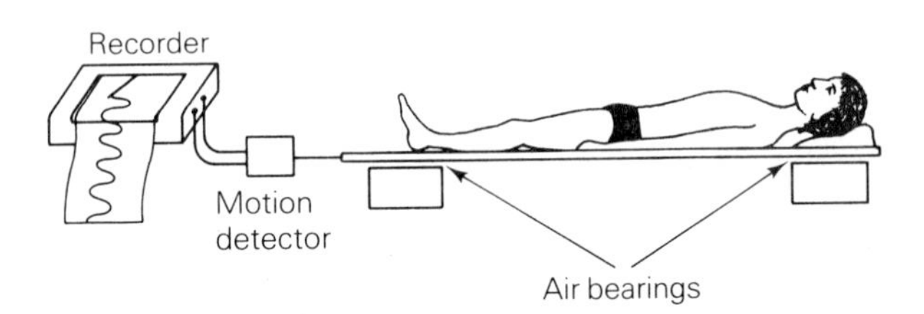

Figure 4-3 The ballistocardiograph.

Now consider what happens as the heart pumps blood up into the aorta toward the head. The heart exerts a force on the blood to start it moving toward the head. But this means that the blood exerts on the heart, an equal but opposite force, which is transmitted to the rest of the body. Thus, as the blood starts to move toward the head, the rest of the body moves a tiny amount in the opposite direction. These small motions of the body and the board are easily measured with modern instruments, and a record of the motions gives the physician a great deal of useful information on blood flow and heart function.

Reaction Forces and Movement

The woman is pushed forward by the reaction force of the ground on her, not by the force of her foot on the ground.

Let's see exactly what force causes a walking person to move forward. The leg muscles produce a force, but it acts on the foot, which in turn, through friction, pushes on the sidewalk. The force of the foot on the sidewalk doesn't make the person go forward, because that force is pushing backward on the sidewalk, not on the person. The force that makes the person move forward is *the reaction force of the sidewalk on the person.* In all forms of locomotion, it is the reaction forces that actually produce the motion.

Similarly, when you jump, the force of your feet doesn't push you up. Your feet push *down* on the ground. You are pushed upward by the equal but opposite reaction force of the ground. You might wonder why the reaction force of the ground doesn't push you up into the air even when you aren't jumping. The answer is that if you just stand without moving, the force of gravity on you is simply transmitted to your feet unchanged. So your feet push down on the ground with a force equal to your weight. The reaction force of the ground pushes up with an equal but opposite force so that the two forces acting on you, gravity and the reaction force of the ground, are equal and opposite. Thus, the net force on you is zero, and you remain on the ground.

When a person jumps, it is the upward reaction force of the ground that actually lifts the person.

To jump up off the ground, you must use your leg muscles to push down on the ground with a force that is greater than your weight. Then the upward reaction force of the ground will be greater than your weight downward, so the net force on you will be upward, lifting you off the ground. Once you leave the ground, it can no longer push on you. Only gravity acts on you, pulling you downward; thus you start to slow down and eventually return to earth.

The reaction force also produces the forward motion in the case of swimming, rockets, jet planes, and the other types of propulsion in which there is nothing solid to push on. These types of propulsion work by throwing mass backward. Whether the material is hot gas, water, or rocks, the principle of operation is the same: The reaction force of the matter thrown backward produces the forward motion. For example, suppose a sailor was carrying a cargo of rocks in a boat when an unfortunate series of events left him in the middle of the lake with the gas tank empty, the oars broken, and no wind blowing. The sailor can still move the boat *forward* by throwing some of the rocks *backward* (Figure 4-4). As the sailor does this, he exerts a force on the rocks toward the rear of the boat. In turn, the rocks will exert an equal and opposite reaction force on the sailor that will cause the sailor and the boat to move forward. Rocket ships and jet planes are propelled in exactly the same way, except that gas molecules are thrown toward the rear instead of rocks. The swimmer is also propelled forward by throwing water backward; again, it is the reaction force of the water that actually propels the swimmer. Of course, the more water the swimmer is able to move backward, the faster the swimmer will move forward. Swim fins are so effective for this reason: They allow the swimmer to move far more water backward than he or she could move with bare feet alone.

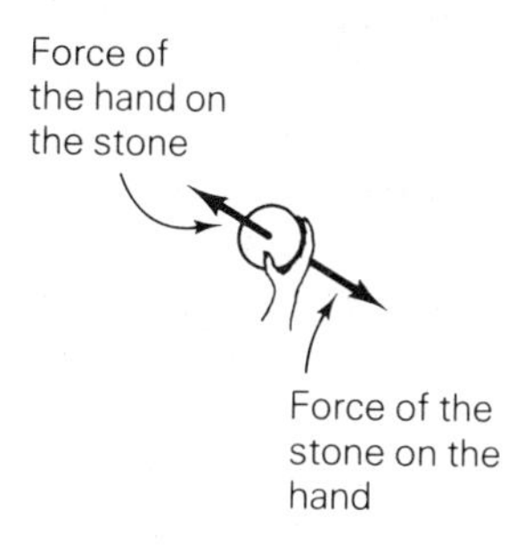

Figure 4-4

Vector Addition

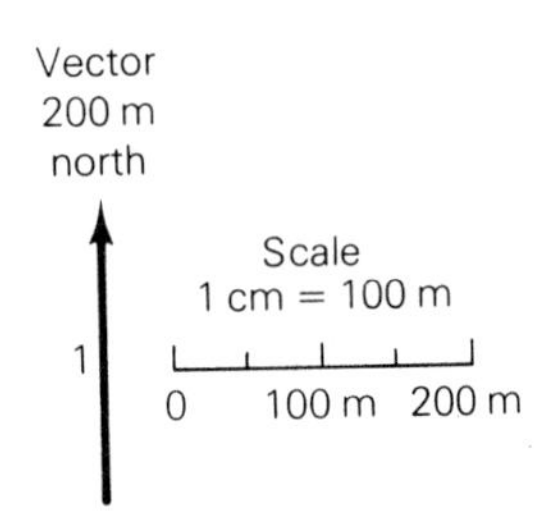

Figure 4-5

Now that we have seen where the forces on objects come from, the next step is to see how forces add to cause or prevent changes in the motion of objects. But before we can do this, we must first find out how to add vectors. This is because forces are vectors, and vectors do not add in the same way as ordinary numbers do.

There are several methods for adding vectors. We will add them by simply making careful scale drawings. This method has the advantages of being straightforward and little math is needed. However, the drawings must be made carefully if we want accurate answers. Of course, sometimes we don't need extremely accurate answers; approximate answers are often sufficient to understand the basic forces in the body. In such cases, we can just sketch the vector addition roughly.

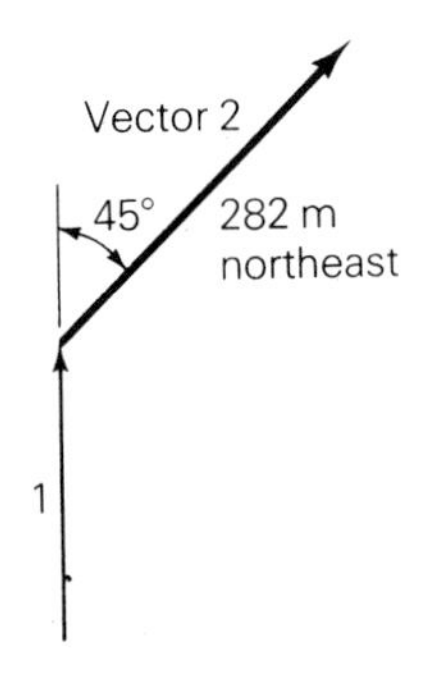

Figure 4-6

Adding vectors by scale drawings has a lot in common with planning a trip on a map. Consider the following problem: Suppose we walked 200 meters north, then 282 meters northeast, and then 400 meters south. Where would we be in relation to our starting point? Note that these three quantities—200 meters north, 282 meters northeast, and 400 meters south—are *vectors,* requiring both size and direction to be described.

One way to find the answer would be to walk outside and follow these directions. This is a perfectly good solution, but it is not very convenient. Another way, which illustrates our method of vector addition, would be to draw a scale map and see where we end up. The first step is to pick the scale of the drawing. We will let 1 centimeter represent 100 meters. The scale we choose is not very important—the only factor to consider is the final size of the drawing. If the scale is too large, the drawing will run off the paper; if it is too small, we won't be able to make accurate measurements. We will let north be at the top of the paper and east to the right—the usual orientation for maps. The vectors will be represented by arrows. The point of the arrow is called the *head,* the plain end the *tail.* We will represent the first vector, 200 meters north, by an arrow 2 centimeters long pointing straight up (Figure 4-5). The second vector, 282 meters northeast, will be an arrow 2.82 centimeters long pointing toward the upper right at a 45° angle. Since this part of the walk begins at the point represented by the head of the first arrow, we will draw the second vector with its tail attached to the head of the first vector (Figure 4-6). Similarly, the third vector, 400 meters south, will be an arrow 4 centimeters long pointing downward with its tail at the head of the second vector (Figure 4-7).

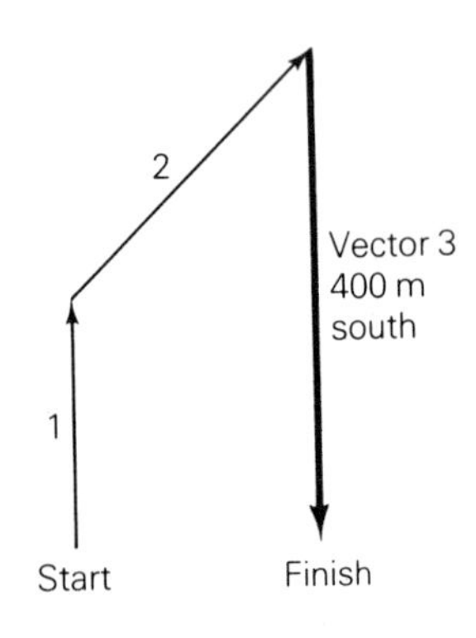

Figure 4-7

If we take a ruler and find the location of the finishing point,

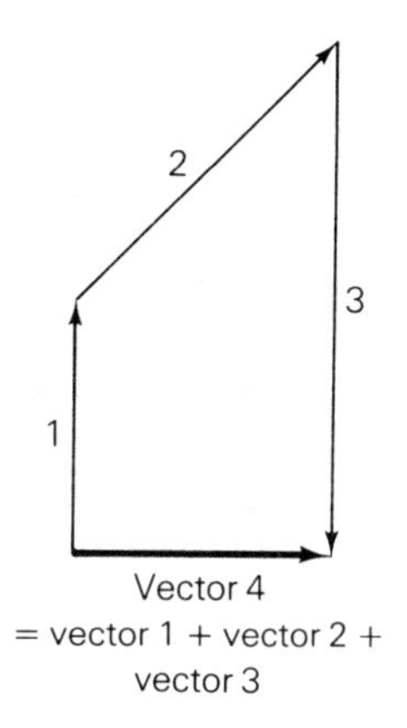

Figure 4-8

we see that it is 2 centimeters to the right of the starting point. Thus, if we had actually made the walk, we would have ended up 200 meters directly east of our starting point. Obviously, we could have gotten to that point simply by walking 200 meters to the east from the starting point. This is shown on the map by the dark vector, Vector 4 (Figure 4-8). Since the end result is the same whether we go to the finishing point by way of Vector 4 or by way of Vectors 1, 2, and 3, we say that Vector 4 is the *sum* of the other three vectors. Note that the sum vector starts at the *tail* of the first vector and has its head at the head of the last vector. This is always true in vector addition.

Interestingly, the order of addition doesn't make any difference (as long as the vectors are correctly added head to tail). The sum is always a vector 200 meters toward the east, as shown in Figure 4-9.

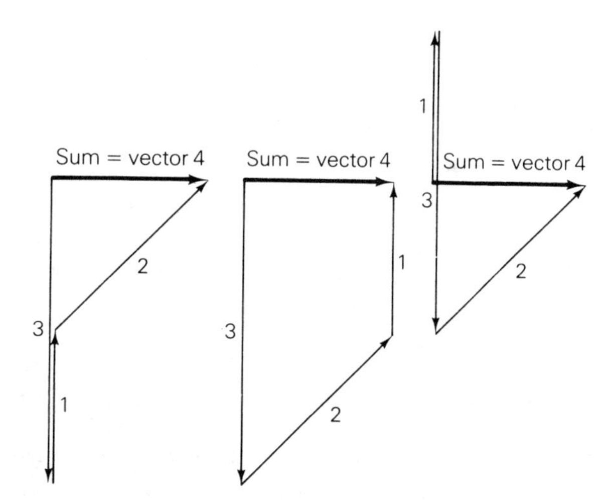

Figure 4-9 The sum of several vectors is the same, regardless of the order of addition.

This method of addition works for all vectors, no matter what they are. For example, suppose a doctor, a nurse, and an orderly were all pulling on a patient in a wheelchair. Suppose the doctor was pulling northward with a force of 200 newtons, the nurse northeast with a force of 282 newtons, and the orderly south with a force of 400 newtons. What is the net force on the patient? That is, what is the vector sum of their three forces? (See Figure 4-10.)

We proceed in exactly the same way as we did in the first example. First, we choose the scale. We will let 1 centimeter represent 100 newtons. Then, we draw the vectors to scale with the tail of the second at the head of the first vector, and so on. Again, the order of addition does not make any difference. When the drawing is done, we see that the net force on the patient is represented by

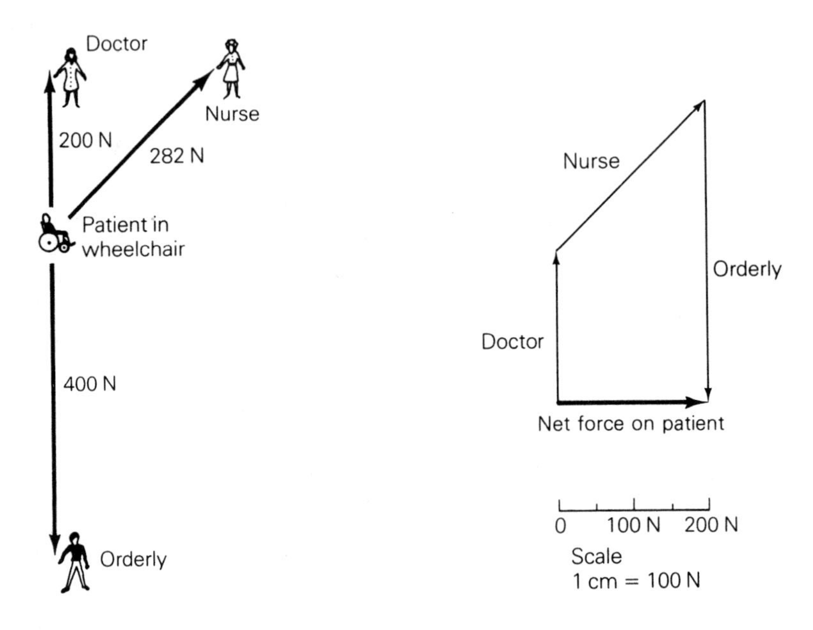

Figure 4-10

Figure 4-11

the dark arrow 2 centimeters long pointing to the right (Figure 4-11). Thus, the net force on the patient would be 200 newtons toward the east, and the patient would start to move toward the east.

EXAMPLE A swimmer dives into a river and heads straight across at 0.4 m/s. However, the swimmer hasn't noticed that the river is flowing toward the right at 0.3 m/s. In what direction and at what speed does the swimmer actually move?

Answer The swimmer's actual velocity is the vector sum of his or her velocity through the water and the water's velocity with respect to the ground. Thus, when the two vectors are added as shown in Figure 4-12, the sum is found to be 0.5 m/s somewhat to the right of the direct path across the river. The exact angle can be measured with a protractor, and it turns out to be 37° (Figure 4-13).

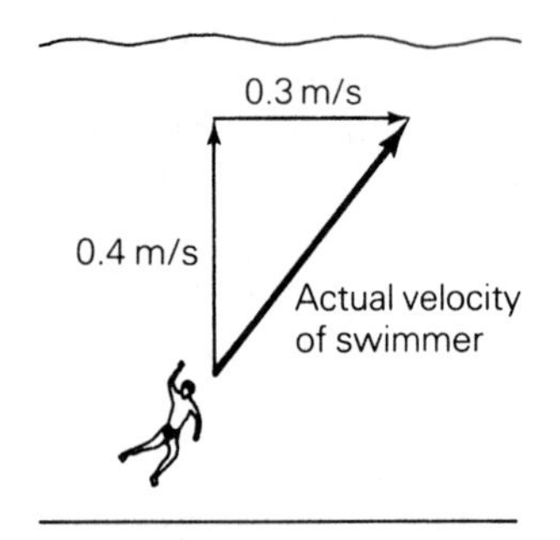

Figure 4-12

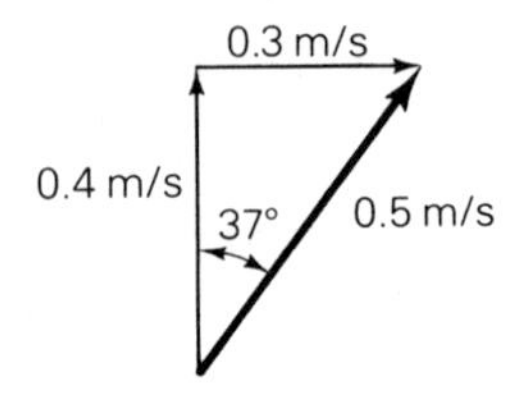

Figure 4-13

EXAMPLE A patient with a dislocated shoulder is placed in traction as shown in Figure 4-14. What is the total force produced by the traction apparatus on the patient's arm?

Answer The traction apparatus applies two forces to the arm: a horizontal force to the left of 2 kilograms × 9.8 m/s² = 19.6 newtons and an upward force of 3 kilograms × 9.8 m/s² = 29.4 newtons. The total force of the traction apparatus on the arm will be the vector sum of these two forces, as shown in Figure 4-15. Adding these two vectors, we find that their sum is 35.3 newtons upward and 34° to the left from vertical (Figure 4-16).

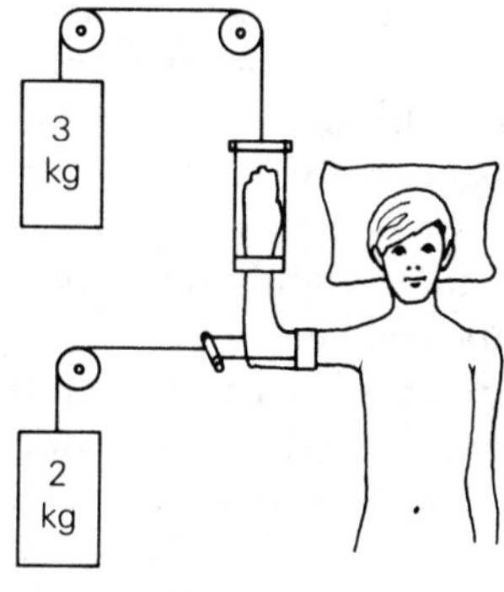

Figure 4-14

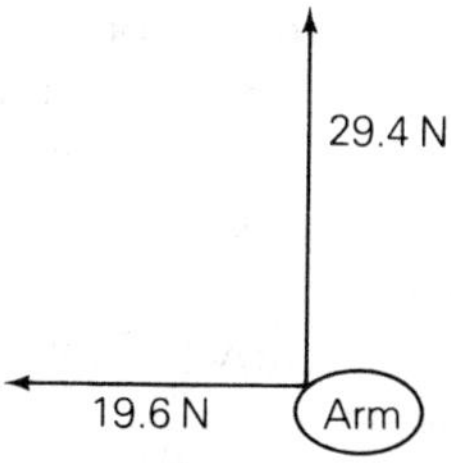

29.4 N
19.6 N
Arm

Figure 4-15

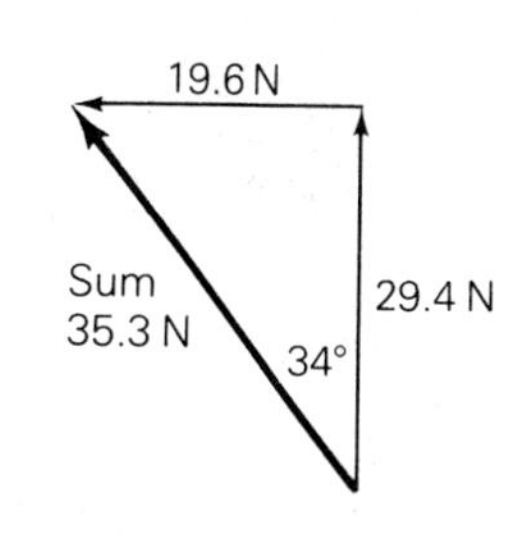

Figure 4-16

EXAMPLE A rope that has a safe working load of 1,000 newtons is stretched between two tall buildings. A tightrope walker, who has a mass of 75 kilograms (and thus a weight of 75 kilograms × 9.8 m/s² = 735 newtons), intends to walk across it. The slack in the rope has been adjusted so that when the tightrope walker is in the middle, the rope will be in the position shown in Figure 4-17. The rope is certainly strong enough to lift the tightrope walker straight up, but is it safe in the position shown?

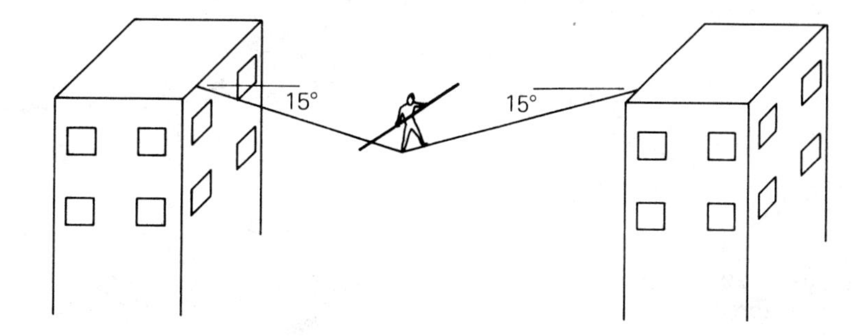

Figure 4-17

Answer The maximum safe tension that the rope can stand is 1,000 newtons. Thus, the maximum force that the rope can exert on the tightrope walker safely is the vector sum of the two 1,000-newton forces shown

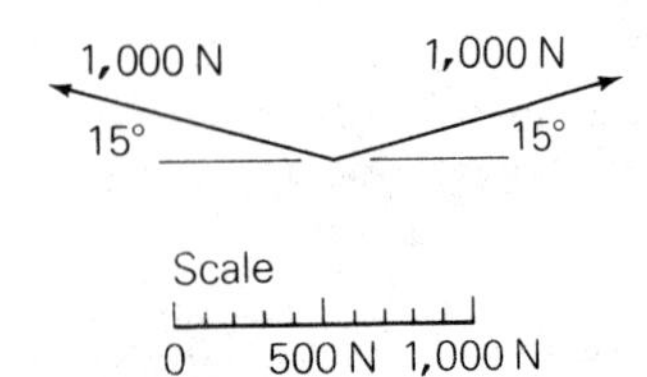

Figure 4-18

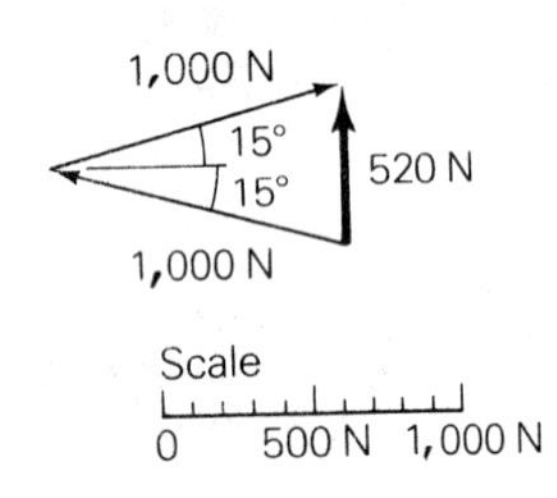

Figure 4-19

in Figure 4-18. When these two forces are added together, we find that their sum is only 520 newtons straight up (Figure 4-19). The rope will therefore be unsafe to use in the position shown.

The more tightly the rope is stretched, the smaller the angles will become and the lighter the load that the rope can safely support. If the rope was stretched so tightly that the angle was only 1°, the rope could safely support only 35 newtons, the weight of a newborn infant.

Mountain climbers sometimes cross a deep chasm by tying a rope across the chasm and then crossing on the rope. The method is known as a Tyrolean traverse. When used in this manner, the rope must not be stretched too tightly. The tension in the rope and the stress at the anchor points is going to be far greater than the weight of the person crossing in any case, but the more tightly the rope is stretched, the greater the tension in the rope will be.

If one wants to produce a very large force with a rope, say, to pull a stuck car out of the mud, the above example shows us how to do it without any pulley systems or other apparatus. Simply stretch the rope between the car and a tree as tightly as possible, then push *sideways* at the middle of the rope. The tension in the rope will be several times greater than the sideways force applied (Figure 4-20).

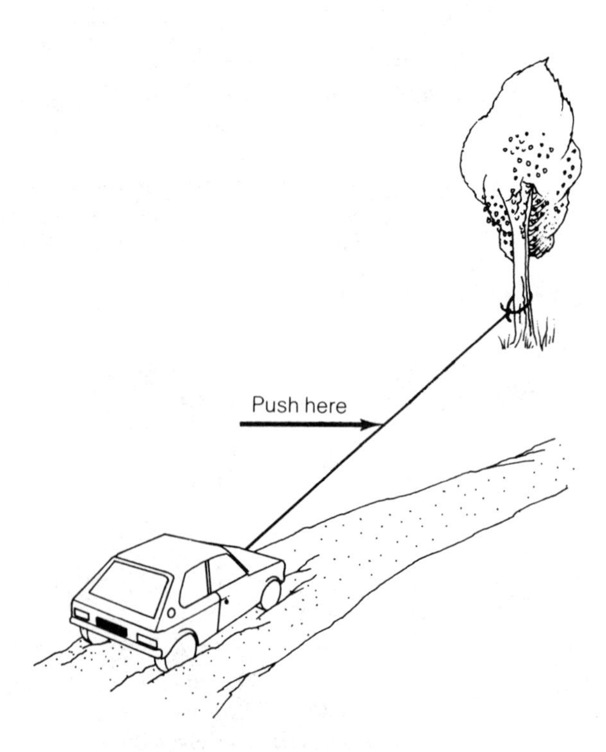

Figure 4-20 To produce a large force to pull a car out of mud, stretch a rope as tightly as possible, then push sideways on the middle of the rope.

During a Tyrolean traverse the tension in the rope is much greater than the climber's weight.

The Arm

The forces in the muscles and bones of the human body are seldom parallel, so vector addition is usually needed to analyze them. Consider, for example, a man who is holding his arms outstretched as shown in Figure 4-21. In that position, the deltoid muscle is the

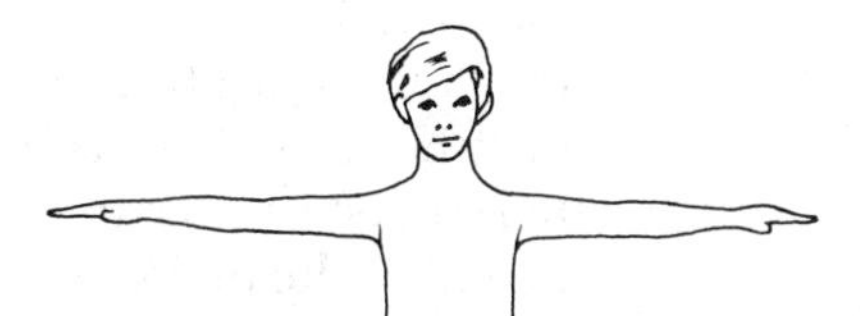

Figure 4-21

primary muscle holding up the arm. This muscle is attached to the upper shoulder at one end, goes over the top of the shoulder, and is attached to the humerus (the upper arm bone) near the elbow at the other end. In this position, the deltoid muscle pulls upward at about a 15° angle to the humerus. The weight of the arm is distributed along its entire length, but for our purposes we can regard the entire weight as being concentrated near the middle of the arm, near the insertion point of the deltoid muscle on the humerus. We will assume that the arm weighs 40 newtons. Figure 4-22 shows our simplified version of the forces on the upper arm.

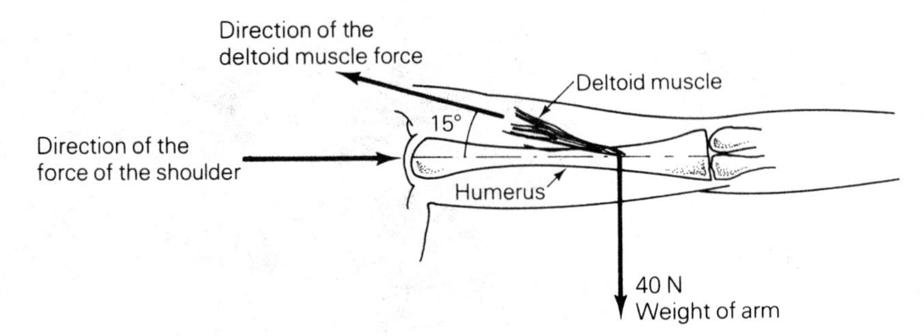

Figure 4-22 The forces on an outstretched arm.

We would like to find the tension in the deltoid muscle and the compressive force on the humerus. Look at the part of the arm around the elbow. The elbow region has three forces acting on it: the weight of the arm (40 newtons) pulling downward, the deltoid muscle pulling to the left and up at 15°, and the humerus pushing horizontally to the right (Figure 4-23). Since that part of the arm is

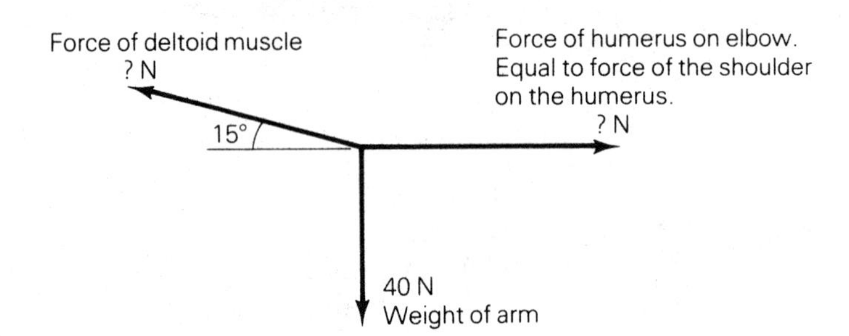

Figure 4-23 The forces on the elbow region.

not moving, we know from Newton's first law of motion that the net force on it must be zero. Thus, the vector sum of these three forces must be zero. Since a vector that is zero has no length whatsoever, there must be no gap between the head of the last force vector and the tail of the first when we add the three force vectors. That is, any three vectors whose sum is zero form a closed triangle, as shown in Figure 4-24.

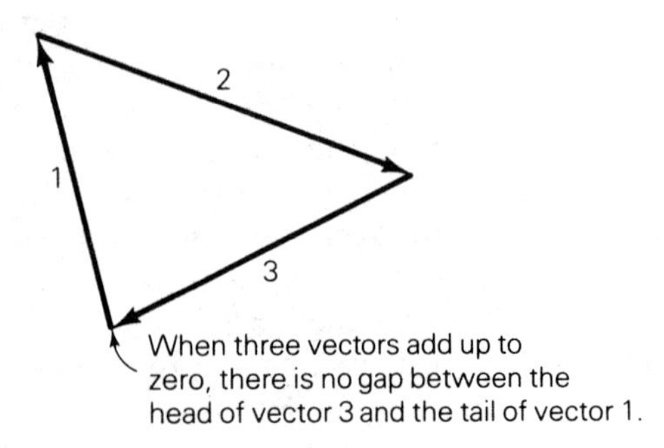

Figure 4-24

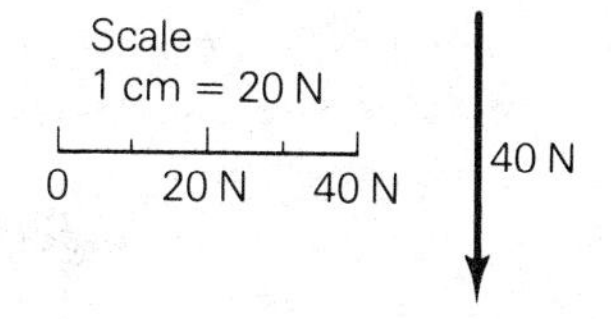

Figure 4-25

We only know one force, the weight, completely. All we know about the other two are their directions. However, this is enough information to solve the problem if we approach it correctly. Start with the weight of 40 newtons, and use a scale where 1 centimeter represents 20 newtons (Figure 4-25). Next, try to add the force of the deltoid muscle to the weight. The tail of the deltoid muscle force vector will be at the head of the weight vector. Although we don't know exactly where the head of the deltoid vector will be, we do know that it will be somewhere along the line going from the head of the weight vector at a 15° angle above the horizontal (Figure 4-26). Next, consider the force of the humerus. If the sum of

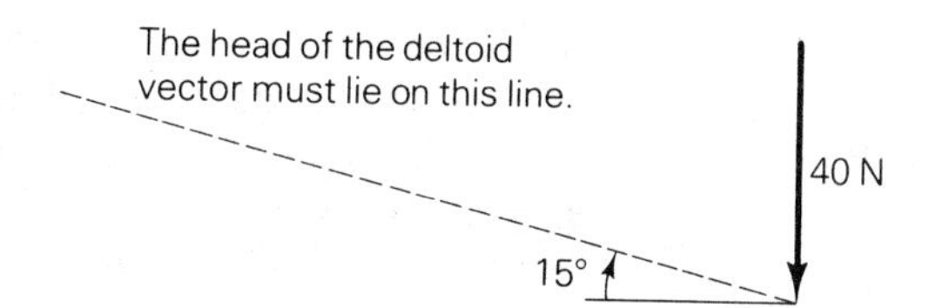

Figure 4-26

the three vectors is to be zero, the head of the humerus vector must be at the tail of the weight vector. The humerus force is horizontal, so the tail of the humerus vector must be somewhere on the horizontal line going to the left from the tail of the weight vector, as shown in Figure 4-27. Now put these last two drawings together.

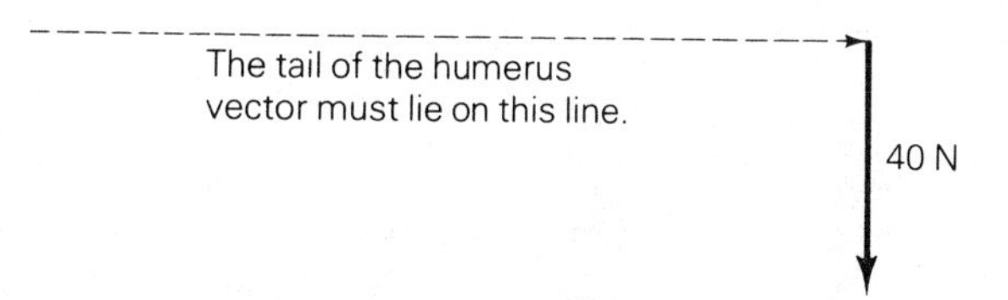

Figure 4-27

The head of the deltoid vector must meet the tail of the humerus vector if the net force is to be zero. Since the head of the deltoid vector is on one line and the tail of the humerus vector is on the other, the only possible meeting place is where the two lines cross: at point *X*, (Figure 4-28). Thus, the humerus vector goes from *X* to the tail of the weight vector, and the deltoid vector goes from the

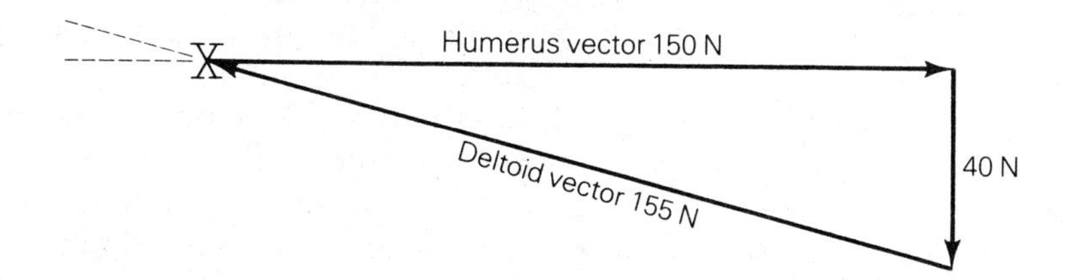

Figure 4-28

head of the weight vector to X. Measuring the vectors, we find that the tension in the deltoid muscle is 155 newtons and the compressive force in the humerus is 150 newtons. This compressive force is transferred by the humerus to the shoulder joint, so the contact force of the humerus with the shoulder is also 150 newtons.

The deltoid muscle is well positioned to produce a wide range of arm movements, but it isn't particularly well located for generating brute lifting strength. For instance, in the example above, the deltoid is pulling toward the side far more than it is pulling upward. Because of the unfavorable lifting angle, the tension in the deltoid is nearly 4 times the weight of the arm.

This example is not an isolated one; on the contrary, the tensions in the muscles and tendons and the contact forces at the joints are usually several times larger than the load applied from the outside. Because of the magnifying effect caused by the unfavorable angles of force application, the stresses inside the body are large even during moderate activity. During vigorous activity, the stresses can easily damage the body if the person is not in proper physical condition. This is one reason why careful training programs are so important in athletics and physical therapy. If a person has not been gradually trained to accommodate increased forces, the muscles and bones may be damaged by the stresses. The damage can range from sore muscles to torn ligaments and broken bones, depending on how badly the body is overstressed.

The Spine

On an evolutionary time scale, humans started to walk erect on two legs only a short time ago. Consequently, some parts of the body are still not perfectly adapted to this position. The back is one of them. The spine is easily damaged by excessive forces, and many people are afflicted with back troubles. Proper positioning of the back when lifting and carrying heavy objects can reduce the strain quite a bit.

Let's compare the forces on the spine in good and bad lifting positions (Figure 4-29). Suppose a nurse was going to lift a child weighing 300 newtons, and the weight of the nurse's upper body was another 300 newtons. In the good position, the weight of the child and the weight of the nurse's upper body are both nearly parallel to the spine. The force of the spine in this case is essentially equal but opposite to the sum of the two weights: 600 newtons (Figure 4-30). Six hundred newtons is a large force, but the spine of a normal, healthy person can carry it with no difficulty.

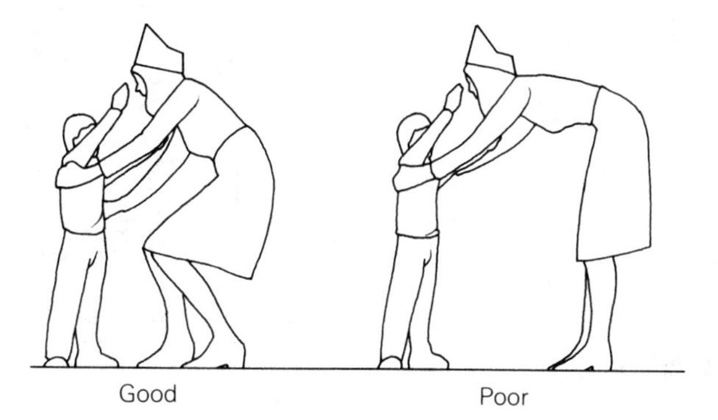

Figure 4-29 Good and poor positions for lifting a patient.

The anatomy of the back in the poor lifting position is shown in Figure 4-31. The body is held in this position primarily by the erector spinae, the two strong muscle groups that run alongside the back of the spine from the pelvis to the neck. In this position, the spine acts like the boom of a crane, while the erector spinae function as cables holding up the boom. Most of the weight of the upper body is near the shoulder region. For our purposes, we can treat the nurse's upper body weight as if it were concentrated at the shoulders. The weight of the child also acts through the nurse's shoulders. Figure 4-32 shows a simplified view of the poor lifting position. The 11° angle between the spine and the erector spinae was determined from skeletal measurements. While the drawing is obviously a simplification, it does present a fairly accurate picture of the major forces on the body in this position.

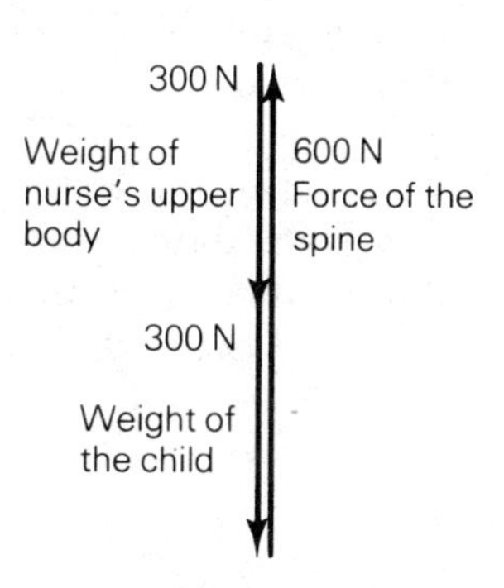

Figure 4-30 The forces on the nurse's shoulder region in the good lifting position.

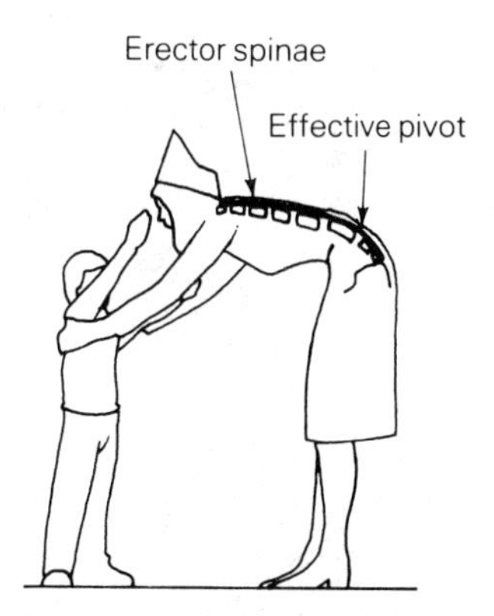

Figure 4-31 Anatomy of the back in the poor lifting position.

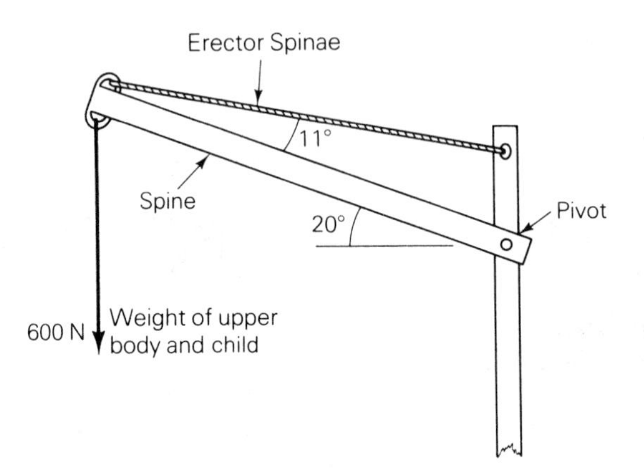

Figure 4-32 Simplified model of the anatomy of the back in the poor lifting position.

There are three major forces on the shoulder region: the force of the spine, the force of the erector spinae, and the sum of the weights. The size of the compressive force in the spine is unknown, but its direction must be along the spine, because the spine is free to pivot at the lower end. The size of the erector spinae force is also unknown, but the direction must be along the muscles, since they can only pull. The sum of the weights is completely known: 600 newtons straight down. We can find the size of the two unknown forces by making use of the fact that the sum of the three forces on the shoulder must be zero when the body is stationary.

The known facts about the three forces are shown in Figure 4-33. We will draw the forces on a scale where 1 centimeter represents 500 newtons. If the vector sum of the three forces is zero, the

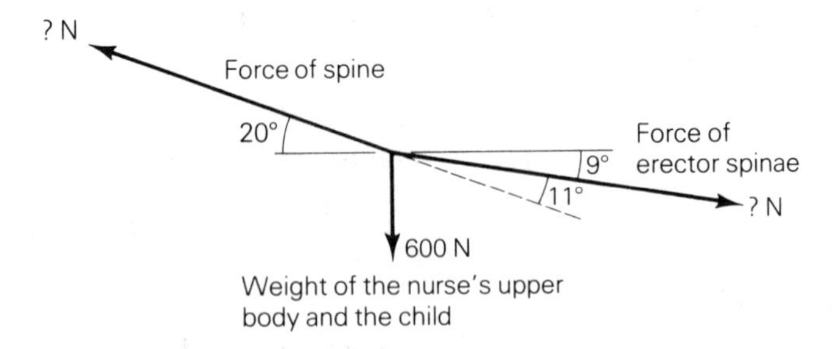

Figure 4-33 The forces on the nurse's shoulder region in the poor lifting position.

tail of the erector spinae vector must lie at the head of the weight vector, and its head must be somewhere on the line going to the right from the head of the weight vector 9° below the horizontal. Similarly, the head of the spine force vector must lie on the tail of the weight vector with its head somewhere on the line going from the tail of the weight vector to the right 20° below the horizontal. Of course, the head of the erector spinae vector must meet the tail of the spine vector. The only possible place is at point X on Figure 4-34.

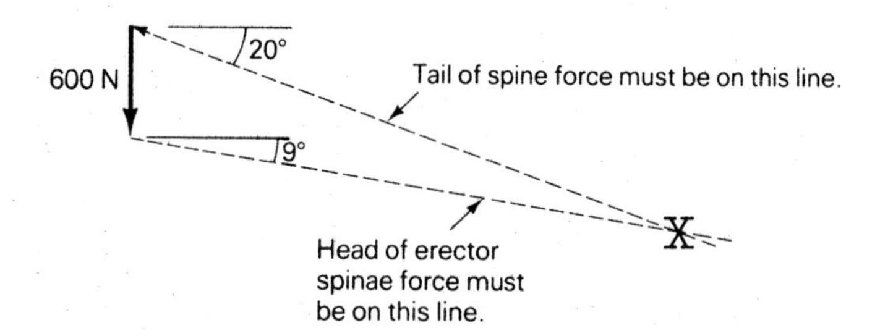

Figure 4-34

Thus, the three vectors must add to zero, as shown in Figure 4-35. Measurement of the two unknown vectors shows that the tension in the erector spinae is 2,950 newtons, and the compressive force in the spine is 3,100 newtons, 5 times the weight being held up. This force is transmitted along the entire length of the spine, all the way to the pelvis. Every vertebra and disk in the nurse's spine is compressed with a force equal to the weight of a horse. So the poor position is the wrong way to lift an object because it can easily overtax the spine with an unnecessarily large force.

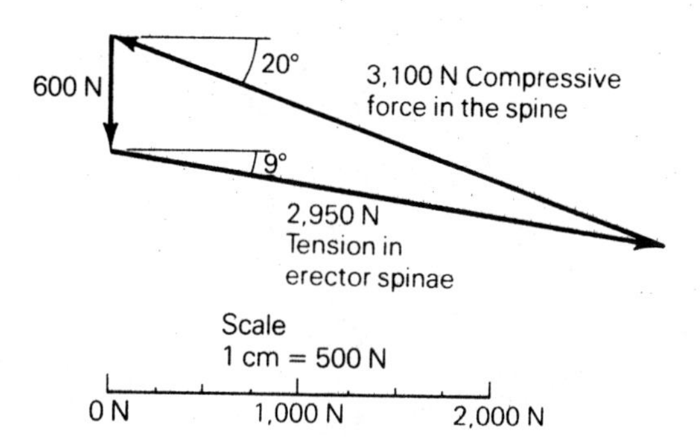

Figure 4-35

Russell Traction Apparatus

The Russell traction apparatus is used for fixation of a fractured femur, the upper leg bone. One of the difficulties in setting a fractured femur is that the powerful muscles of the upper leg pull the two broken sections toward each other. This can shorten the femur as it heals, which would cause the patient to limp. To prevent the contraction of the femur, we need a counterforce parallel to the femur to pull against the force of the leg muscles. At first glance, it is hard to see how the Russell apparatus produces a force parallel to the femur (Figure 4-36). None of the parts of the rope seem to

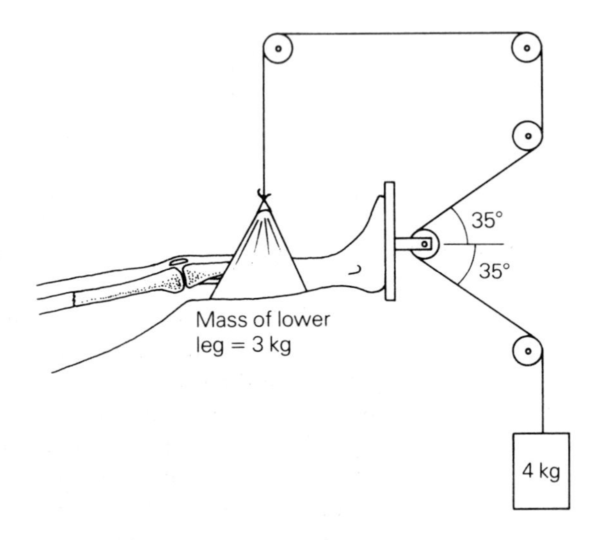

Figure 4-36 Russell traction apparatus.

pull in the right direction. But when we add up the three forces produced by the traction apparatus and the weight of the lower leg (3 kilograms $\times$ 9.8 m/s^2 = 29.4 newtons), the sum of the four forces is 65 newtons and the direction is parallel to the femur (Figure 4-37).

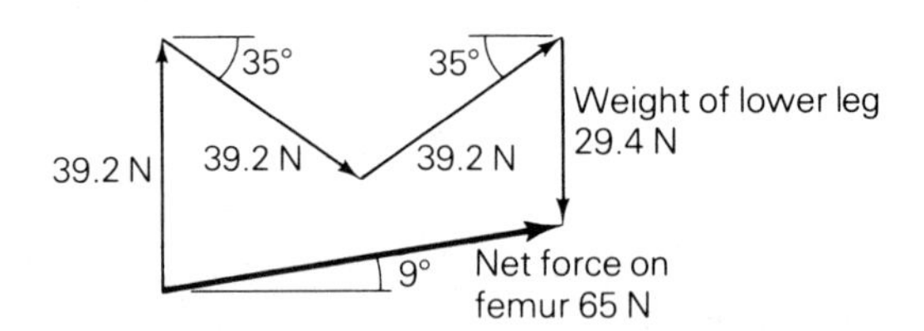

Figure 4-37 Net force of Russell traction apparatus and weight of lower leg.

There is more to designing traction apparatus than just producing the correct force, of course. One important thing to take into account is the behavior of the patient. An orthopedic patient will always wiggle around and change position in bed unless he or she is unconscious. The traction apparatus should be designed so that minor changes in the patient's position don't cause major changes in the applied force. The Russell apparatus is very good in this respect—even moderately large changes in the patient's position don't cause large changes in the applied force. Of course, the patient should still be encouraged to keep the changes to a minimum by not moving around too much. One danger is that the traction weight might reach the floor, in which case the apparatus would exert no force on the patient at all. Therefore, the traction weights should always be placed high enough so that they will not reach the floor when the patient moves. One should also be careful that the weight doesn't become hung up on the bed or nearby furniture. If the weight does catch on the bed, it may drop suddenly and jerk the patient, which can be dangerous as well as painful.

Skin Traction

When large traction forces are needed, a direct mechanical connection to the bones is made surgically; however, for light to moderate traction forces, *skin traction* is often adequate. In skin traction, the force is applied to the skin with adhesive bandages. Let's see how the tape should be placed to transmit the traction forces. If we knew for sure that the force was always going to come from the same direction, one long piece of tape placed as shown in Figure 4-38 would work fine. Of course, in actual practice such an arrangement would never work, because the tape would rip off as

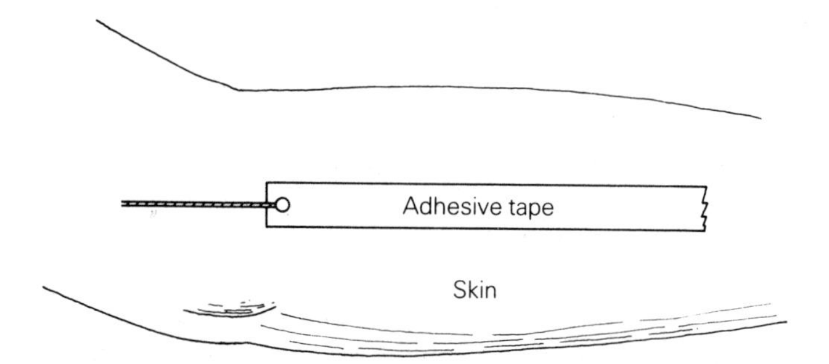

Figure 4-38

soon as the direction of the applied force changed a little bit. Since conscious patients always move around, the direction will change sooner or later.

To prevent the tape from coming off, we must arrange it so that it can tolerate changes in the direction of the force. The arrangement shown in Figure 4-39 will do this. Individually, each tape only pulls in its own direction. But together the three tapes can transfer any force within the dashed lines without coming off. This is because the forces of the tapes can add as vectors to produce any force whose direction is between the dashed lines, providing, of course, that the force does not exceed the strength of the tape.

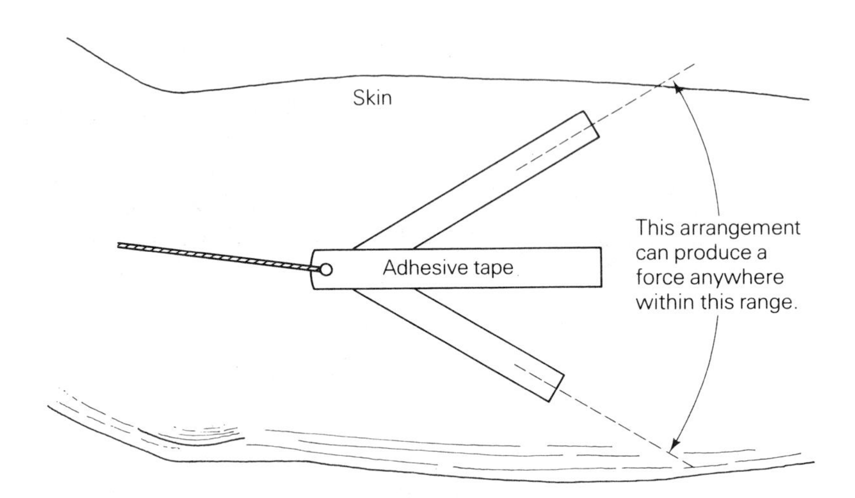

Figure 4-39 Skin traction

This arrangement has the additional advantage of spreading the force out over a larger area of the patient's skin, thus reducing discomfort.

The Ramp

The ramp has been used to raise heavy objects since prehistoric times. To see how a ramp reduces the force needed to lift an object, let's look at the forces on a patient in a wheelchair as he or she is pushed up a ramp. Suppose that a patient is being pushed straight up a ramp at a steady speed. Newton's first law of motion tells us that the net force on the wheelchair must be zero. Three forces are acting on the wheelchair: (1) the weight of the patient and wheelchair, which acts straight down, (2) the force of the person pushing the chair, which would normally be parallel to the ramp, and (3) the reaction force of the ramp. We don't know the size of the reaction force, but we do know that it is perpendicular to the ramp, because the wheelchair rolls with very little friction. In the absence of any significant friction, the ramp cannot exert a force on the chair in the direction parallel to the ramp.

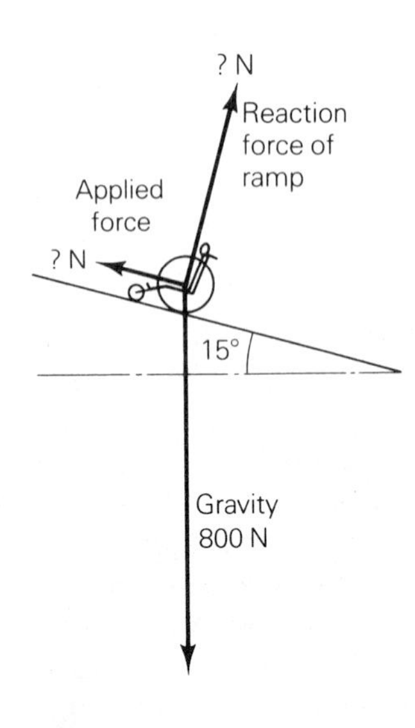

Figure 4-40

Suppose that the patient and wheelchair weigh 800 newtons and that the ramp goes up at a 15° angle to the horizontal. The vector sum of these three forces must be zero, but we are again faced with the problem of not knowing the size of two of the three forces (Figure 4-40). However, we do know their directions, so we can solve the problem in the same way that we solved the earlier examples. Start with the force that is completely known: the weight of the patient and wheelchair. Next add the ramp reaction force to the weight. We don't know where the head of the ramp reaction

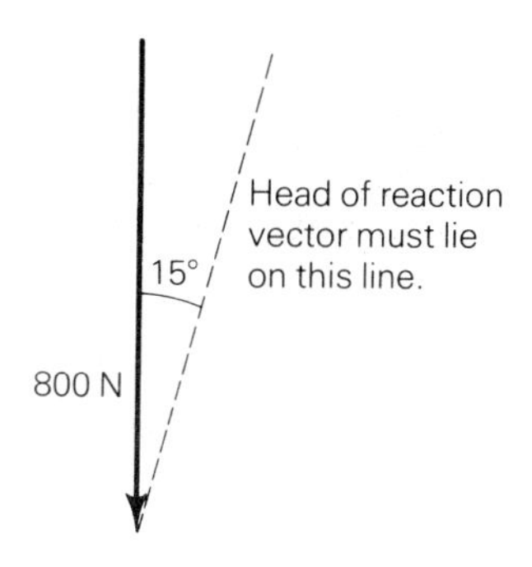

Figure 4-41

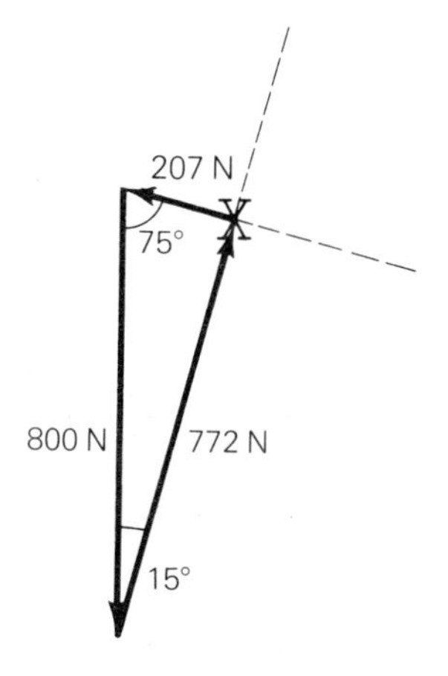

Figure 4-42

vector will be exactly, but we do know that it is somewhere on the line drawn at 15° to the right of the weight vector (Figure 4-41). If the sum of all three forces equals zero, the head of the applied force vector must be on the tail of the weight vector. The tail of the applied force vector lies somewhere along the line drawn at 75° to the right of the weight vector, and it must meet the head of the ramp reaction force vector exactly. Again, this can only occur where the two lines cross, at point *X* (Figure 4-42). Thus, we see that the applied force is 207 newtons—far less than the 800 newtons it would take to lift the patient and wheelchair straight up. Since this ramp makes it possible to lift 800 newtons with a 207-newton applied force, the mechanical advantage of this ramp is 800 newtons ÷ 207 newtons = 3.9. The reaction force (which we are not particularly interested in, since the person pushing the wheelchair doesn't have to generate it) turns out to be 772 newtons.

If we were to draw figures for ramps with other angles, we would quickly see that the steeper the ramp is, the bigger the applied force must be to push the wheelchair up the ramp. For the same reason, cutting tools work much better when they are sharp. The smaller the angle at the edge, the easier the blade can push the sides of the cut apart.

The principle of the ramp, or inclined plane as it is formally known, is put to use in many places. One example is the tilt table described in Chapter 18, which is used by physical therapists to reduce the force on a weakened patient's legs when the patient exercises. Wedges, knives, arrowheads, and freeway ramps are a few other examples. However, the most important example might not be immediately obvious—the screw thread. The screw thread is nothing but an inclined plane wrapped around a cylinder (Figure 4-44). It employs the principle of the ramp to convert the small

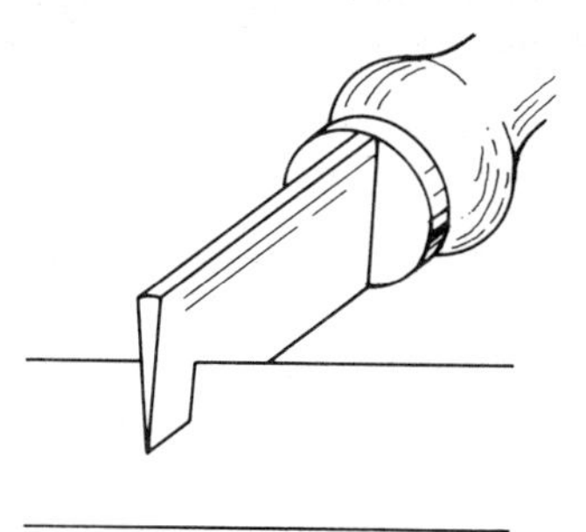

Figure 4-43 Knives use the principle of the ramp to produce a large cutting force.

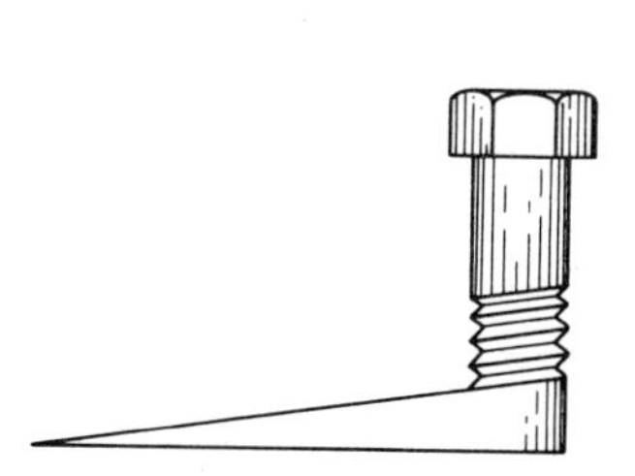

Figure 4-44 Screw threads are ramps wrapped around a cylinder.

twisting force of a wrench, screwdriver, or crank into a very large force. The large force can be used for any number of purposes, such as lifting things with screwjacks, adjusting the position of a hospital bed, and fastening things together with nuts, bolts, and screws.

Concepts

Reaction force
Newton's third law of motion
Internal forces
Vector addition
Ramps

Discussion Questions

1. When you apply a force to a baseball with a bat, what produces the reaction force and what does the reaction force act on?
2. Lean up against a wall. Just what force is it that keeps you from falling over?
3. Once a long jumper is in the air, his or her motions will have only a small effect on the length of the jump. Explain why.
4. A patient in a wheelchair is going up a ramp. Identify the force that is actually pushing the patient and the wheelchair up the ramp.
5. Explain why swim fins make swimming easier and faster.
6. Why don't vectors add in the same way that ordinary numbers do?
7. With the aid of a rough vector diagram, explain why there is a large force on the spine when you lean over to lift something.
8. Give at least one example of a ramp that wasn't mentioned in the book or in your lecture.
9. Why is it harder to climb a steep ramp than it is to climb a gentle slope?

Problems

1. If you push on a patient in a wheelchair with a 27-newton force, what is the force of the wheelchair on you?
2. A woman walks 50 meters down a hall, then turns around and walks 32 meters back toward her starting point. How far away is she from

her starting point? Draw the addition of the vectors (50 meters foward and 32 meters backward).

3. Show that the woman in Problem 2 ends up in the same spot if she first walks 32 meters backward and then 50 meters forward.

4. Can a 7-newton force vector and a 4-newton force vector add to form a 9-newton vector? A 2-newton vector?

5. A nurse leaves her desk and runs 32 meters down the hall. She then turns the corner into a second hall and walks 48 meters to a patient. Assuming that the two halls meet at a right angle, what is the straight-line distance between the nurse's desk and the patient?

6. A woman walks 1 kilometer west, turns and walks 3 kilometers south, then walks 2 kilometers southwest. How far is she from her starting point and in what direction? Add the three vectors in some other order and show that the sum is still the same.

7. Sled dogs are often tied to the sled in a fan hitch, in which each dog is simply tied to its own line and the dogs fan out in front of the sled. Suppose that each dog pulls with a 50-newton force and that the dogs fan out as shown in the illustration. What is the total forward force on the sled?

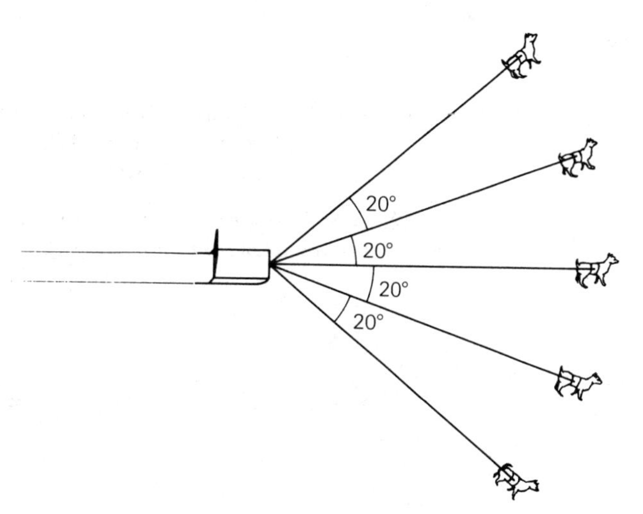

8. A sea captain heads his ship toward the northeast at 14.1 km/hr. Unfortunately, he doesn't know that a strong current is running at 10 km/hr toward the south. In what direction and at what speed will the ship actually move?

9. A coach wants to set up an exercise device in which the force gets larger as the distance moved increases. A spring arrangement would work, but no springs are available. So the coach sets up the pulley arrangement shown below. Find the force needed to hold the movable pulley 0.5, 1, 1.5, and 2 meters away from the wall.

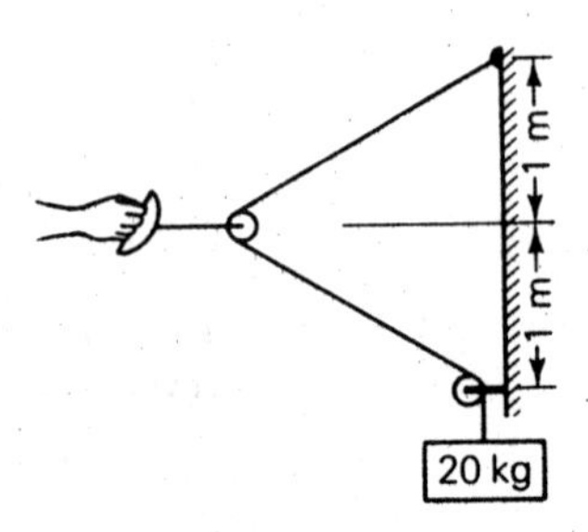

10. A car with a weight of 10 kilonewtons is climbing a steep hill as shown. What is the minimum force that the wheels must develop in order for the car to climb the hill?

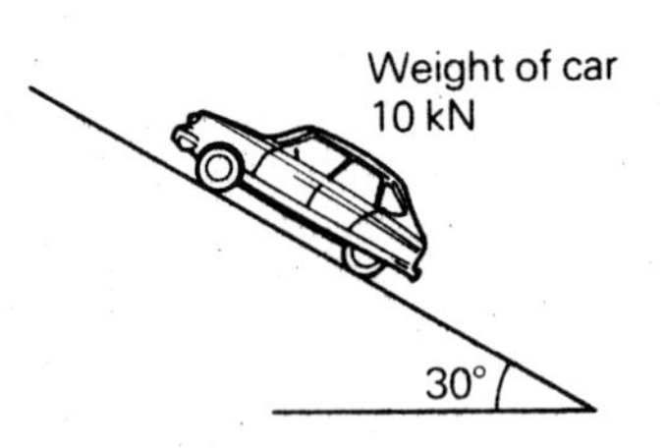

11. During a tug-of-war, the contestants lean in the direction in which they are pulling. Use vector addition to explain why leaning helps the contestants to pull harder.

12. A 50-kilogram woman is crossing a ravine on a rope as shown. What are the tensions in the two sections of the rope?

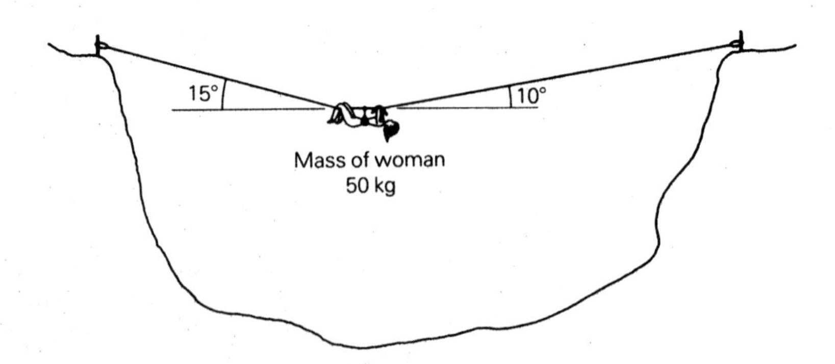

13. If the wheelchair residents of a rest home can all comfortably produce a forward force of 100 newtons and the heaviest patient weighs 1,000 newtons, how steep can the ramps be?

14. Adhesive tape cannot push—it can only pull. Thus, in the type of skin traction arrangement shown below, the direction of the net force of the three tapes always lies between the dashed lines. Keeping in mind that tape can only pull, try out various choices for the three forces and show that their sum always lies between the dashed lines. Choose the size of the three forces by randomly flipping through the text and letting the last digit of the page number be the size of the force. Try at least three examples yourself and then compare your results with those of other students.

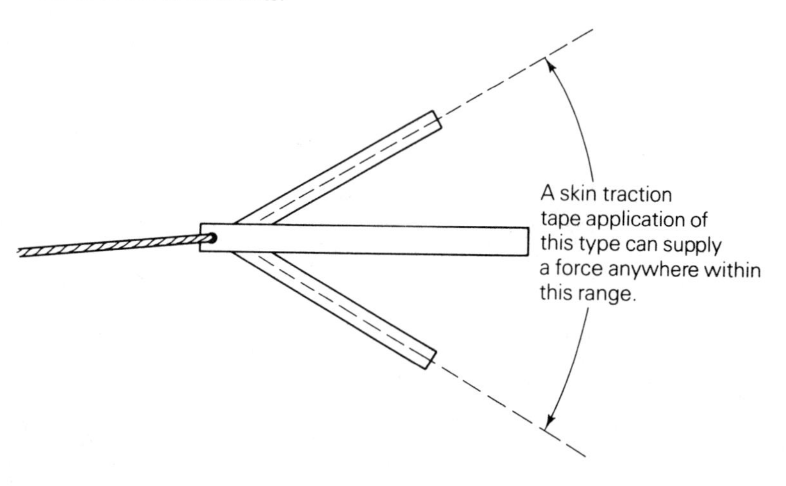

Experiments

1. Go to the middle of a large open field and mark your position with a stone or other object. Then step off the following vectors: 20 meters east, 15 meters northwest, 10 meters west, and 25 meters southeast. Mark your final position. (A good long step is about 1 meter for a person of average height.) Next repeat the process, but change the order of the vectors. Do you still end up in the same place? Do it again in some other order.

2. Hold a magnet in one hand and a piece of iron in the other. Observe that the force on the magnet always feels the same as the force on the iron except that the forces are in opposite directions.

3. You can get a good idea of the tension in a muscle simply by feeling it. Feel the tension in the erector spinae muscles of a friend when the friend is erect, bent over, and bent over lifting an object. The erector spinae run along the back of the spine, one on each side. Of course, your friend should not try to lift too heavy an object.

4. Build a model of a Russell traction apparatus with parts from your old Erector set or Tinkertoys. Verify that the force is indeed along the femur when the traction weights are properly chosen.

5. Tie a light string horizontally between two supports. Tie the string as tight as you can without breaking it. Next, hang weights from the middle of the string until the string breaks. Then show that the same string can easily support these weights when the string is vertical. Explain why this happens.

CHAPTER FIVE

Levers, Balance, and Stability

Educational Goals

The student's goals for Chapter 5 are to be able to:

1. Explain how levers magnify (or reduce) forces and motions.
2. Calculate the forces on a lever.
3. Identify how the bones function as levers.
4. Explain what the center of gravity of an object is and be able to estimate its location without measurements.
5. Identify stable and unstable body positions and be able to explain how to improve the stability of a person or object.

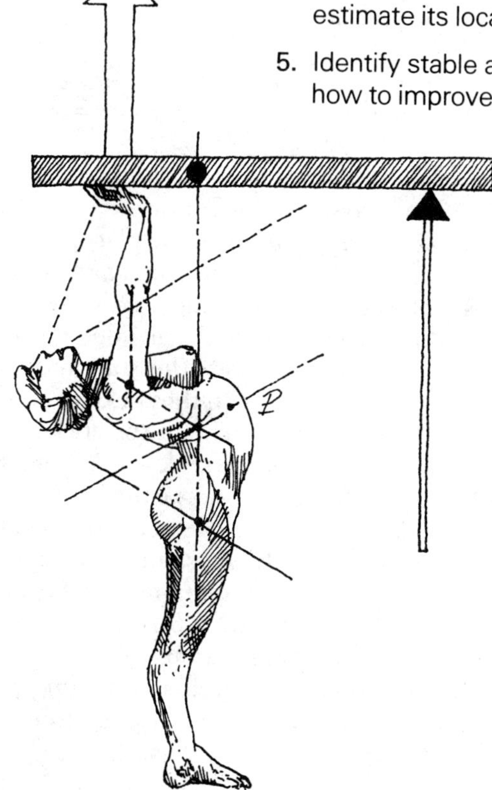

The bones of the body have many functions. They form the body's basic framework, they contain the bone marrow (which produces the red blood cells), and they protect the brain, eyes, and spinal cord from injury. However, the primary function of most of the bones is to act as levers to transmit the forces produced by the muscles. This is obviously the function of the bones of the arms and legs, but most of the other bones act as levers, too, even though providing leverage might not be their primary function. Consider the ribs, for instance. Their main function is to support the thoracic cavity; however, during inspiration they function as levers to magnify the motion of the respiratory muscles and to convert the muscle contraction into an expansion of the thoracic cavity.

There are two reasons why a rigid skeleton is needed for efficient motion. The first is that muscles, like ropes, cannot push very well. A muscle can only produce a large force when it is pulling. The second is that a muscle can only change its length by about 20%; thus, even the longest muscles of the body, the erector spinae, can change their length only by about 20 centimeters. Because of these limitations of muscle tissue, skeletonless animals, such as jellyfish and worms, can move in only a limited number of ways. In order for an animal to have a full range of movement, it must have a skeleton to act as a set of levers, magnifying the motions of the muscles and converting their pulls into pushes.

Rotational Motion

Things move in two different ways. They can move from place to place, and they can rotate while staying in the same place. Newton's first law of motion states that when the net force acting on an object is zero, the velocity of the object does not change. However, Newton's first law applies to the object as a whole; it does not cover rotation of the object. For instance, consider the stationary stick shown in Figure 5-1. The two forces acting on the stick are equal in size but opposite in direction, so the net force on the stick is zero. Therefore, Newton's first law tells us that the stick will not move off to some new place. But the two forces are not acting at the same *point* on the stick, so the stick will begin to rotate about its center. The center will, of course, stay in exactly the same spot, in agreement with Newton's first law. Complicated motions may be a combination of place-to-place motion and rotation; fortunately, the two aspects of motion can always be looked at separately.

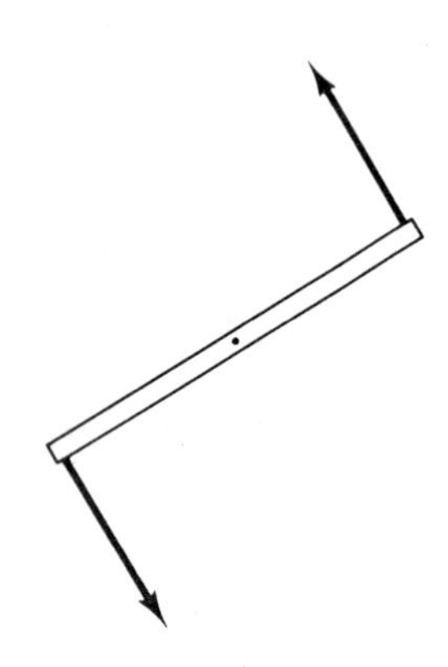

Figure 5-1

Thus, a force can change rotation, as well as changing place-to-place motion. This capability of a force, its twisting strength, is called *torque*. The first law of motion can be extended to include the fact that *the rotational speed of a rigid object will not change unless an outside torque acts upon it.* Note that this only applies to rigid objects—those objects that don't change shape. The rotational speed of a flexible object can change somewhat without any outside forces or torques acting on it. Since the human body is most definitely flexible, its rotational speed can change without any outside torque acting on it. We will look at the rotation of the body later, but right now we are concerned with stationary rigid objects. The above law of motion tells us there are *two* conditions that must be satisfied for a rigid object to remain at rest without rotation: The net force must be zero, and *the net torque on the object must be zero.*

Two other things besides the size of the force determine the twisting strength, or torque, that a force can produce. The first is the distance between the force application point and the center of rotation. The effect of this factor is obvious from our everyday experience. If we want to twist harder on a stuck bolt, we get a longer wrench so that we can move the point of application of the force farther from the center of rotation, thus producing a larger torque on the bolt. For the same reason, door knobs are placed on the opposite side of the door from the hinges so that we can produce the largest possible torque when we pull on the knob. If the knob was on the same side as the hinges, the door would be very hard to open.

The direction of the force is the second thing that affects the torque. In example A in Figure 5-2, the two forces are both directed toward the center of rotation; they produce no torque at all. In example B, the two forces are perpendicular to the lever, producing the maximum possible torque. When the force is perpendicular to the lever, the torque, which we will abbreviate by the Greek letter τ (tau), is just the product of the force F and the distance l from the force application point to the center, which is called the lever arm. Expressed as a formula, the torque is

$$\tau = Fl.$$

If the lever isn't straight, the lever arm is the imaginary line connecting the center and the point of application. In such a case, we would obtain the maximum torque when the force was perpendicular to this line instead of perpendicular to the actual lever. The SI unit of torque is the **newton-meter** (N · m). In the illustration in Figure 5-3, the torque would be

$$\begin{aligned}\tau &= Fl \\ &= 5\text{N} \times 0.5 \text{ m} \\ &= 2.5 \text{ N}\cdot\text{m}.\end{aligned}$$

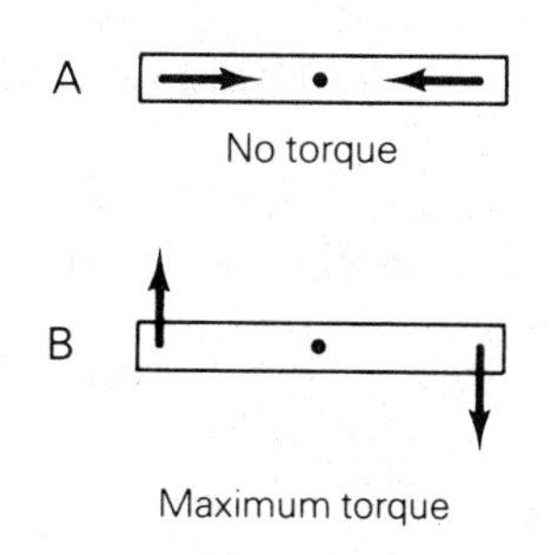

Figure 5-2

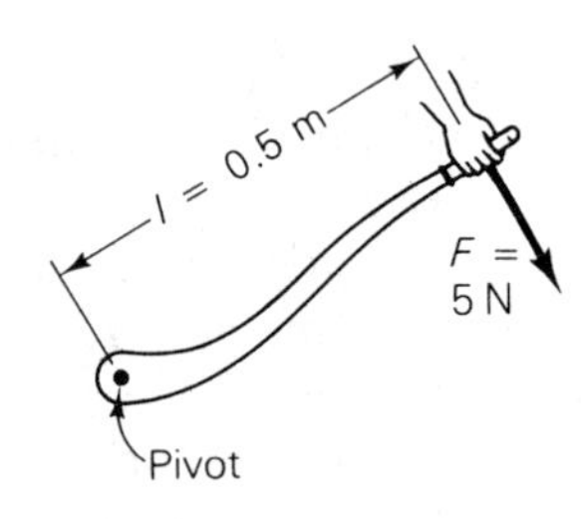

Figure 5-3

It obviously makes a difference whether the force is trying to produce a clockwise or a counterclockwise rotation. We will handle this problem by referring to clockwise torques as positive torques and counterclockwise torques as negative torques. For example, the torque in Figure 5-3 is a positive torque, since the force pushes the lever clockwise. A positive torque would tighten a normal right-hand screw like a bottle top, a toothpaste cap, or a normal bolt; a negative torque would loosen it. By calling clockwise torques positive and counterclockwise torques negative, the numerical sum of two equal but opposite torques is zero. This is as it should be, since two such torques would simply cancel each other.

Effective Lever Arms

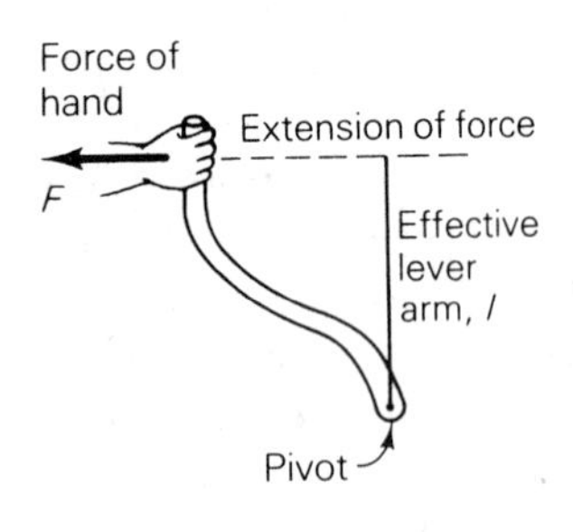

Figure 5-4

When a force does not act directly perpendicularly to the line between the center and the force application point, the effective lever arm is found by first drawing a line along the force to extend whichever end of the vector is closer to the center. Next, a second line is drawn through the center perpendicular to the first line. The length of this second line, from the center to the extension of the force, is the effective lever arm (Figure 5-4). We shall also call it l. Note that the effective lever arm also happens to be the shortest distance from the center to the extension of the force. The torque would again be $\tau = Fl$.

Finding effective lever arms involves careful scale drawings, just as vector addition did. Let's look at an example. Suppose the tendons were pulling on the bones of the lower leg, the tibia and fibula, with a 100-newton force (Figure 5-5). The tendons are connected to the bones 0.06 meter from the pivot point in the knee, but the force is not perpendicular to the lever arm—it is at a 60° angle. To find the effective lever arm, we draw a line along the force (Figure 5-6). Then we draw a second line perpendicular to the first that goes through the pivot point at the knee (Figure 5-7). By measuring our scale drawing, we see that the effective lever arm is 0.052 meter, and the torque is therefore τ = 100 newtons × 0.052 meter = 5.2 newton-meters.

A common error of weekend athletes is to try to go beyond their normal range of movement to get just a "little bit more" on their golf drive, tennis serve, or whatever. This is a mistake because the effective lever arms are small near the ends of most bodily movements, and when the muscular forces act on short effective lever arms, the torques are not very big. By stretching beyond the normal range of movement, the weekend athlete usually loses far more in control than he or she gains in power. (Usually people

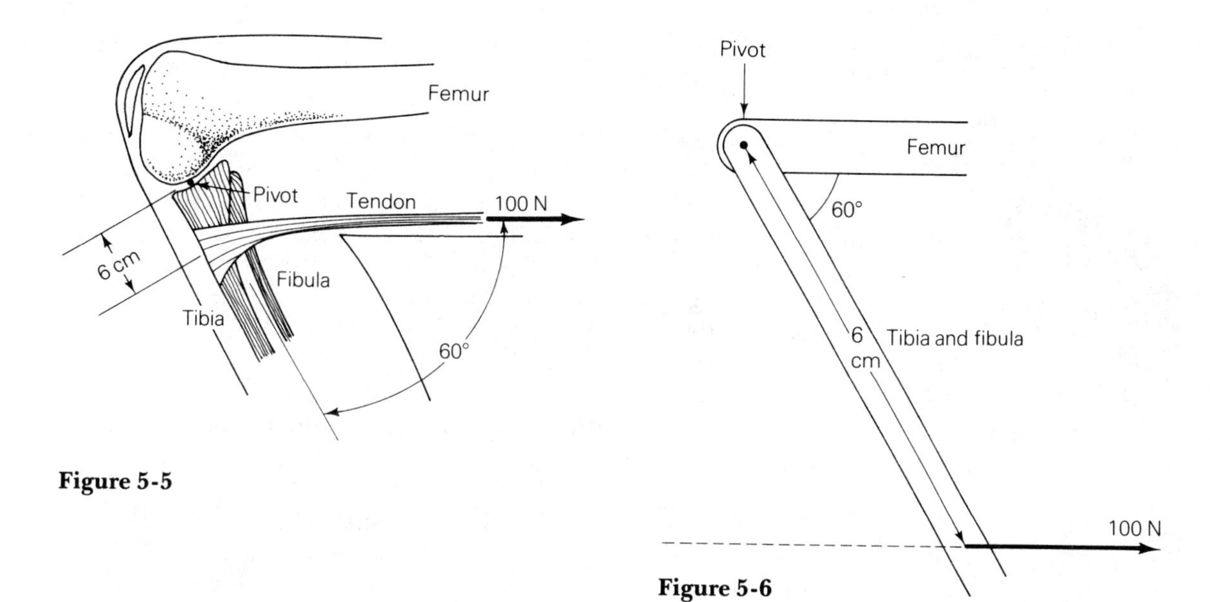

Figure 5-5

Figure 5-6

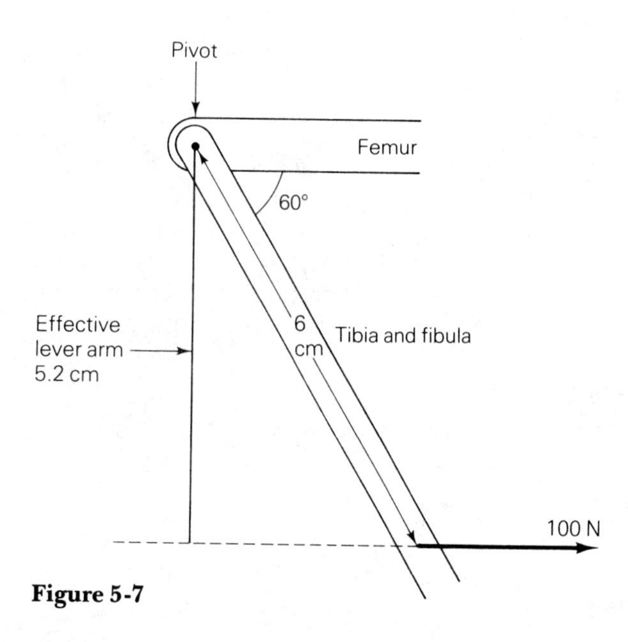

Figure 5-7

think that overstretching is effective because they see pictures of professional athletes taking very full windups. But they forget that the professional is *not* going beyond his or her normal range of movement. The pro, who is constantly working to stay in condition, has a far greater range of movement than the person who exercises once or twice a week.)

Levers

Pry bars, wrenches, artificial arms, baseball bats, steering wheels, bicycle pedals, and most of the bones of the skeleton are levers. One thing that levers are very useful for is transmitting forces from place to place. For example, the bones of the lower arm (the radius and ulna) act as a lever to transmit the forces produced by the biceps and triceps muscles to the hand. Levers can also be used to change the direction of a force. When you push down on a table with your hand, the triceps is pulling up on the ulna, as shown in Figure 5-8.

In addition to transmitting forces and changing their direction, there are four other purposes for which levers are commonly used:

1. *Magnifying motions.* Baseball bats, tennis rackets, and golf clubs are all levers used to magnify the motion of the arms. Indeed, the arm itself functions as a lever to magnify the motion of the muscles. When the arm is flexed, the hand moves about 20 times further than the biceps contract. Because muscles can only change in length by a small amount, magnifying muscular motions to achieve a full range of movement is an important function of many of the bones of the body.

2. *Reducing motions.* There are not many examples of motion reduction by levers in the body. One of the few examples occurs at the three small bones of the middle ear. The hammer, anvil, and stirrup act as a lever system to reduce the size of the sound vibrations as they transmit the vibrations from the eardrum to the inner ear. Outside of the body, control levers are a good

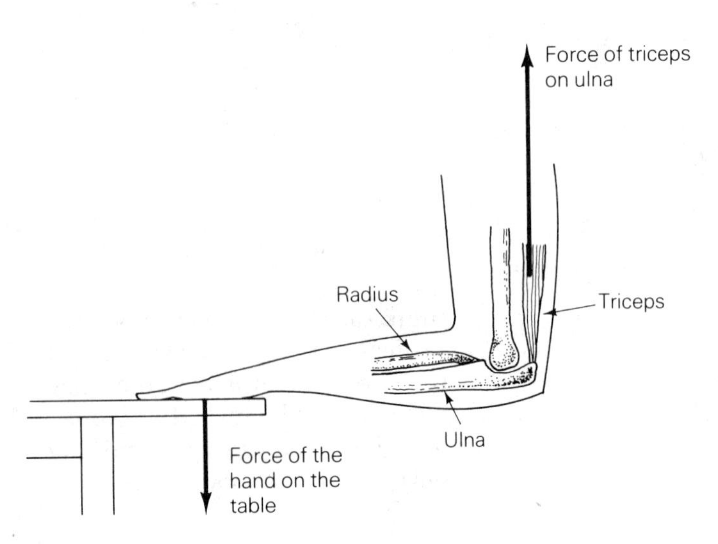

Figure 5-8 The lower arm acting as a lever to change the direction of the force of the triceps muscle.

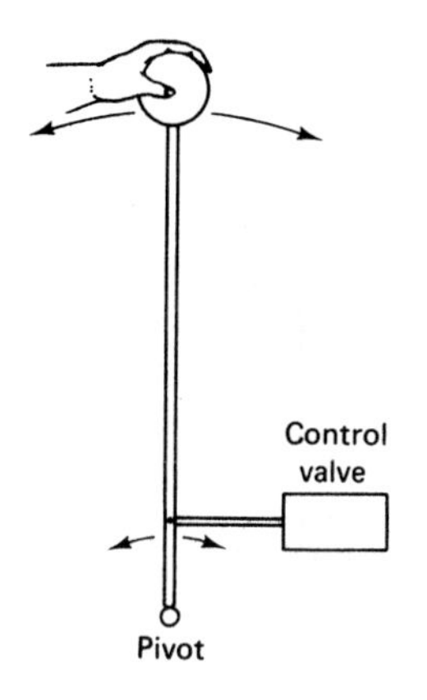

Figure 5-9 Most control levers—like a shift lever on a car—reduce the motion of the hand.

example of levers used to reduce motion. Since we often want to move things with greater precision than is possible by using the fingers alone, levers are often used to reduce the motion of our hands and fingers. The control levers on electronic equipment, the gearshift on a car (Figure 5-9), the handles on fluid control valves, and switch handles are examples of such levers.

3. *Magnifying forces.* This is the purpose that usually comes to mind when we think of levers. Indeed, it is the most common application of levers. Pry bars, jack handles, bicycle handlebars, cranks, pliers, and wrenches are all levers that magnify the forces produced by the human body. Within the body, however, there are only a few examples of force magnification. One instance occurs when a wrestler holds an opponent in a headlock. In a headlock, the lower arm acts as a lever to magnify the force of the wrestler's other hand upon the opponent's head (Figure 5-10).

4. *Reducing forces.* An example of a lever used for force reduction is the seesaw when the two people don't weigh the same. By placing the heavier person closer to the pivot, the upward force on the lighter person can be reduced from what it would be if the two people were the same distance from the pivot. With a few exceptions, such as the headlock hold in wrestling, the bones of the skeleton reduce the forces produced by the muscles.

We might wonder why the bones don't multiply *both* the motion *and* the force of the muscles. After all, both a full movement range and strength are desirable. The reason is part of the basic physics

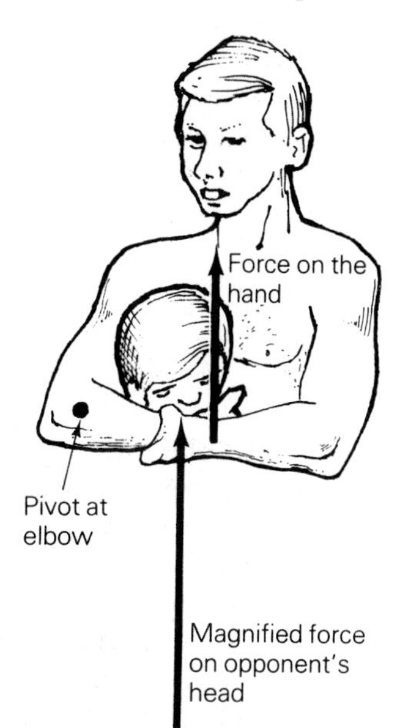

Figure 5-10 In a headlock, the lower arm acts as a lever to magnify the force on the opponent's head.

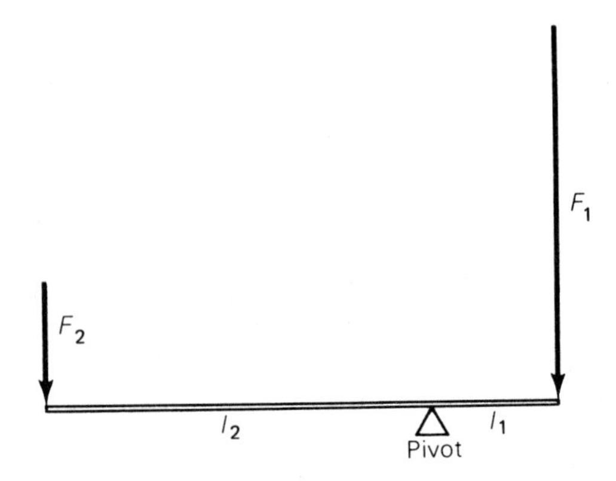

Figure 5-11

of levers: If the motion is magnified, the force will be reduced; if the motion is reduced, the force will be magnified. A lever that would magnify both the force and the motion at the same time would generate energy from nothing. This is impossible, as we will see in Chapter 8. Nor is it possible to build a simple lever that will reduce both at the same time. Since the body can't have both great strength and a wide range of motion, we have the trait that had the greatest survival value to our ancestors—a wide range of motions.

Next let's see how levers magnify (or reduce) forces. Consider the lever shown in Figure 5-11. As long as the lever is stationary or rotating at a constant speed about the pivot, we know that the torque about the pivot must be zero. Therefore,

$$\tau = (F_1 l_1) + (-F_2 l_2)$$
$$= 0.$$

The minus sign indicates that the torque produced by F_2 is counterclockwise about the pivot. We can rearrange the above equation as either

$$F_1 l_1 = F_2 l_2$$

or

$$\frac{F_1}{F_2} = \frac{l_2}{l_1}.$$

So, the ratio of the forces is the *inverse* of the ratio of the lever arms. That is, the large force must occur at the short end of the lever, and the small force at the long end. Of course, we know this from our everyday use of levers—when we pry up something with a lever, we put the short end under the object and apply the force as far out on the long end as possible.

Knowing that the ratio of the forces is the inverse of the ratio of the lever arms, a person can analyze levers as fast as he or she can do the arithmetic. It really is that simple. However, in our examples of levers, we will go into all of the details so that you won't miss any points. But keep in mind that with a bit of practice, these details aren't really necessary. After you go through a few examples in detail, you should try to reanalyze them using the simple relation that the ratio of the forces is the inverse of the ratio of the lever arms.

EXAMPLE If a force of 15 newtons is applied to the handles of the forceps shown in Figure 5-12, what force is produced on the bone chip held by the forceps? Assume that the handles are stationary, so that the net torque on either handle is zero.

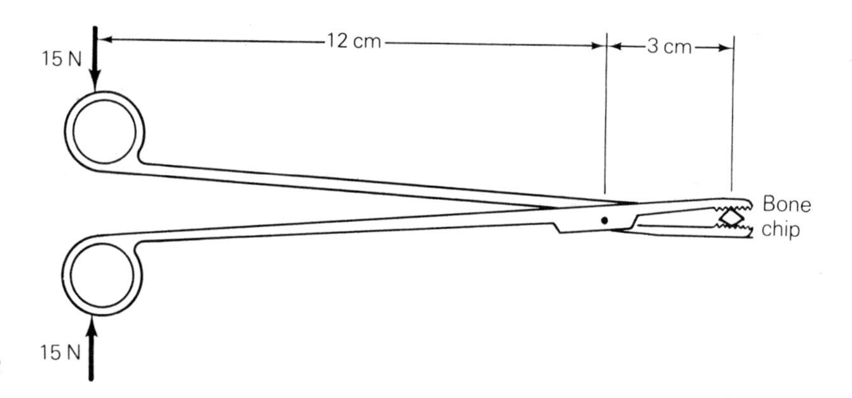

Figure 5-12

Answer Consider the upper handle (Figure 5-13). Call the unknown force on the bone chip F. Call the torque produced by the fingers on the handle τ_h and the torque produced by the force of the bone chip τ_b. The torque on the handle would be

$$\tau_h = -15\ \text{N} \times 0.12\ \text{m}$$
$$= -1.8\ \text{N}\cdot\text{m}.$$

The minus sign indicates that this torque is in the counterclockwise direc-

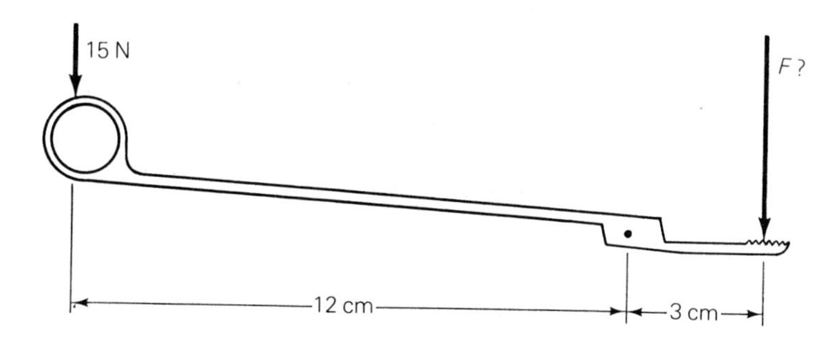

Figure 5-13

tion. The torque caused by the force of the bone chip is

$$\tau_b = F \times 0.03 \text{ m},$$

since the force of the forceps on the bone chip is equal to the force of the bone chip on the forceps. Since the net torque on the upper handle is zero,

$$\tau_b + \tau_h = 0.$$

Rearranging this formula, we see that

$$\tau_b = -\tau_h.$$

Therefore,

$$F \times 0.03 \text{ m} = -(-1.8 \text{ N}\cdot\text{m}),$$

so

$$F = \frac{1.8 \text{ N}\cdot\text{m}}{0.03 \text{ m}} = 60 \text{ N}.$$

This is 4 times greater than the force applied at the handles, so the forceps have a mechanical advantage of 4. We could have arrived at the figure of 60 newtons more rapidly by simply observing that the ratio of the lever arms was 1 to 4, so the ratio of the forces must be just the inverse, 4 to 1. Thus, the force on the bone chip F must simply be 4×15 newtons $= 60$ newtons.

This example illustrates one of the most common uses of levers: multiplication of the force that a person can produce. However, as is true of all nonpower devices that intensify force, the gain in force wasn't accomplished without the loss of something else: The motion of the forceps jaws was only one-fourth the motion of the handles.

One question arises when we look at Figure 5-13. Both the force of the fingers and the force of the bone chip are pushing down on the handle (Figure 5-14). But if the handle is stationary, some other

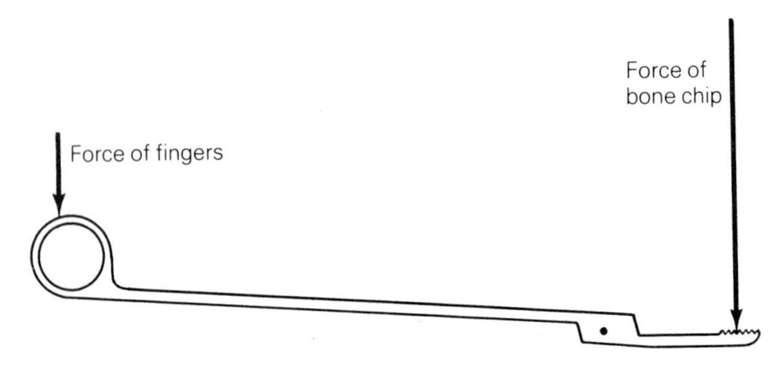

Figure 5-14

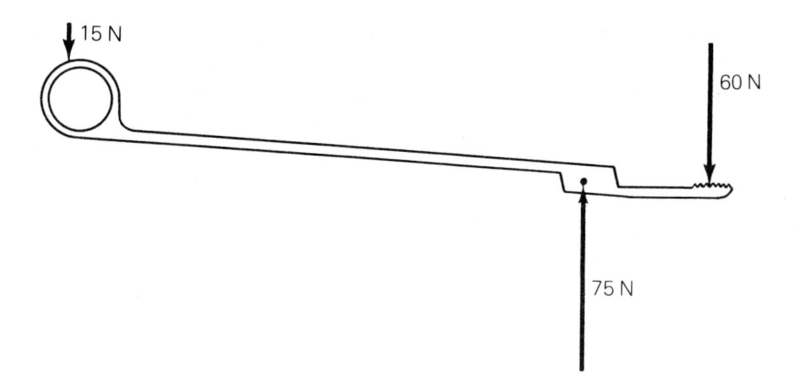

Figure 5-15

force must be acting on it so that the net force is zero. What is that other force, and why did we ignore it in the torque calculation? The other force must act at the only other point where something touches the upper handle, so it must be the force of the lower handle acting at the pivot. In order that the net force on the upper handle be zero, the force on the pivot must be 60 newtons + 15 newtons = 75 newtons acting straight up to balance the two downward forces exactly. We could ignore this force in the torque calculation because it is applied exactly at the pivot point (Figure 5-15). Thus, it *has no lever arm* about the pivot and *produces no torque* about the pivot.

The fact that no torque is produced by a force acting at the pivot can often be used to advantage. We often have some choice in picking our pivot points. For instance, if the lever is stationary—and thus not pivoting about any point—it really doesn't make any difference which point we choose as the pivot. In fact, if the lever is stationary, the net torque about *any* point will be zero. Therefore, if an unknown force is acting at some point, we can save ourselves a great deal of calculation by *choosing the point of application of the unknown force as the pivot.* If we do this, the unknown force is eliminated from the torque calculation. The next example illustrates this procedure.

EXAMPLE There is no reason that a lever must be made from one straight bar. In fact, many common levers are not straight. The jawbone (mandible) is one example of such a lever in the body. The hammer when it is used to pull out nails is another. Suppose the carpenter is pulling on the handle with a force of 400 newtons (Figure 5-16). What force is pulling on the nail?

Answer If we assume that the nail is not moving, we know that the net torque on the hammer is zero. The first question to answer is, Where is the pivot point? Or perhaps we should say, Which point shall we choose as the

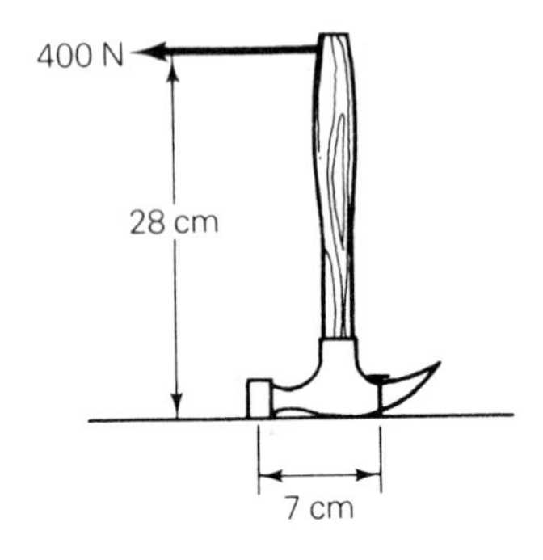

Figure 5-16

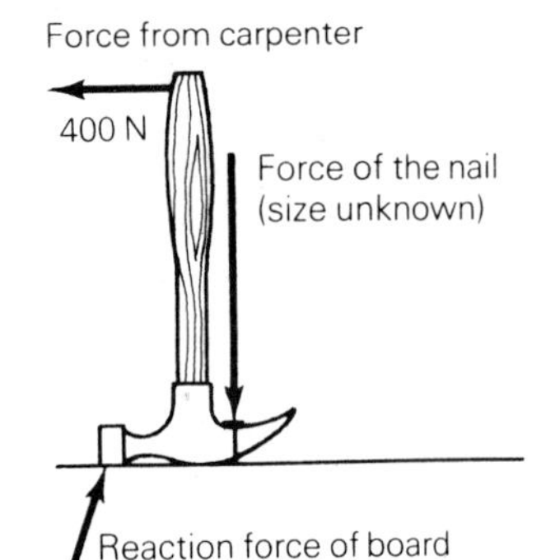

Figure 5-17 The forces on a hammer while pulling a nail.

pivot point? If the hammer is not moving, one point is as good as any other. We should first see if any unknown forces are acting on the hammer. There is one—the reaction force of the board pushing upward and to the right on the head of the hammer. We will pick that point as the pivot point, because the reaction force does not have a lever arm about the head of the hammer. Thus, it does not produce any torque about that point. Let the force on the handle F_h be 400 newtons, and the unknown force of the nail be F_n (Figure 5-17). We will call the lever arm of the handle $l_h = 0.28$ meter, and the lever arm from the pivot to the nail $l_n = 0.07$ meter. Since the hammer is not moving the net torque on it must be zero; that is,

$$\tau_n + \tau_h = 0$$

or

$$\tau_n = -\tau_h.$$

In terms of the forces and lever arms, this equation becomes

$$F_n l_n = -(-F_h l_h).$$

The second minus sign was inserted before the F_h because we know that the force on the handle produces a negative torque. Solving for F_n, we obtain

$$F_n = \frac{F_h l_h}{l_n}$$

$$= \frac{400 \text{ N} \times 0.28 \text{ m}}{0.07 \text{ m}}$$

$$= 1{,}600 \text{ N}.$$

This is the force of the nail on the hammer, but the force on the nail, of course, is equal but opposite to it, according to Newton's third law of motion. So the force of the hammer on the nail is 1,600 newtons straight up. Again, we could have found the force more rapidly by noting that the handle is 4 times longer than the lever arm of the nail puller, so the mechanical advantage of the hammer is 4. Thus, the force on the nail is 4 times larger than the force on the handle. The reaction force on the hammer head is just equal but opposite to the vector sum of the forces on the handle and on the nail puller, since we know that the net force on the hammer must be zero.

EXAMPLE Find the tension in the biceps muscle when a 15-kilogram mass is held in the hand as shown in Figure 5-18. The mass of the lower arm is much less than 15 kilograms, so we can ignore the small extra tension in the biceps caused by its weight. Also, find the contact force at the elbow.

Answer Since we must eventually find all of the forces on the lower arm, either the elbow or the insertion of the biceps on the radius (point of connection of the biceps to the bone in the lower arm) would be a conven-

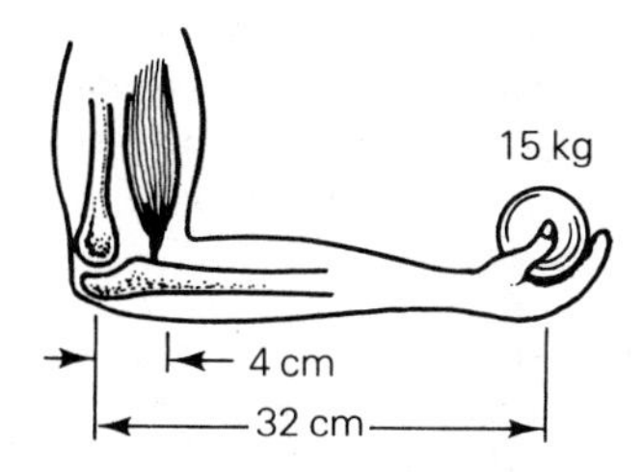

Figure 5-18

ient point to pick as the pivot. We will choose the elbow, but choosing the insertion point would not make the calculation any harder.

The weight acting downward on the hand is

$$W = 15 \text{ kg} \times 9.8 \text{ m/s}^2$$
$$= 147 \text{ N}.$$

Thus, the torque caused by the weight in the hand is

$$\tau_h = W l_h$$
$$= 147 \text{ N} \times 0.32 \text{ m}$$
$$= 47 \text{ N·m}.$$

If we let F_b be the tension in the biceps, the torque produced by the biceps is

$$\tau_b = F_b l_b$$
$$= F_b \times 0.04 \text{ m}.$$

The arm is not moving, so the sum of these two torques must be zero:

$$\tau_h + \tau_b = 0.$$

So

$$\tau_b = -\tau_h$$

and

$$F_b \times 0.04 \text{ m} = -47 \text{ N·m}.$$

Therefore,

$$F_b = -\frac{47 \text{ N·m}}{0.04 \text{ m}}$$
$$= -1{,}200 \text{ N}.$$

The minus sign simply indicates that this force produces a counterclockwise torque.

The contact force at the elbow can be found because it is the only other force acting on the lower arm, and we know that the net force on the lower arm must be zero. The weight of the 15-kilogram object and the force of the biceps acting upward add to

$$1{,}200 \text{ N} - 147 \text{ N} = 1{,}050 \text{ N}$$

acting upward. Therefore, the contact force on the arm at the elbow must be 1,050 newtons acting downward.

EXAMPLE A 75-kilogram woman is off for a brisk walk. Find the tension in the Achilles tendon (tendo calcaneus) and the contact force at the ankle joint

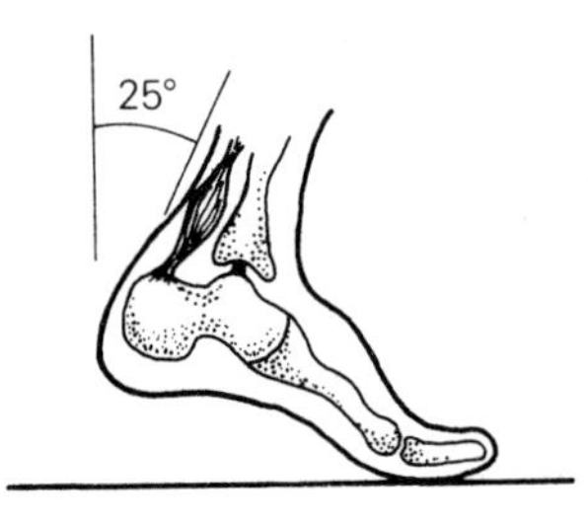

Figure 5-19

when all of her weight is on the foot shown (Figure 5-19). This would be the position of the rear foot just before the forward foot makes contact with the ground. Note that her rear foot is in contact with the ground, and it is neither moving rapidly nor changing its motion rapidly. Thus, both the net torque and the net force on the foot will be nearly zero.

Answer Three forces are acting on the foot. At the toes, the ground reaction force pushes upward with a force that is equal but opposite to her weight: 735 newtons. At the heel, the Achilles tendon is pulling upward and to the right. We don't know its size, but we do know its direction; tendons, like muscles and ropes, can only pull along their lengths. The contact force at the ankle is probably down and to the left, but we really can't say what its size and direction are (Figure 5-20). We know the least about the contact force at the ankle, so that is the best point to pick as the pivot. The effective lever arms of the forces are found by extending the forces where necessary, drawing lines from the pivot perpendicular to the extended forces, and then measuring the lengths of the lines. This is shown in Figure 5-21. Let the tension in the tendon be F_t. The fact that the net torque on the foot is nearly zero leads to

$$F_t \times 0.06 \text{ m} = 735 \text{ N} \times 0.11 \text{ m},$$

so

$$F_t = \frac{0.11 \text{ m} \times 735 \text{ N}}{0.06 \text{ m}}$$

$$= 1{,}350 \text{ N}.$$

The contact force at the ankle is the only other force on the foot. The vector sum of the three forces on the foot must be zero, so the force of the ankle must be equal but opposite to the vector sum of the other two. By measuring the size and direction of the ankle contact force in Figure 5-22, we see that it is 2,000 newtons down and to the left at a 16° angle.

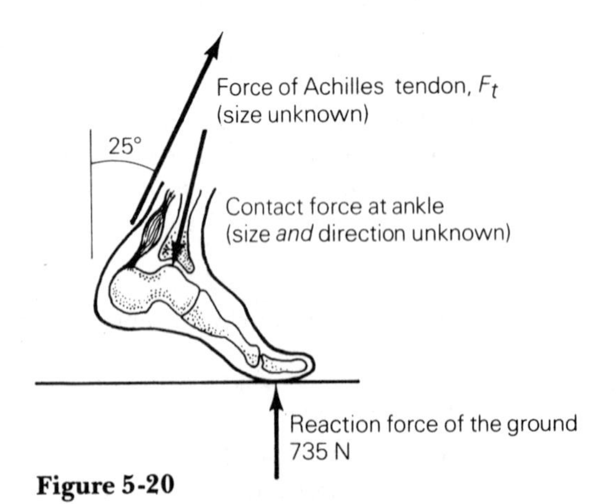

Figure 5-20

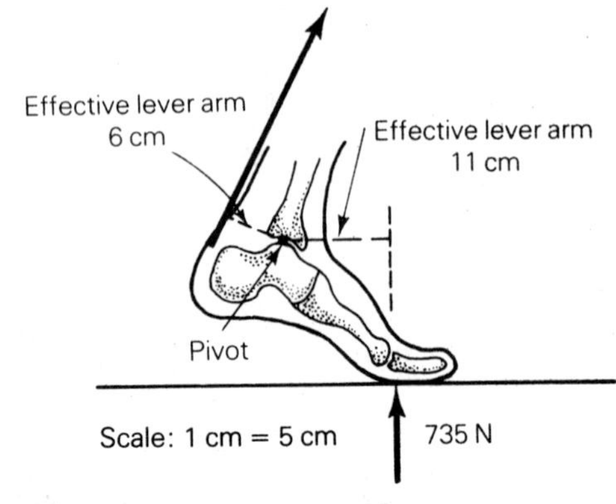

Figure 5-21

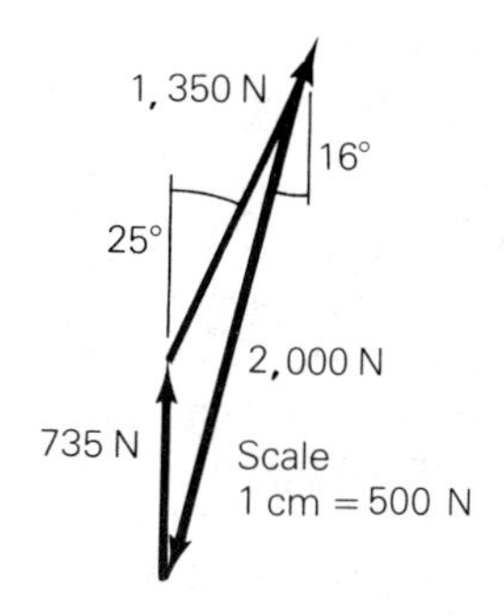

Figure 5-22

The last two examples point out an unavoidable consequence of the body being compact with a wide range of movement: *The stresses in the muscles, tendons, and joints are quite large, even during mild activity.* In the example of a weight held in the hand, we saw that the tension in the biceps was 8 times greater than the weight of the object in the hand. It was also more than the weight of a very heavy person. In the last example, we found that the contact force at the ankle when walking was over 2,000 newtons, about the weight of three average people. If mild activities produce forces this large, it is clear that strenuous activities could produce damaging forces if done carelessly. Coaches, physical therapists, and others who design exercises should keep in mind just how large the forces in the body can be and make sure that the activities that they supervise don't lead to overstraining and injury.

The Center of Gravity

So far we have only considered light levers—levers whose weights are small compared with the outside forces acting on them. For such a lever we can ignore the torque produced by its weight. But many times the weight of the lever is not small and must be taken into account in a torque calculation. How do we go about finding the torque caused by the weight of a lever? Consider an outstretched arm, for example (Figure 5-23). Theoretically, we could calculate the torque atom by atom. The effective lever arm for each atom could be measured and multiplied by each atom's weight; then all the individual torques could be added to find the total torque. Obviously, this is an impractical procedure. However, if the problem does not involve any changes in translational or rotational speed, we can treat the object as if all its mass were concentrated at its balance point. The balance point of an object is usually called the center of gravity or center of mass. We will abbreviate it as CG.

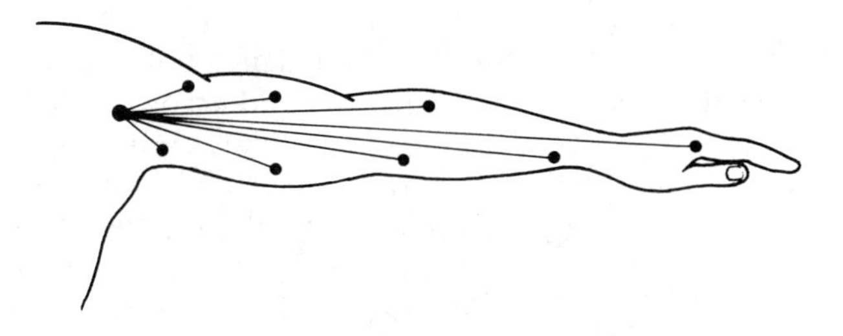

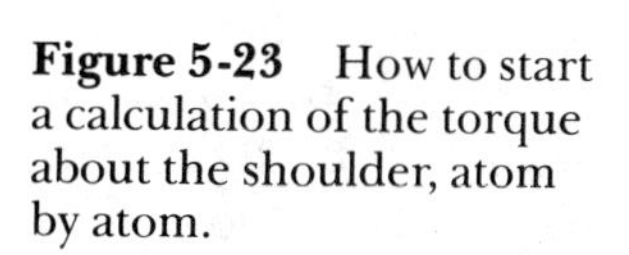

Figure 5-23 How to start a calculation of the torque about the shoulder, atom by atom.

EXAMPLE The lower leg is held straight out as shown in Figure 5-24. Estimate the tension in the quadriceps tendon.

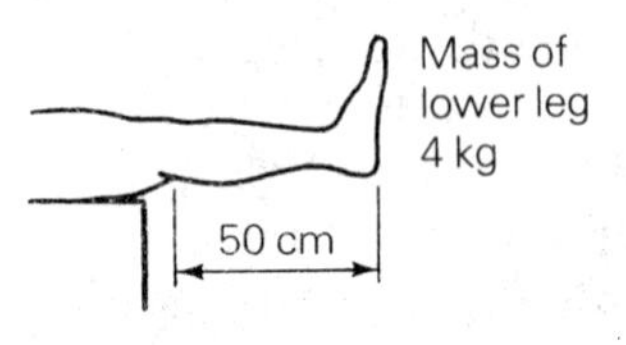

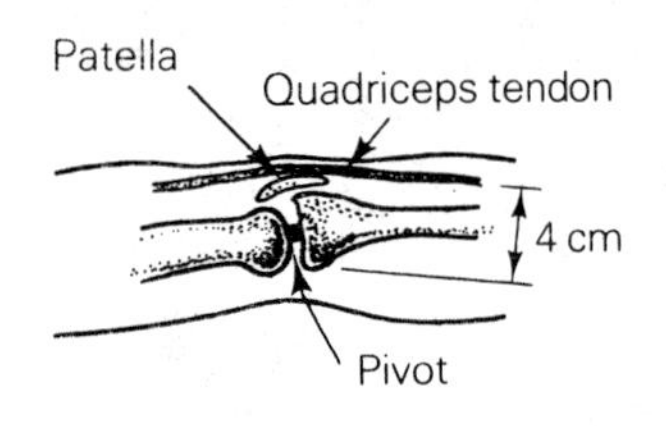

Figure 5-24

Answer First we have to estimate the mass of the lower leg and foot; then we must estimate where the balance point is. A good way of estimating the mass is to place the lower leg in a tank full of water and then measure the mass of the water that overflows. Since the body is mostly water, its density is almost the same as the density of water, so the mass of the water that overflows is nearly equal to the mass of the lower leg. Suppose we find that the mass of the lower leg and foot is 4 kilograms. There is no easy way of finding the exact balance point of the lower leg of a living person, but it must be fairly close to the middle of the lower leg. Actually, the middle is probably a good estimate, since the extra mass of the muscles near the knee is about the same as the mass of the foot. If the lower leg was 50 centimeters long, the CG would be about 25 centimeters from the knee (Figure 5-25). The torque produced by the weight of the lower leg and foot will be

$$\tau_l = 4\text{ kg} \times 9.8\text{ m/s}^2 \times 0.25\text{ m}$$
$$= 9.8\text{ N}\cdot\text{m}.$$

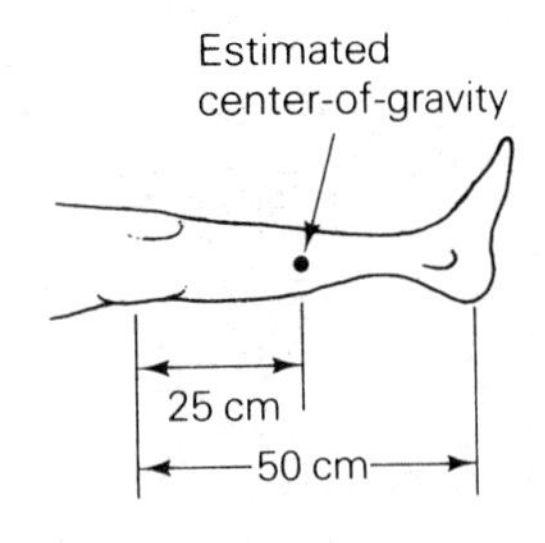

Figure 5-25

The actual pivot in the knee will be chosen as the pivot for the calculation, because the unknown reaction force of the upper leg passes through that point. The torque produced by the quadriceps tendon would be

$$\tau_q = F \times 0.04\text{ m},$$

where F is the tension in the quadriceps tendon. The sum of these two torques is zero, so

$$-F \times 0.04\text{ m} = 9.8\text{ N}\cdot\text{m}.$$

Thus,

$$F = -\frac{9.8\text{ N}\cdot\text{m}}{0.04\text{ m}}$$
$$= -250\text{ N}.$$

Again, the minus sign only indicates that the tension produces a counterclockwise torque.

Finding the balance point of most objects is not very difficult. One way is by trial and error. An especially quick way of finding the CG of long objects like baseball bats and golf clubs is to place the object on your outstretched hands as shown in the photo, then simply bring your hands together smoothly. The CG will be above the point where your hands touch. The CG can also be located by calculation. We just mathematically find the spot where the gravita-

After the hands are brought together, the center of gravity of the club will be directly above the hands.

tional torques add up to zero. Locating the exact CGs of parts of the body is difficult; however, studying the forces within the body usually requires just a careful estimate of the CG locations.

We should keep in mind that the CG of an object is not a place inside the object where all the mass is actually concentrated, nor is it always inside the object. The CG of a doughnut, for example, is at the center of the hole! Many common objects, including pots, cups, tables, and a person bent in the jackknife position, have CGs that are outside of the objects themselves (Figure 5-26).

Figure 5-26 Some objects having a center of gravity outside the object. X is the location of the CG.

The CG of a rigid object is always at the same spot, but this is not true of a flexible object. For instance, the CG of an erect person is inside his or her abdomen, roughly behind the navel. But when the person bends over to pick up something, the CG moves outside the body, somewhere below the abdomen (Figure 5-27). The CG of a flexible object will almost always move when the object is bent. This must be kept in mind when we consider flexible objects—humans and animals in particular. The exact location of the CG depends on the shape of the object at the particular instant under consideration.

The location of the center of gravity depends on the person's build. Muscular people with broad shoulders have higher CGs than the average person; overweight people tend to have lower CGs. Because males tend to have stronger arm and upper body muscles and females have wider pelvises to accommodate child bearing, the

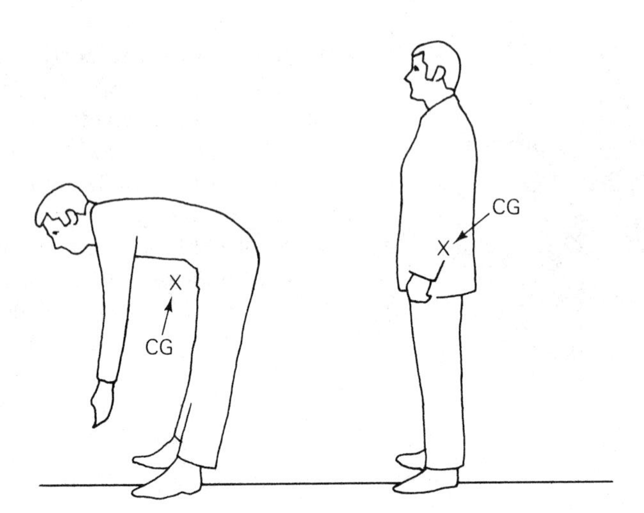

Figure 5-27 Location of the CG of the body in two positions.

CG will usually be a few centimeters higher on a male than on a female of equal height.

During a child's growing years, the CG is constantly moving downward. The newborn infant has a comparatively large head, a small body, and very tiny legs, so its CG is quite high in the body, almost at the armpits. As the child grows, the body and legs grow much faster than the head, gradually moving the CG downward. At age three, the CG is around the bottom of the rib cage; at age eight it is close to the adult position, around the top of the pelvic region (when the legs are outstretched and the hands are at the sides). There are, of course, more shifts as the child grows, reaches puberty, and adds muscle through the teens; however, these shifts are not very large compared with the shifts that occur between infancy and age eight.

The CG of a person can be quickly located by the following simple method, which is based on the principle that the CG of any object is also its balance point. Lay a wooden board on a piece of pipe. Next, have the person lie on the board with his or her navel near the middle of the board. Then roll the board on the pipe until it balances. When the board is balanced, the person's CG is directly above the pipe (Figure 5-28). Another accurate method is described in Experiment 5-5.

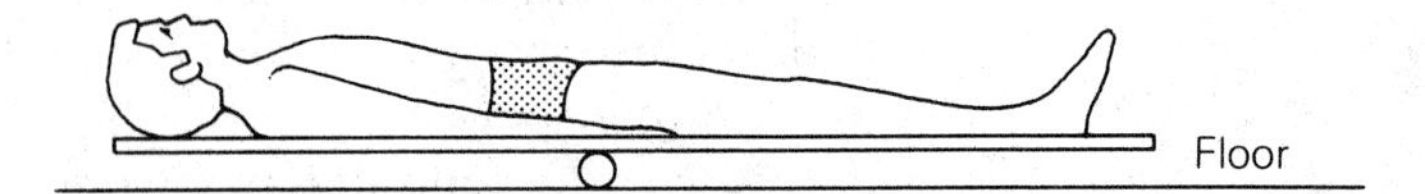

Figure 5-28 The pipe-and-board CG location method.

The Pole-Vaulter

Because no two people are exactly alike, there is no such thing as a standard style that all athletes in a given sport should try to follow. After all, athletes are individuals with different strengths and different weaknesses; what works well for one person might be totally unsuitable for another. During the development of an athlete's style, the athlete and his or her coach must carefully look at the athlete's strong and weak points. Then they must try to create a style that makes maximum use of the athlete's strengths while compensating for the athlete's weaknesses.

To see how the concepts we have been discussing can help in this process, let's determine how high a pole-vaulter should carry the pole while running to make the vault. The extremes found in competitive vaulting range between the cross-body or level carry, where the pole is held nearly parallel to the ground, to the high carry, where the pole is held up at about a 45° angle to the ground. An individual vaulter's style could be anywhere between these two extremes. How should the choice be made? What are the advantages of carrying the pole low? What are the advantages of carrying the pole high?

The principal advantage of a level (low) carry is that it makes controlling the drop of the pole tip into the trough easier, since the tip will only have to drop about a meter. In the high carry the tip will be over 4 meters in the air, and the vaulter will have to control the drop to the trough very carefully. But the high carry has its advantages, too. The most important is that the balance of the pole will be better.

Suppose that the vaulter uses a pole 6 meters long with a mass of 4 kilograms. Also suppose that the vaulter places the rear hand near the end of the pole and the forward hand 1 meter ahead of it. Now let's compare the force on the forward hand in a level and a high carry. We will assume that the vaulter will apply the force of the forward hand perpendicularly to the pole, because this will minimize the force needed. Since we are not looking at the force on the rear hand right now, we will choose the rear hand as our pivot point. We can think of the weight of the pole, 4 kilograms $\times$ 9.8 m/s^2 = 39 newtons, as being concentrated at the CG of the pole, which will be in the middle, 3 meters from either end if the pole is uniform. In a level carry, the situation will be as shown in Figure 5-29. If the net torque on the pole is to be zero, we must have

$$-F \times 1\text{ m} = 39\text{ N} \times 3\text{ m},$$

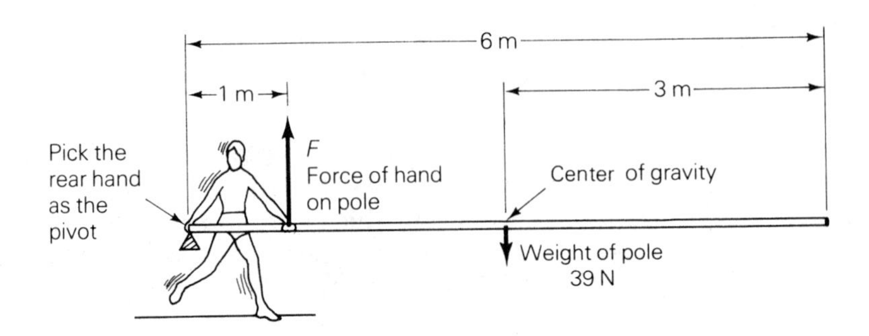

Figure 5-29 The forces on a pole-vaulter's pole during a level carry.

so

$$F = -\frac{39\text{ N} \times 3\text{ m}}{1\text{ m}}$$

$$= -117\text{ N}.$$

In the high carry, we would have the situation shown in Figure 5-30. The effective lever arm for the weight is not 3 meters, because the weight of the pole does not act perpendicularly to the pole. By making a scale drawing, we find that the effective lever arm is only 2.1 meters. Again, the net torque on the pole must be zero, which means that

$$-F \times 1\text{ m} = 39\text{ N} \times 2.1\text{ m},$$

so

$$F = -\frac{39\text{ N} \times 2.1\text{ m}}{1\text{ m}}$$

$$= -82\text{ N}.$$

Thus, the force on the forward hand is 30% less in the high carry than in the level carry, giving the pole better balance and making it easier to carry.

To find the force on the rear hand, we can use the fact that the net force on the pole must be zero; that is, the vector sum of the forces of the two hands must be 39 newtons straight up. This calculation (Problem 5-9) shows that the force on the rear hand is also less in the high carry than in the level carry.

Thus, the choice of pole position during the carry is a compromise between control and balance. If the vaulter is having difficulty with controlling the drop into the trough, a lower pole position might help. If the vaulter is having a problem with pole balance during the approach, a higher pole position might be the answer.

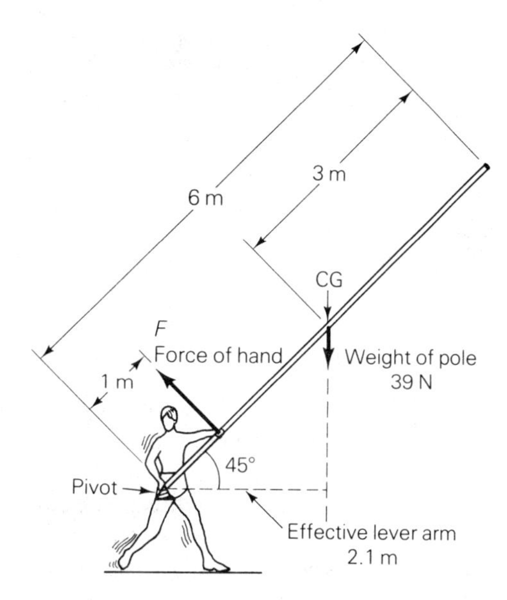

Figure 5-30 The forces on a pole-vaulter's pole during a high carry.

Forces at the Hip and Limping

A person who limps because of an injured hip usually shifts his or her weight *toward* the side of the injured hip (Figure 5-31). Offhand, this might seem strange, since moving the weight toward the injured hip would seem to place a greater strain on it and thus increase the pain. Of course, a person with a bad hip will always try to keep the pain to a minimum, which would require keeping the forces on the injured hip as small as possible. So, in some way, shifting the CG toward the injured hip must reduce the force on the hip.

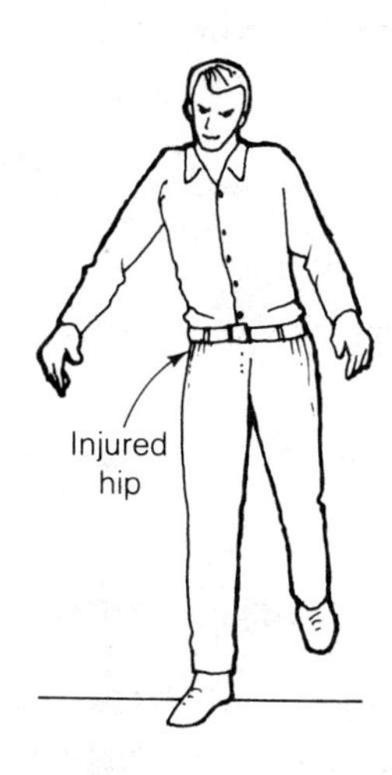

Figure 5-31 A person with an injured hip, limping.

To see how favoring the injured hip reduces the force on the hip joint, we will compare the forces on the leg during normal walking and while limping. The head of the femur (the upper leg bone) is ball-like and fits into a shallow socket (the acetabulum) in the pelvis (Figure 5-32). The advantage of the shallow socket is to allow the leg to have a very wide movement range. The extreme mobility of this joint is most evident in gymnasts, ballet dancers, and football running backs, but even normal walking would be impossible without it. However, the shallow socket has a drawback: The hip joint is easy to dislocate, and the femur must be pulled in (abducted) to the socket by strong abductor muscles and ligaments. Several muscles and ligaments abduct the femur to the pelvis; but during normal walking, when all the weight is on one foot, the gluteus medius and the gluteus minimus are the principal abduc-

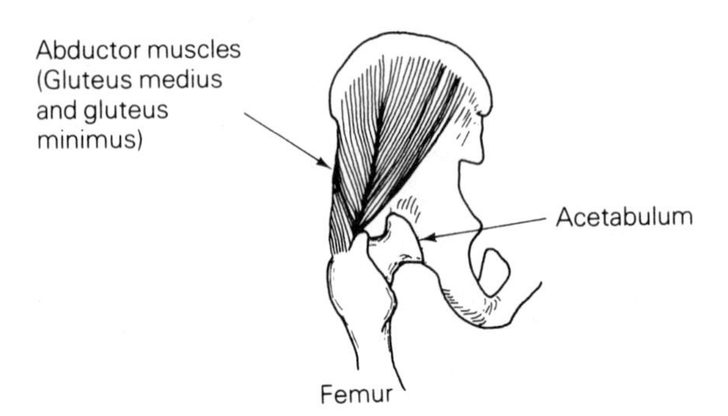

Figure 5-32 The hip.

tors. The location of these muscles and the forces on the lower leg when the person's weight is on one foot are shown in Figure 5-33.

Four forces are acting on the leg: (1) the reaction force of the ground, which is just equal but opposite to the person's weight (the person in the figure weighs 1,000 newtons, so this force is 1,000 newtons straight up; (2) the weight of the leg (150 newtons acting straight down at the CG of the leg, located slightly above the knee, as shown in the figure); (3) the force of the abductor muscles, the gluteus medius and the gluteus minimus (the size of this force is unknown, but the direction must be along the muscles, since muscles can only pull; their direction is about 20° from the vertical toward the pelvis); and (4) the force at the hip joint (this force is unknown in both size and direction; we know that this joint supports the rest of the body, so this force is roughly downward and must be greater than the weight of the remainder of the body). Since we know the least about the force at the hip joint, this joint would be the best place to choose for the pivot in our torque calculation.

The net torque about the hip joint must be zero, so

$$\tau_{\text{abductors}} + \tau_{\text{leg weight}} + \tau_{\text{ground}} = 0$$

$$(F_{ab} \times 0.065\text{ m}) + (150\text{ N} \times 0.01\text{ m}) - (1{,}000\text{ N} \times 0.10\text{ m}) = 0$$

$$F_{ab} = \frac{100\text{ N}\cdot\text{m} - 1.5\text{ N}\cdot\text{m}}{0.065\text{ m}}$$

$$F_{ab} = 1{,}500\text{ N}.$$

The force on the hip joint is now easily found, since it is the last unknown force. Since the net force on the leg must be zero, the force on the hip can be found as shown in Figure 5-34. This force is 2,300 newtons, over twice the weight of the person! This force is so large because the abductor muscles constantly pull the head of the femur into the socket in the pelvis; thus the hip joint carries the abductor muscle force as well as most of the person's weight.

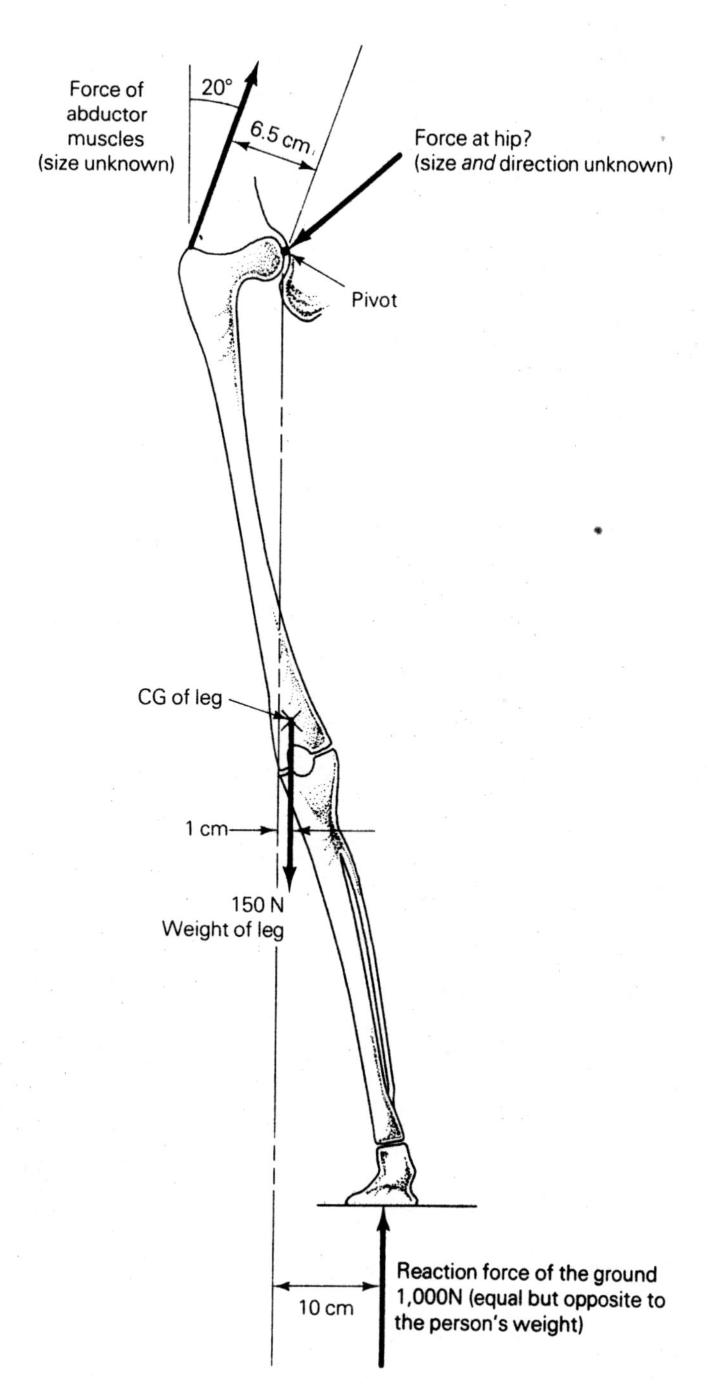

Figure 5-33 Forces on a normal leg when a person's weight is on it.

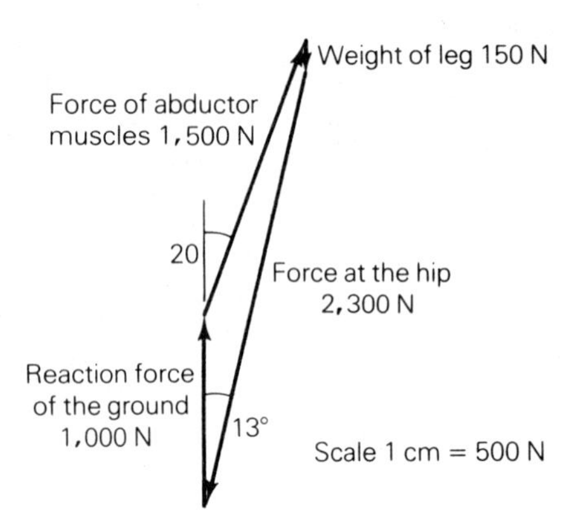

Figure 5-34

$$\tau_{\text{abductor}} + \tau_{\text{leg weight}} + \tau_{\text{ground}} = 0$$

$$(F_{ab} \times 0.07 \text{ m}) - (150 \text{ N} \times 0.02 \text{ m}) - (1{,}000 \text{ N} \times 0.03 \text{ m}) = 0$$

$$F_{ab} = \frac{3 \text{ N·m} + 30 \text{ N·m}}{0.07 \text{ m}}$$

$$F_{ab} = 470 \text{ N}.$$

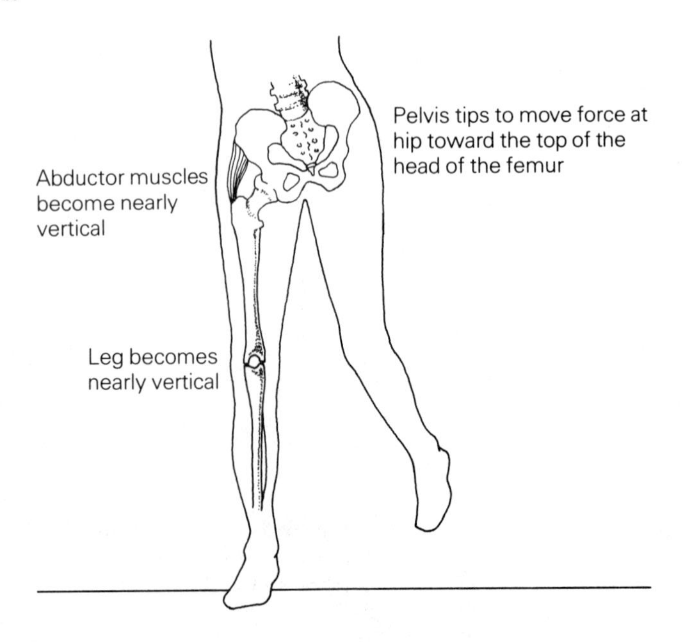

Figure 5-35 Changes of position of the leg and pelvis when limping.

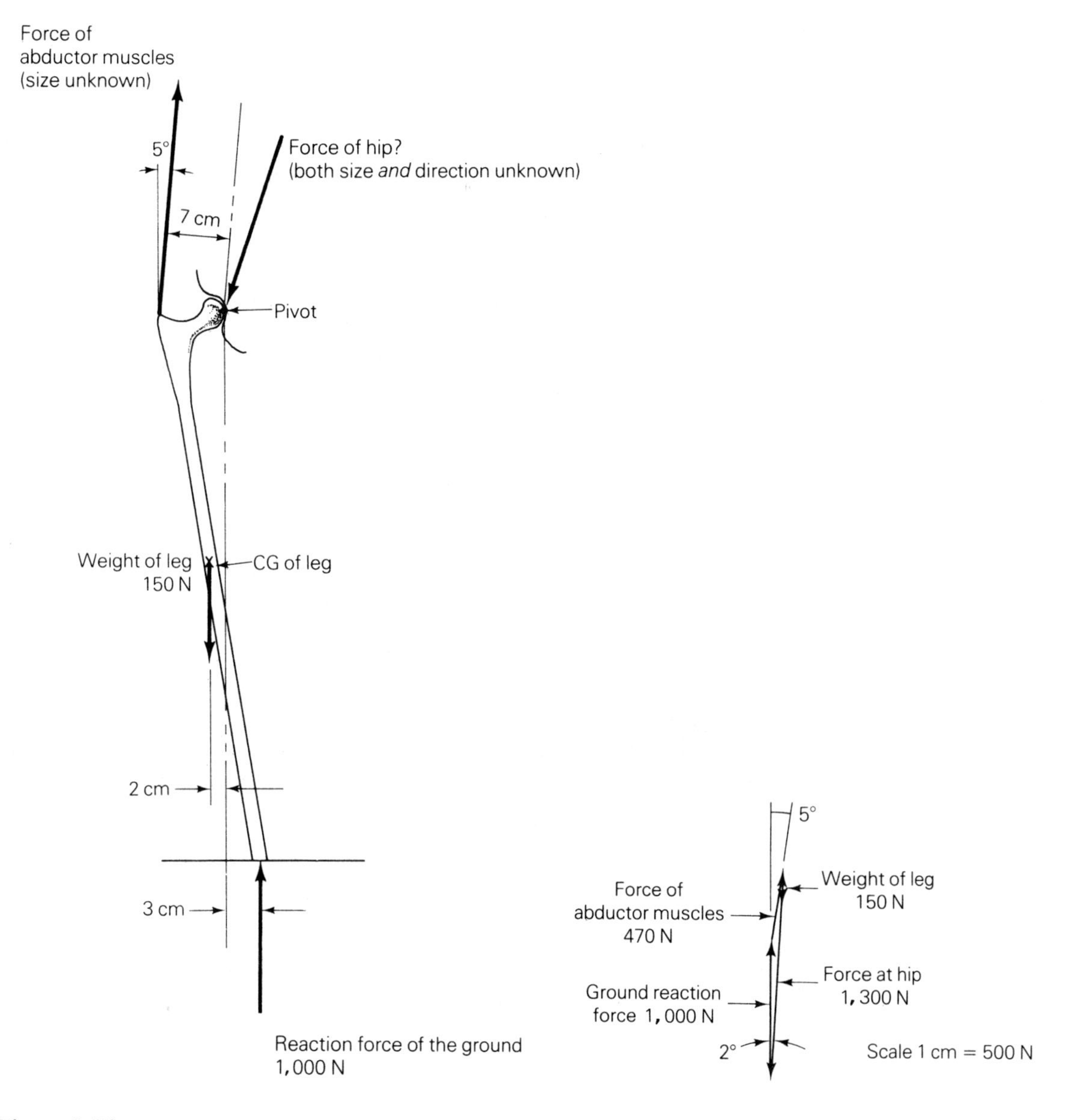

Figure 5-36

Figure 5-37

Since the net force on the leg must be zero, the force on the hip joint is 1,300 newtons, as shown in Figure 5-37. Thus, in limping, the force on the hip joint is 1,000 newtons less than the force on a normal hip. Therefore, limping does reduce the force and discomfort at the injured hip.

Balance and Stability

Balance and stability are important in every human activity. A person's balance can be as precise as the delicate balance of a tightrope walker or as stable as the rocklike stance of an offensive guard in football. Except when we are sitting or lying down, our senses and reflexes are constantly at work to maintain our balance. Motion is even more complicated. A movement like walking involves a deliberate unbalancing, followed by a restoration of balance, then another unbalancing, and so on.

Two conditions must be met for the balance of an object to be stable: The first is that the sum of the gravitational torques acting on the object must be zero. If the net torque is not zero, the object will start to rotate to a new position. The second condition is that *any small change in position must result in a gravitational torque that tends to return the object to its original position.* We can see the importance of this condition by comparing a pendulum hanging at rest with a pencil balanced on its point (Figure 5-38). The pendulum will just hang there while the pencil will fall. That is, the pendulum is stable but the pencil is unstable. Both satisfy the first condition, but only the pendulum satisfies the second. If the pendulum was moved a bit to one side, the gravitational torque would tend to swing it back to the original position. However, if the pencil was disturbed in the slightest, the resulting gravitational torque would cause the pencil to move even farther from its original position and fall over.

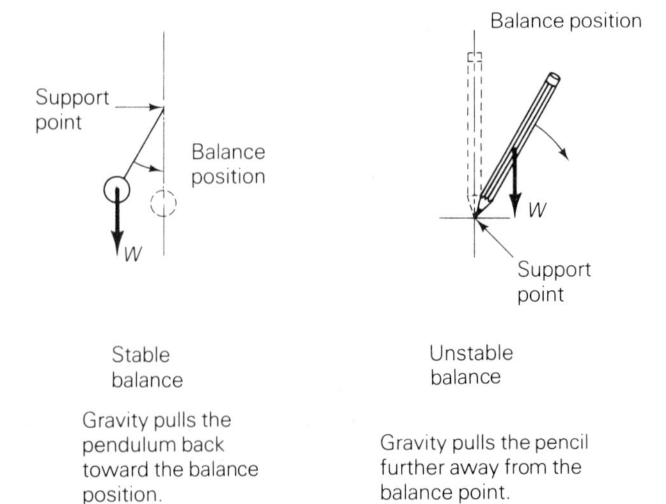

Figure 5-38

If the object is suspended from a single pivot at any point but the CG, the two stability conditions can be satisifed in only one way: The center of gravity must lie directly beneath the pivot point. Thus, any object hung from a string or a free-rotating pivot will, after the swinging stops, come to rest with its CG directly beneath the point of support. It doesn't matter whether the object is rigid or not; when the swinging has stopped, the CG will always be directly under the point of support. Thus, a gymnast hanging on the rings or a single bar always comes to rest with his or her CG directly under the point of support.

Region of Support

Figure 5-39 The regions of support of various objects.

When the center of gravity is above the support, the object won't be stable if it is supported at only one point. To be stable, the object must have a *region of support, and the CG must be directly above an interior point of the region of support.* The region of support includes all the points of the object that are actually in contact with the ground or floor. In addition, if you connect the outside points of contact with straight lines, those points enclosed by the straight lines are also in the region of support. Figure 5-39 illustrates how the region of support is found.

As long as the CG is above an interior point of the region of support, the object will be stable. If the CG is not above the region of support, the object will tip over. If the CG happens to lie exactly over the edge of the region of support, the object will be stable in one direction but easy to push over in the other. The degree of stability depends on the size of the restoring torque produced by tipping the object slightly from the stable position. The restoring torque, in turn, depends on two things: how low the CG is and how far the CG is from an edge of the region of support. In order to attain maximum stability in all directions, the CG should be as low as possible, the region of support should be as large as possible, and the CG should be as close as possible to the center of the region of support.

Stability in Athletics

The stability principles clearly influence an athlete's choice of stance. Normally, a person doesn't worry too much about stability, and the normal stance reflects that lack of concern. The feet are close together, so the region of support is small; the person is erect, so the CG is quite high. In the normal stance, the subconscious brain is always working to keep the CG directly above the region of support. Even so, the person can be toppled by a very small shove if caught off guard. The prizefighter is an example of an athlete desiring greater stability, since he fully expects to have someone disturb his stability. His stance is therefore very wide to produce a large region of support, and he also crouches slightly to lower his CG. The offensive guard in a football game expects violent attempts to disturb his stability. Therefore, his position is on both hands and feet, which creates the largest practical region of support and the lowest practical center of gravity. Of course, absolute maximum stability can be achieved by lying down and spreading the hands and feet far apart. Since the very high stability positions are also low mobility positions, the athlete's choice of stance will be determined by the relative importance of stability and mobility. The maximum mobility position is probably a slight crouch, a position that also offers greater stability than standing erect. This is why the slight crouch is the basic stance in many sports.

Stability in Patient Care

Persons involved in the care of weak patients can take advantage of the stability principles to reduce the strain on the patients and themselves. When helping a patient to walk, try to keep the patient's CG over the region of support at the patient's feet. If this is done, you will only have to apply a small guiding force to assist the patient. If the patient's CG moves outside the region of support, you must quickly apply a larger force to restore the patient's balance; otherwise, the patient will start to fall, in which case you must apply either a large lifting force or at least enough force to lower the patient gently. Either action will put a strain on both the patient and yourself.

Lifting a patient is also easier when the patient's CG is close to the region of support. Take the case of raising a patient to a sitting position. If the patient is lying flat on his or her back, raising the patient takes a lot of effort, because the CG of the patient's upper body is far from the region of support at the buttocks. However, if the bed can raise the patient part way, a much smaller force will bring the patient to an upright position, because the patient's CG will be much closer to the region of support. Incidentally, the hospital bed should not be thought of as just a device that props a patient up in a comfortable position. It is really a machine that can be most helpful in moving and lifting patients.

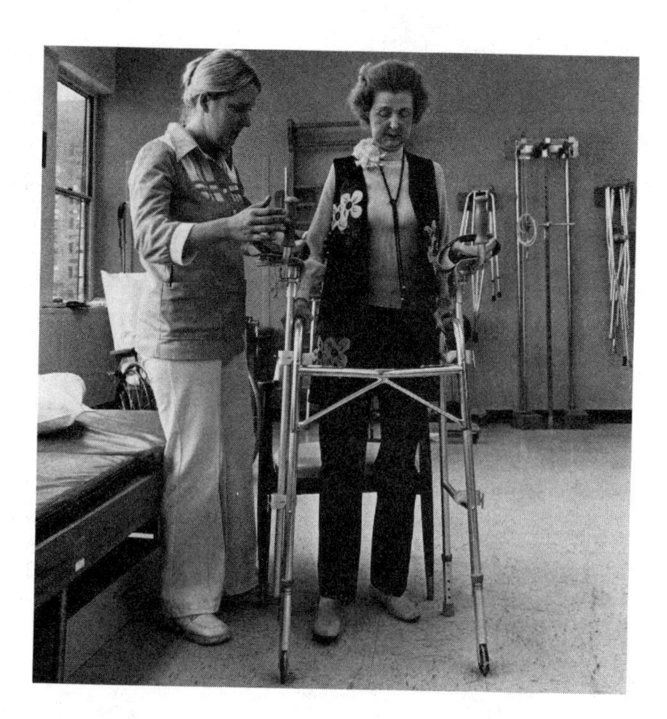

The walking frame enlarges the patient's region of support, thus increasing her stability.

Height and Stability

Keeping the center of gravity low increases the stability in a second way. The lower the CG is, the farther the object can be tilted before the CG goes over the edge of the region of support. This is shown in Figure 5-40, which shows four blocks with the same shape but different CG locations at their maximum tilt angles. Clearly, the lower the CG is, the farther the block can be tilted before it topples.

It is particularly important for vehicles to have good stability. Every vehicle that transports people, from race cars to wheelchairs,

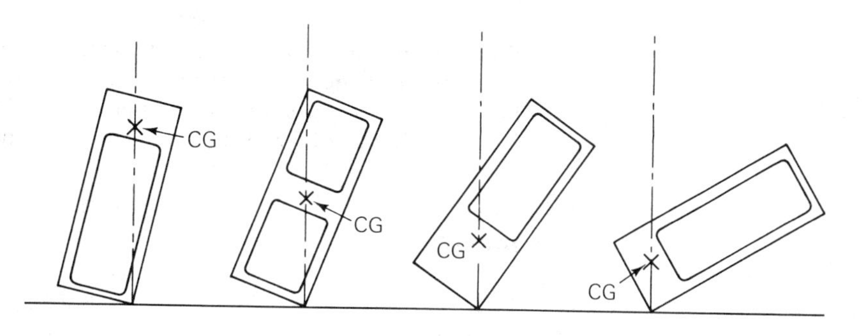

Figure 5-40 Maximum angle at which a block can be tilted without toppling over for various locations of the center of gravity. As long as the CG does not go over the line, the block will return to an upright position when released.

should have as low a center of gravity and as wide a region of support as is practical. A number of things, such as infant high chairs with wheels and high, old-fashioned patient transfer carts, are not as stable as they should be. Such devices should be used with care and moved slowly, especially at corners and at cracks in the floor. Preferably, they should be replaced with safer devices.

EXAMPLE Let's see how these stability principles apply in a specific case. Consider a uniform block of height h, length l, and mass m. Assume that the bottom is square, so its width is also l. What is the minimum force F acting parallel to the floor that will tip over the block?

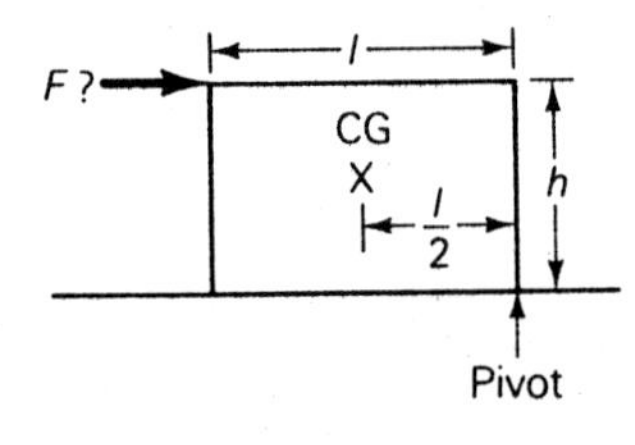

Figure 5-41

Answer The best place to apply the force is at the top of the block, since that position has the longest lever arm. As the force becomes large enough to tip the block, the block will start to pivot about the lower corner on the side opposite to the force (assuming that the friction is great enough to prevent the block from sliding). The block will start to tip when the torque produced by the applied force about this corner just exceeds the opposite torque of the weight of the block. The torque produced by the applied force is

$$\tau = Fh.$$

The torque produced by the object's weight is

$$\tau_g = -mg(\tfrac{1}{2}l) = -\frac{mgl}{2}.$$

When the sum of these two torques is greater than zero, the block will start to tip over; thus, the minimum force that would tip the block over can be found from the equation

$$Fh - \frac{mgl}{2} = 0.$$

Solving for F, we find

$$F = \frac{mgl}{2h}.$$

From this formula, we can see that increasing either the mass or the size of the bottom of the block would increase the force needed to tip over the block. Thus, increasing either or both would increase the stability of the block. Conversely, increasing h would *decrease* the stability.

Concepts

Lever	Center of gravity (CG)
Torque	Balance

Lever arm
Effective lever arm
Pivot point
Stability
Region of support

Discussion Questions

1. Explain why a long-handled crank is easier to turn than a short one.
2. Why do wire cutters, tin snips, and cast cutters have long handles and short blades, while scissors have short handles and long blades?
3. Discuss how the jawbone functions as a lever. Why do the rear molars produce a greater chewing force than the front teeth?
4. Give some examples of levers that (a) magnify motion, (b) reduce motion, (c) magnify force, and (d) reduce force.
5. The arm produces the most pulling force when the lower arm is perpendicular to the upper arm. The arm cannot pull very hard when it is fully extended or fully contracted. Why is this?

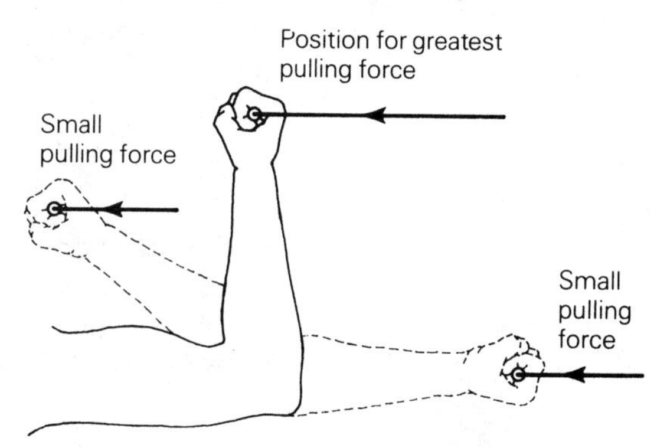

6. Why is it much easier to do a pushup when you rest your knees on the floor?
7. Examine the standard stances of a sport in which you are interested. Which stances offer the most stability? What other factors influence the choice of stance (such as mobility and visibility)?
8. Why do platform shoes greatly increase the danger of ankle injury?
9. Why is this book more stable when it is lying flat on the table than when it is balanced on an edge? When it is on an edge, why is it more stable with the covers fanned out than with the covers closed?
10. The net force needed to lift a long pole is equal (but opposite in direction) to the weight of the pole, no matter how or where you grasp the pole. Nevertheless, it is still easier to pick up a pole at its center than to lift it by one end. Explain why this is the case.

Problems

1. A 50-kilogram boy and a 40-kilogram girl are playing on a seesaw. The girl sits on the board 2.5 meters from the pivot. How far from the pivot should the boy sit to balance the seesaw?

2. What force is needed to lift the handles of the loaded wheelbarrow shown below?

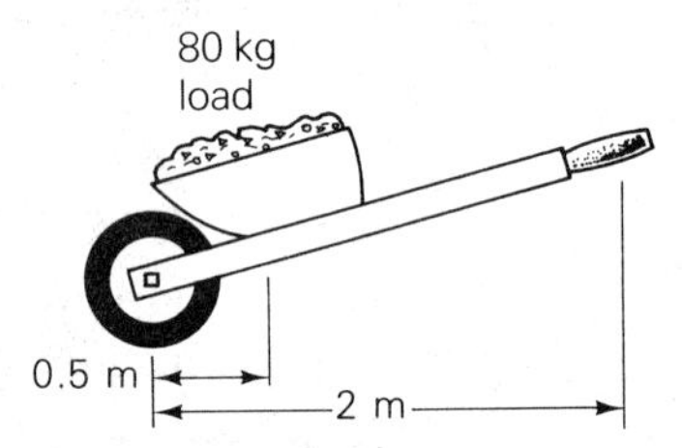

3. Three women are going to carry a heavy mailbag on a 3-meter-long pole. Two women will be at one end of the pole, and the third will be at the other. Where should they put the mailbag to distribute the weight equally among them?

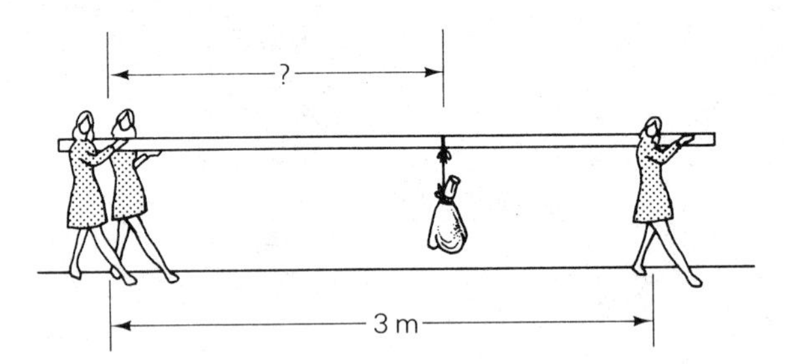

4. Find the tension in the biceps muscle and the contact force at the elbow when a 10-kilogram mass is held in the hand as shown. The weight of the arm and hand is much less than 10 kilograms, so it can be ignored in the calculation. Try to work this problem using the insertion point of the biceps as the pivot point.

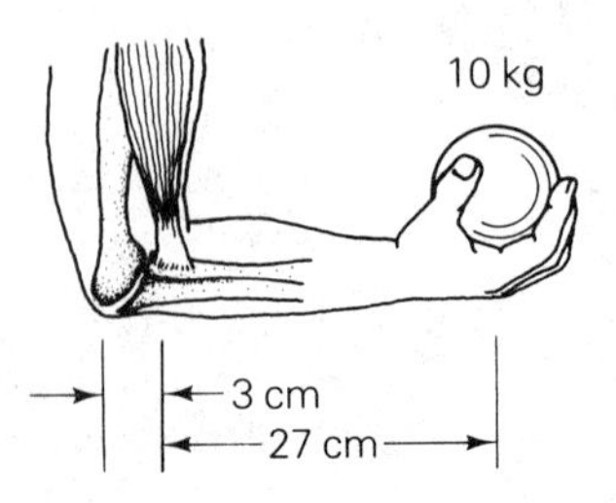

5. Find the tension in the quadriceps tendon in the situation shown.

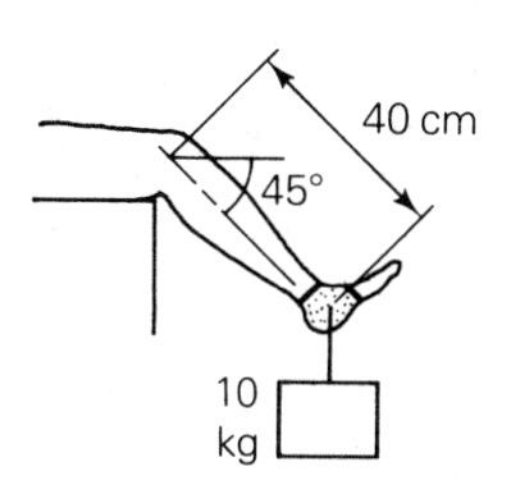

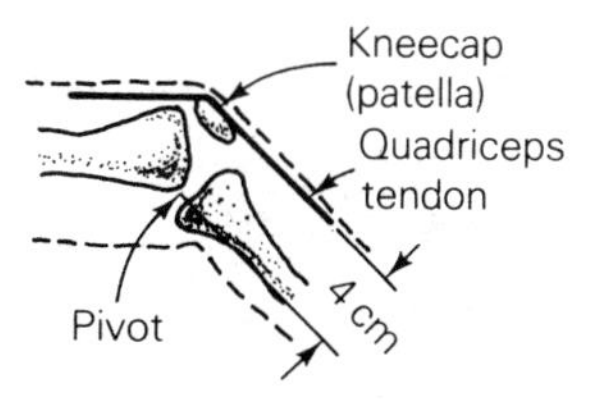

6. Calculate the force at the hands and the force at the toes of the 75-kilogram athlete doing a pushup. What percentage of the athlete's weight is the force at the hands? Repeat the calculation for your size and weight. Assume that your CG is around your navel.

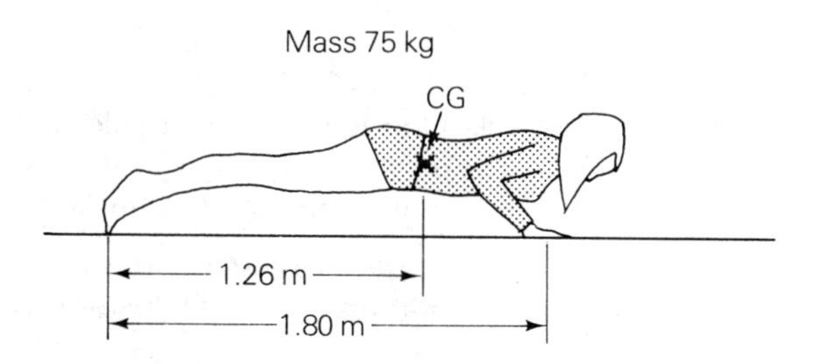

7. What size force would be needed to push over a 75-kilogram man standing erect with his feet close together? His dimensions are shown in the figure. Suppose he widens his stance so that the outer edges of his feet are 80 centimeters apart and his shoulder height drops to 148 centimeters. Now what force would be needed to push him over?

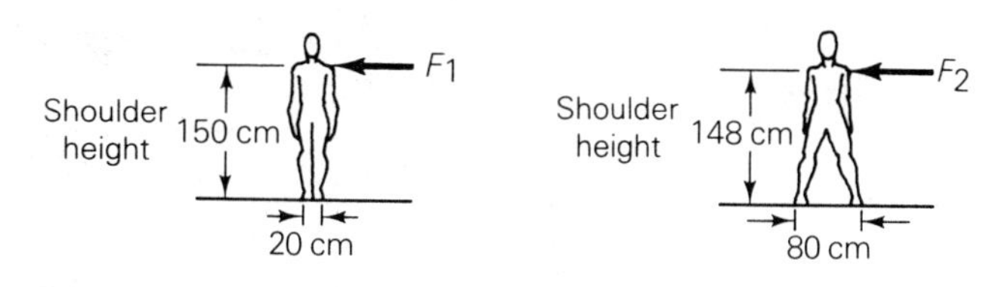

8. What force must the hand provide to hold the pole stationary at angles of 0°, 15°, 30°, 45°, 60°, 75°, and 90°? Plot your results on a graph.

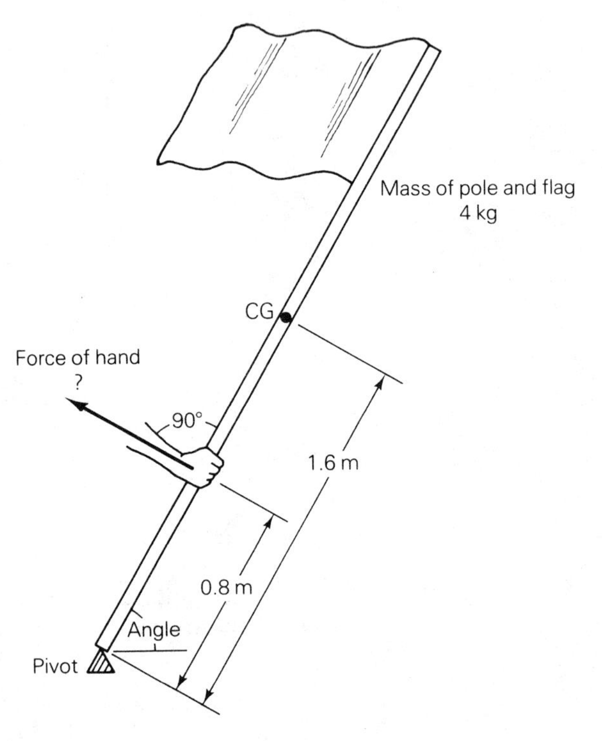

9. Find the force on a pole-vaulter's rear hand during a level carry and during a high (45°) carry. Assume that the hand locations and pole dimensions are the same as those in the text example.

10. If the bicyclist is pressing down on the pedal with a force of 500 newtons, what will the forward force on the bicycle be?

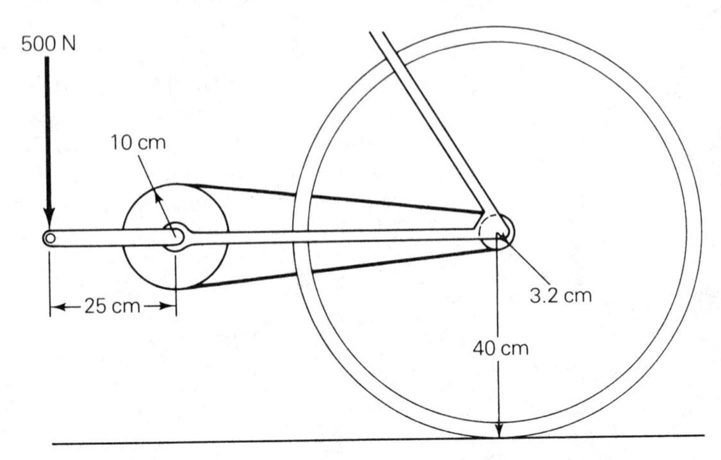

Note: Even though the sprockets are round, they are still basically levers. At any instant, the force at the chain acts at only one point, and the line from that point to the pivot is the lever arm. The dark lines on the sprockets are the lever arms.

11. What would the forward force on the above bicycle be when the pedal is in the position shown?

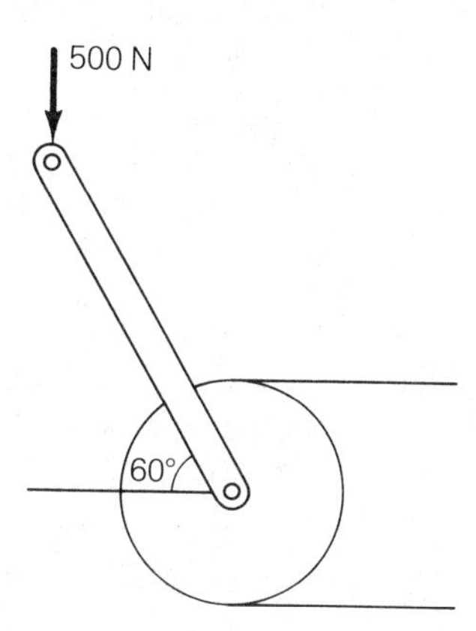

Experiments

1. Using Tinkertoys or an Erector set, construct a lever that has one lever arm twice as long as the other. The pivot should be in between. Add small weights to the short side until the lever balances. Next hang a few weights from the long lever arm and then show that it takes twice as many weights on the short side to bring the lever back into balance. Use identical weights on both sides and hang the weights from the ends of the lever arms.

2. Simply watching a person move and observing the motions of the muscles, tendons, and bones is one of the best ways to learn about human kinesiology. The muscle motions are easiest to see on a thin person, but anyone will do. Watch someone move and try to answer the following questions:

Which muscles produce which motions?
How do the tendons transmit the muscle forces to the bones?
How do the bones act as levers?
Roughly what are the tensions in the muscles and tendons and the contact forces at the joints during various motions?

3. Locate, by any method you choose, the center of gravity of a baseball bat, golf club, tennis racket, pool cue, or any other object of interest.

4. Stand facing a wall with your toes touching the wall. Now try to rise on your toes. Why can't you do it? Stand with your back against a wall with your heels touching the wall, then try to touch your toes. What happens? An interesting thing to observe during these two experiments is the interaction between the muscles and the brain. Because they cannot make the slight weight shift necessary, many people find it extremely difficult to activate the muscles required for the motions. This is an example of the close teamwork (the formal name is synergism) of the muscles.

5. A person's center of gravity can be accurately located by having the person lie on a light board that rests on a pivot at one end and a scale at the other. The roller used as the pivot should be directly beneath the person's feet; the roller on the scale should be directly under the top of the head. Explain how the distance H_{CG} from the person's feet to his or her CG can be found in terms of the person's height h, mass m, and scale reading m_s. Use this method on as many people as you can. How does the relative location of the CG vary with age, type of build, and sex?

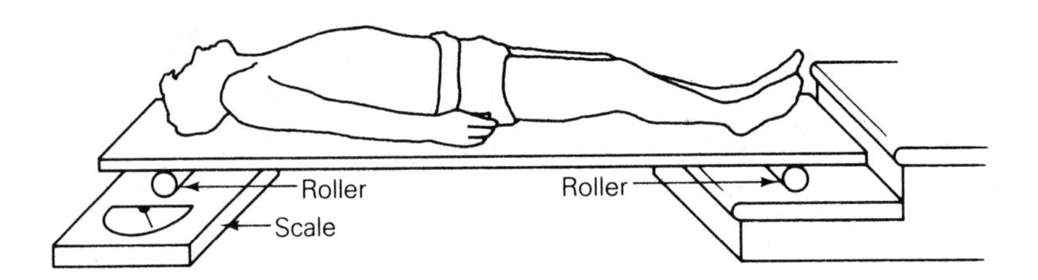

6. Take a ten-speed bicycle and prop it up so you can turn the pedals by hand. Watch the mechanism while you crank it and shift it. Which chain positions produce the greatest speed? Which produce the greatest forward force? Explain why.

CHAPTER SIX

The Effects of Forces on Real Materials

Educational Goals

The student's goals for Chapter 6 are to be able to:

1. Explain how ductility and elasticity influence the behavior of a material under stress.
2. Describe the structure of metals, glasses, and polymers.
3. Describe the structural characteristics of muscle, tendon, bone, and blood vessels.
4. Describe the major sources of friction, the good and bad effects of friction, and how lubrication can be used to reduce friction.
5. Estimate the frictional forces on objects.

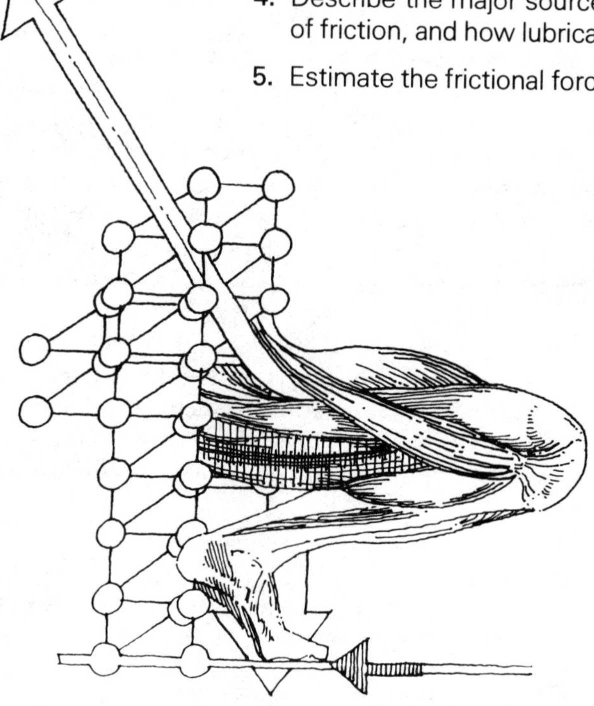

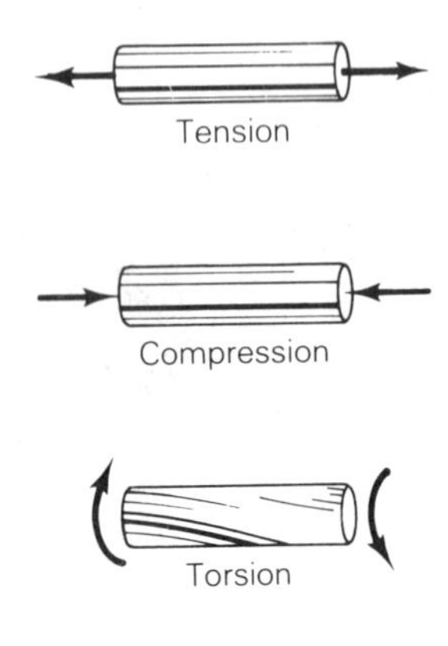

Figure 6-1

Up to now, we have treated solid objects as if they were ideal things that never bent or broke. We have assumed that they are perfectly rigid, that is, that their shape does not change when forces are applied to them. Of course, no object is perfectly rigid. Real materials, including those in the body, can be bent, stretched, compressed, twisted, and broken by the application of forces. In some cases, when either the material is strong or the force is weak, the amount of distortion may be so small that we can ignore it. However, in many cases the effect of the forces on the material must be taken into consideration to understand the resulting motion fully. In this chapter we will look at how the materials of the body and other common materials respond to the application of forces. We will also look into how atomic arrangements influence the properties of materials.

There are three basic ways of applying a force to an object: We can stretch it, compress it, or twist it (Figure 6-1). A force that stretches an object is known as a *tensile* force, a force that compresses an object as a *compressive* force, and a force that twists an object as a *torsional* force. While many materials behave about the same under all three types of stress, some behave drastically differently under the three types. Muscles, tendons, and ropes, for instance, have high tensile strengths, low torsional strengths, and practically no compressive strength at all. Unreinforced concrete has a high compressive strength but only limited tensile and torsional strength.

The distortion produced by a force may be temporary, permanent, or both. If the material springs back to its original shape when the force is removed, it is said to be *elastic*. If the material does not spring back, it is said to have been *plastically* deformed.* Spring steel and bone, if they aren't bent too far, are very elastic materials. Modeling clay and wet plaster are examples of plastic materials. Of course, most distortions are both elastic and plastic, and the material only springs back part way when the force is removed.

Time can also be a factor in the distortion. A piece of wood or bone will be quite elastic if the distortion is released shortly after application. However, if wood or bone are bent for a month or so, the distortion will be permanent. The rate of distortion can also be important. Some materials will be elastic for rapid distortions but

*The materials that are commonly called plastics were given that name because they are in a plastic state while they are being molded. It is really a poor name, because most "plastics" are not plastic in their final form.

completely plastic for slow distortions. Silly Putty is a good example. A ball of Silly Putty will bounce on the floor like a rubber ball, yet if it stands by itself awhile, it will flow and form a puddle.

Thus, the elastic and plastic properties of materials depend on many things. The composition of the material is the primary factor, but the rate and duration of force application can also be factors. Biological materials are particularly sensitive to the duration of a force, since the force may alter the metabolism of the cells. This alteration could in turn change the properties of the material.

How Materials Break

How a material breaks is determined by its elastic and plastic properties. Consider the two extremes of a very plastic material and a material that does not deform plastically at all. The very plastic material will stretch and bend a long way before it breaks; it might even require repeated bendings to break. Such materials are called *ductile* materials. Some ductile metals are lead, gold, copper, and silver. Conversely, a material that doesn't deform at all plastically might bend a bit before it breaks and then snap quickly. A material that does not deform plastically before it breaks is said to be *brittle*. Glass, ceramics, and bones of the elderly are classic examples. Of course, most materials fall somewhere in between these two extremes.

To see why plasticity has such a big effect on fracture characteristics, consider an object with a small crack at the surface, such as the one shown in Figure 6-2. The surfaces of most real objects are covered with such microcracks. Suppose that we apply a force that tends to widen the crack. Because of the leverage of the sides of the crack, the greatest stress will be at the crack tip (Figure 6-3), and the fate of the object will be determined by the behavior of the

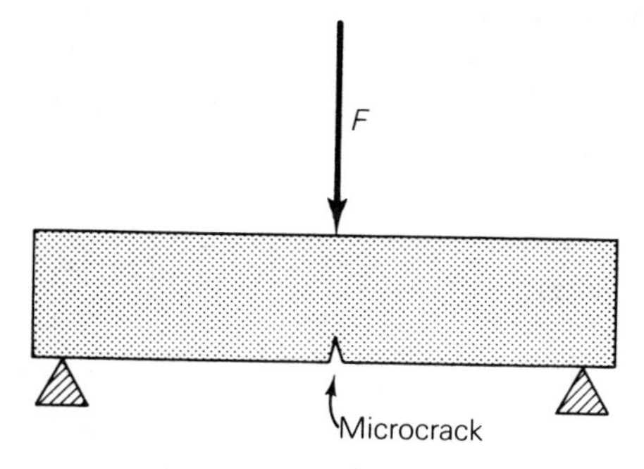

Figure 6-2

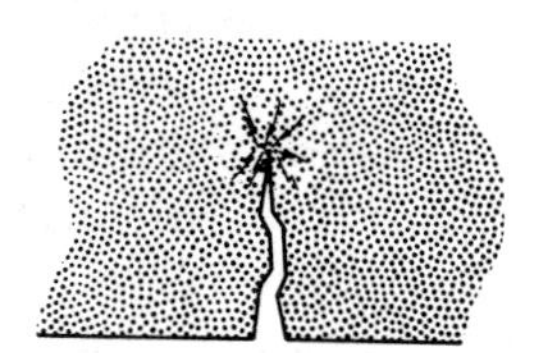

Figure 6-3 The stress caused by an applied force is concentrated at the tip of a crack.

material there. If the material is ductile, the material at the crack tip will flow plastically, thus blunting the tip of the crack (Figure 6-4). As the tip of the crack becomes rounded, the stress at the tip will drop and the crack won't open up further. If the force continues to increase, the object will bend, of course, but it usually will not fracture—at least not until the bending is quite severe. Cracks in soft metals, rubber, and most plastics behave in this way.

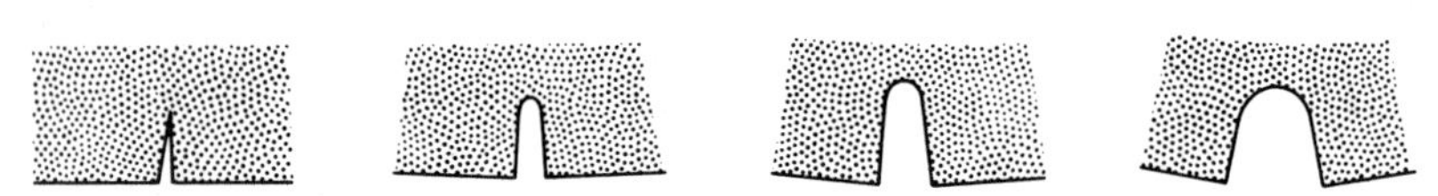

Figure 6-4 A ductile material will flow plastically at the tip of a crack.

Now consider a brittle material in the same situation. Again the stress will be concentrated at the crack tip. However, because the material cannot flow plastically, the stress at the crack tip can only be relieved if the crack grows longer. The deeper the crack penetrates into the material, the weaker the remaining material becomes. Thus, the crack grows faster and faster until the object breaks completely. This is why brittle fracture is such a rapid process (Figure 6-5). It starts without warning and proceeds explosively until the object shatters. For this reason brittleness is seldom desirable in an object. (There are a few exceptions, such as the glass windows covering fire alarm boxes.)

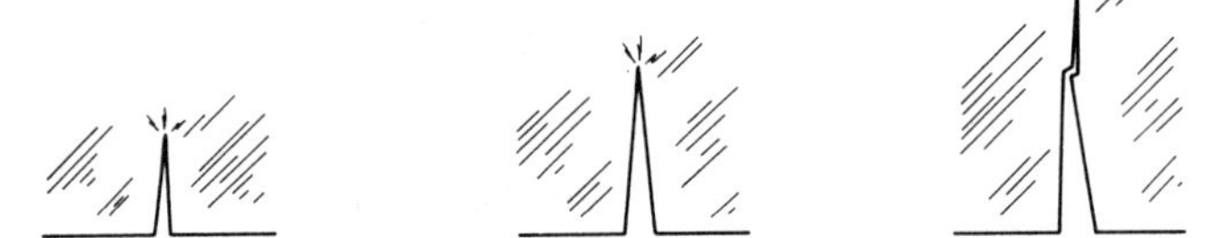

Figure 6-5 A brittle material will not flow plastically at a crack tip. Thus the crack grows rapidly.

Because brittle materials often fail without warning, they are not used where such failure would be hazardous. Ductile materials are generally much safer since they will yield to minor overloads. Unfortunately, ductility and strength don't always go together. Hard, strong materials tend to be brittle, while ductile materials tend to be soft and weak. For instance, high-strength metal alloys are almost always far brittler than the weaker but more ductile metals from which they are made.

Effect of Atom Locations

Because the plastic, elastic, and fracture characteristics of materials are best understood in terms of the behavior of their atoms, let's see how the atoms behave in some common materials.

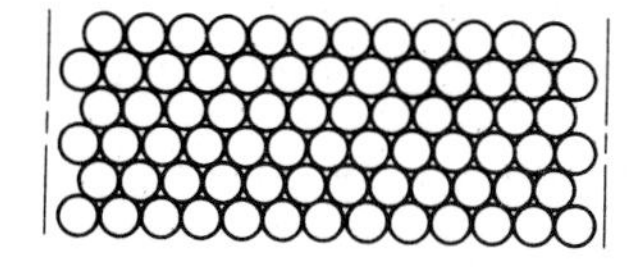

Figure 6-6 Locations of atoms in a crystalline material.

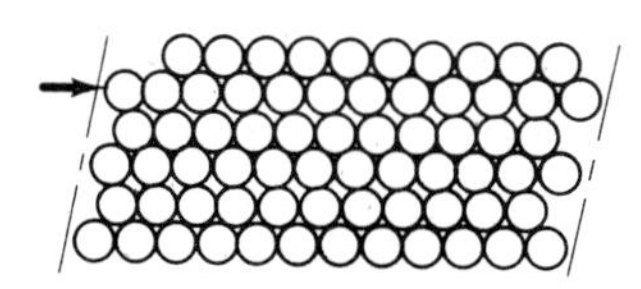

Figure 6-7 Atom locations during an elastic distortion.

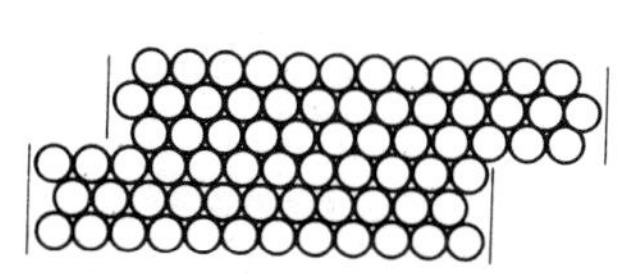

Figure 6-8 Atom locations after slip occurs.

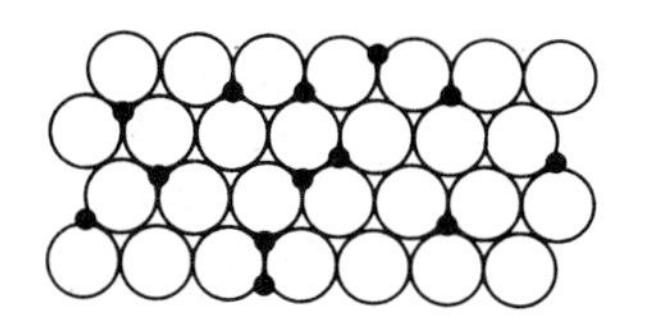

Figure 6-9 A metal strengthened by adding a small amount of impurity atoms.

Metals and Alloys

In most metals, the interatomic forces are strongly attractive, but not very directional. That is, most metal atoms are like sticky, round balls that cling together tightly but roll on each other relatively easily. As a result, most pure metals are ductile and easy to deform.

The atoms of metals are normally stacked in a regular crystalline order, as shown in Figure 6-6.* The arrangement is very similar to a stack of oranges or of other round objects. Now suppose that a force is applied to the metal. If the force is not too large, the metal will just deform elastically. The atomic arrangement will not change permanently, and the atoms will just shift very slightly relative to each other (Figure 6-7). But as the force increases, the atoms will eventually slip along each other to relieve the stress. The location of an individual atom may jump several atomic positions during the slip process. When the slip process stops, the atoms just settle down with their new neighbors—atoms, after all, are identical—leaving no trace of the motion inside of the metal (Figure 6-8).

To strengthen a metal, the slip along atomic planes must be stopped. One commonly used method is exactly analogous to a driver throwing sand between a slipping tire and an icy road. Atoms of another type, preferably quite different from the atoms of the original metal, are mixed in. They might occupy positions between the normal sites as shown in Figure 6-9, or they might replace atoms of the original metal. In any case, they increase the frictional forces between the planes of atoms and prevent slip, just as sand particles stop the slip between a tire and an icy road. Steel is a good example of a material strengthened in this way. Steel is just iron with a small percentage of carbon atoms added. The carbon atoms get in between the iron atoms and increase the resistance to slip, thus strengthening the steel. This is why steel is so much more useful than pure iron.

Alloys (mixtures of different metals) are generally much stronger than the metals from which they are made for the same reason. Since an alloy is a mixture of different atoms, the atomic planes are not smooth and slip is not as easy. Naturally, the ductility of the metal is also reduced, since the increased friction of the atomic planes retards plastic flow.

---

*When discussing atomic arrangements, the word *crystalline* means that the atoms are arranged in a regular pattern—it does not mean that the material is transparent. Some crystalline materials are transparent, but most are not. The type of glass called crystal is not crystalline.

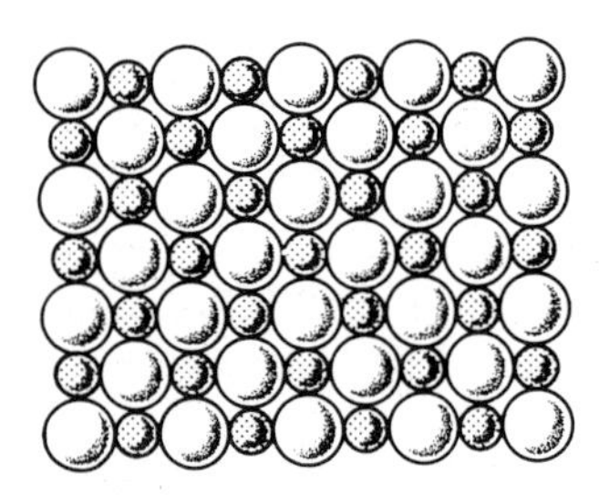

Figure 6-10 A nonmetallic crystal, sodium chloride (NaCl).

Nonmetallic Crystals

The atoms and molecules of nonmetals often have very directional bonds. That is, the atoms or molecules can only be hooked together at certain bond locations (Figure 6-10). Such materials are often quite brittle because slip is almost impossible. Once a bond is broken, it is very difficult to reestablish a new bond with a different atom or molecule, simply because the new bond would have to be at exactly the right spot and at the correct angle. Consequently, once the bonds start to break, they break along an entire crystal plane. Of course, some planes are easier to separate than others; this is why diamond cutters sometimes spend days studying a diamond to determine which planes will be the best to separate.

Glasses and Ceramics

Figure 6-11 Molecular structure of glass.

Glasses and ceramics are similar to nonmetallic crystals except that they do not have any long-range order. That is, their atomic arrangements have little regularity (Figure 6-11). Glass and ceramics are made either by rapidly cooling the molten material or by mixing in a material having atomic bonds that are easily bent. Glasses and ceramics are usually brittle for the same reason that nonmetallic crystals are brittle: The bonds between the atoms are directional; once broken, the bonds are difficult to reestablish with other atoms.

Because glasses and ceramics are brittle, their surface conditions are very important in determining their fracture properties. We tend to think of glass as being weak, but that is not really the full story. Glass is actually strong as long as the surface has no microcracks. As soon as the surface is damaged, the microcracks concentrate the stress, and the glass is considerably weaker.

Polymers

Polymers are materials that have long chainlike molecules. Most plastics and many biological materials fall into this class. A particularly important polymer is DNA, the fundamental genetic material of all living things. The basic part of most polymers is usually formed from hydrocarbons or other carbon-based molecules. A simple example is the hydrocarbon polymer polyethylene, shown schematically in Figure 6-12 (each C stands for a carbon atom and each H for a hydrogen atom).

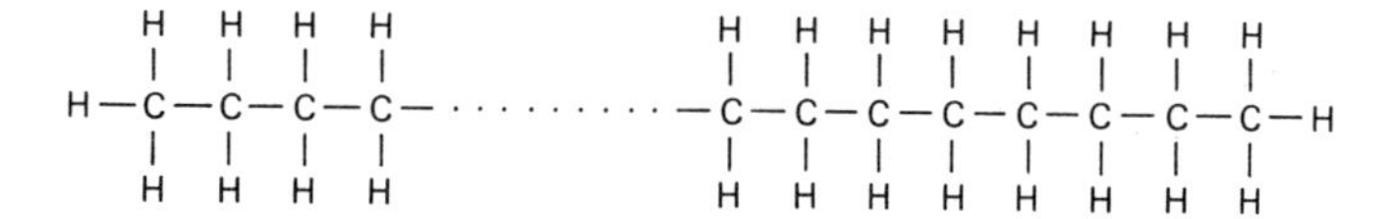

Figure 6-12 Polyethylene.

The elastic and plastic properties of polymers are controlled by two things: the geometrical shape of the molecules, which we will look at later, and the bonds, or *cross-links,* which sometimes form when the molecules touch. Normally, the molecules are only attracted to each other very weakly, and they can slide along each other freely. However, various things—including certain chemicals, radiation, and heat—can bond the molecules together, forming cross-links. The strength, elasticity, and plasticity can be changed by controlling the amount of cross-linking. A lightly cross-linked polymer will be very soft and plastic, perhaps almost fluid. A moderately cross-linked one would be more rigid, but not so rigid that it couldn't absorb shocks without breaking. A highly cross-linked polymer would be very rigid and brittle.

Epoxy resin glues are polymers that are hardened by chemically producing cross-links. The epoxy resin is originally a lightly cross-linked, thick, gummy liquid. It is a liquid because there are few cross-links to stop the molecules from sliding along each other, but it is thick and gummy because the long molecules are tangled up and can only move slowly along each other. A chemical catalyst is then added, which produces cross-linking between the molecules. Depending on the amount of cross-linking produced, the liquid epoxy resin will harden into anything from a moderately cross-linked, rubbery solid to a highly cross-linked, brittle solid.

The white filling materials that dentists use to repair cavities are polymeric resins that are hardened by cross-linking. Some types use a chemical catalyst to produce cross-linking; others are hardened by ultraviolet radiation. Once this type of filling material is in place in the cavity, the dentist shines ultraviolet light on it. The ultraviolet radiation produces cross-links in the resin, which immediately harden the resin to form a durable filling.

The shape of molecules influences the behavior of the polymer in a straightforward manner. For instance, long polymeric molecules would be expected to make excellent fiber materials—with the proper amount of cross-linking, of course. Indeed, most natural and manmade fibers are basically polymers. (The exceptions are such things as steel wool, fiberglass, and asbestos fibers.) When the molecules are long and straight, the fibers are difficult to stretch. Dacron is an example; it is often used in applications where the stretch must be minimized, as in the sails of racing sailboats and

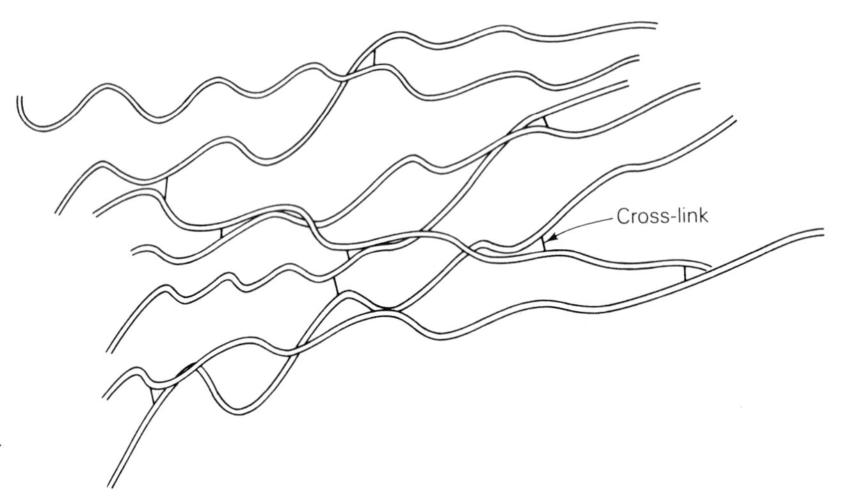

Figure 6-13 Molecules of an elastomer.

in tires. When the molecules are kinky, or coiled like miniature springs, the fibers are very stretchy and elastic. Such materials are called *elastomers* (Figure 6-13). Some examples of elastomers are natural and synthetic rubber, the synthetic elastic polymer Spandex, and the material in the walls of the blood vessels. Elastomers have a multitude of uses, such as absorbing shocks, providing flexibility and freedom of movement in clothes, and providing flexible support, as in the case of an elastic bandage.

Most of the nonliquid parts of the body are polymers. The proteins are a particularly important example, since they are the primary materials of the cells. Proteins are long polymers of the amino acids. They are as complex as life itself, so we won't analyze their behavior; instead, we will just look at a couple of examples of protein cross-linking. One is the coagulation of blood. The blood plasma contains about 8% protein of various kinds. Because these proteins are not normally linked to each other, they are free to move by each other and the plasma stays liquid. The details of the clotting process are far from understood, but it is clear that both foreign surfaces and the chemicals released by injured cells trigger a very rapid cross-linking of the plasma proteins. As the proteins cross-link, they form the gelatinous clot that gradually stops the flow of blood.

Cross-linking also plays an important part in the aging process. The fundamental genetic material, DNA, is an amino acid polymer in the form of a double helix and looks like a twisted ladder. The sides of the ladder are amino acid chains, and the rungs are formed from four types of molecules: adenine, thymine, cytosine, and guanine (Figure 6-14). The genetic code is carried by the ordering

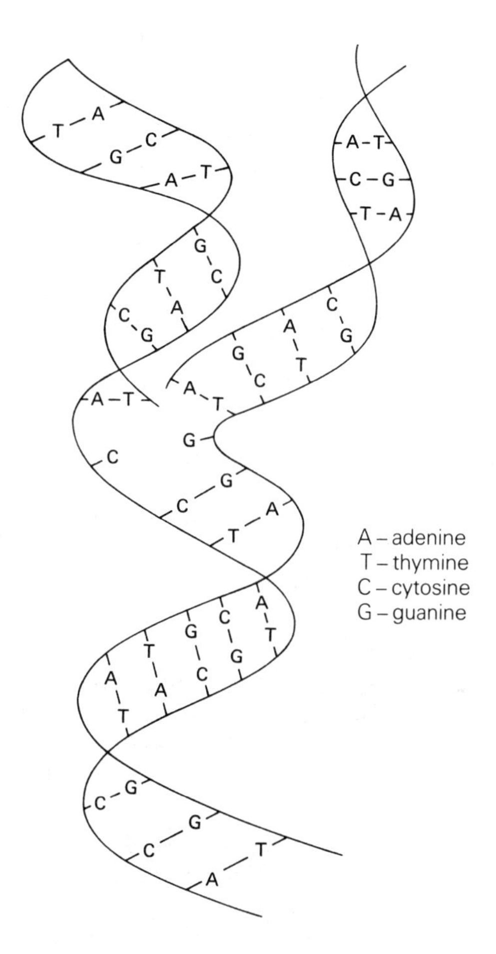

Figure 6-14 The double-helix molecule DNA as it splits to form two identical new molecules.

of these molecules along the ladder in much the same way that the ordering of letters of the alphabet into words and sentences carries information.

During reproduction, the ladder starts to split, and two new ladders are gradually formed as new molecules join the old ones in precisely the proper order to produce two DNA molecules, both identical to the original one—if all goes well. If the DNA strands become cross-linked to neighboring molecules, this will inhibit the movement needed to achieve reproduction and possibly cause incorrect placement of the new molecules—mistakes in the genetic code. Both of these effects cause cells to deteriorate gradually. Whether the cross-linking is the cause or an effect of aging is not yet known, but it seems clear that it is a major factor in the aging process.

Composite Materials

It often happens that no one material will satisfy all of the design conditions that must be met on a particular problem. For instance, concrete has many features that make it a good building material—it is relatively cheap, it can be poured into any shape needed, it has good resistance to weathering, and it is quite strong *in compression*. Unfortunately, it is very weak and unreliable in tension. As a result, concrete by itself is not suitable for use in building bridges, roofs, thin columns, or anything else that might encounter a tensile load. Plain concrete is thus limited to use in sidewalks, foundations, and a few other structures. Steel, on the other hand, is excellent in tension and acceptable in compression, but far more expensive than concrete. The solution in many cases is to reinforce the concrete, which carries the compressive load, with thin steel rods, which carry the tensile load (Figure 6-15). Reinforced concrete is just one example of a composite material that maximizes the advantages and minimizes the disadvantages of the individual materials.

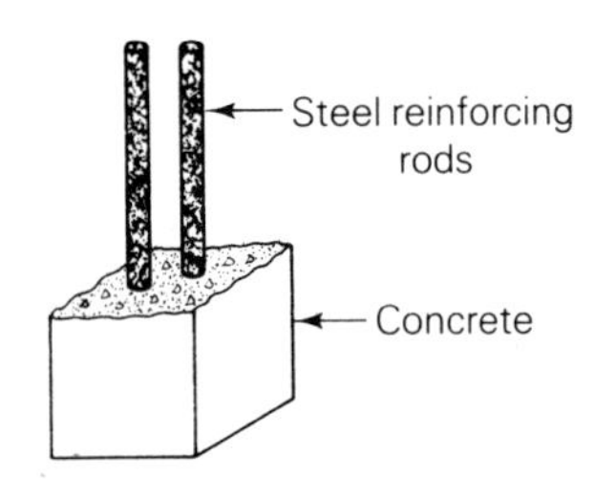

Figure 6-15 Reinforced concrete.

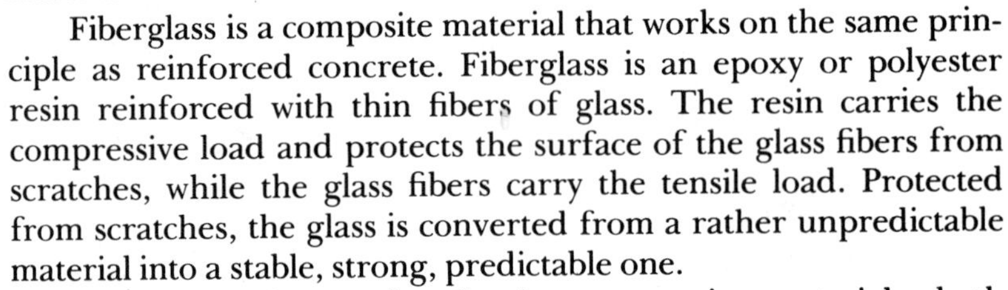

Fiberglass is a composite material that works on the same principle as reinforced concrete. Fiberglass is an epoxy or polyester resin reinforced with thin fibers of glass. The resin carries the compressive load and protects the surface of the glass fibers from scratches, while the glass fibers carry the tensile load. Protected from scratches, the glass is converted from a rather unpredictable material into a stable, strong, predictable one.

There are thousands of other composite materials, both natural and manmade. Some manmade composite materials are galvanized steel, in which steel is covered with a thin layer of zinc to stop rust; tires, which are made from rubber reinforced with polyester, nylon, glass, or steel fibers to limit the stretch of the rubber; wall paneling, in which an inexpensive wood is covered with a thin veneer of expensive wood for good looks; and chrome-plated metals, in which a microscopic layer of chromium prevents rust and provides a shiny surface (at least until the warranty runs out). Of course, many biological materials are composite materials; bone, wood, and blood vessel walls are just a few.

Materials of the Human Body

In many ways the materials of the human body are completely different from nonliving matter. They grow, they can repair themselves after injury or disease, and they can adapt to changes in their environment—things that no piece of inanimate matter can do. The laws of physics, however, apply to both living and nonliving matter equally, so it is not surprising that the materials of the body follow many of the same design principles that man has only recently learned. Of course, so many different materials make up the body that we can only look at the properties of some of the more important ones.

Connective Tissue and Collagen

The main function of connective tissue is to bind the cells together and hold them in place, so it must be well suited to transmitting forces from cell to cell. Connective tissue is found in all organs of the body and is most abundant, naturally, where large forces exist.

The structure of connective tissue is extremely well suited to transmitting forces. Connective tissue is mostly collagen, which is a fiberlike protein molecule. The collagen is arranged in long strands, which in turn are interlaced much like the threads in a fabric. Indeed, the structure of connective tissue is very similar to a high-strength manmade fabric woven from threads made of long polymeric molecules.

The healing of a cut or injury involves the repair of the broken connective tissue. In the first stage of the healing process, which takes about three weeks, a network of collagen fibers is rebuilt over the injured region. Of course, the smaller the gap is between the two sides of a cut, the easier this process will be. Thus, bandages

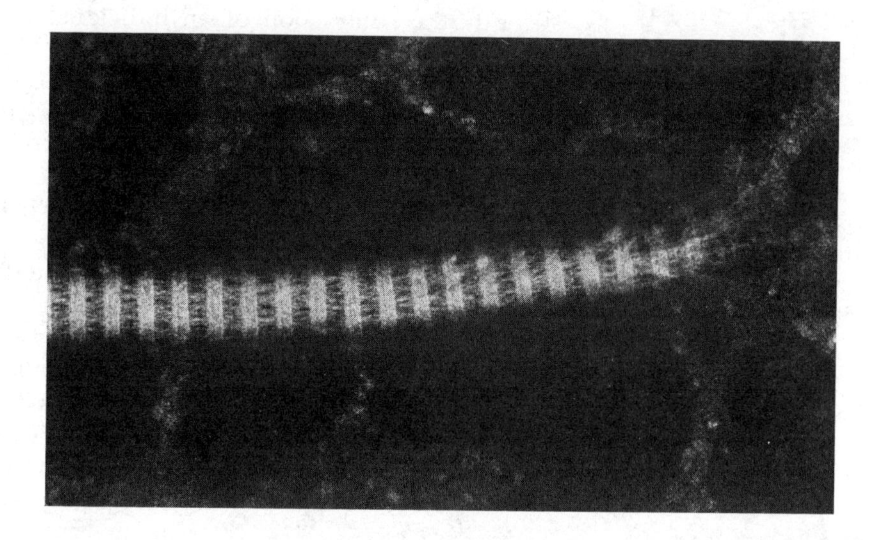

A strand of collagen as seen through an electron microscope.

and, if necessary, stitches are used to keep the cut closed. Although the wound will be closed at the end of this first stage, it will not have regained full strength. The connective tissue takes about a year to rebuild scar tissue that is equal to the surrounding tissue in strength. Evidence suggests that the collagen buildup is most rapid where the stresses are the greatest. For this reason, *gentle* exercise will often be beneficial during the healing process. But the stress must be gentle—overexertion will tear the still weak tissue and severely retard the healing process.

Coaches and trainers must make sure that an athlete's injuries have enough time to heal before being subjected to severe stress. There is always the temptation to send an athlete back into competition when the injury "looks OK" after a few weeks of healing. This is an extremely dangerous practice, since the tissue will still be very weak at that point. Far too many promising athletic careers have been cut short by reinjuring tissue that was still weak from an earlier injury. There really are no substitutes for time and careful reconditioning in the treatment of injuries.

Muscle

The muscle cells are arranged in long fibers whose outstanding feature is that they generate large forces as they shrink in length (and grow in width). There are three major classes of muscle: *skeletal* muscle, the muscles that produce motion of the body and are under our conscious control; *cardiac* muscle, the muscle of the heart; and *smooth* muscle, the muscles of the internal organs and blood vessels, which are not under our conscious control. Apparently the contraction of an individual muscle fiber is an all-or-nothing thing—the individual cell is either relaxed or trying to shrink with the greatest force that it can generate. Fine control is obtained by controlling the *number* of fibers that are stimulated. A muscle produces the greatest force when it is stretched by an outside force in the direction opposite to the normal contraction direction. This is why arm wrestling contests between well-matched opponents can go on for hours: Both opponents can resist a greater force than they can produce, so the contest becomes more a test of endurance than a test of strength.

The details of the contraction process are still not well understood. The *how* of the process is reasonably well-known, in that we know what parts of the cell move in what direction. We also know that an electrical stimulus triggers the contraction. However, *why* the cells can generate such large forces when they contract isn't known and is still a question for future research.

Tendon

The tendons transfer the tensile forces of the muscles to the bones, often over a long distance. For example, the muscles that move the fingers are in the forearm, and their forces are transmitted to the fingers by tendons. Some of these tendons are clearly visible under the skin at the back of the hand. If the finger muscles were actually at the fingers, the fingers would be much bulkier, weaker, and far less dexterous than they are.

Since muscles can only pull, the tendons only transfer tensile loads. They must be very flexible to move around the joints—in much the same way that ropes move over pulleys—yet they must also be extremely strong to transfer the large forces of the muscles. In other words, tendons must transfer forces in the same way that ropes do; in fact, the structure of tendons is very similar to the structure of a high-strength, low-stretch rope. The principal constituent of tendons is collagen, arranged in fiber bundles that are essentially parallel. Thus arranged, the strong fibers of collagen form a material well suited to transmitting the forces between the muscles and bone.

Bone

Bones support the body and act as levers to convert the pulls of the muscles into pushes and pulls. Along with the tendons, they transmit the forces of the muscles to wherever we wish. The bones, therefore, must be stiff—but not too brittle—and very strong in both tension and compression. With these conditions to satisfy, it is not surprising that bone is a composite material. The structure of bone resembles reinforced concrete as one material carries the compressive load and another carries the tensile load. The compressive load is carried by crystals of the calcium compound hydroxyapatite. Hydroxyapatite is like plain concrete in that it can carry high compressive loads but is rather weak in tension. The tensile load is carried by collagen fibers. The resulting composite of hydroxyapatite and collagen is very strong in both tension and compression.

The composite nature of bone can easily be demonstrated by removing either the collagen or the hydroxyapatite. If bone is left out in the sun to dry, the collagen will deteriorate, leaving the white, brittle hydroxyapatite. On the other hand, if the bone is cooked in a mildly acidic solution like vinegar, the hydroxyapatite will dissolve, leaving the collagen. The collagen is so flexible that the "bone" can then be bent into a circle.

The femur.

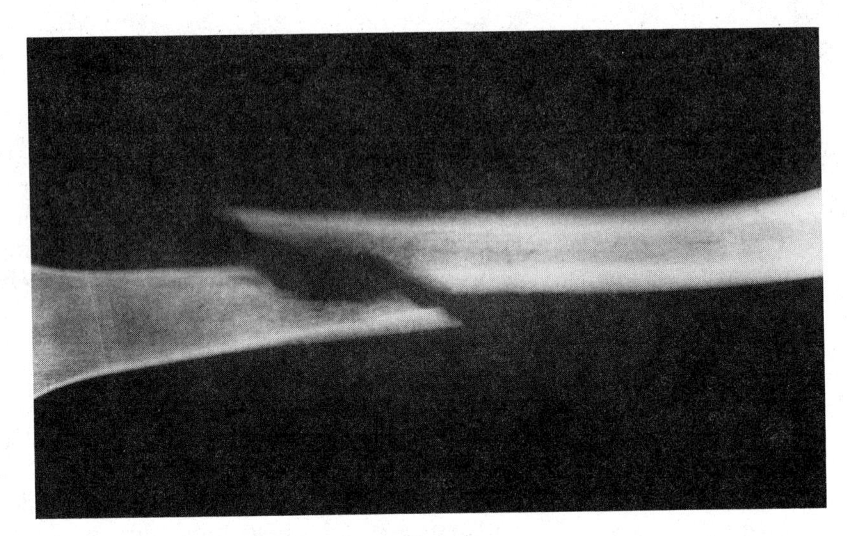

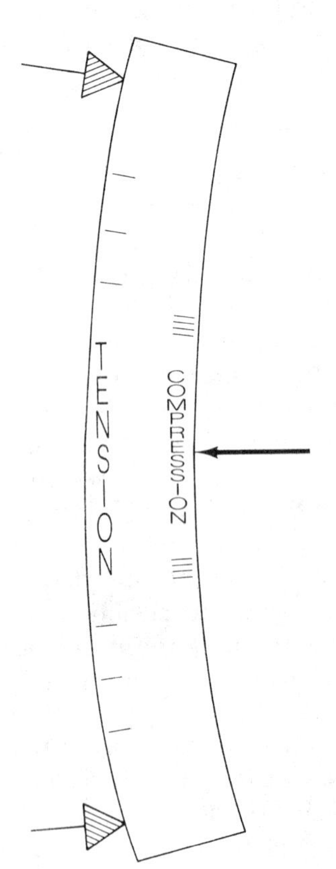

Figure 6-16

The bones are generally not solid and compact throughout. The outer surface is compact bone, but the central region is either open or filled with a spongelike network of bone. Open central regions can either be empty, such as the paranasal sinuses in the skull, or filled with bone marrow, the material that produces the red blood cells. The femur, shown split open above, has both types of central regions—the spongelike structure in the head and an open bone marrow cavity in the shaft. The open center structure is a particularly efficient way of producing a strong structure with a minimum of mass. It is not any stronger than the solid structure in straight tension or compression, but it is far stronger in torsion and bending. The reason is simply that torsion and bending involve rotational movement, and the open center structure has a longer lever arm with which to counteract the outside forces.

The bones of a healthy person are more than strong enough to carry the forces of normal activity; however, accidents often apply abnormally large forces that fracture the bones. The type of fracture that occurs will depend on how the forces are applied, and there are many ways in which compression, tension, and torsion can be combined in an accident.

Let's look at some of the common types of fracture. Often the forces simply bend the bone as shown in Figure 6-16. When something is bent, the material on the side away from the bending force is in tension while the side to which the force is applied is in compression. The break will start on the side away from the force—the side in tension—and in normal adult bone, which is somewhat brittle, the crack will quickly grow all the way across. A fracture of this type is called a *transverse* fracture (Figure 6-17). However, if the

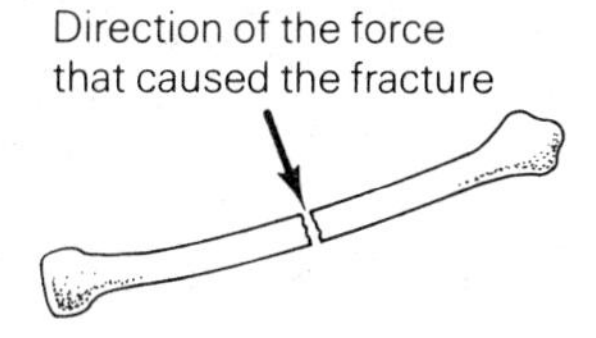

Figure 6-17 Transverse fracture.

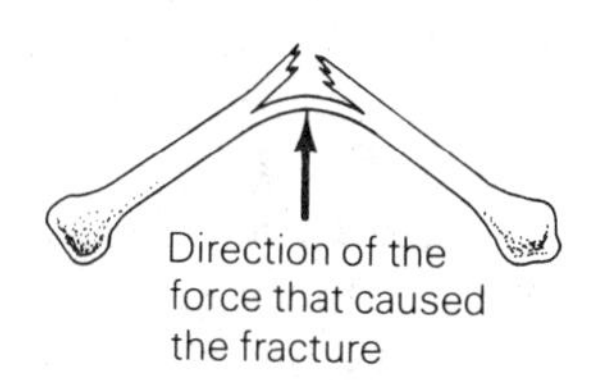

Figure 6-18 Greenstick fracture.

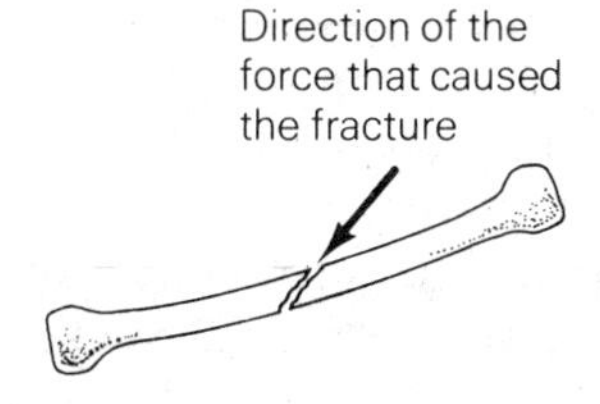

Figure 6-19 Oblique fracture.

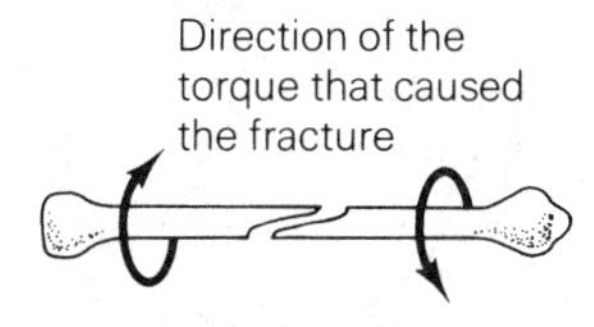

Figure 6-20 Spiral fracture.

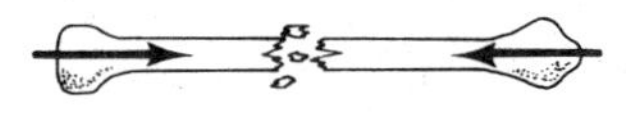

Figure 6-21 Comminuted fracture.

bone is flexible, as it is in children, the break might only go through the region that was initially in tension, and the part that was in compression may just bend. This type of fracture is called a *greenstick* fracture (Figure 6-18), because a green twig of wood breaks in the same way.

If the bending force is combined with a compressional or torsional force, the break may go across the bone at an angle in which case it is called an *oblique* fracture (Figure 6-19). A torsional force alone will cause a crack to move lengthwise as it moves across the bone. The resulting fracture thus takes the form of a spiral and is called a *spiral* fracture (Figure 6-20). Spiral fractures often occur in skiing when the safety bindings don't release soon enough. When this occurs, the long lever arm of the ski can put a large torsional force on the leg, causing a spiral fracture.

When the bone breaks into small pieces instead of breaking cleanly, the fracture is said to be *comminuted.* Such fractures usually result from compressive forces. Comminuted fractures often occur when people try to catch a fall by stretching out their arms or when people jump from too great a height. The initial break doesn't fully relieve the compressive forces; and they push the fractured ends into each other, causing further damage and a comminuted fracture.

Fortunately, the bones will usually repair themselves. The process is slow, but in most cases there is little or no serious disability after the fracture has healed. The first step in the healing process is the formation of a blood clot around the fracture. Next, collagen fibers start to penetrate the clot, gradually rejoining the broken ends with flexible collagen. Then, hydroxyapatite is slowly redeposited in the fracture, and the bone gradually regains its strength. The usual treatment of a fracture is to move the broken pieces back together as close as possible, then fix them in that position long enough so that sufficient hydroxyapatite redeposits to make the bone reasonably rigid. Usually the broken pieces are held in position by a cast, with traction apparatus, or by surgically implanting metal fixtures to hold the bones in place. Experiments using electrical signals to speed up bone growth have shown promise, but the technique is too complicated to use for routine fractures.

The nature of the bones changes with age. The infant's bones are mostly collagen, and they are easily bent but not easily broken. Gradually more and more hydroxyapatite is built up until, in the prime of life, the bones are almost a perfect compromise between flexibility and strength. In later life the calcium is reabsorbed from the hydroxyapatite and the collagen loses its flexibility, so the bones become brittle and easily broken. While the aging process in the bones cannot be stopped, mild exercise and activity seem to slow down their deterioration.

Blood Vessels

The blood vessels must have several properties to carry out their job of transporting blood within the body. First, they must be very strong, because the pressures inside the vessels are high, especially in the arteries. They must be very elastic, both to accommodate the movements of the body and to absorb the shock of the heart pumping the blood in spurts. They must also be able to expand or shrink in response to increases or decreases in the blood flow. In this way, blood can be diverted from regions that have a sufficient supply and sent to regions that need it. Finally, the inside walls of the vessels must be very smooth to prevent damage to the blood cells.

Nature has met these multiple demands with a composite material. The outer layer is mostly connective tissue—collagen—which gives strength to the vessel. The next layer is smooth muscle, which gives the vessel elasticity and the ability to change the diameter of the vessel as needed. The inside layer is formed of epithelial cells—the kind of cells found on the surface of most organs—which provide a very smooth lining. The major difference between the arteries and the veins is that the pressure in the arteries is much higher; thus, the arterial walls must be much stronger. Consequently, the connective tissue and muscle layers are much thicker in the arteries than in the veins (Figure 6-22).

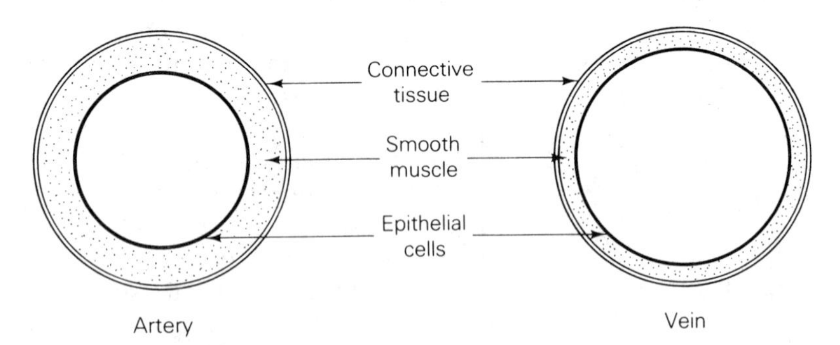

Figure 6-22

Friction

Whenever two surfaces slide along each other, forces will retard or prevent the movement of the two surfaces. Such forces are called *frictional* forces, but there is no single cause of friction. The actual cause depends on the surfaces, and it can vary from the simple abrasion of two rough surfaces to the complex viscous forces in a lubricating oil layer.

Every motion that we make and everything that we build are affected by friction. Enormous amounts of energy are lost to friction that could be reduced with proper lubrication. But even larger amounts of energy are lost to friction that we don't know how to reduce. Friction makes it difficult to move objects, and it can make our joints stiff and painful. Yet friction is not all bad by any means. It is necessary to walk, and it generates the forces needed to steer and stop cars. Nails are held in wood by friction, knots in ropes hold because of friction, and all fabrics are held together by friction. The world would be far different if there were no frictional forces: we would slip when we tried to walk, fabrics would disintegrate, and sand hills would flow and level off like a fluid.

Friction occurs between any two surfaces that are in contact, not just between the surfaces of solids. Friction exists between gases and solids; air friction, for instance, is the major retarding force on a bicyclist at high speeds. There is friction between solids and liquids; such friction occurs when blood flows through our blood vessels and when we swim through the water. There is also friction between gases and liquids: The wind on the water produces waves through gas-liquid friction.

Although frictional forces are always present, they are perhaps the least understood common force. In fact, not one single type of friction is well understood. The best that we can say about existing theories of friction is that they are useful in some cases, even if they are not complete. But when we need exact information about frictional forces, we must actually measure them. For instance, if we wanted to find out how long a certain car takes to stop on a particular type of pavement, we would have to find out through actual trials—no calculations will give us an accurate figure. Even then, slight changes in the road or the tires could have a large effect on the frictional forces. The most that theoretical calculations can provide is a rough estimate of the forces caused by friction.

Figure 6-23
Microscopically, most smooth surfaces are very rough. When two such surfaces slide over each other, the roughness causes friction.

In this chapter we will look only at solid-solid friction in detail. On a microscopic level, the surfaces of solids are almost never smooth. Even though the surfaces might look and feel flat, a microscope will show that they are quite rough (Figure 6-23). When two solid surfaces are in contact, the peaks of one surface drop into the valleys of the other, and vice versa. Obviously, frictional forces will result when we try to slide one surface along another, because the peaks bump into each other. In addition, the two surfaces will actually weld at the points of contact. Since the surfaces will actually make contact at only a very few places, the stresses at those

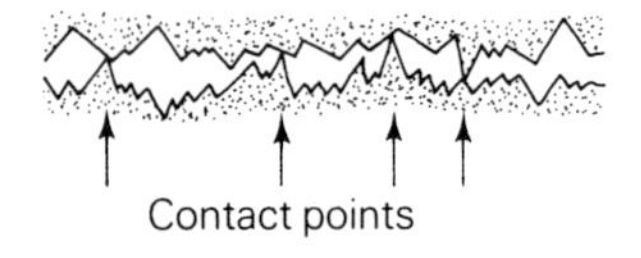

Figure 6-24 The ultrahigh pressure at contact points produces pressure welding.

points will be so enormous that the two materials will flow and mix together, creating pressure welds (Figure 6-24).

It is much harder to *start* an object sliding than it is to keep it going because two stationary surfaces have had time to settle together, making the pressure welds stronger. When we try to start the object moving again, we must first break the two surfaces apart. Thus, we usually make a distinction between static and moving friction: Static frictional forces occur when the two surfaces are stationary, and moving frictional forces occur when they are sliding. Static friction is always at least as large as moving friction and often much larger.

Even when we try to move things smoothly, we often get a rough motion accompanied by creaking or squealing. This type of motion, called *stick-slip* motion, results from the static friction being larger than the moving friction. What happens is that the force builds up to a high level to start the motion. But once the motion starts, the object jumps ahead some distance because the moving friction is smaller than the static friction. If the force that is pushing the object cannot keep up with this sudden jump, the object stops and sticks again. Then the motion doesn't start until the force builds up enough to break the static friction, at which point the object suddenly jumps forward again, and the process is repeated. Creaking doors, squealing brakes and tires, chalk squeaking on a blackboard, and gnashing teeth are all examples of stick-slip motion. However, stick-slip motion doesn't produce only ugly sounds. The sounds of a violin are produced by stick-slip friction of the bow with the strings. In fact, rosin is used on the bow to increase the static friction.

Lubrication

When we put a layer of lubricating fluid between two surfaces, the friction is reduced in two ways. First, the lubricant keeps the peaks of one surface from settling so deeply into the valleys of the other. Second, it prevents pressure welding at points of contact by introducing a nonweldable layer between the two solid surfaces. Even a layer of air can reduce welding at points of contact.

The effectiveness of a lubricating oil layer strongly depends on its thickness. A narrow oil layer reduces the friction somewhat, but the higher peaks of the surfaces can still contact to produce significant frictional forces. Of course, when the peaks collide during sliding, they often break off and the surface gradually wears away. If we want to cut down friction and wear as much as possible, we must maintain a lubricating layer that is so wide that the actual solid surfaces never come into contact. One way to do this is to place a layer of thick, heavy grease between the surfaces. If the grease is thick enough, it will not be squeezed out from between the surfaces, and it will prevent solid-solid contact. This prevents wear very effectively, but unfortunately the thick grease itself retards sliding, so the friction is still rather high. If we want to reduce both friction and wear, we must use a wide layer of a thin lubricating oil (thin in the sense of being nonviscous and easily flowing). However, the thin oil will run from between the two surfaces very quickly—so quickly that either the oil will have to be sealed in (a very difficult thing to do) or pumps will have to be used to replenish the oil layer continuously, as in a car engine.

Lubrication is one of the purposes of the various oils and lotions used on the skin. A good backrub would be painful without the lubrication of a lotion. Similarly, baby powder reduces the friction between the clothes and the skin to cut down chafing.

The joints of the human body are lubricated with a wide layer of lubricating fluid, the synovial fluid. The design of the joints is far more efficient than anything man has designed in that the synovial fluid is encapsulated by membranes that hold the fluid in the joints. In this way a wide lubricating layer is retained without continuously replenishing the fluid. The synovial fluid is an extremely good lubricant; it lowers friction in the joints to about 1/100th of what it would be for bone on bone. If the quality of the joint or the synovial fluid deteriorates, as can happen in a number of diseases including rheumatoid arthritis, the friction in the joints increases sharply. This results in pain, stiffness, and sharply increased wear on the joints.

Estimating Frictional Forces

There is no one theory of friction, and the theories that explain special cases are not highly accurate. For this reason the ideas in this section will help you to estimate frictional forces, not to find them exactly. Nevertheless, a rough estimate of the forces and a basic understanding of the factors that cause friction are enough to understand many motions of the body.

We will only consider the case of one solid sliding on another with at most a narrow lubricating layer. Our treatment does not apply to fluid friction, such as air or water friction; nor to cases where the lubricating layer is wide, which are instances of solid-liquid friction.

We can find out what factors are important in determining friction by considering an object sliding on a flat surface. The frictional force will depend primarily on two things: the surface conditions—the materials and their smoothness—and the force pushing the object and the surface together.

Surprisingly, the area of contact is not very important,* because it influences the frictional force in two opposing ways: One way tends to increase the friction, and the other tends to decrease it. A large area of contact has many more points of contact, which tend to increase the friction. But the force pushing the two surfaces together is also spread out over more contact points; thus the stress at each point of contact is less. Consequently, the peaks do not penetrate into the valleys as deeply, causing less deformation and pressure welding; this tends to reduce the friction. Because these two effects essentially cancel each other, the area of contact doesn't have much effect on the frictional force.

The speed at which the surfaces slide doesn't have much effect on the friction, either. Once static friction is overcome and the object is moving, the speed at which the peaks and valleys bump into each other doesn't have much effect on the frictional force. (This is only true for solid-solid friction; fluid friction increases rapidly with speed.)

It has been found experimentally that the frictional force for a particular set of surfaces increases uniformly as the force pushing the two surfaces together increases. That is, if you push the surfaces together twice as hard, the frictional force will also be twice as large. Expressed as a formula where F is the contact force pushing the two surfaces together, the friction force F_f will be

$$F_f = \mu F,$$

where the constant μ (mu) is known as the *coefficient of friction.* The coefficient has no units because it is simply a ratio of the two forces.

The value of the coefficient of friction depends entirely on the two surfaces—the materials they are made from, their roughness, and the lubrication between them. Since the coefficient of static friction is generally larger than the coefficient of moving friction, we will distinguish between the two by denoting them as μ_s and μ_m, respectively.

Up to now we have assumed that the force pushing the two surfaces together was perpendicular to the surfaces. If this is not the case, only the part of the force that acts perpendicular to the surface is involved in determining the frictional force. The portion

---

*While the area of contact is never very important, there are cases in which it does play a minor part. One case is rubber sliding on a road. On a dry road a wide rubber tire with a large contact area holds slightly better than a narrow tire. However, on a wet road a narrow tire holds better.

of the force that acts parallel to the surface does nothing to push the surfaces together, so it doesn't affect the friction at all.

EXAMPLE A 2-kilogram steel block is sitting on a wooden table (Figure 6-25). What force is needed to start it moving and what force is needed to keep it moving? The coefficient of static friction for steel on this particular wood is $\mu_s = 0.5$ and the coefficient of moving friction is $\mu_m = 0.3$.

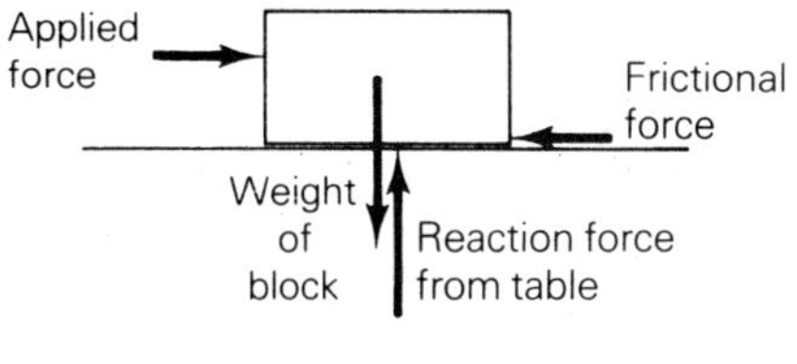

Applied force
Frictional force
Weight of block
Reaction force from table

Figure 6-25

Answer The force pushing the two surfaces together is just the weight of the block: $F = W = mg = 2$ kilograms $\times$ 9.8 m/s² $= 19.6$ newtons. The force required to start the block is equal to the maximum force that static friction can produce, which is

$$
\begin{aligned}
F_f &= \mu_s F \\
&= 0.5 \times 19.6 \text{ N} \\
&= 9.8 \text{ N}.
\end{aligned}
$$

To keep the block moving requires a force equal to the force of moving friction, which is

$$
\begin{aligned}
F_f &= \mu_m F \\
&= 0.3 \times 19.6 \text{ N} \\
&= 5.9 \text{ N}.
\end{aligned}
$$

Note that when the block is moving at a constant speed, the net force on the block is still zero. The frictional force is equal in size but opposite in direction to the force applied by whatever is pushing the block.

The coefficients of friction for various materials are given in Table 6-1. However, keep in mind that these values are only approximate. They strongly depend on the precise surface conditions. For instance, a little bit of water can easily drop the friction 50% or more. Your estimates will be much more accurate if you actually measure the coefficients of friction for the actual surfaces under consideration.

Table 6-1 Coefficients of Friction

Surface	*Coefficient*	
	μ_s	μ_m
Steel on steel (dry)	0.2	0.1
Steel on steel (oiled)	0.04	0.03
Metal on wood	0.5	0.3
Rubber on dry concrete	1.0	0.7
Rubber on wet concrete	0.7	0.5
Wood on wood	0.4	0.3
Rope on wood	0.5	0.3
Teflon on steel	0.04	0.04
Shoes on ice	0.1	0.05
Leather-soled shoes on carpet	0.6	0.5
Rubber-soled shoes on wood floor	0.9	0.7
Mountain climbing boots on rock	1.0	0.8
Bone lubricated with synovial fluid	0.016	0.015

EXAMPLE A 100-kilogram wooden box is sitting on a flat stone floor. It takes a 490-newton force to get it started and a 390-newton force to keep it moving. What are the coefficients of static and moving friction for these surfaces?

Answer The two surfaces are pushed together by the weight of the box:

$$\begin{aligned} F &= W \\ &= mg \\ &= 100\ \text{kg} \times 9.8\ \text{m/s}^2 \\ &= 980\ \text{N}. \end{aligned}$$

We know that

$$F_f = \mu_s F,$$

so

$$\begin{aligned} \mu_s &= \frac{F_f}{F} \\ &= \frac{490\ \text{N}}{980\ \text{N}} \\ &= 0.50. \end{aligned}$$

Similarly,

$$\mu_m = \frac{F_f}{F}$$

$$= \frac{390 \text{ N}}{980 \text{ N}}$$

$$= 0.40.$$

The above example illustrates a good practical way of finding coefficients of friction. An object made of one material is simply dragged over a level surface made of another while you measure the horizontal force needed to start the object moving and the horizontal force needed to keep the object moving at a constant speed. The force pushing the two surfaces together is just the weight of the object, so the coefficient of static friction is equal to the horizontal force required to start the object moving divided by the weight, and the coefficient of moving friction is equal to the horizontal force needed to keep the object moving at a constant speed divided by the weight.

Friction and Movement

How we walk is directly influenced by the friction between our shoes and the surface we are on. For instance, a hill that is normally climbed with ease might be impossible to climb when it is wet or icy. Our choice of stride depends on the amount of friction; our stride on a slippery surface is completely different from our usual stride. Friction also determines how fast we can stop and how sharply we can turn.

Consider hill climbing. When a person stands on a slope, the person's weight acts in two different ways. The weight pulls the person back down the hill parallel to the slope, but it also pushes the person's shoes perpendicularly into the surface of the hill. To see exactly how the coefficient of friction limits us when climbing hills, we must look at the details of these two different effects of the person's weight. We know from vector addition that a particular force may be equal to the sum of two or more other forces. Suppose we think of the weight as the sum of a force acting parallel to the surface of the hill and a force acting perpendicular to the hill. Call them $F_{\parallel}$ and $F_{\perp}$, respectively. Figure 6-26 shows how the weight can be broken down into the sum of $F_{\parallel}$ and $F_{\perp}$ for different slopes. Note that the steeper the slope is, the larger $F_{\parallel}$ is and the smaller $F_{\perp}$ is. Only $F_{\perp}$ pushes the two surfaces together, so the maximum static frictional force is $F_f = \mu_s F_{\perp}$. Thus, the frictional force also decreases as the slope becomes steeper. Eventually, as the steepness is increased, $F_{\parallel}$ (the part of the weight pulling the person down the slope) becomes larger than F_f, and the person slides back down the

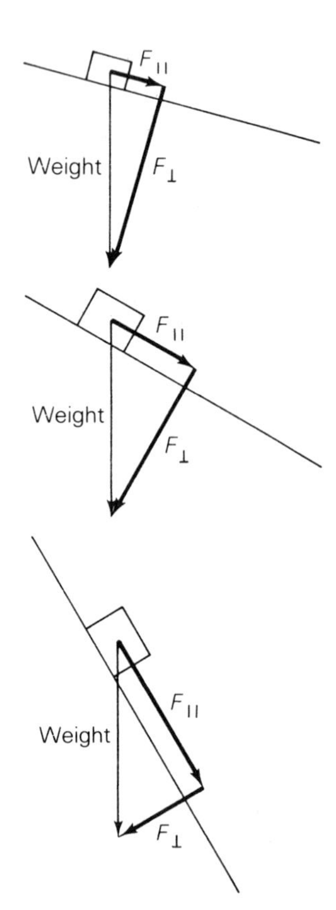

Figure 6-26 The weight of the block can be thought of as the sum of a force parallel to the slope and a force perpendicular to the slope.

hill. The larger the coefficient of friction is, the larger the frictional force will be. The larger the coefficient of friction is, the steeper the slope a person can stand on without sliding down. The coefficient of friction needed to climb a particular slope can be found by comparing $F_{\parallel}$ and $F_{\perp}$ for that slope. In order not to slide down, F_f must be greater than $F_{\parallel}$. But $F_f = \mu_s F_{\perp}$, so $\mu_s F_{\perp}$ must be greater than $F_{\parallel}$, too. This means μ_s must be greater than $F_{\parallel}/F_{\perp}$; thus, the smallest coefficient of friction that can be used to climb the slope would be

$$\mu_s = \frac{F_{\parallel}}{F_{\perp}}.$$

The chart in Figure 6-27 shows the results of such calculations.

One rule of beginning mountain climbing is "never try to sit down when you become frightened on a steep slope." This is a good example of how the coefficient of friction can influence technique. A beginning climber who becomes scared on a steep slope will have a very strong urge to sit down. This urge can easily lead to a fall, because modern climbing boots have soles with a very high coefficient of friction, usually somewhere in the range of 1.0, so slopes as steep as 45° can be climbed without slipping. Unless one sews rubber patches on the seat of his or her pants, the coefficient of friction between the pants and the rock will be much less than 1.0, usually around 0.3. With a coefficient of friction this small, a person would slip on a mere 20° slope. Thus, a person is far less secure sitting down than standing on a steep slope.

A particularly simple way to measure the coefficient of friction is to place an object made of one material on a board coated with another. One end of the board is raised until the object starts to slip. This angle is measured, and the coefficient of static friction can be found on the chart in Figure 6-27. This method can also be

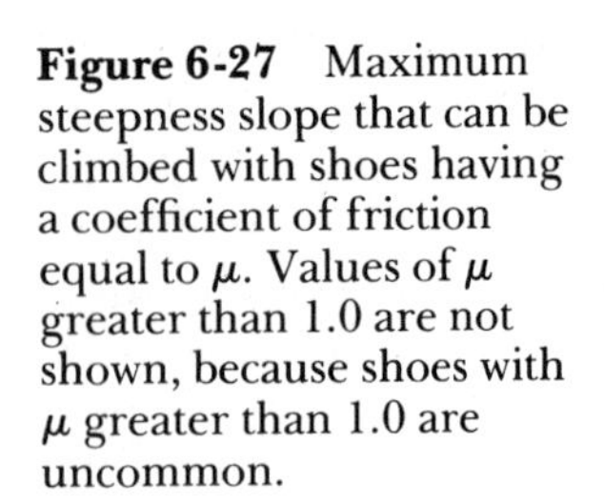

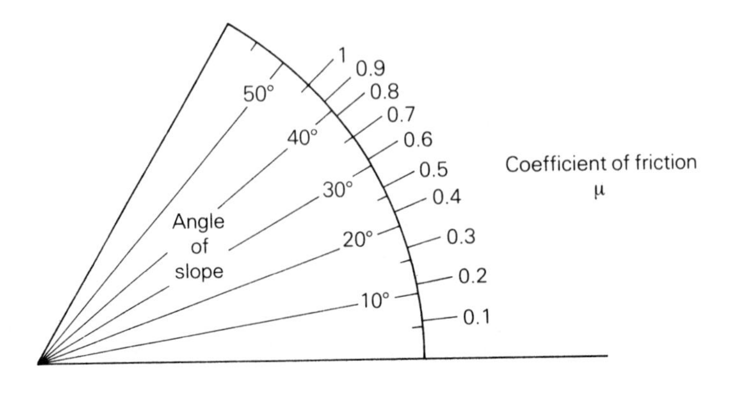

Figure 6-27 Maximum steepness slope that can be climbed with shoes having a coefficient of friction equal to μ. Values of μ greater than 1.0 are not shown, because shoes with μ greater than 1.0 are uncommon.

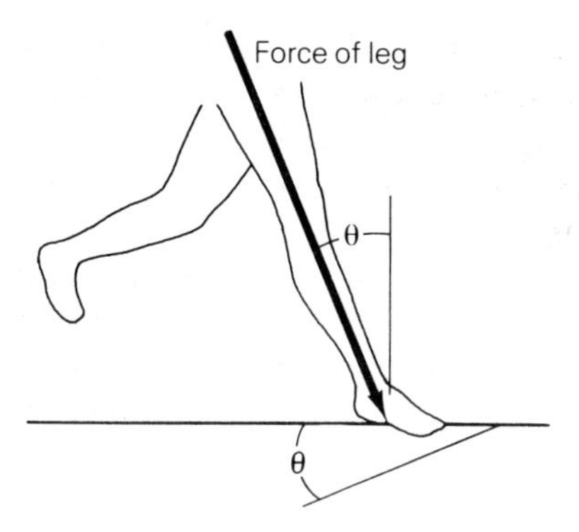

Figure 6-28

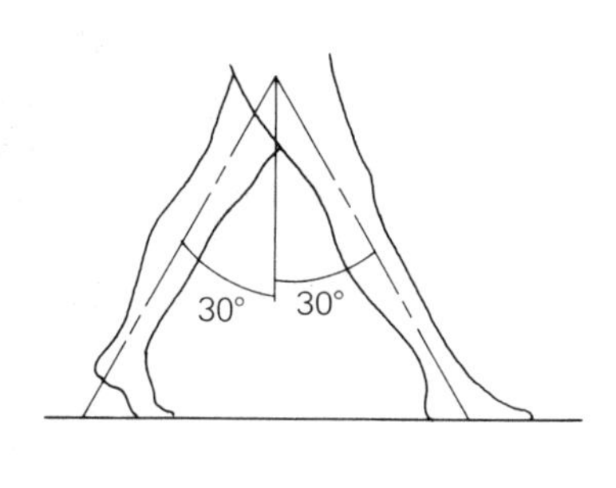

Figure 6-29
Maximum stride.

used to find the coefficient of moving friction, but you have to keep nudging the object to start it moving as the board is raised. The angle at which the object moves down at a constant speed after the nudging corresponds to the coefficient of moving friction.

When a person walks at a steady pace on level ground, the force of the leg is normally directed parallel to the leg (Figure 6-28). If you turn the book so that the force is pointed straight down and the ground is sloping up at an angle θ above the horizontal, this case looks exactly the same as the case of an object sitting on a hill. The force again acts in two ways: Part of it pushes the shoe and the ground together, and the other part acts parallel to the ground. The force can again be broken down into parallel and perpendicular components, $F_{\parallel}$ and $F_{\perp}$, and the person will slip if $F_{\parallel}$ exceeds the maximum frictional force, $F_f = \mu_s F_{\perp}$. This analysis is exactly the same as the case of a person on a hill, so the same chart can be used to determine the maximum angle that the leg can make with the ground without slipping. For instance, on a slippery surface having a coefficient of static friction of 0.2, the maximum angle that the leg could make without slipping would only be 11°. This is why people take very small steps on slippery surfaces. The actual stride length can be measured on a scale drawing of the person. If the person's legs were 80 centimeters long, the maximum stride when $\mu_s = 0.2$ would only be 30 centimeters. If the average person's maximum stride is roughly equal to his or her leg length, a scale drawing of the legs in this position shows that the angle between each leg and a vertical line is 30° (Figure 6-29). Thus, the chart tells us that the coefficient of friction must be greater than $\mu_s = 0.6$ in order to walk at full stride without slipping.

When we are walking, our stride tends to be rather conservative, because we know from experience that a slip of the foot is very hard to stop and can often cause a nasty fall. Often a person who falls while walking lands very awkwardly; injuries to the legs, hip, spine, arms, and head are common. Hip fractures are particularly common in icy weather. A slip of the foot is so difficult to control because once slipping has started, the friction is no longer static friction—it is *moving* friction, which is considerably smaller. Thus, the person does not have the full force of static friction available to control his or her motion; the smaller force of moving friction is usually too small to restore control. The person falls—and falls hard—because moving friction doesn't even provide enough force to control the position of the body.

Falls while walking are dangerous to everyone—even athletes in peak condition have been seriously injured in such falls. However, they are particularly hazardous for the elderly and the sick. Consequently, walking surfaces in hospitals, nursing homes, and the homes of the elderly should never be slippery. Wet areas like bathtubs and showers are especially dangerous. If they have slippery surfaces, various types of tapes and stick-ons are available to eliminate the hazard.

Concepts

Tensile force
Compressive force
Torsional force
Elastic materials
Plastic materials
Strength
Ductile materials
Brittle materials
Metals
Alloys
Crystalline materials
Nonmetallic crystals
Glasses and ceramics
Polymers
Cross-linking
Elastomer
Composite materials
Collagen
Muscle
Tendon
Bone
Transverse fracture
Greenstick fracture
Oblique fracture
Spiral fracture
Comminuted fracture
Friction
Coefficient of friction
Lubrication
Static friction
Moving friction

Discussion Questions

1. Give three examples of substances or objects that behave about the same under tension as under compression. Give three examples that behave very differently under tension and compression.

2. Give examples of elastic materials, plastic materials, brittle materials, and ductile materials.

3. Why are spiral fractures of the lower leg bones common in skiing?

4. Discuss the role of cross-links in the hardness of polymers.

5. Explain how epoxy resin glues work.

6. In what ways is bone similar to reinforced concrete?

7. Explain how tendons are similar to ropes in composition and in function.

8. Why are backrubs seldom given without applying a lotion?

9. Most athletic shoes have high coefficients of friction so that the athlete can stop and turn quickly. Can you think of any sports or physical activities where the shoes do not have a high coefficient of friction?

10. Mountain and rock climbers should always be belayed by a safety rope when climbing dangerous slopes. However, a safety rope will be of value only if the person holding it at the other end, the belayer, can actually stop and hold the fallen climber. Often the belayer just sits down on a rock without making any effort to find a secure point. Could a belayer who is sitting down on a rock actually hold a climber's fall?

11. In Chapter 5, in the section on center of gravity, we saw how the CG of a long object can be easily located by placing it on your outstretched hands and then bringing your hands together smoothly. The CG is then directly above where the hands meet. Explain why the friction between the object and the hands makes this method work.

Problems

1. The head of a certain x-ray machine is made of steel, and it slides horizontally on a dry steel track. If it takes a force of 15 newtons to start the head sliding, how much force will be needed to keep the head moving at a steady speed?

2. If the track described in Problem 1 was oiled, how much force would it take to start the head moving and how much would it take to keep it moving?

3. A 55-kilogram girl wearing rubber shoes and standing on dry concrete has a tug-of-war with a 95-kilogram boy who is wearing shoes and standing on ice. Who is the most likely winner?

4. If you were wearing shoes that had a coefficient of static friction of 0.8 on rock, how steep a rock slope could you climb?

5. If the length of your leg was 85 centimeters from your hip to your heel, what would be your maximum safe stride on a surface that had a coefficient of friction of 0.3 with your shoes?

Experiments

1. Study how different materials bend and break. Choose several different materials, such as iron wire, a plastic rod, a wooden stick, and a piece of chalk. Then bend them until they break. Study the broken pieces with a magnifying glass if you have one. Describe how each sample broke. Note: Do not try this experiment with any material that might be dangerous, such as glass.

2. The best way to learn how different types of bone fractures are caused by different types of force application is to get some bones from a small animal and break them. In particular, look at fractures caused by bending, compression, and twisting. Fresh bones will show you the most about how human bones fracture, but cooked bones, or even green twigs or blackboard chalk will give you a good idea of what happens when the forces on a bone exceed its strength. Chalk broken by torsion is especially instructive. Make a drawing showing how you applied the forces and what the fractures looked like.

3. Measure the coefficient of friction of some of your shoes on a wooden board by tilting the board until the shoes slide. Measure the angle of the board with a protractor and then use the chart in Figure 6-27 to find the coefficient of friction. Which types of shoe sole have the largest coefficients? Which types have the smallest?

CHAPTER SEVEN

The Dynamic Behavior of Matter

Educational Goals

The student's goals for Chapter 7 are to be able to:

1. Calculate the acceleration of an object when it changes its speed or direction or both.
2. Find the acceleration of an object when the force on the object is known.
3. Use the concept of momentum to explain the results of collisions between objects.
4. Describe the path of falling and thrown objects.
5. Explain why a centripetal force is needed to make an object move in a circle.
6. Estimate the effect of forces on human movement.

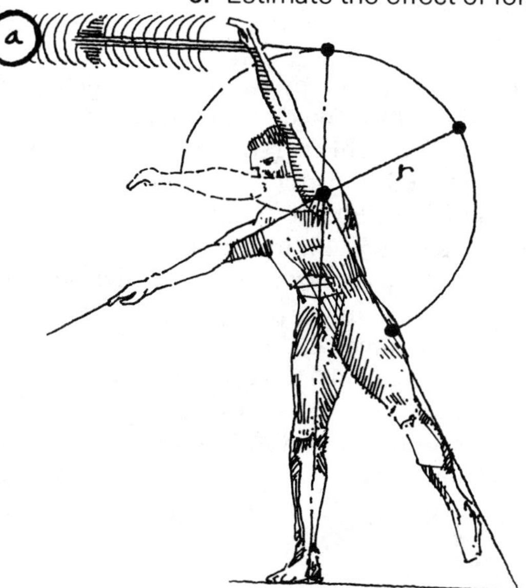

In our earlier study of matter, we saw that forces produce *changes* in the motion of objects. However, we were then mainly interested in cases where the net force was either zero or so close to zero that we could ignore any small change in motion. The next step in our study of matter and motion is to find out exactly *how* the motion of an object changes when the net force is not zero. But before we start, let's first look into how we describe changes.

Describing things that are changing is harder than describing things that are stationary because there are many ways to describe changes. We could talk about the *total change* that occurs. For instance, if Joe had been on a diet, we could describe the total change by saying, "Joe lost 10 kilograms when he went on a diet." Or we might talk about the *rate* of loss by saying, "Joe lost 2 kilograms per month on his diet." Assuming that his diet took five months, both the total loss of 10 kilograms and the rate of loss of 2 kg/month accurately describe the change in Joe's mass.

Which description we choose depends on our interests. Joe might be most concerned about the *total change.* But the health professionals guiding his diet would be more concerned with the *rate of loss.* They don't want the rate to be so fast that Joe's health is injured, nor do they want the rate to be so slow that Joe gets discouraged and quits. Although most changes can be accurately described in several ways, one way may be far better than others for understanding a particular change. This is definitely true in the case of motion.

Describing Acceleration

The change in motion caused by a force is best described by giving the *rate* at which the velocity changes, which is called the *acceleration* of the object. We know from everyday activity that force, mass, and acceleration are related and that large forces are needed to produce large accelerations. For instance, the harder we throw something, the faster it goes when it leaves the hand. We also know that light objects are easier to accelerate than massive objects. Throwing a pebble is much easier than throwing a large rock.

Acceleration is defined as the *rate* at which the velocity of an object changes. Since velocity is a vector quantity, its rate of change, the acceleration, will also be a vector quantity. Thus, not only is the size of the acceleration important, but so is its direction. We can find the acceleration in the same way that we find all other rates: We divide the change by the time needed to make the change.

Thus,

$$\text{acceleration} = \frac{\text{change in velocity}}{\text{time needed to make the change.}}$$

The change is found by subtracting the old velocity from the new velocity, so we can also write

$$\text{acceleration} = \frac{\text{new velocity} - \text{old velocity}}{\text{time needed to make the change.}}$$

For example, consider a person who is driving down a straight road at 10 m/s. The person steps on the gas and 3 seconds later is going 19 m/s. The acceleration would be

$$a = \frac{19\ \text{m/s} - 10\ \text{m/s}}{3\ \text{s}}$$

$$= \frac{9\ \text{m/s}}{3\ \text{s}}$$

$$= \frac{9\ \text{m}}{\text{s}} \times \frac{1}{3\text{s}}$$

$$= 3\ \text{m/s}^2.$$

The units in the last example, m/s^2, look a little odd. To see what they mean, let's look more closely at what rates mean in general. Consider a person's rate of food consumption. How much food, on the average, does that person consume in a day? To obtain an accurate estimate of the rate of consumption, we would measure the total consumption over several days and then divide the total food intake by the number of days. In other words,

$$\begin{matrix}\text{rate of food}\\ \text{consumption}\end{matrix} = \frac{\text{food consumed}}{\text{time required to consume the food.}}$$

Speed is another common rate. Speed is just the *rate* at which distance is traveled, which is found by dividing the distance traveled by the time required to travel that distance. Thus,

$$\text{speed} = \frac{\text{distance traveled}}{\text{time required to travel that distance.}}$$

The units of speed will be distance units divided by time units, such as m/s, km/h, or miles per hour. (The word *per* is equivalent to *divided by.*)

Since acceleration is the rate of change of velocity, the units of acceleration will be velocity units divided by time units, such as (m/s)/s = m/s^2. We can write (m/s)/s as m/s^2 because units follow the same mathematical rules as numbers do. (Recall that the rule for dividing fractions is to invert the divisor, then multiply. This, followed by a bit of algebraic fiddling, will show that both ways of writing the units of acceleration are exactly equivalent.)

The fact that acceleration is a rate of something that is a rate itself sounds confusing. However, rates of rates are actually quite common. For instance, consider the effect of inflation on rents. Rents themselves are rates: so many dollars *per* month. So when we discuss the rate of change of our rent, we are talking about the rate of change of our rate of payment. Suppose that the rent was \$200/month last month and \$220/month this month. Thus,

$$\begin{aligned}\text{rate of rent increase} &= \frac{\text{new rent} - \text{old rent}}{\text{time for the change}} \\ &= \frac{\$220/\text{month} - \$200/\text{month}}{1 \text{ month}} \\ &= \frac{\$20/\text{month}}{\text{month}} = \$20/\text{month}^2.\end{aligned}$$

We normally would call the above rate of rent increase "\$20 per month per month." "Twenty dollars per month squared" is perfectly correct mathematically, but it doesn't sound as logical. For the same reason, m/s^2 is normally called "meters per second per second" rather than "meters per second squared." Of course, both terms are correct.

Acceleration is a vector, but its direction doesn't have to be the same as the direction in which the object is moving. In fact, whenever an object moves in a curved path, the acceleration is in a different direction from that of the velocity. The acceleration can be any direction, but two directions are particularly important. The first is the simple case where the acceleration is parallel to the velocity; that is, it is either in the same direction as the velocity or directly opposite to it. When the acceleration is in the same direction as the velocity, the object's speed increases and the direction of the motion remains the same. When the acceleration is directly opposite the velocity, the object slows down without changing direction. We speak of this as a deceleration, and mathematically we refer to it as a negative acceleration. Thus, when the acceleration is parallel to the velocity, the object either speeds up or slows down, without changing its direction.

Just the opposite occurs in the second important case where the acceleration is perpendicular to the velocity. Here the direction of motion constantly changes, but the size of the speed remains the same. This type of motion occurs whenever we go around a corner without changing the size of our speed. Uniform motion in a circle is another example. An object going around a circle is *always accelerating* even though the size of the speed is always the same, because the object's direction constantly changes. A change in direction involves acceleration, just as a change in the size of the speed does.

When the acceleration is neither exactly parallel nor exactly perpendicular to the velocity, both the size and the direction of the velocity change simultaneously. Most accelerated motions are of this type. However, the mathematical details are more complicated than either of the two simple types.

EXAMPLE A motorcyclist is traveling down the road at 25 m/s when she sees a dog in the road. She hits both brakes and 2 seconds later is going 5 m/s. What was the acceleration of the motorcycle during the time when the brakes were on?

Answer We can find the acceleration directly from its definition:

$$a = \frac{v_n - v_o}{t},$$

where v_n and v_o are the new and old velocities and t is the time required to change the velocity. Putting the velocities and time into this formula, we find

$$a = \frac{5 \text{ m/s} - 25 \text{ m/s}}{2 \text{ s}}$$

$$= -10 \text{ m/s}^2.$$

The minus sign reflects the fact that the motorcycle was slowing down; negative acceleration simply means deceleration. Incidentally, 10 m/s^2 is about the best deceleration that can be obtained with a rubber-tired vehicle. If the road was at all slippery, it would be impossible to stop this rapidly.

EXAMPLE A boy is swimming eastward at 0.75 m/s. He then gradually turns southward without changing the size of his speed of 0.75 m/s. He takes 25 seconds to make the turn. What is his acceleration while turning?

Answer We can again find the acceleration from its definition. However, since the change in velocity is a change in direction, we must compute the change using vector subtraction. To subtract a vector, we simply add the negative of that vector. The negative of any vector is simply a vector of the same size pointing in the opposite direction to the original vector. Thus, to

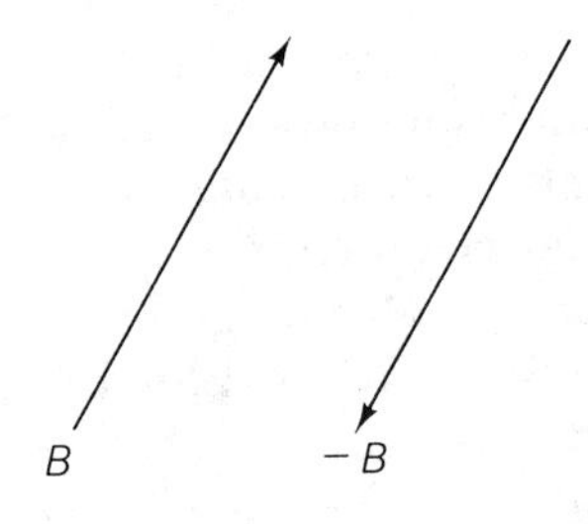

Figure 7-1 The negative of a vector is simply a vector of the same length in the opposite direction.

change a vector into its negative, we only have to switch its head and tail (Figure 7-1). If we want to subtract Vector B from Vector A, we simply note that

$$A - B = A + (-B).$$

So to subtract B from A, we simply switch the head and tail of B to get $-B$, then add A and $-B$ in the normal way. This is illustrated in Figure 7-2.

In our example, the first step is to find the change in the velocity, $v_n - v_o$. We must do this graphically, since the two velocities are in different directions. When we do this as shown in Figure 7-3, we find that the change in velocity, $v_n - v_o$, is 1.1 m/s toward the southwest. Dividing the change by the time, we get

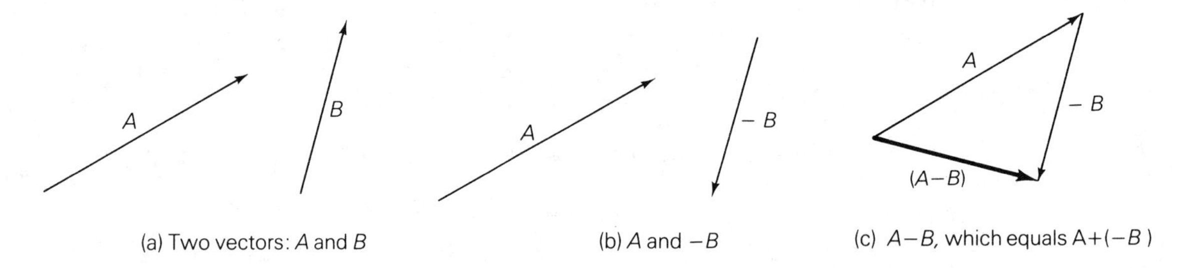

Figure 7-2 A − B, which equals A + (−B).

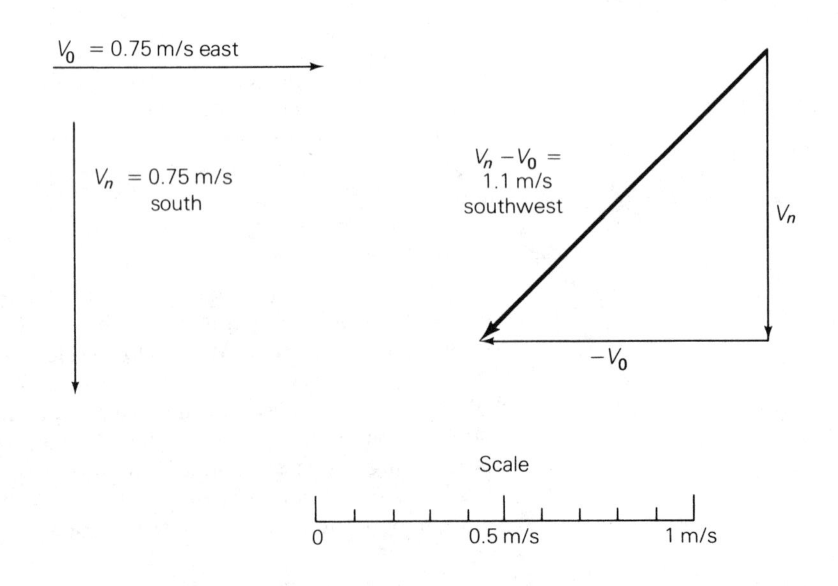

Figure 7-3

$$a = \frac{1.1 \text{ m/s}}{25 \text{ s}}$$

$$= 0.044 \text{ m/s}^2.$$

Thus, during the turn the boy was accelerating at 0.044 m/s^2 toward the southwest.

EXAMPLE Suppose a sprinter starts out from rest and accelerates at a rate of 5 m/s^2. How fast will the sprinter be going 2 seconds later?

Answer First we must rearrange the formula for acceleration,

$$a = \frac{v_n - v_0}{t},$$

to find the new velocity. Moving the t and switching sides on the equation gives

$$v_n - v_0 = at;$$

thus,

$$v_n = at + v_0.$$

In this case the sprinter starts from rest, so v_0 is zero. So

$$v_n = (5 \text{ m/s}^2 \times 2 \text{ s}) + 0$$

$$= 10 \text{ m/s}.$$

Forces and Acceleration

When an unbalanced force acts on an object, the object's motion changes. In other words, the object accelerates. People have always known that forces change motions, but it took thousands of years to figure out precisely what aspect of the change was most closely related to the force. In the seventeenth century, Newton realized that the force was directly related to the acceleration. Like most great discoveries, the relationship is really quite simple: *The acceleration is directly proportional to the net force.* That is, if we were to double the net force on an object, the acceleration would double; if we were to triple the net force on an object, the acceleration would triple; and so on. Note that we are *not* saying that the speed doubles or triples; we are saying that *the rate at which the speed changes* doubles or triples.

The only other factor that influences an object's acceleration is its mass. We know that the more massive an object is, the more difficult it is to change its motion. So the relationship between an object's mass and its acceleration must be an inverse relation. That is, for a particular force, the acceleration must decrease as the mass increases. The exact relationship between mass and acceleration is simple, too: *The acceleration is inversely proportional to the mass.* That is, if the same force acted on an object with twice the mass, the acceleration would only be half as large; if the object had 3 times the mass, the acceleration would be one-third as large; and so on.

The relationship between force, mass, and acceleration is known as Newton's second law of motion:

Newton's Second Law of Motion

When an unbalanced force acts on an object, the net force produces an acceleration that is directly proportional to the force and in the same direction as the force. The acceleration is inversely proportional to the mass of the object.

Expressing this as a formula is easier than saying it. In formula form, the second law of motion is

$$a = \frac{F}{m},$$

where a is the acceleration, F is the net force, and m is the mass of the object. Of course, we can rearrange the formula into any of its equivalent forms. The most common form is $F = ma$, because it is easy to write and easy to say.

Because many motions of the body are too complicated to analyze exactly, we won't emphasize exact calculations with the second law. Instead, we will mainly be interested in seeing how the second law can be used to help understand human motion. For instance, whenever a person makes rapid changes in his or her motion, the second law tells us that large forces must be involved. When a person makes small, gradual changes, we know that the forces causing these changes must be small.

The second law explains why smoothness of motion is such a desirable goal. When a motion is jerky, it changes irregularly; in other words, there are large, rapidly changing accelerations. Where there are large, irregular accelerations, there must be large, irregular forces. The muscles and nerves aren't very well suited to producing such forces, so a jerky motion is hard to control. In addition, a jerky motion leads to unnecessary strain on the body

The continuous circular motion of the feet while pedaling is much more efficient than the start-stop motions of walking and running.

and a waste of muscular energy. In a smooth motion, the muscular forces and the stresses on the joints are held to a minimum. The forces change slowly and the muscles are easier to control. Most importantly, the effort is concentrated on the motion, not on rapid, unproductive changes.

Running and walking aren't especially efficient forms of propulsion, because the legs are continually starting and stopping. One instant the foot is stationary on the ground, the next it is yanked forward ahead of the body, only to be stopped abruptly and placed on the ground again. Since the foot and leg are almost constantly being accelerated, the forces of the leg muscles are always large. These forces are mostly wasted in starting and stopping the leg. Bicycling is far more efficient than running or walking; most people can bicycle twice as fast as they can run. One reason is that the feet and legs move continuously while bicycling. Thus, the accelerations are smaller and less energy is wasted.

Why are good runners usually thin? The second law gives the answer: Thinner people have less mass in the leg to accelerate; thus, they can accelerate their legs more rapidly. Large leg muscles don't help because larger muscles have more mass. In fact, the mass increases faster than the strength of the muscle, so additional muscle often does a runner more harm than good. This line of reasoning also explains why deer can run much faster than humans. In a human, the leg muscles are rather low in the leg and move with the leg. Consequently, the force needed to accelerate the mass of the leg is large. In a deer, the muscle is concentrated near the top of the leg; the lower part of a deer's leg is almost nothing but skin, bone, and tendons. Since the muscle is close to the pivot point of the lever formed by the leg, the muscle in a deer's leg moves far more slowly than the hoof of the leg. Thus, the acceleration of the deer's leg muscles is relatively small; more of the force goes into moving the deer forward and less is wasted starting and stopping the leg.

The second law of motion also explains why lighter people are usually more agile than heavier people. The less mass a person has, the more quickly the person can change his or her motion. This is why the athletes in sports where great agility is needed, such as gymnastics and figure skating, often have small, light builds. In sports where a balance between strength and agility is needed, athletes with medium builds are usually most successful. Football demands players who are either very difficult to move, like linemen, or difficult to stop moving, like a running back, so football players usually have large builds. Great running backs are very rare because they must have both a heavy build and great agility. The second law tells us that such people are not going to be common.

EXAMPLE How fast can a person accelerate if his or her shoes have a coefficient of friction of 0.3? How long would it take to reach a speed of 6 m/s starting from rest?

Answer Since the mass of the person isn't given, call it m. Maybe we won't need to know it. The maximum forward force F that a person can produce is just equal to the coefficient of friction times the force pushing the person toward the ground. That force is simply the person's weight. So

$$F = \mu W$$

$$= \mu m g.$$

But by the second law, F is also equal to ma; thus,

$$ma = \mu m g.$$

Because the mass cancels on both sides of the equation, we don't need to know it. So

$$a = \mu g$$

$$a = 0.3 \times 9.8 \text{ m/s}^2$$

$$= 3 \text{ m/s}^2.$$

The time is found by rearranging the formula for acceleration:

$$a = \frac{v_n - v_0}{t}$$

$$t = \frac{v_n - v_o}{a}$$

$$= \frac{6 \text{ m/s} - 0 \text{ m/s}}{3 \text{ m/s}^2}$$

$$= 2 \text{ s}.$$

EXAMPLE The air friction, or drag, of a parachute will hold the falling speed of a 75-kilogram parachutist at about 5 m/s. Assuming that the legs absorb the shock of landing over a distance of 0.5 meter, estimate the force of the ground on the parachutist's body during the landing.

Answer The first step is to estimate the time required to stop. At a steady 5 m/s, a person would cover 0.5 meter in

$$t = \frac{d}{v}$$

$$= \frac{0.5 \text{ m}}{5 \text{ m/s}}$$

$$= 0.1 \text{ s}.$$

However, during the landing, the speed isn't a steady 5 m/s: It starts out at 5 m/s and ends up at zero. So the average speed will be somewhere between zero and 5 m/s, probably about 2.5 m/s. If we assume that the average speed during the landing is 2.5 m/s, it would take

$$t = \frac{d}{v}$$

$$= \frac{0.5 \text{ m}}{2.5 \text{ m/s}}$$

$$= 0.2 \text{ s}$$

to stop.

Once we know the time it takes to stop, we can find the acceleration from

$$a = \frac{v_n - v_o}{\text{t}}$$

Because the parachutist stops, v_n will be zero. So

$$a = \frac{0 \text{ m/s} - 5 \text{ m/s}}{0.2 \text{ s}}$$

$$= -25 \text{ m/s}^2.$$

The minus sign just signifies a deceleration. That is, the acceleration is upward, opposite to the velocity. We will drop the minus sign at this point and just remember that the acceleration, and thus the force on the parachutist, is upward during the landing. This net upward force is found from Newton's second law:

$$\begin{aligned} F &= ma \\ &= 75 \text{ kg} \times 25 \text{ m/s}^2 \\ &= 1{,}900 \text{ N}, \end{aligned}$$

compared with his or her weight of

$$\begin{aligned} W &= mg \\ &= 75 \text{ kg} \times 9.8 \text{ m/s}^2 \\ &= 740 \text{ N}. \end{aligned}$$

Thus, during the landing, the net force on the parachutist is well over twice his or her weight. But the situation is even worse than this. Remember that the net force F on the body is the vector sum of two forces, the parachutist's weight W, which acts downward, and the force of the ground F_g, which acts upward. That is,

$$F = F_g - W.$$

(The minus sign occurs because F_g and W are in opposite directions.)

$$\begin{aligned} F_g &= F + W \\ &= 1{,}900 \text{ N} + 740 \text{ N} \\ &= 2{,}600 \text{ N}. \end{aligned}$$

So the force of the ground on the parachutist is well over 3 times the parachutist's weight. The legs must transmit almost all of this force to the main part of the body; thus, the force acting on the legs is over 3 times the normal load. Obviously, the bones and muscles are under quite a strain during the landing. Normally, the legs can take this strain for a short time, but a slight slip could easily cause injury. Like all people who participate in highly strenuous activities, parachutists should be carefully and gradually trained to take the strain.

The Sense of Acceleration

The body has very sensitive organs for detecting accelerations: the utricle and the semicircular canals, both located in the inner ear. The utricle detects accelerations in a straight line, and the semicircular canals detect rotational accelerations. The utricle is a small sac that has a heavy membrane inside, which is supported by fine hairs connected to nerve endings. When the body is accelerated, the force needed to accelerate the membrane is transmitted through the hairs, bending them slightly. The nerves detect the bending

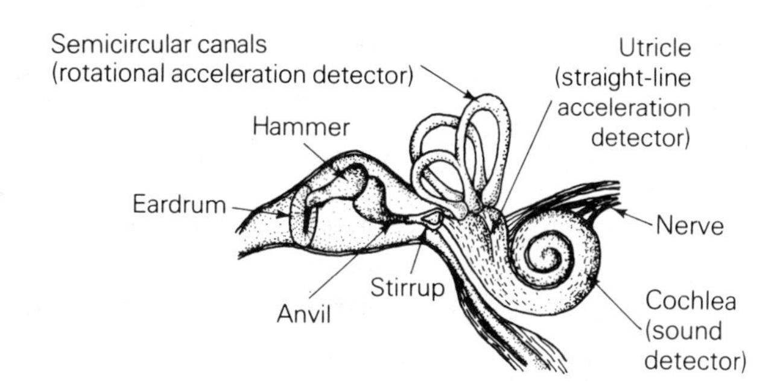

Figure 7-4 Location of the acceleration and sound detectors of the inner ear.

and transmit the signal to the brain. The utricle can detect accelerations as small as 0.1 m/s^2, 1% of the acceleration due to gravity. The semicircular canals detect rotational accelerations in a similar way. They are filled with a fluid, the endolymph. When the head is twisted, the canals move but the fluid tends to remain stationary. The flow of the fluid signals the brain that a rotational acceleration is occurring.

Both acceleration detectors play a much bigger role in our lives than is generally realized. Most nonvisual information about our motions comes from these organs. In diving and gymnastics their information is crucial, and we all depend on them for balance and stability. Walking on two legs would be difficult without them. The sense of acceleration is also important to our mental well-being. Humans go to great lengths to satisfy and amuse their senses, and the sense of acceleration is no exception. Carnival rides, skiing, and car racing are exciting largely because they involve large changing accelerations. Indeed, much of the pleasure of sport and physical activity comes from the stimulation of the acceleration-detecting organs. Of course, the acceleration sense can be overstimulated, just as the other senses can. Dizziness and seasickness both result from excessive changing accelerations that overwhelm the brain's ability to keep track of them.

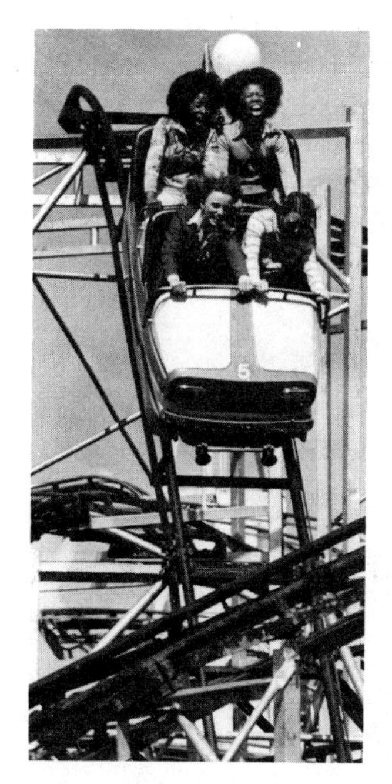

Most of the thrills of a carnival ride come from stimulation of the acceleration detectors of the inner ear.

The sense of acceleration is more important to our well-being than our sense of smell; a person who couldn't keep his or her balance would have a much harder time than someone who couldn't smell. Yet the sense of acceleration isn't even included in the usual list of the five senses, undoubtedly because people figured out what an acceleration is only within the last 300 years. The sense of acceleration is definitely a major sense: The acceleration detectors occupy as much of the middle ear as the sound detectors do. Of course, our ability to move is crucially dependent on them. We also have a strong drive to stimulate these organs—not as strong as hunger or sex, but probably as strong as the need to listen to music or see interesting sights. Physical activity will satisfy this drive; this mental benefit of physical activity can sometimes be as important as the physical benefits. Certainly anyone working in recreation or mental health programs should keep in mind the benefits of stimulating the sense of acceleration. In addition to the general benefits, such stimulation can be effective in combating boredom and depression. Since the acceleration detectors are very sensitive, carnival rides or contact sports aren't necessary to stimulate them; almost any activity will work, even walking.

Padding and Impact Protection

People seldom realize how large the forces in collisions actually are. For example, when a bowling ball is dropped from the waist to the floor, the force of the impact with the floor will be more than the weight of a large car. Even low-speed collisions, like those in contact sports and falls, can cause severe injury when proper safety padding isn't used. Consider the force on the head when a person runs into a brick wall at a normal running speed of 7 m/s. (This is equivalent to falling onto a concrete floor from 2.5 meters.) The head has very little padding of its own, only a few millimeters of skin. Assume that the head stops in around 4 millimeters. The average speed during the collision will be about ½ × 7 m/s = 3.5 m/s, since the head is slowing down from 7 m/s to zero during the collision. Thus, the impact takes only

$$t = \frac{d}{v} = \frac{0.004 \text{ m}}{3.5 \text{ m/s}} = 0.0011 \text{ s}.$$

We can calculate the head's acceleration from

$$a = \frac{v_h - v_0}{t} = \frac{0 \text{ m/s} - 7 \text{ m/s}}{0.0011 \text{ s}} = -6{,}400 \text{ m/s}^2,$$

over 600 times the acceleration of gravity. (Again, we will drop the minus sign, which just indicates that the head is slowing down.) The mass of the head is around 5 kilograms, so the force on the head during the impact will be

$$F = ma$$
$$= 5 \text{ kg} \times 6{,}400 \text{ m/s}^2$$
$$= 32{,}000 \text{ N},$$

which is about the weight of a loaded pickup truck! Obviously, that person is going to be severely, perhaps fatally, injured.

A good way to reduce the danger of such collisions is to use proper padding in safety helmets, car dashboards, athletic pads, and wherever else impacts are likely. With 4 centimeters of padding instead of 4 millimeters of skin, the force could be reduced to 3,200 newtons, one-tenth of what it was in the example. This is still a hard blow to the head, but it would not seriously injure most people. But 4 centimeters of just any kind of padding won't do. The padding must be carefully selected to absorb the shock at the proper rate. Padding involves more than just slapping on a bit of foam rubber here and there with adhesive tape.

Consider safety helmet design, for example. The main problem is that there isn't much room for padding, a few centimeters at most. If the padding was much thicker, the helmet would be too cumbersome to wear. Ordinary foam rubber might seem to be a good padding material, but it isn't because it is much too soft. Soft foam rubber, as it compresses, only produces a small, comfortable force that is unable to slow the head down very much. The foam quickly becomes completely compressed; the head then slams against the inside of the hard shell of the helmet. This is only slightly better than hitting a brick wall; again, serious injury is likely. In fact, soft foam rubber padding gives people a false sense of security, encouraging them to take needless chances. For maximum protection, the padding should produce *the maximum force that the head can safely stand for the entire time that the padding is being compressed.* Then the padding will make the most of the rather short distance available to stop the head. Because the maximum force that the head can endure is far above the threshold of pain, such a padding will not be comfortable. In fact, it will seem rather hard. Even when such a padding works properly, it can cause severe pain, because it is designed to prevent injury and death, not discomfort.

How hard should padding be? People vary in their ability to withstand acceleration, but tests indicate that most people will survive something over 50 times the acceleration of gravity for short times without permanent damage—something in the range of 500 m/s^2. If we assume that the mass of the head is 5 kilograms, helmet padding should not yield until the force is around

$$F = ma$$
$$= 5 \text{ kg} \times 500 \text{ m/s}^2$$
$$= 2{,}500 \text{ N},$$

which is about the weight of two or three people. Certainly, this padding would not feel soft, but it would work well to prevent serious injury. Exactly how hard the padding in helmets should be for various activities is still an open question. Hard padding would be the choice for car or motorcycle racing where maximum protection is needed. But for a sport like football, where many minor bumps occur, the padding might be a compromise between comfort and maximum safety. However, even for contact sports, the possibility of a severe collision exists, and the compromise should be made on the side of ultimate protection—not mere comfort.

Soft grass is one of the best cheap padding materials available. It doesn't "give" much for small forces, but it starts to compress at about 1,000 newtons when a person's head hits it. It will compress for several centimeters at this rate—thus absorbing much of the shock of a fall. If possible, sports where falls are common should be played on grass or a field having safety properties that are at least as good as those of grass. A piece of ordinary carpet on a hard concrete floor is a poor substitute for natural grass. Of course, some artificial surfaces are even better than grass; unfortunately, many are not.

Momentum

The calculations aren't always easy, but if we know the force on an object, we can always use $F = ma$ to find out how the object will move. The condition "if we know the force" is a big if, however. Forces are very difficult to measure, particularly if the object's motion is changing rapidly. Since the most interesting examples involve rapid changes, we might wonder if we can learn anything about an object's motion without knowing the force. Actually, we can learn quite a bit if we combine $F = ma$ with the fact that forces come in action-reaction pairs.

Suppose that a little girl and a little boy were playing on a friction-free surface, such as a very slippery ice skating rink. Then, as little children often do, they started to fight and one shoved the other (Figure 7-5). Regardless of who did the shoving, Newton's third law of motion says that the force on the boy F_b was exactly equal but opposite to the force on the girl F_g. That is,

$$F_g = -F_b.$$

Figure 7-5 A girl and boy playing on the ice. The force on the girl is equal but opposite to the force on the boy.

The minus sign indicates that the two forces were in opposite directions. From $F = ma$, we know that $F_g = m_g a_g$ and $F_b = m_b a_b$, so substituting for the forces in the equation above, we find

$$m_g a_g = -m_b a_b.$$

Note that the force, which we didn't know, is no longer in the equation. Next, multiply both sides of the equation by t, the time the shove took.* The time must be the same for both children, since the length of time they were in contact was the same for both. This gives

$$m_g(a_g t) = -m_b(a_b t).$$

Now look at the terms in parentheses: $(a_g t)$, for instance. $(a_g t)$ is the girl's acceleration multiplied by the time it took for her velocity to increase. This is just her velocity, since she started out from rest, so

$$a_g t = v_g.$$

The same reasoning holds for the boy,

$$a_b t = v_b.$$

Now, substituting these into the previous equation gives

$$m_g v_g = -m_b v_b.$$

The quantity mv is called the *momentum* of the object. By finding this equation, we have eliminated the force of the shove, which

*The purpose of multiplying by t is certainly not obvious at this point in the calculation. Newton was the first person to realize what could be learned by doing this manipulation.

would have been very difficult to measure. The equation relates mass and velocity, which are much easier to measure.

What else does the equation tell us? It says that the two children have equal but opposite momentums, which means that they move off in opposite directions, the heavier child moving more slowly than the lighter child. The equation also lets us solve for any one of the quantities, provided the other three are known.

Let's see how the net momentum of the two children changed during the shove. They weren't moving before the shove, so the net momentum was zero then. After the shove, their momentums were equal but opposite, so the net momentum was still zero. Thus, the net momentum didn't change at all. In fact, the net momentum of any system of objects always remains the same, as long as there are no forces from outside the system acting on the objects. This concept is very useful for analyzing motions:

Principle of Conservation of Momentum

If no outside forces act on the objects, the net momentum of a set of objects doesn't change.

This principle holds true no matter how many objects there are, how they interact, or how they were moving at the start.

Since momentum is the product of mass and velocity and since velocity is a vector, momentum is also a vector. Its direction is the same as the direction of the velocity. When the objects move in different directions, vector addition must be used to find the net momentum.

EXAMPLE A little boy and girl run directly toward each other, grab on to each other, and fall over. The boy has a mass of 20 kilograms and was running at 4 m/s from the left; the girl has a mass of 25 kilograms and was running at 3 m/s from the right (Figure 7-6). Which way do they fall over?

Answer We will call the right the positive direction and the left negative. The boy's momentum will be

$$\begin{aligned} m_b v_b &= 20 \text{ kg} \times 4 \text{ m/s} \\ &= 80 \text{ kg·m/s.} \end{aligned}$$

The girl's momentum will be

$$\begin{aligned} m_g v_g &= 25 \text{ kg} \times -3 \text{ m/s} \\ &= -75 \text{ kg·m/s.} \end{aligned}$$

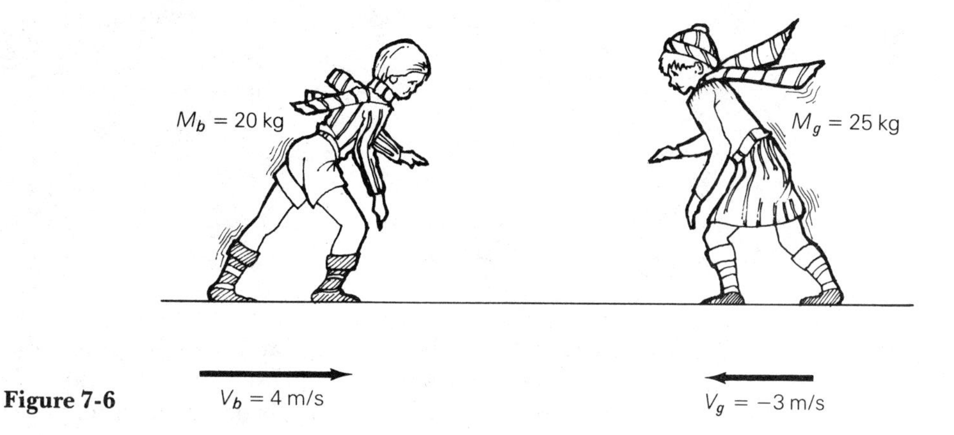

Figure 7-6

The minus sign indicates that she is moving toward the left. Thus,

$$\text{net momentum} = 80 \text{ kg·m/s} - 75 \text{ kg·m/s}$$
$$= 5 \text{ kg·m/s.}$$

The net momentum stays the same during the collision. Because it is positive, the children must be moving toward the right after they collide. Thus, they fall toward the right—the direction in which the boy was going. Note that the boy isn't heavier than the girl. They fall toward the right because he had greater *momentum;* the *product* of his mass and velocity was greater than the product of hers.

EXAMPLE A 100-kilogram ball carrier runs straight down the football field at 8 m/s. He is tackled by a 120-kilogram player running directly across the field at 5 m/s. In what direction and at what speed do they move before they hit ground?

Answer The momentum of the ball carrier was

$$m_b v_b = 100 \text{ kg} \times 8 \text{ m/s}$$
$$= 800 \text{ kg· m/s}$$

straight down the field. The momentum of the tackler was

$$m_t v_t = 120 \text{ kg} \times 5 \text{ m/s}$$
$$= 600 \text{ kg·m/s}$$

across the field. When we add their momentums as vectors, as shown in Figure 7-7, the sum is 1,000 kg·m/s at a 37° angle away from the downfield direction. Since the combined mass of the two players is 220 kilograms, their speed after the tackle is

$$v = \frac{\text{net momentum}}{\text{total mass}}$$

$$= \frac{1{,}000 \text{ kg}\cdot\text{m/s}}{220 \text{ kg}}$$

$$= 4.6 \text{ m/s}$$

at 37° away from the downfield direction.

When a quarterback is hit by a heavy guard, the quarterback is far more likely to be injured than the guard. The same thing is true in car accidents. When a large car and a small car collide, the small car is nearly always more severely damaged. The principle of conservation of momentum explains why the small car or person always suffers more in a collision. Since the net momentum is the same before and after the collision, whatever momentum is lost by the large object must be gained by the small one. If we put this into a formula, we find

$$m_l v_l = -m_s v_s,$$

where l stands for the large object and s for the small one. The v's are the changes in velocity. If m_l is bigger than m_s, then v_s must be greater than v_l to conserve momentum. No matter what their speeds were, no matter how they strike each other, no matter what directions they go in after the collision, the small object's velocity will change more than that of the large one. In other words, the small object accelerates more rapidly during the collision; thus, the internal forces within the small object will be bigger. As a result, the damage to the small object will be greater, unless the small object is made of a much stronger material. Since this fact is the result of a physical law, it cannot be avoided. Heavy guards will nearly always get the better of light quarterbacks, and cars will continue to be demolished by trucks when they collide.

The principle of conservation of momentum can help us to obtain maximum efficiency in swimming, rowing, and canoeing. These methods of propulsion work by throwing mass backward with the reaction force of the mass propelling us forward. Clearly we should generate larger forces on the mass thrown backward to increase our forward speed, but how can we do this most effectively? If we look at the problem from a momentum viewpoint, the answer is clear. To increase our forward momentum, we must increase the momentum of the material thrown backward. Because momentum equals mv, we can do this by throwing more mass backward, throwing it faster, or doing both. Which option we take de-

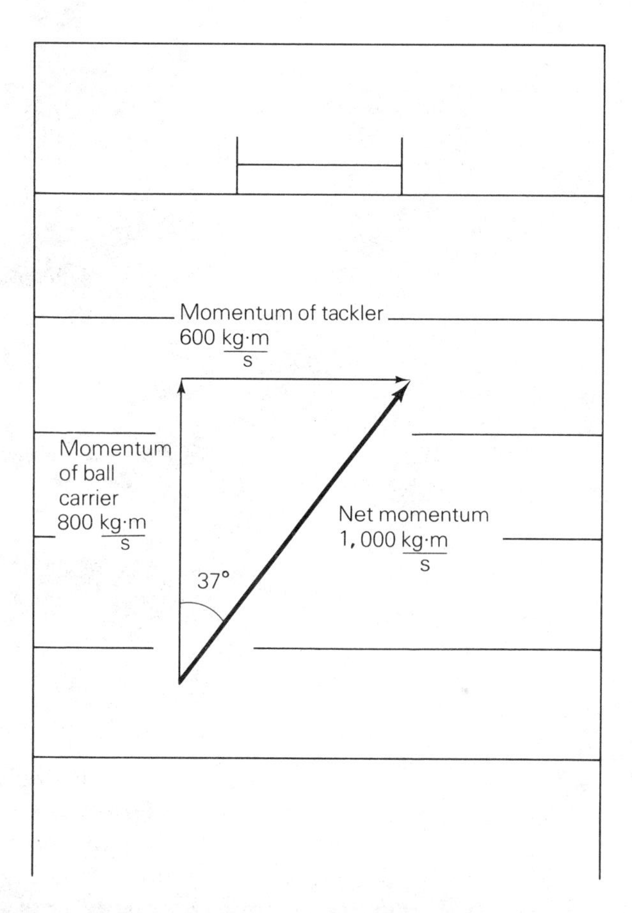

Figure 7-7

In a car-truck collision, the velocity of the car will change more than that of the truck. Consequently, the car will be damaged more than the truck in most accidents.

Racing-canoe paddles are large in order to move the maximum amount of water backward.

pends on the situation. For instance, in rocket travel the amount of mass is limited; therefore, throwing the mass backward at a faster rate is the better technique. Swimming and canoe paddling present the opposite situation. Increasing the speed of the hands or paddle increases the amount of energy lost to friction; however, the amount of mass (water) to throw backward is unlimited. Therefore, in swimming the hands and feet should be positioned to move the most water possible. Swim fins are so helpful because they throw much more water to the rear per stroke than our feet do. For the same reason, a canoe paddle should be as big as possible without being too heavy or cumbersome to handle efficiently.

Motion and the Center of Gravity

When a person jumps into the air, he or she is isolated from all outside forces except gravity. Although the person's motion in the air can be quite complicated, the motion of his or her center of gravity turns out to be simple, because the motion of a system's center of gravity cannot be changed by its internal forces. This is a result of conservation of momentum, which is easy to see in the simple case of two people initially at rest on a frictionless surface. If

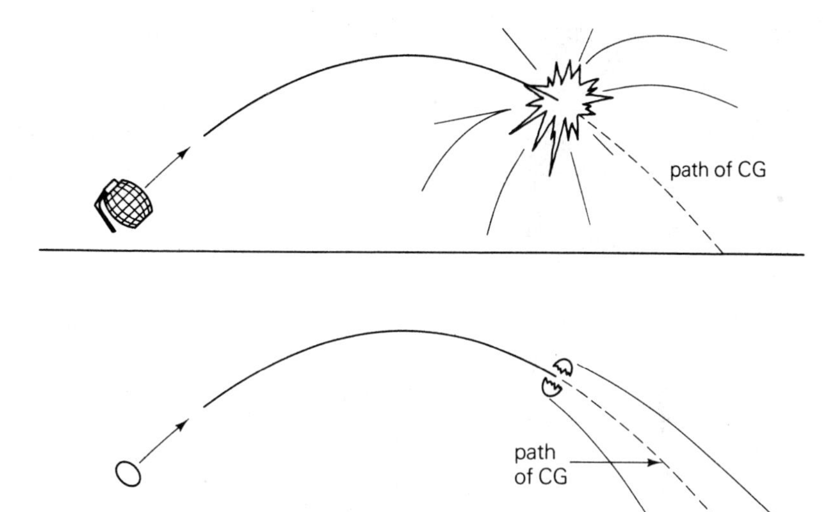

Figure 7-8 The path of the center of gravity of an object that explodes or breaks in flight is the same as the path that the object would have followed if it had not broken.

one shoves the other, they will move away from the starting point in opposite directions. To conserve momentum, the light one will move away from the starting point faster than the heavy one. If we calculate where they are at any instant and then determine where the balance point is, we will find that it remains at the starting point. Thus, the center of gravity remains stationary. The same reasoning will show that the motion of the CG of any system cannot be changed by internal forces or internal motions. For instance, suppose we throw a cracked rock that splits in two in flight. There is no way of telling where either piece will land, but their center of gravity will follow exactly the same path that the rock would have followed if it had not split. Similarly, when a hand grenade explodes in the air, the center of gravity of all the pieces follows the same path that the grenade would have followed (Figure 7-8).

Except for skydivers, skijumpers, and people who dive from high cliffs, the force of the air on a person is quite small. Once a person leaves the ground or a diving board, the person becomes an isolated system. No matter what wild movements he or she makes, absolutely nothing that the person can do will change the path of his or her center of gravity. For example, after leaving the board, nothing that a diver can do will change the point where he or she will enter the water. The entry point is fixed the instant the diver leaves the board.

The same principle applies to high jumpers, too. The moment the high jumper leaves the ground, the path of his or her center of gravity is fixed. Any motions that the jumper makes after leaving the ground will not influence how high the CG gets. Then why do

The high jumper's center of gravity is at almost the same level as the bar in most advanced high jumping styles.

high jumpers perform all sorts of wild twists during the jump? High jumps aren't scored on where the center of gravity goes; they are scored on where the body goes. As we discussed in Chapter 5, the CG isn't a stationary point inside the body, and it isn't even always inside the body. The goal of all high jump styles is to get the part of the body that is passing over the bar as high above the center of gravity as possible. In the simple jump style, where the body stays erect and the feet are lifted to clear the bar, the CG must clear the bar by about 40 centimeters. In most competitive jumping styles, the body is close to horizontal as it clears the bar, and the center of gravity is only a few centimeters above the bar. In some of the more advanced high jumping styles like the "Fosbury flop," the center of gravity might actually go *under* the bar.

Thus, there are two main considerations in developing a winning high jump style: First, the legs must be positioned at takeoff to obtain maximum upward thrust. (Actually more than the legs are involved in producing the largest possible upward force. The motion of the rest of the body, the arms in particular, are also important, as is explained in the discussion on jumping in Chapter 18.) Second, in the air the objective is to get the part of the body passing over the bar as high above the CG as possible. To do so, the jumper must keep the parts of the body that are not passing over the bar as low as possible. Keeping the arms and legs down before and after they pass over the bar will keep the rest of the body as high as possible.

Gravity

The force due to gravity, one of the most important forces in our lives, is one of the simplest to understand. On or near the surface of the earth, the force due to gravity is constant. Therefore, gravitational accelerations are also constant, which means that the velocity of a falling object always changes at a constant, uniform rate.

Freely Falling Objects

In the absence of air friction, only one force acts on a falling object—the force due to gravity. That is, the net force on a freely falling object is equal to its weight. Since we know the force, finding the acceleration is easy. For a freely falling object,

$$F = W.$$

But we also know that for any object,

$$F = ma,$$

and, as we saw in Chapter 3,

$$W = mg.$$

When we put these values in for F and W, we find

$$ma = mg.$$

Thus,

$$a = g.$$

This is why g is called the acceleration due to gravity: Freely falling objects accelerate at $g = 9.8\ \text{m/s}^2$. That is, a freely falling object will pick up 9.8 m/s of speed every second that it falls. Thus, an object dropped from rest will be going 9.8 m/s at the end of 1 second, 19.6 m/s after 2 seconds, 29.4 m/s after 3 seconds, and so on.

Of course, air friction reduces the acceleration of a falling object. The amount of air friction depends on several things; the size, shape, mass, and speed of the object are the most important factors. Air friction varies widely. A feather's fall is dominated by air friction in a fraction of a second. However, most things, including humans, won't be influenced much by air friction for the first 3 or 4 seconds of fall. If we were to remove the air, all things—feathers, humans, cannonballs, everything—would fall with exactly the same acceleration.

EXAMPLE How fast will a coin be going 1 second after it is dropped into a deep wishing well? How fast will it be going after 4 seconds? The air friction is quite small in both cases.

Answer We know that the acceleration of a falling object is a constant 9.8 m/s^2 as long as the air friction is small. Therefore, the object's speed increases by 9.8 m/s every second. The definition of acceleration is, as we saw before,

$$a = \frac{v_n - v_o}{t},$$

where v_n is the new velocity, v_o is the old velocity, and t is the time. For objects dropped from rest, $v_o = 0$, so

$$a = \frac{v_n}{t},$$

or

$$v_n = at.$$

But $a = g$ for freely falling objects, so if the object was dropped from rest,

$$v_n = gt.$$

Thus, we see that at the end of 1 second, the coin will be going at 9.8 m/s^2 × 1 second = 9.8 m/s, and at the end of 4 seconds the velocity will be 9.8 m/s^2 × 4 seconds = 39.2 m/s.

How does an object fall when it is thrown to the side? Consider an object that is thrown horizontally off a tall building. Although it is going sideways, gravity is still the only force acting on the object, and it still acts straight down. Thus, gravity accelerates the object downward, which means that the object's downward speed will increase constantly. However, there is no horizontal force, so the horizontal speed doesn't change, in accordance with Newton's first law of motion. Putting these two motions together, we see that the object's motion is a combination of a steady horizontal speed and a continually increasing downward speed (Figure 7-9). Thus, the path will curve downward, becoming steeper and steeper as time goes on. This type of a path is called a *parabola.* When an object is thrown upward and to the side, the path is also a parabola. In this latter case, the object's horizontal speed remains constant while its vertical speed decreases until the object reaches the top of its path.

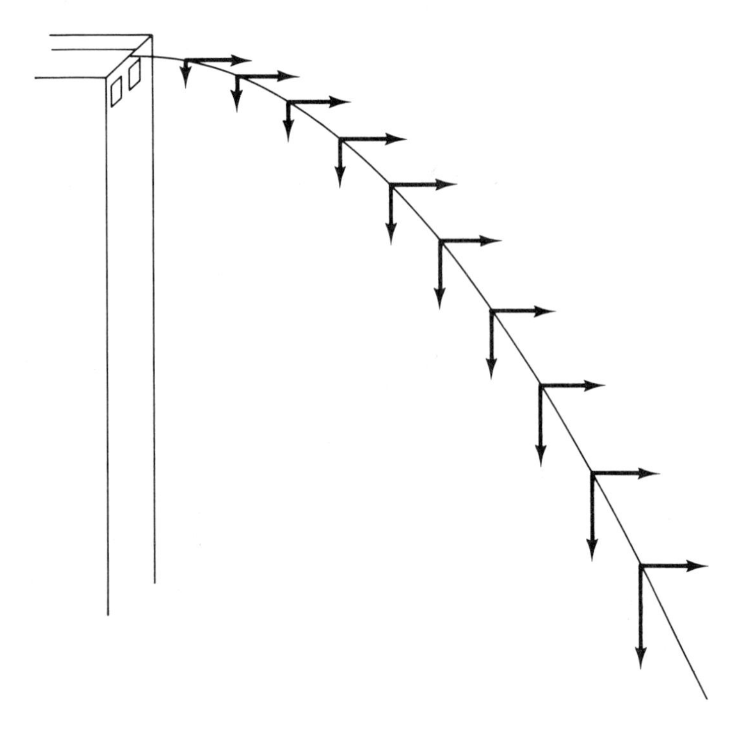

Figure 7-9 Parabolic path of an object thrown horizontally. The horizontal speed stays constant while gravity increases the vertical speed at a constant rate of 9.8 m/s^2. The arrows represent the vertical and horizontal parts of the object's velocity at the points shown.

The javelin is thrown at a 45° angle for maximum range.

At the top, it moves horizontally for an instant, then starts on the downward part of the parabola. The paths for various throwing angles are shown in Figure 7-10. The initial speed is the same, 25 m/s, in all the cases; this is a normal throwing speed. Air friction is ignored. The object will go farthest over level ground when it is thrown at a 45° angle.

For maximum distance, the long jumper wants a good balance between high horizontal speed and high upward takeoff speed. One without the other, as you can see in Figure 7-10, would get the jumper nowhere. A long jumper cannot attain the same upward speed as horizontal speed, because the jumper has several strides to build up horizontal speed but only one stride to build up vertical

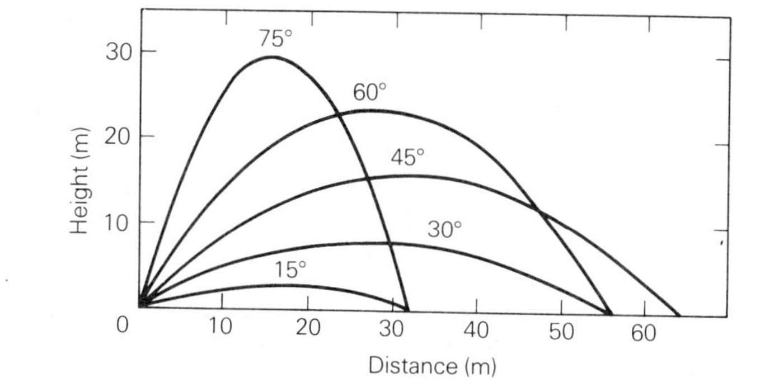

Figure 7-10 Path of an object thrown at 25 m/s for various throwing angles. The maximum range on level ground of 64 meters occurs when the angle is 45°.

speed. Thus, the maximum-range angle of 45° is not practical. Because the horizontal run is considerably easier than the upward jump, the last takeoff stride is a key element in long jumping. The greater the upward thrust, the greater the distance will be. Once in the air, nothing that the jumper can do will change the path of his or her center of gravity. Of course, jumping isn't scored on where the CG hits the ground; the important thing is how far the feet go before landing. So jumpers try to get their feet out as far ahead of their CG as possible without losing their balance during the landing.

Terminal Velocity

The faster an object falls, the greater the air friction. Eventually, as the speed increases, air friction will increase to the point where its force upward is equal to the weight of the falling object. At that point, the net force on the object will be zero and its speed will no longer increase. The velocity at this point is known as the *terminal velocity* of the object.

When an object is dropped in air, its acceleration will be 9.8 m/s^2 at the start. But as its speed increases, the air friction also increases, reducing the object's acceleration. Note that we are not saying that the object slows down; we are saying that the *rate* at which its speed increases becomes smaller. The speed continues to increase but not as rapidly as it did at the beginning. Gradually the acceleration diminishes until it finally becomes zero. After that, the velocity doesn't increase any more. The object has reached its ter-

Table 7-1 Table of Terminal Velocities for Various Objects Falling in Air

Object	*Terminal Velocity in m/s**
Baseball	40
Cannonball	250
Falling bullet	100
Feather	0.4
Human	55
Human with parachute	5
Paper	0.6
Stone (large)	200
Stone (small)	75

*All of these values depend on the details of the shape and size of the object, so they are only approximate. If you would like to convert these velocities to miles per hour, multiply by 2.24. For example, 100 m/s = 224 mph.

minal velocity, which is maintained until the object hits the ground.

We often want to know whether we can ignore air friction in a problem. Naturally, the answer depends on the amount of error that can be tolerated in the answer, but we can usually ignore air friction if it is less than 20% of the object's weight without causing any serious error. For most objects, the air friction will not exceed 20% of the object's weight until the object's speed is about one-half its terminal velocity. A rough rule of thumb is to disregard air friction until the object moves faster than one-half its terminal velocity.

EXAMPLE How long does it take before air friction retards a skydiver's fall? In other words, for what length of time can we use the free-fall formulas to find the skydiver's speed and still trust the answer?

Answer The terminal velocity of a falling person is about 55 m/s. (This depends on the person's position. A person curled up in a ball will fall faster, while a person who is spread-eagled will fall more slowly.) If we ignore air friction until the skydiver reaches one-half of his or her terminal velocity, we can treat the skydiver's fall as free-fall until his or her speed is about 28 m/s (½ × 55 m/s). In free-fall, a person will reach that speed in

$$t = \frac{v}{g} = \frac{28 \text{ m/s}}{9.8 \text{ m/s}^2} = 2.9 \text{ s}.$$

Thus, the fall will be nearly free for the first 3 seconds or so.

Sedimentation Time

The terminal velocity is closely related to the time required for dust and smoke particles to settle out of the air. The important factor here is the size of the particles: The smaller the particles are, the smaller their terminal velocity will be and the longer it will take for them to settle out of the air. Larger sandlike particles will settle out in a few seconds, while most dust and smoke will settle out in a few hours. However, very small particles will have such a small terminal velocity that the slightest air current can keep them airborne indefinitely. This effect can be seen on any sunny day by watching the dust in a narrow sunbeam shining into a room. This effect is also why sneezing spreads contagious diseases. The sudden airflow breaks up the germ- and virus-laden fluid into extremely small drops that stay aloft for a long time. Fine aerosol mists are better than coarse sprays for getting medicine into the lungs because of the small terminal velocity of the mist. Of course, if we don't want the spray to get into the lungs, the coarser the spray is, the better. For this reason, whenever you have a choice between using a hand pump spray bottle and a pressurized spray can to spray a toxic product, the hand spray is safer.

In liquids, the terminal velocity is much smaller than in air because liquid friction is much greater than air friction. Sand particles that settle out of the air in a few seconds take several hours to settle out of water. In a thick oil they might take days. The time that it takes particles to settle out of a particular liquid is called the *sedimentation time.* Of course, the smaller the terminal velocity, the longer the sedimentation time will be.

A good test for the presence of certain infections in the body is the sedimentation rate of blood cells. The blood cells are denser than the blood plasma, and they will gradually sink to the bottom of a test tube if an anticoagulant is added. Some infections alter the cells and plasma, increasing the terminal velocity of the cells. Thus, an abnormally rapid blood sedimentation rate is a good indicator of an infection somewhere within the body. This test is called the erythrocyte sedimentation rate, or ESR (red blood cells are also known as erythrocytes).

Rotation

When an object goes around in a circle, the direction of its velocity is constantly changing. Thus, an object moving in a circle is always accelerating, even when the size of the velocity doesn't change. The second law of motion, $F = ma$, thus tells us that a force must be acting on that object. When the revolution is uniform, the force will be directed toward the center. Think about whirling a ball on the end of a string. The string is the only thing supplying the force needed to keep the ball going in a circle. But strings can only pull, so the force must be along the string, pointing toward the center of rotation. The change in velocity is calculated in Figure 7-11. As you can see, the direction of the change in velocity, which is the same as the direction of the acceleration and the force, is toward the center. Because the force on the object is *toward* the center, it is called the *centripetal* force, which means center-seeking force.

The size of the centripetal force F_c can be found from Figure 7-11 and the equation $F = ma$. However, we will not go into the

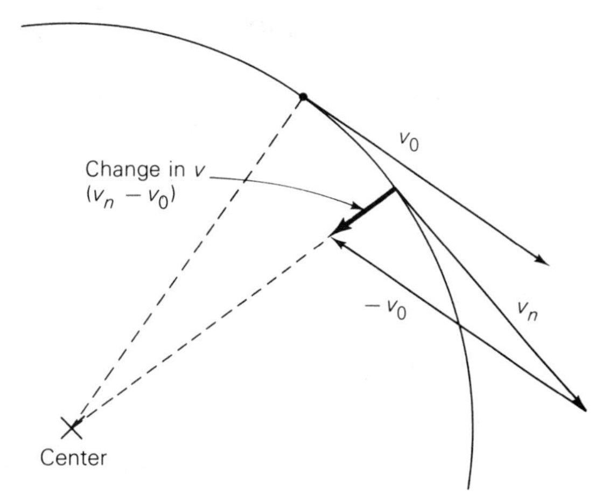

Figure 7-11 The change in velocity of an object going around a circle uniformly always points toward the center of the circle.

details; it turns out that

$$F_c = \frac{mv^2}{r},$$

where m is the mass of the object, v is the velocity, and r is the radius of the circle. The formula agrees with our experiences in driving and running around corners. The centripetal force increases when either the mass or the speed of the object is increased. As the corner becomes sharper, r becomes smaller, causing an increase in F_c.

When we run or drive around a level corner, the necessary centripetal force comes from friction with the road surface. Of course, friction is limited; it can't exceed $F_f = \mu mg$, the coefficient of friction times our weight or the car's weight. If we try to go around the corner too fast, the centripetal force needed will be more than friction can supply. So we will slip and go off the road toward the outside of the corner, according to Newton's first law of motion.

We can go around banked turns faster without slipping because the reaction force of the road's surface can provide some or all of the centripetal force (Figure 7-12). The steeper the banking is, the faster we can go around the corner. However, banking corners is not a cure-all; if a corner is banked too steeply, a person will slip down the road if he or she goes around it too slowly. This is unlikely to occur on a dry day, but it could happen when the road is slippery.

You have probably heard the term *centrifugal force* used in connection with rotation. For instance, when you slide against the side

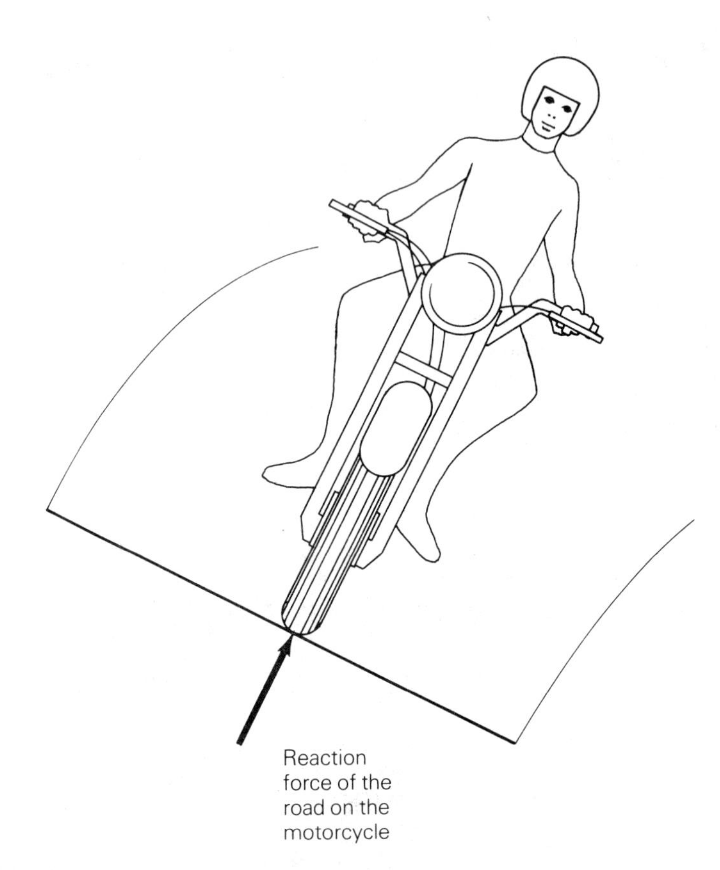

Figure 7-12 On a banked turn, the reaction force of the road on the motorcycle furnishes some or all of the centripetal force needed to turn the motorcycle.

of a car when the car goes around a sharp turn, someone might say, "Centrifugal force pushed you across the seat." If you try to figure out where the centrifugal force came from, you will see that this explanation is wrong. The centrifugal force didn't come from the car, it couldn't have come from the road since the road didn't touch you, and it couldn't be some mysterious force coming from the trees alongside the road. There is no object that could have produced a centrifugal force on your body.

To see what really happens, let's watch a car and driver go around a sharp right turn. As the car starts to turn, the road exerts a centripetal force on it toward the right, accelerating the car toward the right. But the seat doesn't exert much of a force on the driver, so the driver continues to move in the same direction in which he or she was originally going. What really happens is that *the centripetal force of the road on the car moves the car to the right* to hit the driver while the driver continues to move in a straight line (Figure 7-13).

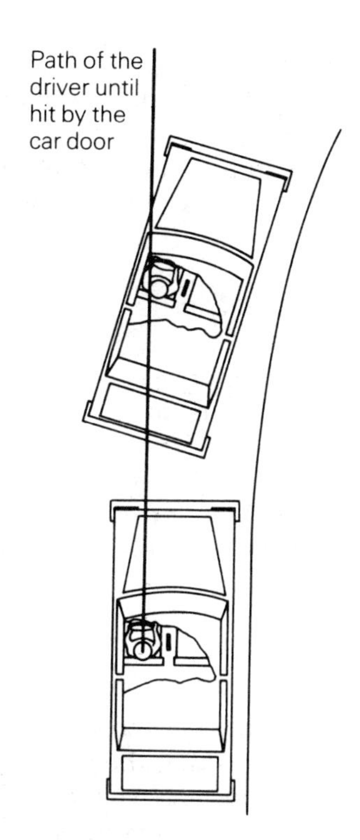

Figure 7-13 As the car turns the corner, the driver continues to move in a straight line until hit by the door. The driver isn't pushed toward the door, the door is pushed toward the driver by the centripetal force of the road, which causes the car to turn.

The Centrifuge

The centrifuge is an extremely useful piece of laboratory apparatus that uses rotational motion to speed up the sedimentation of cells (or particles) mixed in a liquid. The test tube containing the mixture is held in the centrifuge with the mouth of the test tube pointing toward the center and the closed end pointing toward the outside of the centrifuge (Figure 7-14). When the centrifuge is spun, the cells tend to settle toward the closed end of the test tube for the same reason that you slide toward the side of a car as it goes around a corner—the cells (and you) tend to remain in motion in a straight line.

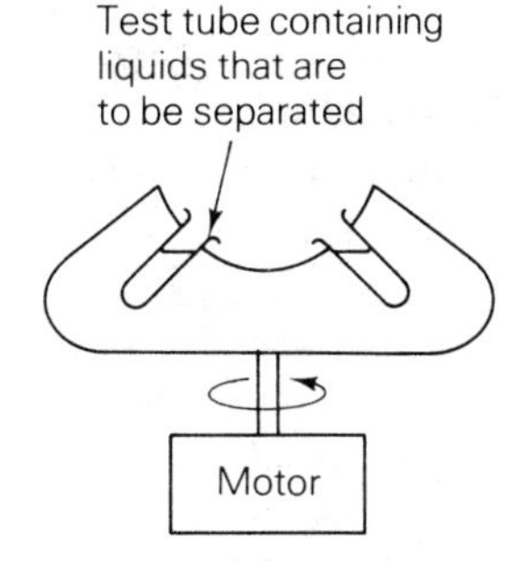

Figure 7-14 A centrifuge.

If more than one type of cell or particle is mixed in the liquid, the centrifuge can often separate the different cells. The heavier, denser cells will be found near the closed end, while the lighter cells will be closer to the mouth of the tube. Naturally, the faster the centrifuge spins, the faster the cells will settle and the better the

The figure skater pulls in her arms to decrease her radius and thus increase her rotational speed.

different substances will be separated. For this reason, research centrifuges spin extremely rapidly—in excess of 100,000 revolutions per minute in many cases. At these high rotation speeds, it is very important that nothing breaks and that no liquid escapes, because anything that escapes will be sprayed over the room.

Angular Momentum

Spinning objects keep spinning in the absence of friction, because the centripetal force acts perpendicularly to the velocity of the various parts of the object. As a result, the force only changes the direction of their velocity and not the size. We call this tendency to stay spinning the object's *angular momentum.* The angular momentum (L), of an object going in a circle is

$$L = mvr,$$

where m is the mass of the object, v is its velocity, and r is the radius of the circle.

With Newton's laws and a little work (perhaps a lot of work) we can show that the angular momentum of an object stays the same when no external torque is applied. This concept is called the principle of conservation of angular momentum. The roots of this principle lie in the same ideas used earlier in this chapter to show that ordinary straight-line momentum is conserved.

For rigid objects, the fact that mvr is a constant just says that v must stay constant, because neither m nor r can change. So rigid objects with no external torques on them just spin without changing speed. However, internal forces can change the shape—and thus the radius r—of nonrigid objects. Of course, internal forces do not produce external torques, so the angular momentum doesn't change. If r is decreased, the only way the angular momentum can stay constant is for v to increase. This effect is easily seen by whirling a ball on a string, then pulling in on the string. The ball speeds up to conserve the angular momentum. (Pulling in the string doesn't exert any torque on the ball, since the string pulls directly toward the center.)

Dancers, gymnasts, divers, and figure skaters all use this principle to control their spins and turns. Figure skaters start spins with their arms outstretched, then build up their rotational speed by pulling their arms in to decrease r. Since r decreases, their speed increases to keep their angular momentum constant. This effect is also important in diving. Once a diver leaves the platform, his or her angular momentum is fixed because there are no torques to change it. To increase rotational speed, a diver must decrease r by

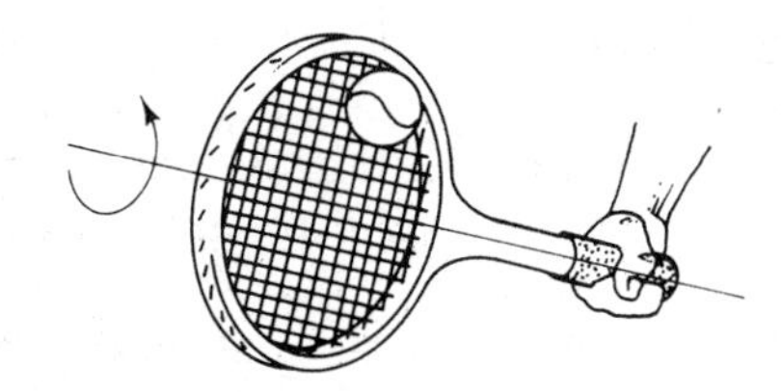

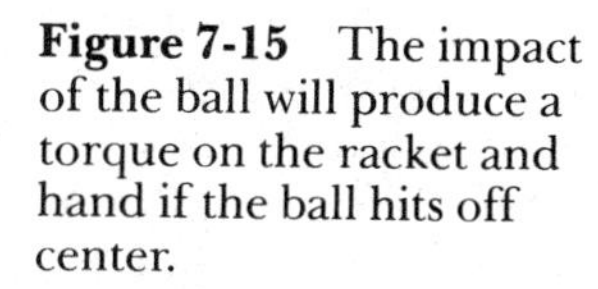

Figure 7-15 The impact of the ball will produce a torque on the racket and hand if the ball hits off center.

pulling his or her body into a more compact position. To slow down the rotation, a diver extends his or her body to increase r. Controlling r is the only way that a diver can control his or her rotation.

Recently, tennis rackets and golf clubs changed drastically when the principles of rotational motion were applied to their design. Consider what happens when the ball strikes off center toward the side of a tennis racket. The force twists the racket in the player's hand; in other words, the impact produces a torque about the racket's axis (Figure 7-15). Naturally, the more the racket twists, the poorer the return will be. Controlling the direction of the return will be harder, and energy will be wasted twisting the racket in the hand. To see how the twisting can be reduced, look at how the racket's angular momentum changes during the impact. The torque of the ball's impact gives the racket a certain amount of angular momentum L. We know that

$$L = mvr,$$

and we also know that we want the racket to twist as slowly as possible; that is, we want to keep v to a minimum. There isn't much we can do about L, which is determined primarily by the torque of the impact. Therefore, the only way we can keep v small is to make

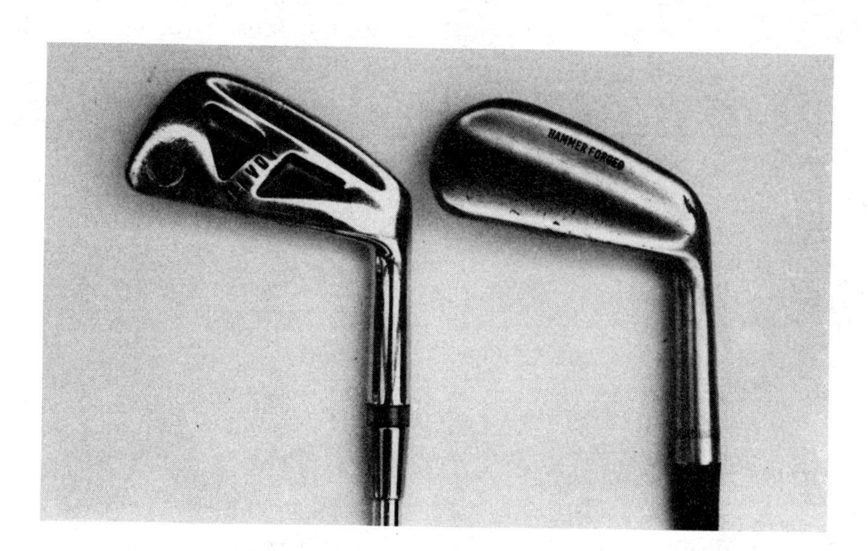

The mass of the modern golf club on the left has been concentrated near the edges to increase the size of the "sweet spot."

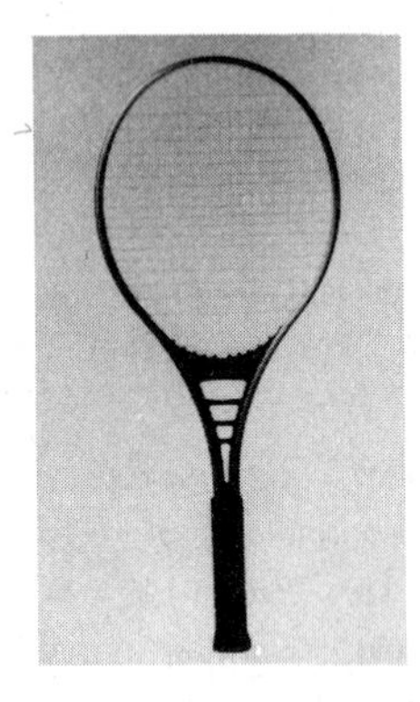

m or r (or both) large. Adding mass is easy, but it is counterproductive, because the extra mass makes the racket heavy and hard to swing. But if we make the racket wider, the increase in r drastically reduces the amount of twist when the ball strikes off center without increasing the "swing weight" of the racket. Thus, compared with the conventional racket, the ball can strike farther from the center of the wide racket without seriously twisting it. Modern golf clubs also use this principle, as you can see in the photograph. The total mass of the modern club is the same as the conventional club, but the mass has been moved to the edges of the clubhead. The increase in r increases the size of the "sweet spot," the region where the impact of the ball doesn't twist the clubhead much.

The Pendulum—Limitations on Walking and Running

Why can people run faster than they can walk? Surprisingly, the reason is directly related to how pendulums work. A pendulum is anything hanging from a pivot that is free to swing. A rock hanging from a string, a stick swinging on a hinge, and our legs are pendulums.

One important fact about a pendulum is that a small swing and a large swing take the same amount of time to go back and forth (Figure 7-16). Tie something on the end of a string and try it. (This fact is why pendulums are used in clocks.) The reason large and small swings take the same time is that the greater distances and the greater speeds of larger swings are exactly balanced. The farther the pendulum swings from the center, the bigger the force that gravity produces to accelerate it back toward the center. But because the pendulum must cover a greater distance to get back to the center, a large swing takes the same time to swing back as a short swing. Thus, a pendulum has a natural swing time (or period), which is determined by the pendulum's length. The longer the pendulum is, the smaller the force pushing the pendulum toward the center will be. Thus, the swing time increases with the length of the pendulum.

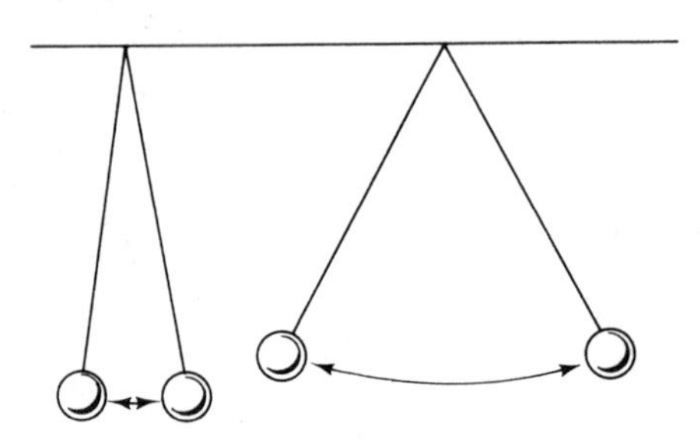

Figure 7-16 The pendulum. The time for one swing is the same for large and small swings.

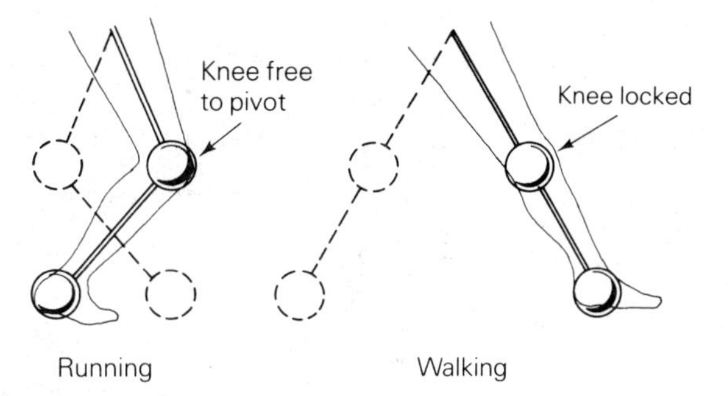

Figure 7-17 When running, the leg acts like two short pendulums. When walking, the leg acts like one long pendulum.

To try to move a pendulum faster than its natural swing rate is extremely inefficient. Take a tennis racket, golf club, or a stick, and let it hang straight down. Now start it swinging back and forth at its natural swing speed. It takes almost no effort to keep it moving. Next try to swing it back and forth twice as fast. It takes an extreme effort to do so. Since the legs are pendulums, too, they are very hard to swing faster than their natural swing speed. Normally, we increase our walking speed by increasing the length of our stride, not by trying to speed up the swing rate. If we are already taking as long a stride as we can and then try to increase the swing rate, walking will become extremely inefficient and tiring.

Returning to our original question, why is the running motion faster than the walking motion? When we run, the lower leg is free to pivot at the knee. Thus, the leg acts like two short pendulums: The upper part of the leg is a short pendulum pivoting at the hip, and the lower leg is another pendulum pivoting at the knee. When we walk, the knees are usually straight and the entire leg swings as one long pendulum with a much longer swing time than while running—almost twice as long, in fact. Thus, the leg motions while running are naturally much faster than while walking (Figure 7-17).

Concepts

Rates
Acceleration
Vector subtraction
Newton's second law of motion
Utricle
Semicircular canals
Impact protection
Momentum
Conservation of momentum
Motion of falling objects
Motion of thrown objects
Terminal velocity
Sedimentation time
Rotational motion
Centripetal force
Centrifuge
Angular momentum
Conservation of angular momentum
Pendulum

Discussion Questions

1. How does shaking your wet hands get rid of excess water?
2. Explain the difference between acceleration and velocity.
3. Why is a steel hammer far better for driving nails than a rubber hammer of equal mass?
4. Why are jerky body movements less efficient than smooth ones?
5. What organs of the body detect accelerations? How do these organs work?
6. When a light person and a heavy person collide, why is the light person more likely to be injured?
7. Matching the mass of a tennis racket, baseball bat, or golf club to an individual's swing is not an easy task. What are the advantages and disadvantages of a light racket, bat, or club? Of a heavy racket, bat, or club?
8. Why is it important to slow down when you turn a corner with a patient on a stretcher cart?
9. A centrifuge is a machine that whirls test tubes in a circle at very high speeds to speed up the settling of suspended particles. Explain why the particles will settle out of the liquid faster when it is spun in a centrifuge than when it is stationary.
10. Why is running faster than walking?
11. A figure skater usually starts a spin with the arms outstretched; then the arms are drawn in close to the body. Why do these actions speed up the spin?

Problems

1. If your telephone bill averaged $30 per month last year and $42 per month this year, what was the average rate of increase?
2. A bicyclist takes 4 seconds to go from rest to 12 m/s. What was his acceleration?
3. A jet plane starts from rest and accelerates at 20 m/s^2. How fast is it going at the end of 15 seconds?
4. A car going 20 m/s slows down to 4 m/s over 5 seconds. What was the acceleration of the car?
5. A ship is heading straight at a rock at 4 m/s. The captain calls for full speed astern, which results in an acceleration of -0.02 m/s^2. How fast

is the ship going at the end of 100 seconds? 200 seconds? 300 seconds?

6. A diver enters the water going straight down at 10 m/s. Once in the water she quickly levels off, and 0.25 second later she is going forward at 7 m/s. What was her acceleration as she changed her velocity?

7. A jogger is going westward at 5 m/s. He comes to a corner and heads north at 7 m/s. If he took 2.5 seconds to turn the corner, what was his acceleration?

8. What force would be needed to accelerate a 48-kilogram hospital bed at a rate of 2.7 m/s^2?

9. A nurse applies a horizontal force of 55 newtons to a patient and wheelchair that have a combined mass of 79 kilograms. How fast does the patient accelerate?

10. What force would be needed to accelerate a 1,500-kilogram car from zero to 30 m/s in 6 seconds? Compare this force with the car's weight.

11. If the coefficient of friction of a basketball player's shoes with the floor is 0.8, how long would it take the player to stop from a speed of 7 m/s?

12. If the coefficient of friction of a car's tires with the road is 1.0, how much time will it take the car to stop from a speed of 49 m/s? How much time would it take to stop if the coefficient of friction was only 0.1? Note: $\mu = 1.0$ is a typical value for good tires on a clean, dry concrete road; $\mu = 0.1$ is typical of ice.

13. Suppose a pitcher throws a baseball toward the batter at 30 m/s and the batter hits a hard line drive toward center field at 50 m/s. If the impact of the ball and bat takes 3 milliseconds, what is the average force on the ball at impact? A baseball has a mass of 145 grams.

14. A 10-gram bullet is fired from a 4-kilogram gun at a velocity of 1,000 m/s. What is the recoil velocity of the gun?

15. A woman jumps out of a balloon that is high enough so that she doesn't have to open her parachute for 1 minute. How fast will she be dropping after 2 seconds? After 3 seconds?

16. If the coefficient of friction of your shoes on the sidewalk is 0.5, how fast could you run around the corner shown in the figure without slipping? The sidewalk surface is level.

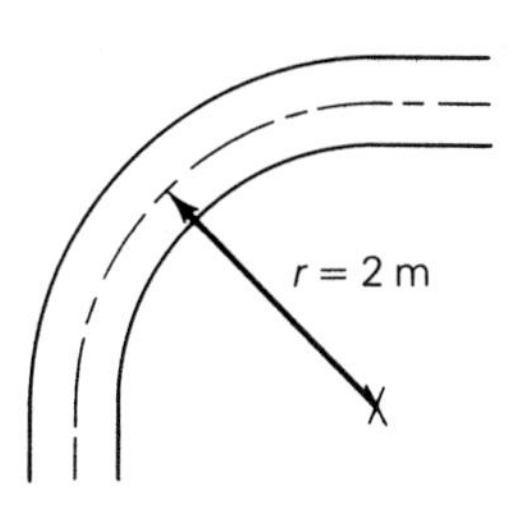

Experiments

1. Test various types of padding by dropping eggs from 2 or 3 meters onto foam rubber, plastic foam packing material, and any other padding material that is available. Which does the best job of stopping the eggs? (Use hard-boiled eggs. There won't be any mess and you can eat the eggs afterwards.) Note: The best material for stopping eggs isn't necessarily the best material for stopping human bodies, so don't use eggs to test helmets or body pads.

2. Put some water in a bucket, then whirl it rapidly in a vertical circle. Why doesn't the water fall out of the bucket when the bucket is overhead? (Note: The answer is not because of centrifugal force.) The first time you try this experiment, do it outdoors in a bathing suit.

3. Take a tennis racket, golf club, or a stick, and swing it back and forth at its natural swing speed. Make a mental note of the effort needed to keep it swinging. Next try to make it swing back and forth twice as fast. Note that this takes far more effort.

4. To see why running is faster than walking, take two things of approximately equal mass and tie them on a string as shown. The lengths of the two sections of string should be about equal. Compare the swing times when the two masses are swinging opposite each other (like the leg does in running) and swinging together (like the leg does when walking).

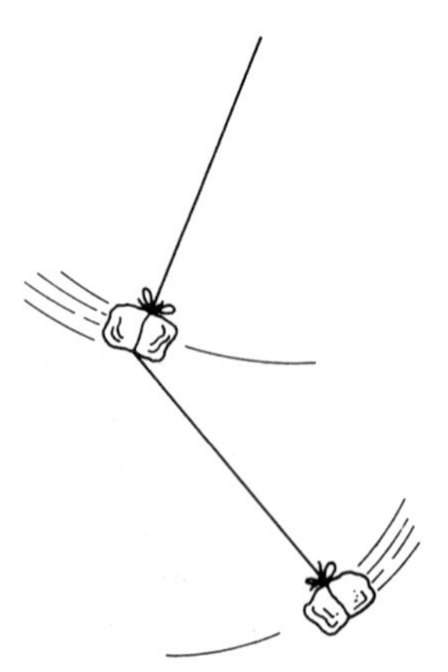

CHAPTER EIGHT

Energy, Work, and Power

Educational Goals

The student's goals for Chapter 8 are to be able to:

1. Trace the path of the energy flowing through a complex system like the human body.
2. Calculate the kinetic energy of a moving object and the potential energy of an object that has been raised to a certain height.
3. Use the principle of conservation of energy to solve problems in human motion.
4. Explain what power is and how it differs from energy.

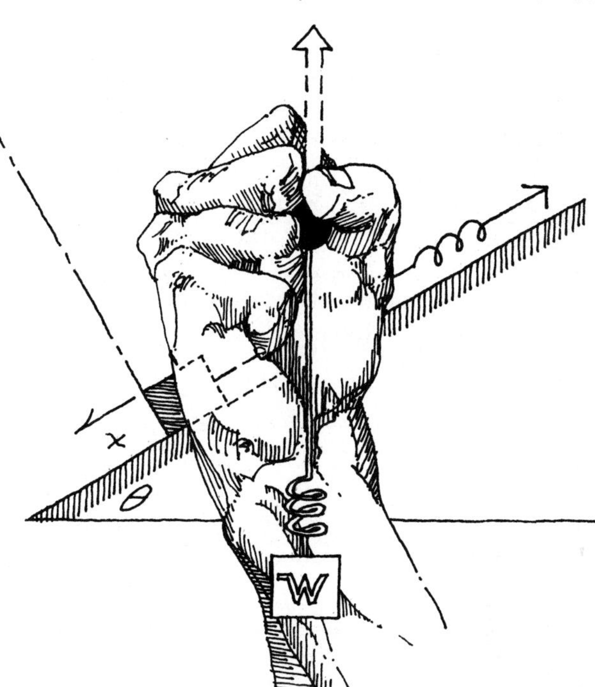

Of all the physical concepts we will discuss, energy is the most useful for analyzing complex systems. Quite often the best first step in investigating a complicated system is simply to follow the energy flow within it. Find out where the energy comes from, how it is transformed in the system, what the efficiencies of the transformations are, and where the energy ultimately ends up. This information will provide a very good foundation for understanding the system, whether the system is a machine or a living thing. It even helps to understand a very complicated system. Take world politics and economics, for example. For most of this century, the issues of political ideology, race, religion, and ancient rivalries have been trivial compared with the question of controlling the world's energy resources. As the supply of fossil fuels is consumed, energy's importance will grow even more. Every geopolitical decision of the future will be influenced, if not determined, by energy considerations.

The flow of energy is central to all life processes. When there is no energy flow, there is no life. One of the best ways to start the study of any organism is to look at its metabolic processes—the processes by which the organism obtains energy, stores it, converts it to motion and heat energy, and then disposes of the waste energy. Weight control is a good example. For normal people with no glandular disorders, weight control is simply a matter of balancing energy intake with energy use. If a person takes in more food energy than he or she uses to produce heat and motion, the person stores the extra energy as fat. If a person uses more energy than he or she eats as food energy, the person loses weight. In this case, the missing energy comes from fat if it is available and from muscle if it isn't. Weight control has a multitude of problems, involving psychological and medical questions. But in the end, weight control comes down to this: If you eat more energy than you use, you get fat; if you eat less energy than you use, you lose weight.

Saving energy can be important to an elderly person or a person weakened by disease. The last chapter of this book includes a discussion of how an occupational therapist can use the concept of energy to help such patients save their limited energy.

Energy

What exactly is energy? Energy is the *ability to move against a force*. *Work** is a quantity that is closely related to energy. Work occurs

*The usage of the word *work* in physics is similar to, but not exactly the same as, the everyday usage of the word. For example, both a nurse lifting a patient into bed and

when we move against a force, so energy can be thought of as the *ability to do work.* Note that energy is *not* the ability to produce a force; a brick in a wall can produce an enormous force on the other bricks, but it isn't producing energy or doing work because it is *not moving the other bricks.* However, an ant crawling up the wall *is* doing work because it is moving against the force of gravity.

The directions of the force and motion are also important; moving sideways to a force doesn't take energy, because the motion is not against the force. In order to do work, the force and the motion must both be in the same direction. When the force and motion are perpendicular, as they are when a ball is whirled at a uniform speed on the end of a string or when we push a frictionless wheelchair across a level floor, no work is done and no energy is needed or produced. Only that part of the force that is parallel to the motion requires energy.

Energy occurs in many different forms. Chemical energy is the energy stored in the chemical makeup of a material. Food, petroleum, and chemical explosives all contain chemical energy. Heat is the energy contained in hot objects. Nuclear energy is the energy stored in the nucleus of the atom, which can be released in nuclear reactors and nuclear explosives. We obtain electrical energy from batteries and generators. There is mechanical energy, which comes in two basic forms: kinetic energy and potential energy. Kinetic energy is the energy associated with an object's motion, like the energy of a moving car. Potential energy is the energy that is stored in a stationary system by winding it up, lifting it to a higher position, or otherwise elastically bending or compressing it. For instance, the energy stored in a wound-up spring is potential energy, as is the energy stored in an object sitting on a high shelf and the energy stored in the compressed gas of a spray can.

Conversions from one type of energy to another happen constantly. The chemical energy in a barrel of oil might be converted into heat by burning the oil. This heat could then be used to generate steam for running an engine, thus converting the heat energy into mechanical energy. If the engine is hooked up to a generator, the mechanical energy could be converted to electrical energy, which might go to an oven to be converted into heat again. Energy transformations are also occurring all the time in the body. We take in chemical energy in the form of food, converting it into heat to keep us warm and into mechanical energy to move about. A little bit is even converted into electrical energy for the brain and nerves.

---

a writer sitting in a chair thinking would be working in the ordinary sense of the word. In the physical sense, only the nurse would be doing work, because only the nurse is moving against a force. In contrast, the force of the writer on the chair is *not* moving the chair.

One very common transformation is the conversion of mechanical energy into heat by friction. Rubbing our hands together to warm them is an example. We often say that the mechanical energy is "lost" to friction, but it isn't lost in the sense of being destroyed—the energy still exists as heat energy. We say that it is lost because it is impossible to reconvert all the heat back into mechanical energy.

All of the various types of energy are fundamentally the same; energy is never destroyed when it is converted into another form. However, in some forms—heat in particular—the energy is far less useful to us than it is in some other forms. Once the energy has been transformed into heat energy, it can no longer be reconverted into the other forms with 100% efficiency.

According to the precise definition, work is the product of the distance moved and the force needed to make the motion. That is,

$$\text{work} = \text{force} \times \text{distance},$$

or simply

$$w = Fd.$$

The energy needed to do the work is exactly equal to the work done, so we can write

$$\text{energy needed} = \text{work done} = \text{force} \times \text{distance},$$

or

$$E = w = Fd.$$

Only that part of the force that is parallel to the motion does work. For instance, consider a boat as it is pulled slowly through the water. Three forces act on the boat (assuming that the motion is slow enough so that water friction can be ignored): Gravity pulls the boat downward, the balancing force of the water pushes upward, and the horizontal force pulls the boat through the water. The only force that takes energy is the horizontal force, because it is the only one that is parallel to the motion (Figure 8-1). There is no motion up or down, so neither gravity nor the water does any work.

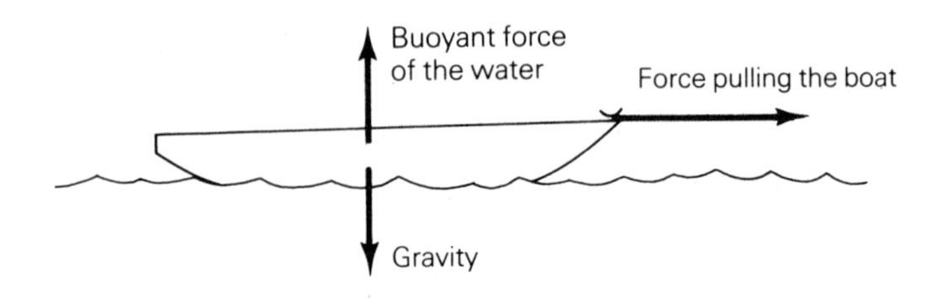

Figure 8-1 Only the horizontal force pulling the boat takes energy, because that is the only force parallel to the motion of the boat.

The units of work and energy are the product of force units and distance units. Thus, they would be newton-meters* in the SI system. Since the energy unit is used so often, it has a simple, short name; the newton-meter is given the name **joule** (J):

$$\text{SI energy unit} = \text{SI work unit} = \text{newton-meter} = \text{joule}.$$

EXAMPLE Suppose we push a child on a sled 100 meters across a frozen pond (Figure 8-2). The coefficient of friction between the sled and the ice is 0.05, and the total mass of the child and the sled is 20 kilograms. How much energy is needed to push them this distance? How much energy would be needed if the ice was covered with sand, raising the coefficient of friction to 0.5?

Answer Friction is the only force opposing the motion of the sled, and it is

$$\begin{aligned} F &= \mu W \\ &= \mu m g \\ &= 0.05 \times 20\ \text{kg} \times 9.8\ \text{m/s}^2 \\ &= 9.8\ \text{N}. \end{aligned}$$

Thus, the energy needed to push the sled across the pond would be

$$\begin{aligned} E &= Fd \\ &= 9.8\ \text{N} \times 100\ \text{m} \\ &= 980\ \text{J}. \end{aligned}$$

If the coefficient of friction were raised to 0.5 by spreading sand on the ice, the force needed to move the sled would be increased to

$$\begin{aligned} F &= \mu m g \\ &= 0.5 \times 20\ \text{kg} \times 9.8\ \text{m/s}^2 \\ &= 98\ \text{N}, \end{aligned}$$

---

*You may recall that torque is also measured in newton-meters. This is just a mathematical coincidence that results from the manner in which angles are measured. Torque and energy are, of course, two very different things—even though they are accidentally measured in the same units. To help distinguish these two concepts, we will always use joules when referring to energy and newton-meters when referring to torque.

and the energy required to move the sled across the pond would be

$$\begin{aligned} E &= Fd \\ &= 98\ \text{N} \times 100\ \text{m} \\ &= 9{,}800\ \text{J}. \end{aligned}$$

This example shows why keeping friction at a minimum is so important when we are moving things. Any increase in the friction increases the amount of energy needed to move the object. Since this extra energy seldom serves any useful purpose—and energy almost always costs money—unnecessary friction often results in unnecessary economic waste.

EXAMPLE How much energy is needed to lift a 100-kilogram barbell from the floor to a position overhead, 2.0 meters above the floor (Figure 8-3)?

Figure 8-3

Answer The force that the weightlifter must overcome is the weight of the barbell, which is $W = mg$. The distance moved is simply the height h to which the barbell is lifted. Thus, the energy needed to lift the barbell is

$$\begin{aligned} E &= mgh \\ &= 100\ \text{kg} \times 2.0\ \text{m} \times 9.8\ \text{m/s}^2 \\ &= 1{,}960\ \text{J}. \end{aligned}$$

Potential Energy

The amount of energy stored in a compressed, stretched, or raised object depends on the details of the force that the object produces as it is decompressed, unstretched, or lowered. The first two cases are a bit complicated, because the force changes as the object releases the energy. However, the case of a raised object is simple, because the object produces a force just equal to its weight, mg. As the object is lowered, it can exert this force over a distance equal to whatever height h it had been lifted. Therefore, the object's potential energy is

$$E = mgh.$$

Note that this is exactly the same as the energy it took to lift the object in the first place.

Kinetic Energy

All moving objects have kinetic energy contained in their motion. To convince ourselves of this, we need only watch a moving object collide with a stationary one. The work done bending and distorting both objects must have come from energy associated with the moving object's motion—its kinetic energy.

Two things determine the amount of kinetic energy that a moving object has: the object's mass and its velocity. We know that the kinetic energy of a fast object is greater than that of a slow one because a fast collision produces more damage than a slow one. We also know that the kinetic energy of a heavy object is more than that of a light one (it hurts much more to be hit by a baseball than by a Ping-Pong ball going at the same speed).

We won't go into the technical details of finding the kinetic energy of a moving object, because the details are somewhat involved. We will just state the formula: It is

$$\text{energy} = \frac{1}{2} \times \text{mass} \times (\text{velocity})^2,$$

or

$$E = \frac{1}{2}mv^2.$$

This relation agrees with our experience; the kinetic energy increases with both increasing mass and increasing velocity. The surprising thing is how rapidly it increases with velocity. The velocity is *squared* in the formula, which means that if the velocity is doubled, the kinetic energy will be *quadrupled.* If the velocity is tripled, the kinetic energy will be *9 times* larger. The effect of this relationship can be vividly seen in car accidents: An accident at 40 km/hr will be *4 times* as destructive as a similar accident at 20 km/hr; an accident at 100 km/hr will be *25 times* more destructive than a similar one at 20 km/hr; and an accident at 200 km/hr will be *100 times* more destructive than a similar accident at 20 km/hr.

Conservation of Energy

We have seen many examples of how energy can be transformed from one form to another, then into a third form, or perhaps back to the original form. The obvious question is, what finally happens to the energy? Does it eventually disappear, or does it just keep on transforming from one form into another? When we take all forms of energy into account, no energy is lost whatever. The total amount of energy always remains the same, as stated in the following principle:

Law of Conservation of Energy

Energy cannot be created or destroyed; it can only be changed from one form to another. The total amount of energy is always exactly the same.

We might wonder why there is a shortage of energy if it cannot be destroyed. The reason is that not all forms of energy are useful to us. For instance, the high-grade energy originally contained in fossil fuels ends up as low-grade heat energy at the end of its useful life. Although there is just as much energy, it is not in a very useful form. Low-grade heat merely warms up the environment by a fraction of a degree; then eventually it is radiated into outer space. During the entire process—as the energy goes from the chemical energy in the fossil fuel to the low-grade heat radiating into space—the total amount of energy is always the same. However, whether the energy is stored in barrels on earth or diffused 5 light-years away from us in interstellar space makes a big difference. Nevertheless, when we use the law of conservation of energy to solve a problem, we must be sure to take all forms of energy into account, not just those forms that are useful.

Let's look at an example of how energy is transformed. Suppose a woman consumes the energy stored in food, uses that energy to climb up a diving platform, and then dives into the water. What forms does the energy take during this process? First, the energy is stored in the food in the form of chemical energy. In fact, the *calorie** is a unit of energy used for measuring heat energy and the energy in food;

1 food calorie = 4,188 joules.*

The woman's body transforms the food energy into mechanical energy as she climbs the ladder to the platform. The conversion isn't 100% efficient, because some of the food energy is changed into heat as she climbs. When she is on the diving board, she has potential energy, which is then gradually transformed into kinetic energy as she drops toward the water. As she slows down in the water, her kinetic energy is changed into heat, which is quickly lost to the environment. Thus, the energy that starts out as food energy ends up as low-grade heat energy. All of the energy is still there, but not in a very useful form. No energy was destroyed at any time during the process.

The law of conservation of energy can be used to solve many mechanical problems quickly. In fact, many of the problems involving pulleys, levers, and ramps can be solved in about half the time by using the law of conservation of energy instead of force analysis methods.

EXAMPLE If a car traveling 50 km/hr (14 m/s) takes 10 meters to stop on dry concrete, how far will a car traveling at 100 km/hr (28 m/s) take to stop? (Note that these values are typical of a car with good tires and good brakes on a dry road. The distances for cars with poor brakes or cars on slippery roads will be much longer.)

Answer In both cases, the kinetic energy of the car is transformed into heat by the frictional forces of the brakes and tires. Because there is no other place to which the kinetic energy can go, the energy absorbed by the frictional forces must be equal to the initial kinetic energy of the car. The coefficient of moving friction doesn't change much with the speed of the object, so the frictional force F will be about the same in both cases. The energy that the brakes absorb will be the frictional force times the stopping distance d:

$$E = Fd.$$

---

* Unfortunately, there are two different calories in common use, and people seldom state which one they are using. The calorie used for measuring food energy is the amount of energy needed to raise the temperature of 1 kilogram of water by 1°C. The other calorie is the amount of energy needed to raise the temperature of 1 *gram* of water by 1°C. Thus, these two calories differ by a factor of 1,000.

As we said above, this energy is also equal to the initial kinetic energy of the car,

$$E = \frac{1}{2} mv^2.$$

Thus,

$$Fd = \frac{1}{2} mv^2;$$

solving for d,

$$d = \frac{1}{2} \frac{m}{F} v^2.$$

The frictional force F and the mass of the car m remain the same. Since the formula says that the stopping distance increases as the *square* of the car's speed when all other things are the same, doubling the car's speed to 100 km/hr will quadruple the stopping distance to 40 meters (about half the length of a football field).

When driving, it is worth remembering that *both* stopping distance and crash severity increase as the *square* of the speed of the car. Therefore, doubling your speed means that it will take *4 times* longer to stop; if you don't stop in time, the accident will be *4 times* as damaging. Both these effects call for increased caution at high speeds.

EXAMPLE A person with a strong arm can throw a baseball about 33 m/s. Approximately how high could such a person throw a baseball straight up?

Answer This is a good example of a problem that is hard to solve by using Newton's laws directly but easy to solve using the law of conservation of energy.

If we assume that air friction is small, conservation of energy says that the kinetic energy of the ball as it leaves the person must be equal to the potential energy at the top of its flight, where it stops for an instant before it starts back down.

$$E_{\text{kinetic}} = \frac{1}{2} mv^2$$

$$E_{\text{potential}} = mgh$$

But these two energies are equal, so

$$mgh = \frac{1}{2} mv^2.$$

Solving for h,

$$h = \frac{v^2}{2g}$$

$$= \frac{(33 \text{ m/s})^2}{2 \times 9.8 \text{ m/s}^2}$$

$$= 56 \text{ m}.$$

The law of conservation of energy can be a big help in estimating muscle forces in the body. Using conservation of energy, we only need to know the length of the muscle contraction to find the average force. A reasonable estimate of the contraction can be made by observing the body. If a better estimate is needed, the contraction distance can be measured on a skeleton or model skeleton. Either estimate is far easier to obtain than measuring all the bone lengths and joint angles needed for lever calculations. However, estimating the force with the law of conservation of energy does have one drawback: We can only find the average force during the contraction; the exact force at any specific position must still be found with lever methods.

EXAMPLE What is the average tension in the biceps during a chin-up? Suppose that the person has a mass of 75 kilograms. During the chin-up, the person's center of gravity will be raised by about 0.40 meter. From skeletal measurements, we estimate that the biceps contracts by 4 centimeters (0.04 meter) during the chin-up.

Answer The energy needed to lift the person's CG is

$$E_{\text{lifting}} = mgh$$

$$= 75 \text{ kg} \times 9.8 \text{ m/s}^2 \times 0.40 \text{ m}$$

$$= 294 \text{ J}.$$

If the biceps muscles in both arms share the work equally, each biceps must produce 147 joules. The energy produced by one of the biceps muscles is

$$E_{\text{biceps}} = Fd,$$

where F is the tension in the muscle and d is the length of the contraction (the change in the length of the biceps, not the length of the biceps). Solving for the tension in the biceps (F),

$$F = \frac{E}{d}$$

$$= \frac{147 \text{ J}}{0.04 \text{ m}}$$

$$= 3{,}700 \text{ N}.$$

This is the average tension in the biceps during the contraction. However, the tension doesn't change too much during the contraction, so it is a good estimate of the tension at any time during the contraction.

Energy and Human Performance

The records for human performance are broken almost routinely, and no doubt they will continue to be broken. However, no one ever expects to see more than just small improvements. As new records are set, it gets harder and harder to break them, because improvements are limited by the structure of the body. Muscle and bone must follow the laws of physics just as all other materials must.

Most of the limitations on human performance are closely related to energy production in the body. High jumping is a case where the relationship is especially close. The energy needed to jump to a height h is simply that height times the weight of the jumper:

$$E = mgh.$$

The variable h is not the height of the bar above the ground, it is the change in the height of the jumper's center of gravity. Since the height of the jumper's CG will be about 0.5 meter at the start of the jump and just about equal to the height of the bar at the top of the jump, the height of the bar above the ground will be about 0.5 meter more than h (Figure 8-4).

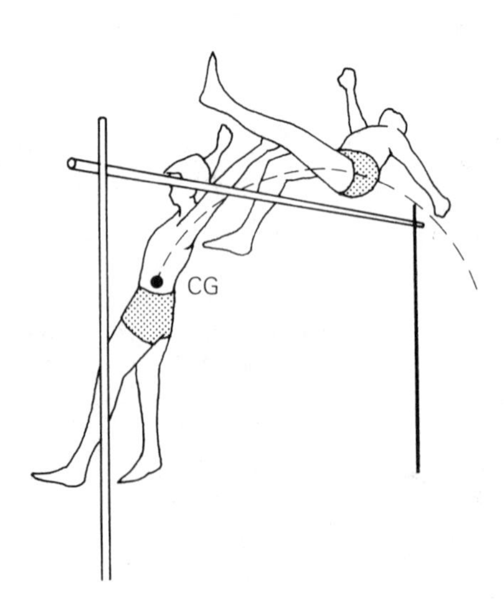

Figure 8-4 Path of a high jumper's center of gravity.

The leg muscles supply almost all of the energy for the jump. That energy will be

$$E = Fd,$$

where F is the force of the legs and d is the distance that the legs extend during the jump. For a person of average build, d is about 0.5 meter. We can estimate F by looking at the maximum weight that people can lift with their leg muscles. On the average, people can lift about twice their weight with their legs, and the exceptional person can lift around 3 times his or her weight. Since we are interested in the limitations on record jumps, we will suppose $F = 3 \times W = 3mg$. Because very little energy is lost to joint friction, almost all the energy goes to lifting the jumper upward. Thus, the energy produced by the legs is equal to the energy needed for the jump, so

$$\begin{aligned} mgh &= Fd \\ &= 3mgd. \end{aligned}$$

Solving for h,

$$\begin{aligned} h &= 3d \\ &= 3 \times 0.5 \text{ m} \\ &= 1.5 \text{ m}. \end{aligned}$$

This calculation explains why the high jump record is just a bit more than 2 meters. (A 2-meter high jump corresponds to about a 1.5-meter change in the height of the jumper's CG.) As we have seen, a 1.5-meter change means that the jumper must produce a force equal to 3 times his or her weight over a distance of 0.5 meter. For the high jump record to be increased substantially, the person would either have to be far stronger than any person has ever been or have far longer legs so that d would be substantially increased.

Power

Quite often we are interested in the *rate* of energy flow in a system. The rate of energy production, consumption, or flow is called the *power.* Unfortunately, the everyday usage of the word *power* is far broader than its technical usage. In ordinary conversation, power can mean anything from *ability* to *strength.* To avoid confusion, the meaning of *power* in a technical conversation should be restricted to its technical meaning: *the rate of energy production or use.* Even in a technical discussion, it would be correct to say that a horse is more powerful than a human, but not because the horse is bigger or stronger. A horse is more powerful than a human because a horse can produce energy *more rapidly* than a human can.

Power is found in the same way that other rates are. The energy is divided by the time needed to produce or use that energy. That is,

$$\text{power} = \frac{\text{energy}}{\text{time}}.$$

Expressed as a formula,

$$p = \frac{E}{t},$$

where p is the power. Power is measured in energy units divided by time units, so the SI unit of power would be a joule/second. Of course, because the unit of power is used so often, it has a short name of its own, which is the **watt** (W):

$$\text{SI unit of power} = \frac{\text{joule}}{\text{second}} = \text{watt}.$$

The watt is the same unit that we encounter in electrical appliances. For instance, a 100-watt light bulb uses 100 joules of electrical energy every second; a 1,350-watt coffee pot uses 1,350 joules of energy every second; and so on.

EXAMPLE A 60-kilogram athlete runs up to the top of a 30-meter-high building in 20 seconds. What was the athlete's power during the climb?

Answer Power is the rate of energy production or use, so

$$\begin{aligned} p &= \frac{E}{t} \\ &= \frac{mgh}{t} \\ &= \frac{60 \text{ kg} \times 9.8 \text{ m/s}^2 \times 30 \text{ m}}{20 \text{ s}} \\ &= 880 \text{ W}. \end{aligned}$$

This figure is around the maximum power that a human can maintain for any length of time; it is close to the old English unit of power, the horsepower. Only people in good condition can keep this power level up for more than a minute. Of course, for very short times, say the time needed to swing a hammer, higher power levels are possible.

EXAMPLE A 75-watt light bulb is left on overnight for 12 hours. How much energy goes to waste? How much does the wasted energy cost if electrical energy costs 3¢ per megajoule?

Answer Since

$$p = \frac{E}{t},$$

$$E = pt$$
$$= 75 \text{ W} \times 43{,}200 \text{ s}$$
$$= 3{,}240{,}000 \text{ J}.$$

(Remember that quantities must always be expressed in SI units, and the SI unit of time is the second.)
The cost would be

$$3.24 \times 3¢$$

or 10¢.

EXAMPLE If the engine in a 2-tonne car can deliver 100 kilowatts of energy to the wheels, how long will the car take to accelerate from rest to 30 m/s on a level road?

Answer Since we know the power of the engine, the maximum amount of energy that it can produce in a time t can be found from

$$E = pt.$$

As long as the friction is small, which it is on a paved road, all the energy produced by the engine is transformed into the kinetic energy of the car, which is

$$E = \frac{1}{2}mv^2.$$

Thus,

$$pt = \frac{1}{2}mv^2.$$

Solving for t, we find

$$t = \frac{mv^2}{2p}$$

$$= \frac{2{,}000 \text{ kg} \times (30 \text{ m/s})^2}{2 \times 100{,}000 \text{ W}}$$

$$= 9 \text{ s}.$$

The Metabolic Rate: The Power of the Body

A person's metabolic rate—which is simply a person's power level—is closely related to the activity of the thyroid gland. In fact, the body's metabolic rate is controlled primarily by the amount of secretion of this gland. Too much thyroid secretion causes a person to be overactive, nervous, and underweight; too little causes the person to be sluggish and overweight. (However, most overweight people do not have a thyroid problem; they simply eat too much food.) Because of the close relation between thyroid activity and a person's metabolic rate, the metabolic rate is measured to determine the condition of the thyroid.

However, the metabolic rate also depends on a person's level of physical activity, and it is also larger for larger people. In order to eliminate these variables, a person's metabolic rate is measured while a person is resting but awake—the so-called *basal* conditions. To eliminate the size of the person as a factor, the basal power level is divided by the total skin area of the person. The result is a good indicator of thyroid activity alone. This number, the person's resting power level divided by his or her skin area, is called the person's *basal metabolism rate* (BMR). The average person has a basal power level around 90 watts and a skin area around 1.5 square meters, so the average BMR is about

$$\frac{90\ \mathrm{W}}{1.5\ \mathrm{m}^2} = \frac{60\ \mathrm{W}}{\mathrm{m}^2}.$$

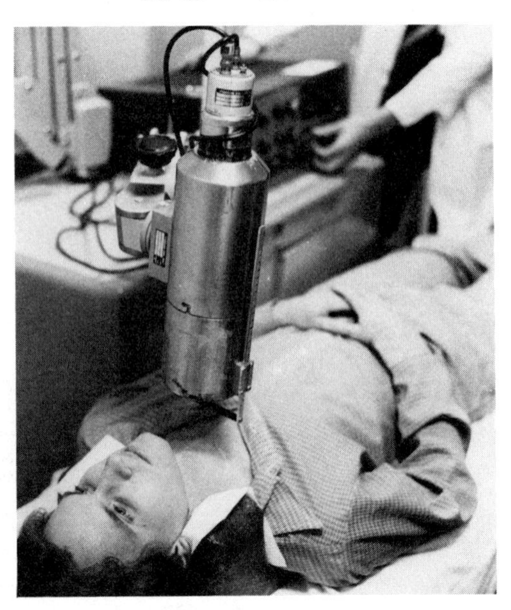

Radioactive iodine is often used to measure thyroid activity.

Concepts

Energy
Work
Heat
Chemical energy
Kinetic energy
Potential energy
Conservation of energy
Power
Metabolic rate

Discussion Questions

1. When you rub your hands together rapidly, they get hot. Where does the heat come from?
2. What is the difference between force and work?
3. Give examples of chemical, electrical, heat, kinetic, and potential energy.
4. A gymnast drops from a platform onto a trampoline, then bounces from the trampoline back up to the platform. Discuss the changes in the energy, from potential to kinetic and so on, that occur.
5. Why is a collision at 80 km/hr far more than twice as dangerous as a similar collision at 40 km/hr?
6. Give a brief description of the flow of energy within your body. Where does the energy come from, how is it transformed within the body, and in what forms does it leave the body?
7. A ball thrown straight up at 30 m/s will go 4 times higher than one thrown at 15 m/s. Why?
8. Although exercise is a very important part of any fitness program, by itself it is not too helpful in controlling a person's weight. Why is the person's diet generally a much bigger factor in weight control than exercise?
9. Do you think that anyone will ever build a perpetual motion machine that produces energy from nothing at all? Why or why not?
10. How is power different from energy?

Problems

1. How much energy is used by a 100-kilogram person climbing the Washington Monument (which is 150 meters high)? Neglect the energy that is lost to heat and friction. Give your answer in both joules and food calories.

2. How much energy is needed to drag a 50-kilogram table 5 meters across the floor when the coefficient of friction between the table and floor is 0.3?

3. A 50-kilogram woman rises on her toes, lifting her center of gravity by 6 centimeters. The muscle group attached to the Achilles tendon (tendo calcaneus) furnishes most of the energy for this motion, and during the motion it contracts by 3 centimeters. What is the average tension in her Achilles tendon as she rises on her toes? Assume that she distributes her weight equally on her legs.

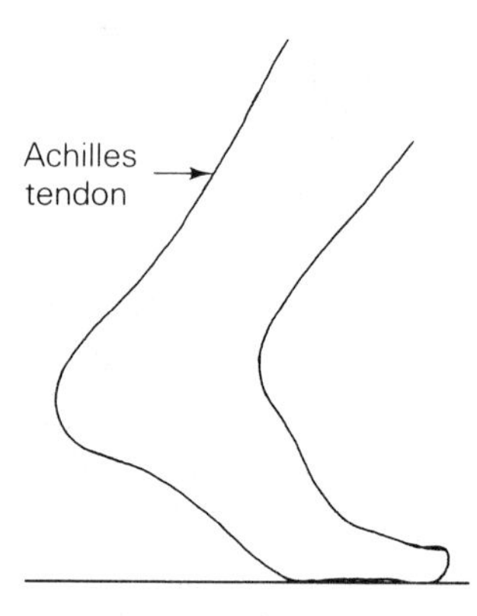

4. If you dropped a stone off a 20-story building (about 60 meters high), how fast would it be going when it struck the ground?

5. Compare the stopping distance of a car going 50 km/hr on dry concrete to the stopping distance of a car going 100 km/hr on a slippery, slush-covered road. The coefficient of friction for tires is about 1.0 on dry concrete and about 0.2 on a slush-covered road.

6. How fast will a skier be going at the bottom of a 50-meter-high hill if no energy is lost to friction?

7. How many sit-ups would you have to do to wear off the energy in a 55-calorie slice of bread? Assume that you use 4 times more energy than is actually used to lift the body. Assume that the mass of the upper body is 35 kilograms and that the CG of the upper body is raised by 45 centimeters during the exercise.

8. How much energy is needed to get a 200-tonne airplane up to an altitude of 10,000 meters at a speed of 720 km/hr? These numbers are typical of a fully loaded 747 at normal cruising altitude and speed. How much does it cost to get the plane to this altitude and speed if the fuel costs 5 cents per megajoule?

9. How many times would you have to lift a 50-kilogram barbell to wear off a kilogram of fat? Assume that you lift the barbell 2 meters per lift and that no energy is lost to heat within the body. In practice, you would only have to lift it around one-fourth as many times because

about three-fourths of the energy would be lost to heat in the body. One kilogram of fat contains about 7,000 calories of energy.

10. How long would it take a person to lose 5 kilograms if his or her diet was cut by 1,000 calories per day and the level of physical activity was not changed? Suppose the person also increased his or her physical activity by doing 250 push-ups every day. How long would it take to lose the 5 kilograms? Assume that the person's mass is 75 kilograms and that the center of gravity is raised by 30 centimeters during a push-up. Also assume that a push-up burns up 4 times more energy than is actually used to lift up the body. (See Problem 9 for the energy content of fat.)

11. A person who just sits around most of the time needs about 2,000 calories per day. What is that person's average power level in watts?

12. The average person uses about 150 calories per hour when walking at a normal speed. What is the power of such a person in horsepower? (The horsepower is the English unit of power, which is equal to 746 watts.) The horsepower is also roughly equal to the power that a strong workhorse can maintain for an eight-hour day. Incidentally, the term *horsepower* was first used to describe the power of steam engines by James Watt, the inventor of the modern steam engine. But the horses have gotten even—in the SI system we measure the power of a horse, or anything else, in watts.

13. A 73-kilogram person runs up a 4-meter-high flight of stairs in 3 seconds. What was the person's power level during the climb?

14. How long would it take a 800-watt electric coffee pot to heat up 2 liters of water from 20°C to 85°C? Recall that the food calorie is the amount of energy needed to raise the temperature of 1 kilogram of water by 1°C and that 1 calorie equals 4,188 joules.

15. If a 1-tonne car is to accelerate from zero to 72 km/hr in no more than 4 seconds, what power must the engine deliver to the wheels?

Experiments

1. Keep a record of the food you eat for one day, then calculate the energy content of the food in calories and in joules. (Calorie tables for common foods are available in most drugstores and bookstores.) Assume that 25% of the energy was used in physical activity, and calculate how high a ladder you could have climbed with that energy.

2. Measure the height of a flight of stairs and then run up them as fast as you can while a friend times you. Calculate the energy used and your power during the climb. Convert your power in joules to horsepower (1 horsepower = 746 joules).

3. Choose an exercise that is tiring but not so tiring that you can't do at least 50 repetitions of the exercise. If you are in very good condition, you might do sit-ups or push-ups. A simple one-step climb is a good choice for the average person. If you have weightlifting equipment available, any exercise that involves lifting a known weight through a known distance would be suitable. Estimate the energy needed for one cycle of the exercise by finding the distance your CG (or the CG of the weight) is raised during the exercise and then multiplying by your weight (or the weight of the barbell) in newtons. Next investigate the effect of fatigue on your power. Have a friend time you while you do ten repetitions of the exercise, followed by ten more repetitions, and so on, until you are tired. Then rest ten minutes and do ten more repetitions to see how a short rest affects fatigue. Calculate your power during each set of exercises and record them in a table. Some questions to look into are: How fast did your power drop? Did your power drop gradually or all at once? What percentage of your original power did you recover after a ten-minute rest? Note: Don't choose an exercise that will strain you or lead to overexertion. A mildly tiring exercise is just fine for this experiment.

CHAPTER NINE

Fluids—Liquids and Gases

Educational Goals

The student's goals for Chapter 9 are to be able to:

1. Describe the solid, liquid, and gaseous states of matter.
2. Calculate and/or measure an object's density.
3. Clearly explain the difference between pressure and force.
4. Explain how fluids transmit pressure.
5. Explain how the pressure in a fluid increases with depth and how this pressure increase causes objects to be buoyed up by the fluid.
6. Relate the basic principles of fluids to blood circulation, pregnancy, elimination, and the other functions of bodily fluids.
7. Estimate the percentage of fat in a person's body by measuring the density of the person.
8. Understand the role of surface tension and capillary action in the body.

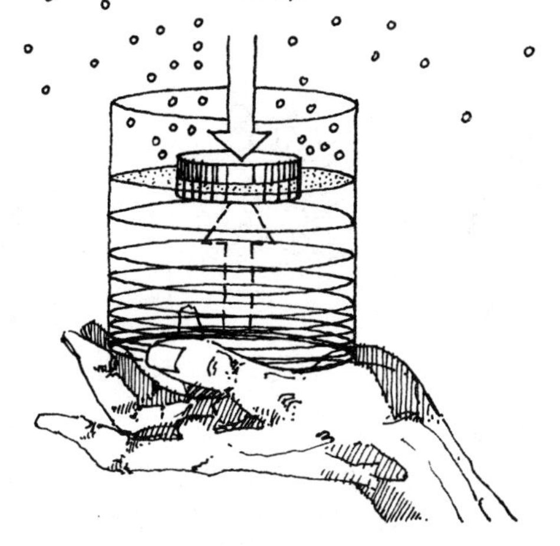

According to Darwin's theory of evolution, our ancestors moved from the sea to the land millions of years ago, and in many ways our bodies still reveal that ancient origin. Water is the most common compound in our bodies, and the environment of most of our cells is nearly identical to the warm, slightly salty sea in which life began. The major difference between ourselves and the early forms of life is that nutrients and wastes are no longer delivered and removed by random currents and the ebb and flow of the tides; we now have a highly organized system of pumps, valves, tubes, and filters to move fluids within our bodies. In this chapter we will look into how the laws of physics apply to fluids, particularly fluids in the human body.

The States of Matter—Solids, Liquids, and Gases

At normal temperatures a substance is either a solid, a liquid, or a gas (Figure 9-1). The form that a particular substance takes depends on the strength of the forces between the molecules. If the molecular binding forces are very strong, the molecules are held tightly in place, vibrating a bit but rarely moving to a new position. Such a substance is a solid. In liquids and gases, the two fluid states of matter, the intermolecular forces are too weak to keep the molecules from moving. Thus, a fluid has no shape of its own, and it simply adopts the shape of its container.

Although the molecular forces are too weak to hold the molecules of a liquid in one spot, they are still strong enough to hold the molecules close together. That is, the molecules of a liquid behave like a set of oily magnetic balls: they are very difficult to pull apart, but they slide over and around each other easily and mold into any shape we want. Since pulling one molecule away from the

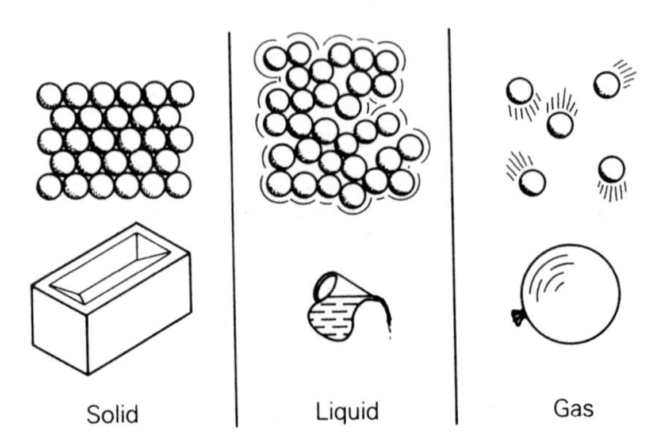

Figure 9-1

others is difficult, a liquid has a well-defined surface. Thus, when a liquid is poured into a container, it takes on the shape of the container up to a level top surface, provided that the liquid doesn't fill the container.

Of course, the molecules of a liquid are not little magnets. The intermolecular forces are actually electrical in character, like the static electrical forces that cause clothes to cling. The forces between the molecules are similar to the chemical bonds that hold the molecule together, but they are much weaker.

Since the intermolecular forces in a liquid hold the molecules close together, there is very little space between the molecules. This makes liquids very hard to compress. Thus, the volume of a liquid—like the volume of a solid—doesn't change much when either the temperature or the pressure changes.

In gases, the intermolecular forces are so weak that the molecules rarely stick to one another. Thus, the molecules of a gas move around randomly with vast amounts of totally empty space between them. They only contact each other during brief collisions. Therefore, gases have no surface, and they completely fill whatever container they are in. If the container has a hole in it, the molecules will flow out quickly. Gases are highly compressible because the molecules are so far apart. You can always pump more gas into a container unless the container bursts or the molecules are forced so close together that the gas becomes a liquid.

How Fluids Are Described

Since the molecules of a fluid move randomly, describing the motion of a fluid is difficult. In theory we could analyze the motion of every individual molecule, but this would be impossibly complicated—and we really don't care where each molecule goes anyway. We want to see how *averages* of quantities such as force and mass affect the overall motion of a fluid. Two especially important average quantities are density and pressure.

Density

Mass plays an important part in fluid motion, just as it does in all other types of motion. However, neither the mass of an individual molecule nor the total mass of a fluid is a very useful quantity. What we need is some sort of average of the mass. The *density* is that quantity. The density of a substance is the amount of mass contained in one unit of volume. For a specific quantity of fluid (or solid), the density is found by dividing the mass of the material by its volume. In formula form,

$$\text{density} = \frac{\text{mass}}{\text{volume}},$$

or

$$D = \frac{m}{V}.$$

The unit of density will be a mass unit divided by a volume unit; therefore,

$$\text{SI density unit} = \frac{\text{kilogram}}{\text{meter}^3} = \text{kg/m}^3.$$

As with all other SI units, this is the best density unit to use in problems. When other density units are used, the answer will not have the proper SI units.

Another commonly used density unit is the gram per cubic centimeter (g/cm ^{3}, or g/cc). While the g/cm ^{3} is not an SI unit, it is a convenient size for most everyday density measurements. We can easily convert kg/m ^{3} to g/cm ^{3} by using the following information:

$$1 \text{ kg/m}^3 = 1 \frac{\text{kg}}{\text{m} \cdot \text{m} \cdot \text{m}}$$

$$= 1 \frac{1{,}000 \text{ g}}{(100 \text{ cm}) \cdot (100 \text{ cm}) \cdot (100 \text{ cm})}$$

$$= 0.001 \frac{\text{g}}{\text{cm} \cdot \text{cm} \cdot \text{cm}}$$

$$= 0.001 \text{ g/cm}^3$$

Thus, when you want to change the density from kg/m ^{3} to g/cm ^{3}, just multiply by 0.001; to change from g/cm ^{3} to kg/m ^{3}, multiply by 1,000. For example, 3.2 g/cm ^{3} = 3,200 kg/m ^{3}.

Materials have a wide range of densities. Water has a density of 1,000 kg/m ^{3} (1 g/cm ^{3}). (The size of the kilogram was chosen so that this relationship would exist.) Normal air has a density 1,000 times smaller, about 1 kg/m ^{3} (0.001 g/cm ^{3}). The density of most common liquids and solids is between 500 kg/m ^{3} and 8,000 kg/m ^{3} (0.5 g/cm ^{3} and 8 g/cm ^{3}). Very heavy metals, such as gold and tungsten, have densities around 20,000 kg/m ^{3} (20 g/cm ^{3}). The densities of many common substances are given in Table 9-1.

Table 9-1 Densities and Specific Gravities of Some Common Substances

Substance	*Density* g/cm^3	kg/m^3	*Specific Gravity**
Solids			
Aluminum	2.7	2,700	2.7
Bone (compact)	1.6	1,600	1.6
Brass	8.4	8,400	8.4
Copper	8.9	8,900	8.9
Cork	0.25	250	0.25
Glass	2.6	2,600	2.6
Gold (24 karat)	19.3	19,300	19.3
Ice	0.92	920	0.92
Lead	11.3	11,300	11.3
Plastics†	1.1	1,100	1.1
Steel	7.7	7,700	7.7
Stone†	4.0	4,000	4.0
Tungsten	19.1	19,100	19.1
Wood†	0.8	800	0.8
Liquids			
Alcohol (ethyl)	0.79	790	0.79
Blood (whole)	1.05	1,050	1.05
Blood plasma	1.03	1,030	1.03
Gasoline	0.69	690	0.69
Mercury	13.6	13,600	13.6
Oil†	0.8	800	0.8
Water	1.0	1,000	1.0
Water (seawater)	1.03	1,030	1.03
Gases			
Air (dry)	0.0013	1.3	1.0
Air (humid)	0.0012	1.2	0.92
Carbon dioxide	0.002	2.0	1.5
Cyclopropane	0.0019	1.9	1.5
Ether	0.0033	3.3	2.5
Helium	0.00018	0.18	0.14
Hydrogen	0.00009	0.09	0.07
Nitrogen	0.00125	1.25	0.96
Nitrous oxide	0.002	2.0	1.5
Oxygen	0.0014	1.4	1.1

Note. The densities of the gases are given at normal atmospheric pressure at 20°C (approximately room temperature).

* Solids and liquids are compared to water; gases are compared to air at the same temperature.

† Varies widely depending on type.

Because the density is the average of the mass per unit volume, it does not depend on either the size or shape of an object; it depends only on the type of *material* from which the object is made. For example, a gold filling in a tooth has exactly the same density as a gold bar, a gold ring, or any other gold object. Since the density is a property of the material and not the object, it can be very helpful in identifying materials.

People sometimes confuse the density of a fluid with its viscosity, which is the resistance to flow. They mistakenly think that dense liquids should also be viscous and hard to pour. Actually, density and viscosity are completely unrelated; some dense fluids are very runny, while some very light fluids are viscous and flow very slowly. Part of the confusion comes from using the words *thick* and *thin* to describe both density and viscosity. To avoid confusion, do not use these two words to describe fluids.

Specific Gravity

Specific gravity is a concept used to compare the density of a substance with that of a standard substance. We will use the usual standards of water for solids and liquids and of air for gases.

The specific gravity of a substance is just the ratio of its density to the density of the standard. For instance, a liquid that was twice as dense as water would have a specific gravity of 2.0. If we had a solid with a density of 7,200 kg/m^3 (7.2 g/cm^3), we would determine its specific gravity as follows:

$$\begin{aligned}\text{specific gravity} &= \frac{D}{D_w} \\ &= \frac{7{,}200\ \text{kg/m}^3}{1{,}000\ \text{kg/m}^3} \\ &= 7.2.\end{aligned}$$

Note that when the density of a solid or liquid is given in g/cm^3, the specific gravity will be numerically equal to the density, because water has a density of 1 g/cm^3. For example, the specific gravity of bone is 1.6, since the density of bone is 1.6 g/cm^3. This relation does not apply to gases, because the density of air is not equal to 1 in either common system.

Later in the chapter we will see that the specific gravity is the key to determining whether a substance will float or sink in the standard substance.

Density of the Body

The density of the body is closely related to the percentage of fat. Fat is less dense than muscle and bone, so the density of the body goes down as the percentage of fat increases. Thus, a measurement of the density of the body can be used to estimate the percentage of fat.

Knowing the percentage of body fat can be very helpful in planning a fat reduction program. For instance, a mass loss goal can be especially helpful in motivating a patient. But how does one set a safe goal? The standard methods are not very accurate. A person's appearance can be deceiving, and the charts of average mass versus height make little allowance for a person's build and no allowance for variations in muscularity. (For instance, the tables say that a 180-centimeter-tall person who has a mass of 100 kilograms is fat. Yet many football players and weightlifters have exactly that height and mass without having any extra fat.) To set a safe mass loss goal, you must know the percentage of fat in the patient's body. Then you can set the mass loss goal to bring the fat percentage down to between 10% and 15%, which is generally considered normal.

The fat percentage is also a valuable guide in athletic conditioning. Measuring an athlete's mass will not tell you whether the athlete has gained 5 kilograms in muscle or in fat, but measuring the athlete's fat percentage will. We will see exactly how the fat percentage is found later in the chapter.

Density of Body Fluids

Most bodily fluids are water based, and their densities are near that of water. However, the small changes in density that do occur can be helpful in determining the condition of the fluids. For instance, blood cells are denser than blood plasma; thus, the density of whole blood drops as the cell count drops. (However, cell-counting methods are superior to the density test for this purpose, so the blood density test is not used as often as it once was). The density of the urine indicates the amount of dissolved material. This quantity can be helpful for maintaining the proper fluid balance in a patient and for checking the waste concentrating ability of the kidneys.

Pressure

Now that we have seen how density is used to describe the average mass of a fluid, let's see how pressure is used to describe the average force of the molecules of a fluid. To do this, we will first look into the effect of forces on surfaces.

What determines whether a force will distort or damage the surface of an object? The size of the force plays a part, but the *area* over which the force is distributed is just as important. When the force is spread over a wide area, the surface distorts only a little bit. But if the same force is concentrated on a small area, the surface can be easily damaged. For instance, when you push on someone with your finger, not much happens to the person. But if you apply the same force to the edge of a scalpel blade, the scalpel easily cuts into the person.

When we choose a mattress, we want one that spreads the weight of the body over as large an area as possible. The force of the mattress on us will be the same in all cases; it is just equal but opposite to our weight, according to Newton's third law of motion. However, the quality of the mattress depends on how evenly it spreads this force over the body. A waterbed distributes the force especially evenly, and a good mattress is almost as comfortable. If the mattress is lumpy, it will concentrate the force on the tops of the lumps, and we will not sleep well. If we tried to sleep on a bed of nails, the force would be concentrated on a very small total area, and we would be quite uncomfortable.

Pressure is the physical quantity that describes how concentrated a force is on a surface. It is simply the force divided by the area over which the force is distributed. Expressed as a formula,

$$\text{pressure} = \frac{\text{force}}{\text{area}},$$

or

$$P = \frac{F}{A}.$$

For a particular force, the pressure will be small when the force is spread out over a large area and large when it is spread out over a small area. The units of pressure will be force units divided by area units; the name **pascal*** (Pa) was picked as the name for the N/m^2.

$$\text{SI pressure unit} = \frac{\text{newtons}}{\text{m}^2} = \frac{\text{N}}{\text{m}^2} = \text{Pa}.$$

Several other non-SI pressure units are commonly used. Many are based on how high up against gravity a particular pressure will

*Named for Blaise Pascal (1623–1662), the French philosopher, physicist, and mathematician, who was one of the founders of the science of fluids and the author of the quotation, "The heart has its reasons which reason knows nothing of."

push water or mercury. For example, blood pressures are commonly given in millimeters of mercury (mm of Hg). A blood pressure of 100 mm of Hg could push a column of mercury up 100 millimeters against the earth's gravity. The conversion factor from mm of Hg to pascals is

$$1 \text{ mm of Hg} = 133 \text{ Pa.}$$

So to convert a pressure in mm of Hg to pascals, multiply by 133. For instance, a systolic (peak) blood pressure of 125 mm of Hg is equal to

$$P = 125 \text{ mm of Hg} \times 133 \frac{\text{Pa}}{\text{mm of Hg}} = 16{,}600 \text{ Pa.}$$

When a fluid presses on us, it seems to exert a steady force. However, this is not the case. What we actually feel is the total effect of all the individual molecular impacts. Each impact lasts for only a brief moment, but the impacts are so numerous and each impact is so tiny that we don't feel the separate collisions. Instead, we feel the average of all the impact forces. The same is true for the walls of any container or the surface of any object immersed in a fluid—the forces of the individual collisions are too rapid and too small to be detected individually. Thus, the pressure is the average force of the molecular impacts divided by the area of the surface under consideration.

The force that a stationary fluid exerts on a surface is always exactly perpendicular to the surface (Figure 9-2), because the molecules of a stationary fluid are free to move parallel to the surface; thus, they can't push in any direction that is parallel to the surface. The smallest shove will move a stationary boat, because the water does not exert a significant force opposing the motion until the boat builds up speed.

If we could watch the molecules collide with a surface, it would be clear why the force of a stationary fluid is always perpendicular

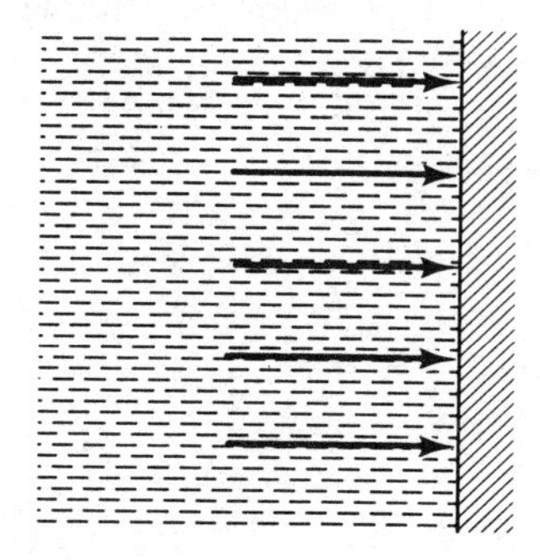

Figure 9-2 The force of a stationary liquid on a surface is always perpendicular to the surface.

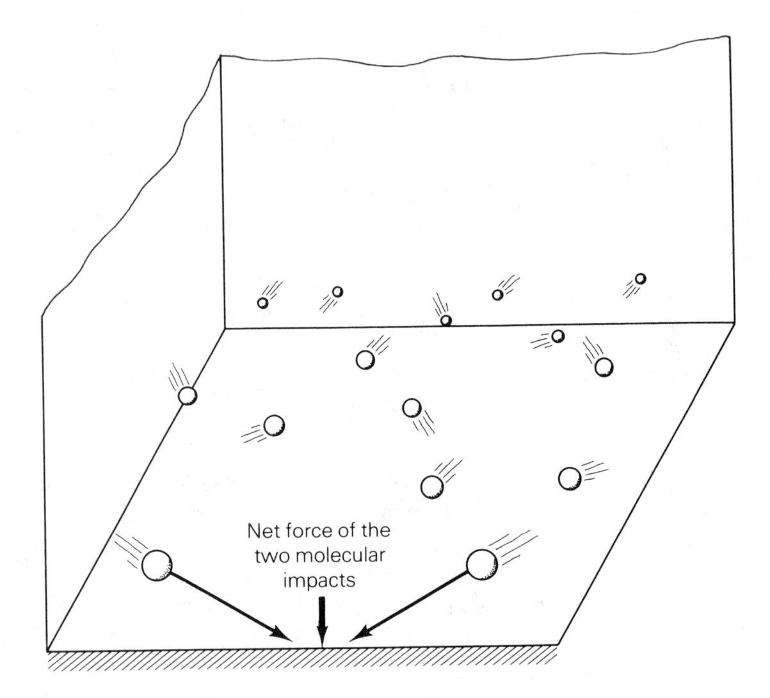

Figure 9-3 On the average, the number of molecules striking from the right is the same as from the left. Thus the average force of the molecular impacts is perpendicular to the container wall.

to the surface. Consider the horizontal surface shown in Figure 9-3; the same number of molecules strike the surface from the right as the left. Thus, the sideways forces cancel, and the net force of all the molecular impacts will point straight toward the surface.

Let's see what the pressures are in a few common situations.

EXAMPLE What is the pressure under the feet of a 50-kilogram girl standing barefoot in the sand?

Answer Typically, the foot of a 50-kilogram girl is about 25 centimeters long and about 6 centimeters in width. Thus, the area of each foot would be approximately 0.25 meter × 0.06 meter = 0.015 m^2. The total area of both feet would be twice that, 0.03 m^2. The force on her feet is just her weight, 50 kilograms × 9.8 m/s^2 = 490 newtons. The pressure is simply that force divided by the area of the bottom of her feet;

$$P = \frac{F}{A}$$

$$= \frac{490 \text{ N}}{0.03 \text{ m}^2}$$

$$= 16{,}000 \text{ Pa.}$$

Even though this pressure is numerically large, it isn't very large compared with the pressures of normal activities. For instance, the pressure under a fingertip when we touch someone with a light force of 1 newton is around 10,000 pascals.

EXAMPLE Suppose that the same 50-kilogram girl puts on a pair of the very narrow high heels that were in style in the fifties (and that return from time to time), and she then stands with all her weight on one of the heels. If that heel is 1 centimeter on a side, what is the pressure under the heel?

Answer The area of the heel is

$$0.01 \text{ m} \times 0.01 \text{ m} = 0.0001 \text{ m}^2,$$

so

$$P = \frac{490 \text{ N}}{0.0001 \text{ m}^2}$$

$$= 4{,}900{,}000 \text{ Pa}.$$

This is a high pressure! In fact, it is bigger than the compressive strength of many common flooring materials, such as linoleum and asphalt tile. This is why even light people sometimes make dents in floors when they wear tiny high heels.

Everytime you buy an airline ticket, part of the money pays for the cost of those tiny high-heeled shoes of the fifties. They came into style at the same time that the first of the modern jet airliners was being designed. The high pressures under the heels dented the floors in the test models rather badly, so the floors were strengthened. To carry the heavier floors, the airframes had to be heavier, as a result, bigger engines were needed and fuel economy was a little bit poorer. Today's planes follow those same basic designs, and all their floors can support the smallest heels—but at the cost of higher plane fares. Per ticket, the costs don't amount to much, but over the years those heels have cost millions of dollars.

EXAMPLE What is the pressure under a 400-kilogram dune buggy with four very wide tires, each having a rectangular region of contact 40 centimeters by 30 centimeters? Assume that each of the four wheels supports one-fourth of the weight of the buggy.

Answer The weight of the buggy is

$$W = mg$$

$$= 400 \text{ kg} \times 9.8 \text{ m/s}^2$$

$$= 3{,}900 \text{ N}.$$

The total contact area is

$$A = 4 \times 0.40\ \text{m} \times 0.30\ \text{m}$$
$$= 0.48\ \text{m}^2.$$

Since $F = W$, the pressure under the tires is

$$P = \frac{W}{A}$$
$$= \frac{3{,}900\ \text{N}}{0.48\ \text{m}^2}$$
$$= 8{,}100\ \text{Pa},$$

which is only about half the pressure under the feet of a girl standing barefoot in the sand. Because of the large contact area under the wide tires, the pressure that a dune buggy exerts is small. Thus, the tires don't sink very far into the sand.

EXAMPLE A person bites down hard on a tough piece of meat. If the force is 40 newtons, and it is applied to a single front tooth that has a top surface 6 millimeters by 1 millimeter, what is the pressure on the biting edge?

Answer The area of the biting surface would be

$$0.006\ \text{m} \times 0.001\ \text{m} = 0.000006\ \text{m}^2.$$

The pressure is

$$P = \frac{40\ \text{N}}{0.000006\ \text{m}^2}$$
$$= 6{,}700{,}000\ \text{Pa}.$$

This is also quite a high pressure, but it is less than the compressive strength of healthy tooth enamel. Nevertheless, it is amazing that teeth can stand up to these pressures day after day and still be in reasonable condition in old age.

The sensation we feel when a force is applied to our skin depends more on the pressure than the size of the force. When the force is spread out over a large area, the pressure is small and we feel only the sensation of touch. But if the force is concentrated on a small area, like the point of a needle, the pressure is large and we feel pain. The pressure that causes pain varies from person to person—and from place to place on the body—but the borderline between pain and discomfort is around 1 million pascals. Fortunately, because pain usually starts just before damage starts, pain usually warns us of possible danger.

A fluid's weight produces a pressure, of course. The force of gravity pulling downward on all the fluid above a given level must be balanced by an equal but upward force from the fluid below that level. Therefore, the deeper we go into a fluid, the greater the pressure will become, because the deeper we go, the more fluid there will be above us to support. We live at the bottom of an ocean of air, whose pressure at sea level is about 100 kilopascals. The higher one goes into the atmosphere, the less air there is above us to support; thus, the atmospheric pressure decreases. At an altitude of 5 kilometers, the pressure is only half of the sea level pressure.

Pressure, Volume, and Respiration

The volume that a particular amount of gas fills depends on the pressure. If the temperature is constant, the higher the pressure is, the smaller the volume into which the gas can be squeezed. Conversely, when a gas expands into a larger container without changing its temperature, the pressure will drop. The reason is simple: When the molecules go into a larger container, they take longer to travel across the container. Therefore, they do not hit the walls as often. Because there are fewer impacts per second, the average force and therefore the pressure on the walls go down.

As long as the amount of gas and the temperature don't change, the product of the pressure and the volume will always be the same. That is, under these conditions,

$$PV = \text{constant}.$$

Thus, if a gas expands into a container that is 10 times larger in volume, the pressure will drop to one-tenth of the original value, and so on.

Since

$$D = \frac{m}{V},$$

the density also drops to one-tenth of its original value when the gas expands into a volume that is 10 times larger. So the percentage change in the pressure and the percentage change in the density of a gas will be exactly the same as long as the temperature stays the same. This is why breathing at high altitudes is difficult. For instance, at an altitude of 5 kilometers, where the pressure is one-half normal, the density is also one-half normal. Thus, a lung full of air at 5 kilometers contains only half as much oxygen as a lung full at sea level. At 5 kilometers a person would have to breathe twice as often to take in the same amount of oxygen as at sea level.

Patients who have difficulty breathing are sometimes helped if they are placed in hyperbaric chambers. (Roughly translated, hyperbaric means high pressure.) The air pressure in the hyperbaric chamber is kept at 3 to 4 times normal, and the percentage of oxygen is often increased above the normal 20%. These conditions greatly increase the amount of oxygen reaching the lungs and drastically reduce the effort that the patient must use to breathe. Hyperbaric chambers range in size from a single-patient unit to a full operating room. At high pressure, the blood plasma will carry a substantial amount of dissolved oxygen to supplement the oxygen carried chemically by the hemoglobin of the red blood cells. Patients who are poor surgical risks because of low blood counts can often be operated on successfully in hyperbaric operating rooms. Carbon monoxide poisoning is another condition that can often be overcome by using a hyperbaric chamber.

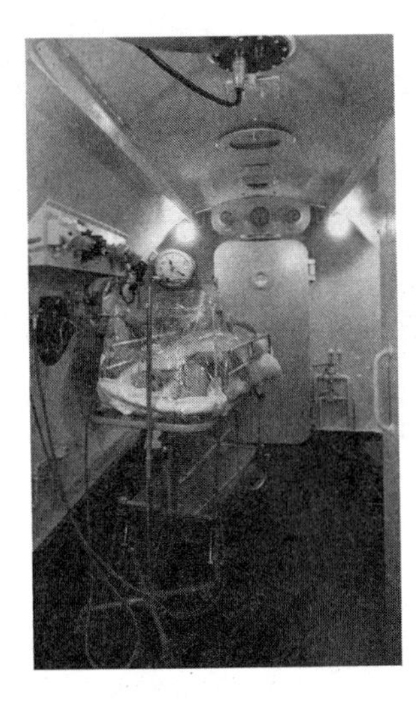

A hyperbaric operating room.

Respiration

When we breathe in, we increase the volume of the chest (thoracic) cavity by expanding the rib cage and dropping the diaphragm (the flat muscle below the lungs). The increase in volume causes the pressure in the lungs to drop below atmospheric pressure. The greater outside pressure then forces air into the lungs. To breathe out, we reverse the process. We contract the chest volume to increase the pressure in the lungs, thus pushing the air in the lungs out.

Pressure and Motion of Fluids

Pressure is a key concept in studying fluid motion. For instance, flow rates are very closely related to pressure differences. With one exception, fluids will flow from a high-pressure region to a low-pressure region. (The exception is the pressure increase with depth caused by the weight of a stationary fluid. This pressure increase just balances the weight and thus does not cause a flow. We will look at pressures caused by a fluid's weight later in the chapter.)

To prove that the flow will be from a high-pressure region to a low-pressure one, consider the fluid in a horizontal pipe connecting two tanks, 1 and 2. Suppose that the pressures in the two tanks are different, P_1 and P_2, respectively, and that the fluid in the pipe is

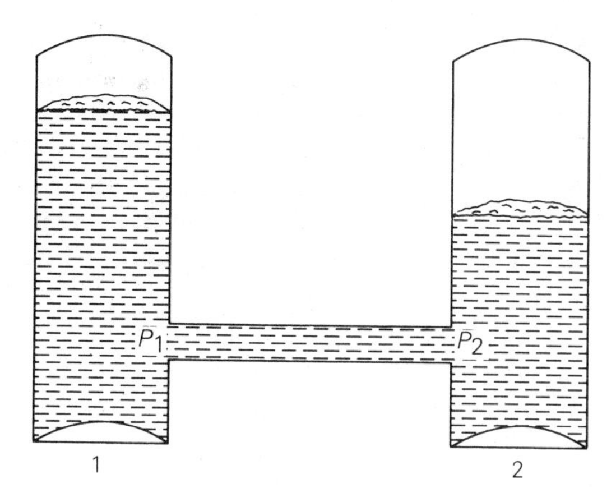

Figure 9-4

initially stationary. Now look at the forces on the fluid in the pipe, particularly those that might cause the fluid to go to the left or the right. The pipe wall does not push to the left or the right, because the force of the pipe on the stationary fluid must be perpendicular to the surface of the pipe. Thus, the net force of the pipe must be up or down. It can be shown that the force of the wall just supports the weight of the fluid. This leaves the forces at the two ends to push horizontally. The force on the left end, at tank 1, pushing toward the right will be

$$F_1 = P_1A,$$

where A is the cross-sectional area of the pipe. Similarly, the force at the other end due to the pressure in tank 2 is

$$F_2 = P_2A,$$

and this force will be toward the left. The net force on the fluid will be the difference between these two forces:

$$\begin{aligned} F_{\text{net}} &= F_1 - F_2 \\ &= P_1A - P_2A \\ &= (P_1 - P_2)\,A. \end{aligned}$$

Thus, if the pressures are different, there will be a net force on the fluid pushing toward the region of lower pressure. This unbalanced force would start the fluid flowing toward the low-pressure region, and the flow would continue until the two pressures were equal. (When P_1 is greater than P_2, $P_1 - P_2$ is positive, which means that F_{net} is toward the right. When P_1 is less than P_2, $P_1 - P_2$ is negative, and F_{net} is toward the left. That is, F_{net} is always toward the low-pressure region.)

If the two pressures were equal, the net force on the fluid would be zero. Thus, in accordance with Newton's first law of motion, the fluid would stay stationary. The first law also tells us that the opposite must be true: If the fluid is stationary, the net force must be zero; thus, the pressure must be the same everywhere in the fluid (except for those pressure differences that just support the weight of the fluid.) This idea is called Pascal's principle.

Pascal's Principle

The pressure in a stationary fluid is the same everywhere except for the pressure differences due to the weight of the fluid.

If the pressure was not the same everywhere, the fluid would not stay stationary—it would flow toward the low-pressure region.

Pascal's principle is exactly true only for stationary fluids. However, it is approximately true for a slowly moving fluid in a wide pipe if the fluid is not too viscous. (As you will recall, a viscous fluid is one that is hard to pour.) As long as the fluid isn't too viscous, doesn't move too rapidly, and doesn't flow into a narrow channel, the pressure will not change too much from place to place in the fluid.

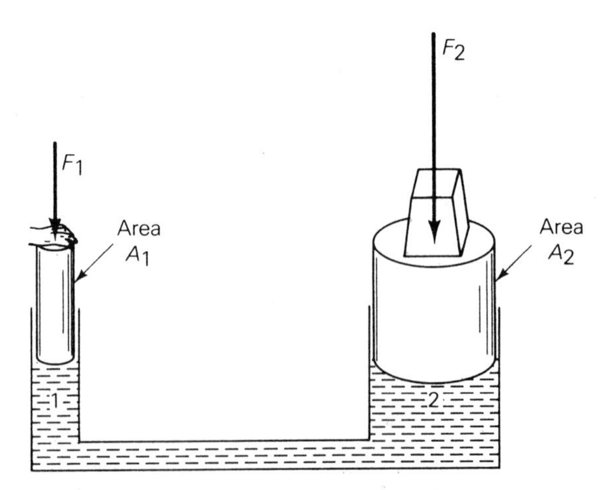

Figure 9-5 The basic hydraulic jack.

Hydraulic jacks, used for all types of heavy lifting, make use of Pascal's principle to efficiently multiply the force that we produce with our arms. Many patient lifts use hydraulic jacks, and jacks are commonly carried by emergency rescue teams to pry open smashed cars so that victims can be removed. The basic hydraulic jack is shown in Figure 9-5. The person pushes on the small piston, number 1. Often a lever is used to multiply the force of the hands further. Let the force on piston 1 be F_1 and its area be A_1. The small cylinder is connected to a much larger cylinder (number 2) with a pipe or hose. The large cylinder is placed under the object that we want to lift or wherever we want the large force to be applied. Let the area of the large piston be A_2. We want to find the force F_2 that the large piston produces. We know from Pascal's principle that the pressure is the same everywhere in the fluid. So the pressure at piston 1 is the same as that at piston 2. Since pressure is the force divided by the area,

$$P = \frac{F_1}{A_1},$$

and since the pressure is the same everywhere,

$$P = \frac{F_2}{A_2},$$

also. Therefore,

$$\frac{F_1}{A_1} = \frac{F_2}{A_2}.$$

If we rearrange this to find F_2, we find

$$F_2 = \left(\frac{A_2}{A_1}\right) F_1.$$

Thus, the force is multiplied by the ratio of the areas of the two pistons.

The hydraulic jack is a very efficient device for multiplying the force that a person can produce. The frictional losses are very small, because the hydraulic oil lubricates the pistons in addition to transmitting the pressure.

EXAMPLE The hydraulic jack shown in Figure 9-6 is the type carried by emergency rescue teams to pry open smashed cars so that victims can be removed. If the rescue worker can exert a force of 300 newtons at the handle, how large a force will the jack produce?

Answer We first note that the force on the small piston F_1 will be 10 times larger than the force on the handle, because the handle is 10 times longer than the distance from the pivot to the small cylinder. Thus, $F_1 = 3{,}000$ newtons. The area of the small cylinder is

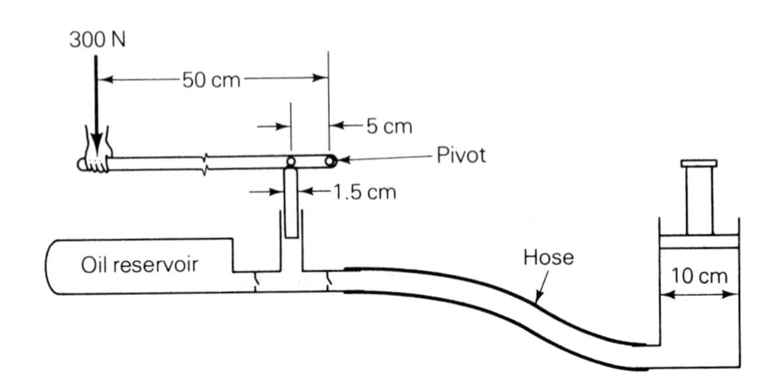

Figure 9-6

$$A_1 = \frac{\pi d_1^2}{4}$$

$$= \frac{3.14 \times (0.015 \text{ m})^2}{4}$$

$$= 0.00018 \text{ m}^2.$$

Similarly, A_2 is

$$A_2 = \frac{\pi d_2^2}{4}$$

$$= \frac{3.14 \times (0.1 \text{ m})^2}{4}$$

$$= 0.0079 \text{ m}^2.$$

Therefore, the force of the large cylinder F_2 is

$$F_2 = \left(\frac{A_2}{A_1}\right) F_1$$

$$= \frac{0.0079 \text{ m}^2}{0.00018 \text{ m}^2} \times 3{,}000 \text{ N}$$

$$= 130{,}000 \text{ N},$$

enough to lift a 13-tonne truck and more than enough to pry open a car.

There is almost no limit to the size of the force that a person can generate with a hydraulic jack. There are jacks that enable a single person to lift the largest of airplanes—but not very rapidly. The jack multiplies the force that the person produces, but it does not increase the person's power or energy. The small piston moves much farther than the large piston, so just as in the case of the lever, the work done by the large cylinder is just equal to the work done by the person pushing the small cylinder minus any frictional

Accident victims can be quickly freed with this hydraulic cutter.

losses. To prove this, consider the fact that the hydraulic fluid is incompressible; that is, when we push a given volume of fluid out of the small cylinder, an equal volume must flow into the large cylinder. If the small cylinder goes in a distance d_1 and the large cylinder goes out a distance d_2, we can say that

$$d_1A_1 = d_2A_2.$$

This equation expresses algebraically the fact that the volume pushed out from 1 is equal to the volume pushed into 2. Next, multiply both sides by F_1 to get

$$F_1d_1A_1 = F_1d_2A_2$$

and then divide by A_1 to get

$$F_1d_1 = \left(F_1 \frac{A_2}{A_1}\right) d_2.$$

From our study above of how a hydraulic jack multiplies forces, we know

$$F_2 = \frac{F_1A_2}{A_1}$$

when friction is small. That is, the quantity in parentheses is F_2, so

$$F_1d_1 = F_2d_2.$$

As is always the case with unpowered tools, the force multiplication is gained at the expense of moving a greater distance.

Hydraulic automobile brakes work in the same way that hydraulic jacks do. When we push on the brake pedal, we generate a pressure in a small cylinder that is transmitted through the pipes and hoses to large cylinders that operate the brakes on the wheels. One of the big advantages of hydraulic brakes is that the braking force on both sides of the car is the same because the pressure in the brake fluid is the same on both sides, in accordance with Pascal's principle.

Pressures Due to the Weight of a Fluid

The weight of a fluid produces a pressure that simply holds up the fluid. This pressure will always be in addition to any pressures caused by pistons or other outside means. The pressure due to the weight increases as we go down into the fluid because the deeper we go, the more fluid there is above us to support. This increase with depth occurs in both gases and liquids, but it is not easy to calculate in gases because the density of a gas changes with pressure. In liquids, density hardly changes with pressure increases because liquids are so hard to compress. As a result, a liquid's pressure at a particular depth is easy to find.

Consider a container with straight sides and a bottom of area A. The container is filled to a height h with a liquid of density D (Figure 9-7). Because the walls cannot exert an upward force on the liquid, the weight of the liquid must be completely supported by the bottom. So the force of the liquid on the bottom is just the weight of the liquid. The mass of the liquid is the density of the liquid times its volume,

$$m = DV,$$

and the volume is the height times the area, so

$$m = DhA.$$

Thus, the weight of the liquid, which equals the force on the bottom, is

$$W = F$$
$$= DhAg.$$

Again we find the pressure by dividing the force by the area,

$$P = \frac{W}{A}$$
$$= \frac{Dh\not{A}g}{\not{A}}$$

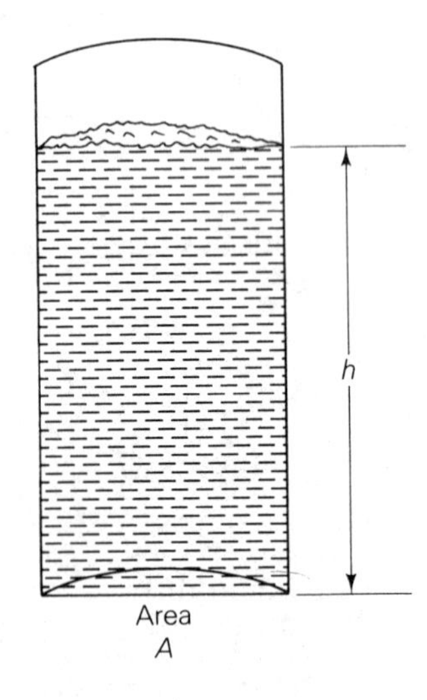

Figure 9-7

$$= Dhg.$$

This relationship holds true for any container shape.

EXAMPLE A scuba diver is working 30 meters below the surface of the ocean. What is the water pressure there, and how does it compare with the normal atmospheric pressure of 100 kilopascals?

Answer The density of seawater is 1,030 kg/m^3, so the water pressure 30 meters below the surface is

$$P = Dhg$$
$$= 1{,}030 \text{ kg/m}^3 \times 30 \text{ m} \times 9.8 \text{ m/s}^2$$
$$= 303{,}000 \text{ Pa},$$

more than 3 times normal atmospheric presure.

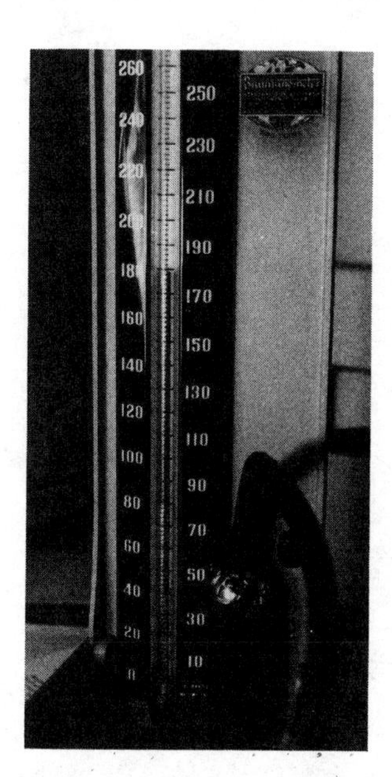

A sphygmomanometer

As we mentioned earlier, pressure is often measured by seeing how far it will lift a liquid in a tube. Such a pressure gauge, called a *U-tube manometer,* is shown in Figure 9-8. This gauge consists of a glass tube bent into a U that is partially filled with a liquid, such as mercury or water. The *sphygmomanometer* used to measure blood pressure uses a mercury manometer. While U-tube manometers are still widely used, they are cumbersome, easily spilled, and easily broken. Spillage and breakage are serious problems when the manometer is filled with mercury, because mercury vapors are extremely toxic.*

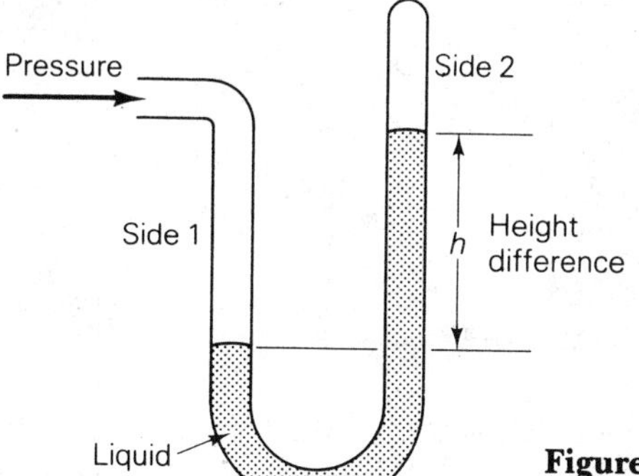

Figure 9-8 A U-tube manometer.

*The Mad Hatter in *Alice in Wonderland* and the phrase "mad as a hatter" both stem from the wide use of mercury compounds by eighteenth- and nineteenth-century hatters to make felt from fur. Hatters often became insane from mercury poisoning. Mercury evaporates fast enough that the air near any exposed mercury is a serious health hazard. Mercury must be stored in tight containers, and any spilled mercury must be quickly cleaned up by a person trained in mercury decontamination procedures.

The difference in the liquid height h on the two sides of the manometer is directly proportional to the pressure difference between the two sides. For instance, if there was a vacuum on side 2 so that the pressure was zero above the liquid on that side, the pressure on side 1 would be

$$P_1 = Dgh,$$

where D is the density of the liquid in the manometer. If there was a pressure P_2 on side 2, then the height would be proportional to $P_1 - P_2$. The formula would be

$$P_1 - P_2 = Dgh.$$

When there is a vacuum above the fluid on side 2, the manometer measures the actual pressure on side 1. The actual pressure is often called the *absolute pressure.* Gauges and manometers that measure absolute pressure are called absolute pressure gauges (Figure 9-9). The barometer, used to measure the atmospheric pressure, is an example of an absolute pressure gauge.

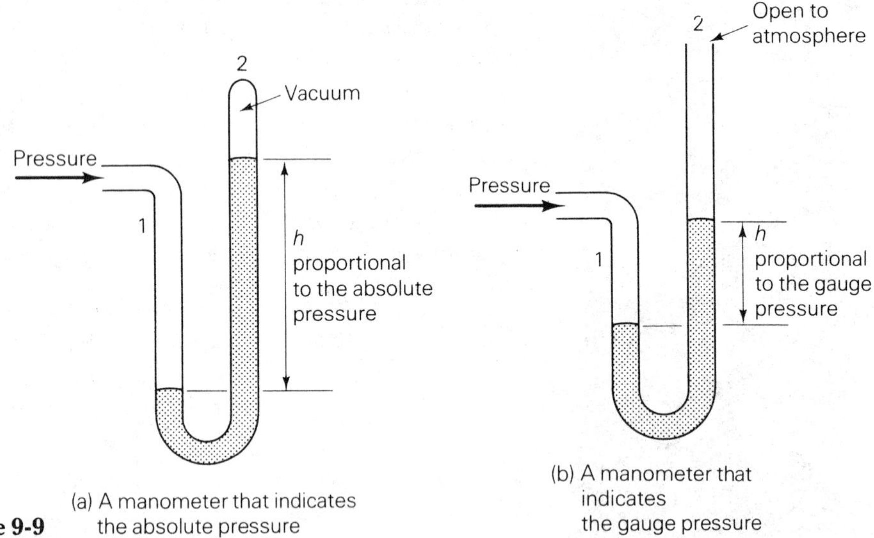

Figure 9-9 (a) A manometer that indicates the absolute pressure (b) A manometer that indicates the gauge pressure

Most manometers (and gauges) do not have a vacuum on side 2; instead, they are just left open to the atmosphere. In this case the indicated pressure is not the absolute pressure on side 1 but the difference between the pressure on side 1 and atmospheric pressure. A manometer of this type is also shown in Figure 9-9. This pressure difference is called the *gauge pressure;* expressed as a formula,

$$\text{gauge pressure} = \text{absolute pressure} - \text{atmospheric pressure}.$$

Blood pressures, most liquid pressures, and the pressures in tires are all commonly given as gauge pressures. If you need to find the absolute pressure, as you would in any calculation involving gas volumes and amounts, just add the atmospheric pressure to the gauge pressure.

For many centuries manometers were the only accurate pressure gauges, so pressures were given in such units as mm of Hg and centimeters of water (cm of H_2O), since these units could be easily read off the manometer with a ruler. These units are still used often; however, they have the serious disadvantage of being hard to use in calculations of such things as the volume of a gas or the force on the walls of a container.

EXAMPLE The pressure of the atmosphere will lift a column of mercury about 760 millimeters. How high will it lift a column of water?

Answer The pressure at the bottom of a column of liquid is

$$P = Dhg.$$

Since P is the same for both the mercury and water columns,

$$D_w h_w g = D_m h_m g.$$

Thus,

$$\begin{aligned} h_w &= \frac{D_m}{D_w} h_m \\ &= \frac{13{,}600 \text{ kg/m}^3}{1{,}000 \text{ kg/m}^3} \times 0.760 \text{ m} \\ &= 10.3 \text{ m}. \end{aligned}$$

Because the height of a water-filled manometer changes more than the height of a mercury-filled manometer, a water-filled manometer is more sensitive to small pressure changes.

Fluid Pressures in the Body

Our bodies are more than half fluid; most of the fluids are water based, but a few, like the air in our lungs and the oils in our fat cells, are not. The interiors of the cells are mostly fluid, the regions between the cells are filled with fluid, the cells are supplied with food and oxygen by fluids, and cell wastes are taken away by fluids.

The balance of the various bodily fluids is very delicate. Even small changes are quite noticeable. When a little bit of fluid collects in our feet, we quickly notice the slight swelling (edema). When we are slightly short of water, we feel thirsty long before there is any serious danger of dehydration. Any large change in the fluid balance immediately changes the body. If the body loses 10% of its water, as it can during severe vomiting or diarrhea, the pulse will weaken, the skin will lose its elasticity, pain will be common, and, if much more water is lost, the person will lose consciousness. When the fluid balance is restored, either orally or intravenously, the person will rapidly improve. This improvement is so rapid that it was regarded as a near miracle by the physicians who first treated the severe dehydration that accompanies many diseases.

Blood Pressure

Let's follow the blood as it flows through the circulatory system (Figure 9-10), which takes about 25 seconds on the average. We will start at the heart, but keep in mind that the heart is really two separate pumps: The blood flows first through one side of the

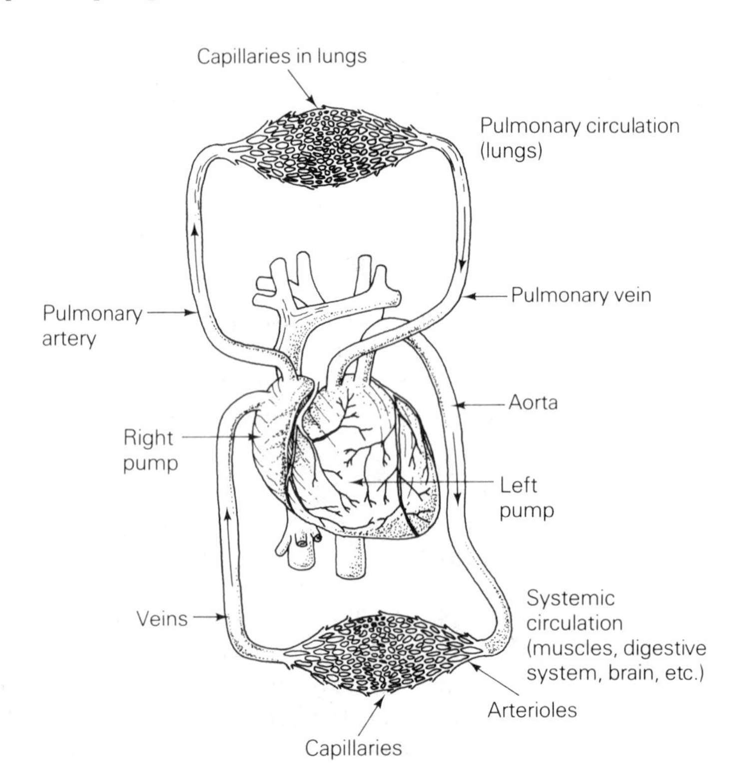

Figure 9-10 The circulatory system.

heart and later through the other side. We will start in the left side of the heart, where the freshly oxygenated blood arrives from the lungs. As the heart contracts, the pressure rises and the blood is forced into the main artery, the aorta, to start its trip around the body.

Since the heart pumps in beats, the pressure rises and falls in the aorta. The maximum, or *systolic,* pressure in the aorta occurs as the heart contracts; the minimum or *diastolic,* pressure occurs as the heart relaxes to refill with blood from the lungs. The values of the two pressures vary widely, depending on a person's activity and state of health. In a relaxed, healthy person, the systolic pressure is about 17 kilopascals (128 mm of Hg), and the diastolic pressure about 11 kilopascals (83 mm of Hg).

The aorta branches into smaller arteries, with some going to the head and the upper extremities, some to the abdomen, and others to the lower extremities. The arteries are all fairly large, so the blood flows through them easily, just as the traffic flows easily on a wide highway in the country. Because there is only a small frictional loss at the walls of the large, smooth arteries, the pressures are about the same everywhere in the arterial system. Because the pressure drops only slightly in the arteries, the pressure pulses can be felt in the wrist, neck, and other spots where the arteries are close to the surface of the body.

The smallest branches of the arteries are the arterioles. The arterioles are narrow, and there are not very many of them for the amount of blood they carry. Thus, just like a narrow road that carries more traffic than it was designed to handle, the resistance to the flow is high in the arterioles, and the major blood pressure drop occurs here as can be seen in Figure 9-11. The arterioles branch further into the microscopic capillaries, some of which are so thin

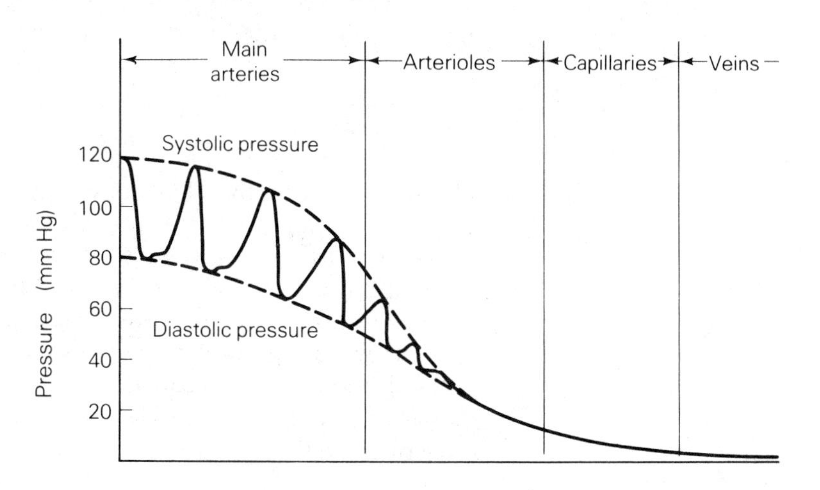

Figure 9-11 Typical circulatory system pressures.

that the blood cells pass through single file and folded over like a piece of bread wrapped around a hot dog. Although the capillaries are narrow, the pressure doesn't drop very much in them because they are so numerous. Each individual capillary carries such a small flow that the whole capillary system doesn't present much resistance to the flow. The region around the capillaries is where the greatest exchange of oxygen, carbon dioxide, food, and wastes takes place.

The capillaries rejoin to form larger and larger veins, which return the blood to the right side of the heart. However, when the blood gets to the veins the pressure has dropped so much that the blood needs help to get back to the heart. For this reason, the entire venous system functions as a pump. Most of the veins have valves that allow the blood to flow only one way, toward the heart (Figure 9-12). Whenever the muscles around the vein contract, they squeeze the blood into the next section of the vein, gradually pumping it back toward the heart. If there is little or no muscle activity, the blood will pool up in the veins and return much more slowly. This is why people whose jobs require them to stand without moving develop varicose veins and other circulatory problems more often than active people do. This is also why maintaining proper circulation in bed-ridden patients can be difficult.

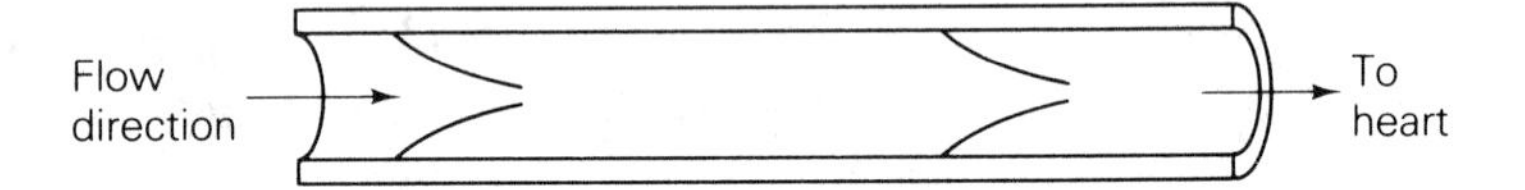

Figure 9-12 One-way valves in the veins.

The blood returns from the veins to the right side of the heart, which pumps it through the pulmonary artery into the lungs. There the waste carbon dioxide leaves the blood and fresh oxygen is absorbed. The blood then returns through the pulmonary veins to the left side of the heart to start a new cycle.

Vessel Size

The thickness of a blood vessel wall depends on both the pressure in the vessel and the diameter of the vessel. The arteries must have thicker walls than the veins to withstand the much greater pressures. But the aorta has a thicker wall than the smaller arteries, not because the pressure is greater, but because a small-diameter tube can withstand a greater pressure than a large tube with the same wall thickness. The reason is that a small-diameter tube has less wall area than a big one; thus, the same pressure will produce less force on its walls.

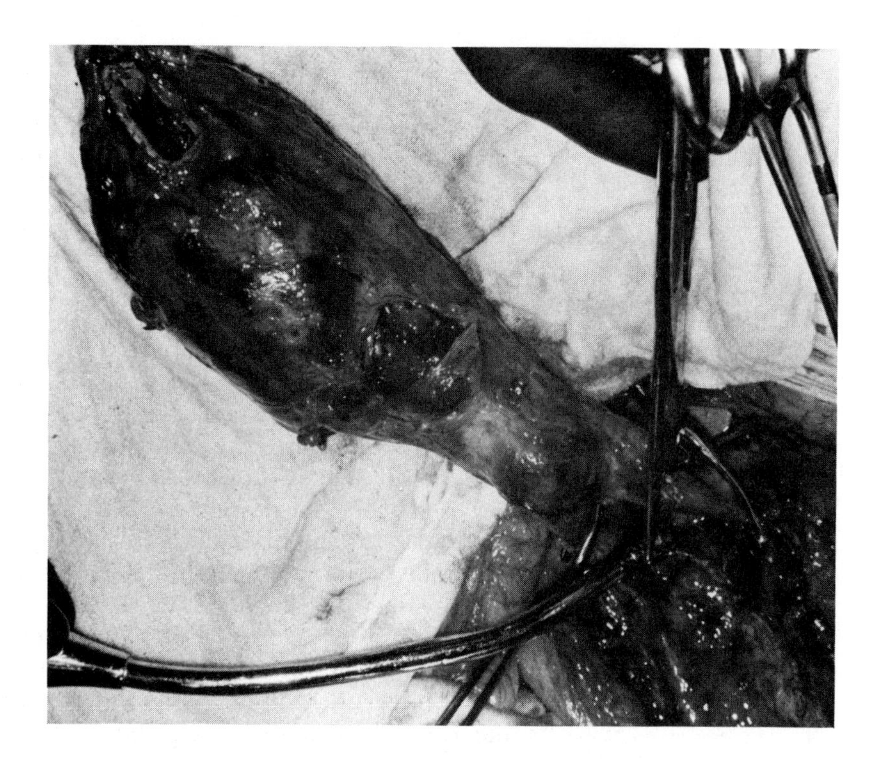

The walls of the bulged-out aneurysm are under great stress and could rupture without notice.

Any disease that weakens the arterial walls is extremely dangerous. Often a weakened artery will bulge out just like an old, worn-out hose does. The bulge is called an *aneurysm.* An aneurysm is very dangerous because as an artery expands, its wall becomes thinner. This thinner wall is then under greater stress, both because the wall is thinner and because the diameter of the artery is larger.

Infection and Pressure

The swelling and inflammation that often accompany an infection are caused by an increase in the fluid pressure between the cells. The infected cells release a chemical that causes the capillaries to enlarge (dilate). This speeds healing by increasing the blood supply to the infected region, but it also increases the pressure of the blood in that region by reducing friction at the capillary walls. The higher pressures cause the capillary walls to leak more fluid and infection-fighting materials. Consequently, the pressure rises throughout the infected region, resulting in swelling, pain, and inflammation.

Blood Pressure Measurement

The sphygmomanometer used to measure arterial pressure consists of an inflatable air cuff connected to a hand air pump and a pressure gauge. The cuff is wrapped around the patient's upper arm, air is pumped into the cuff, and the pressure is measured on the attached gauge. A stethoscope is used to listen for flow sounds in the brachial arteries below the cuff. The pressure in the cuff is raised until it stops all blood circulation in the arm. When the flow is cut off, all flow sounds stop, too, and you will hear no sounds in the stethoscope. The pressure is then gradually dropped until the sounds of flow just start again. This pressure is the systolic (or peak) pressure. Since the blood is only flowing for the very short time that the blood pressure is greater than the cuff pressure, the flow is in very short spurts, which sound something like sharp drumbeats. As the pressure is further lowered, the spurts become longer, because the time that the blood pressure is greater than the cuff pressure becomes longer (Figure 9-13). The sounds of the spurts are called Korotkoff sounds. When the cuff pressure drops below the lowest blood pressure (diastolic pressure), the flow is no longer interrupted and again you will hear no flow sounds. When interrupted flow sounds stop, the diastolic pressure is equal to the cuff pressure. (The flow sounds don't stop suddenly as the pressure is lowered, however; they stop gradually. Thus, it takes some practice and judgment to make accurate diastolic pressure readings.)

The venous pressure at the intake of the heart—the central venous pressure (CVP)—is much harder to measure. The one-way valves in the veins are closed most of the time, so the pressure in the arm veins is not the same as the pressure in the veins near the heart. In order to measure the CVP, either the pressure gauge must somehow be inserted directly into the heart, or a tube (catheter) must be inserted through the veins of the arm into the heart. Since the tube doesn't have any valves and has no flow through it, Pascal's principle applies and the pressure at the outside of the catheter tube is the same as the CVP. Of course, the CVP is not very big, so the outside end of the catheter must be at exactly the same height as the heart to prevent errors in pressure due to the weight of the fluid in the tube.

The CVP drops when the amount of blood is too small to maintain proper circulation; thus, the CVP is a good indicator of abnormal blood losses. Knowing the CVP is also useful when giving blood transfusions to someone who has lost a lot of blood; the CVP will rise when the proper amount of blood is back in circulation. Of course, inserting a catheter into a patient's heart is too complicated

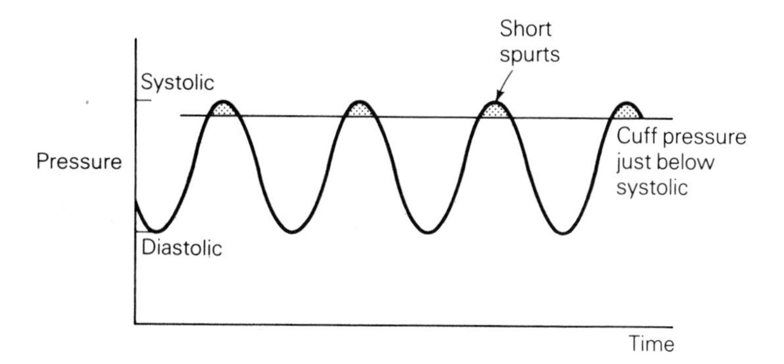

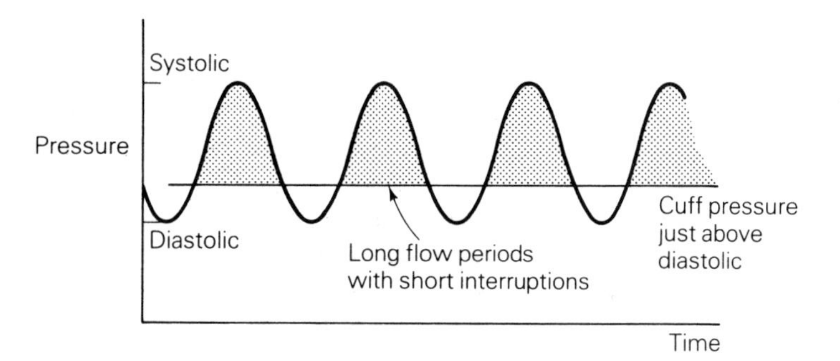

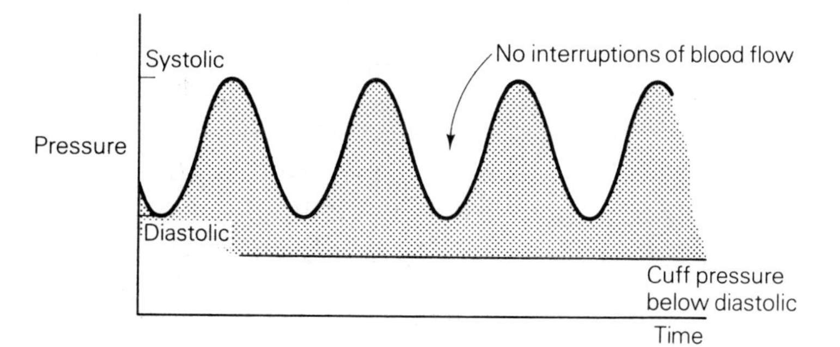

Figure 9-13

to do on a routine basis, and as with all invasive procedures,* there is some risk involved. For these reasons, the CVP is only measured when the value of the information outweighs the risks and complications involved.

---

*Invasive procedures are those involving the invasion of the body by surgical instruments or drugs.

Outside Forces and Their Effect on Circulation

A steady force on the body can easily produce a pressure that cuts off the flow of blood to that part of the body. For instance, when we sit too long on the edge of a hard chair, the pressure can stop the blood flow to the leg and the leg "falls asleep." That is, there is a tingling sensation in the leg, resulting from an inadequate blood supply. Normally, this is no problem; a healthy person will wiggle and change position as soon as the slightest discomfort is felt. But a bed-ridden or paralyzed patient might not be able to move or feel the discomfort. Of course, such patients must be moved regularly to prevent bed sores from developing.

Outside forces can also be applied to the body to control bleeding. Generally, even severe bleeding can be stopped by applying pressure with a cloth dressing or bandage. If this doesn't work, then a piece of surgical tubing—or a piece of cloth or tape in an emergency—can be used as a tourniquet to apply pressure. The tourniquet is tightened around the bleeding arm or leg and stops the flow. However, the flow cannot be stopped for a long period; this would kill all of the tissue beyond the tourniquet. The tourniquet must be loosened periodically to permit enough blood to flow to keep the arm or leg alive. Thus, tourniquets are so tricky to use that they are used in first aid work only as a last resort. In surgery, where the pressure can be carefully measured and controlled, tourniquets are useful for controlling blood flow and bleeding.

Tourniquets are also used to make veins visible so that intravenous needles can be inserted. Ordinarily, the pressure in the veins is so low that they are hard to see. But if a tourniquet is put on an arm or leg with just the right pressure—big enough to stop the blood flow out of the veins but not so big that blood can't flow in through the arteries—the veins will swell and become visible. However, if the tourniquet is put on so tight that it stops the blood from flowing in through the arteries, the arm or leg will simply turn blue.

How Worms Move

Since muscles can't push, we might wonder how boneless animals such as worms can move forward. What the worm does is to pressurize the fluids inside its body by squeezing down on one part of its body. This pressure is transmitted to the rest of the body—Pascal's principle again—causing the rest of the body to expand. By alternating the parts that contract and expand, the worm moves forward. However, this forward motion is not nearly as efficient as the backward motion, a motion that is mainly a muscular contraction. A worm can snap back into its hole 100 times faster than it can move forward—as anyone who has hunted worms knows.

Pregnancy and Pressure

The fetus is remarkably well protected by the amniotic fluid that surrounds it inside the uterus (Figure 9-14). You might wonder how a fluid can protect against outside blows. Remember that the amniotic fluid is sealed within the uterus—it cannot flow out rapidly. Suppose that the mother slips and falls against a table, as shown in Figure 9-15. As the force of the table compresses the uterus, the increased pressure generates a counterforce, stopping the intrusion of the table into the uterus. So, even though the force of the table is concentrated on the uterine wall, the pressure in the amniotic fluid is uniformly spread out over the fetus, in accordance with Pascal's principle. The fetus only feels a small, uniform pressure increase. Since this pressure increase only lasts a short time, it doesn't harm the fetus. Of course, there is a limit to the protection that the fluid can provide; a large force will injure the fetus. But the fetus can withstand quite a bit of bumping before any serious damage is done. In most accidents, the fetus is far less likely to be injured than the mother.

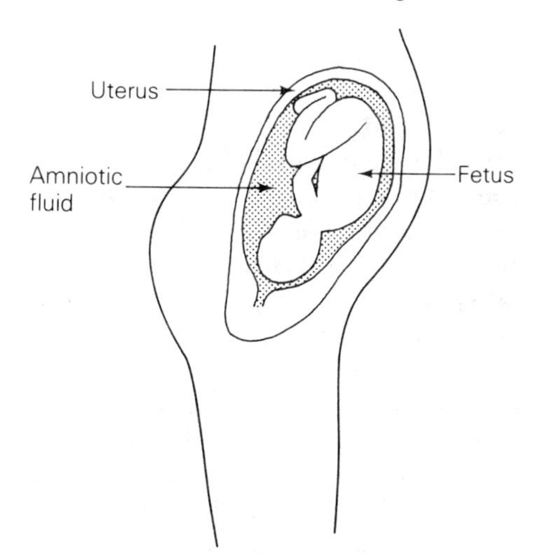

Figure 9-14 The fetus is surrounded by the amniotic fluid in the uterus.

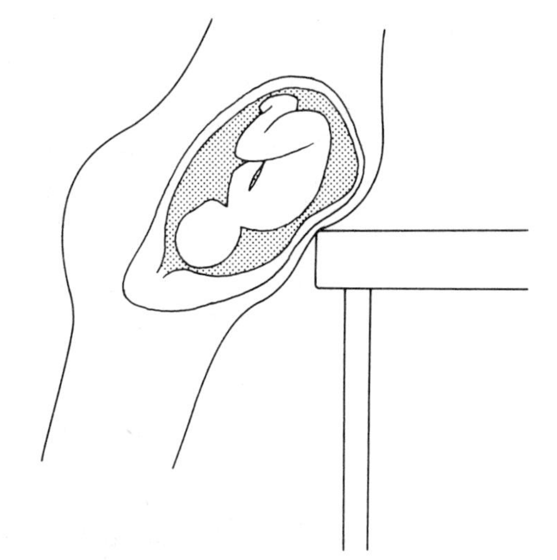

Figure 9-15 The impact of the table is distributed over the fetus by the amniotic fluid.

Although the fetus is not usually harmed by momentary pressure increases, a long-term pressure increase, even a small one, can be dangerous. The higher the pressure is inside the uterus, the harder it is for the mother's blood to flow into the uterus. Thus, the higher pressure reduces the nutrient supply, and it can also alter the fluid balance. Of course, both of these changes can seriously harm the growing fetus. For these reasons, pregnant women should not wear tight clothing. Even a tight belt or girdle can easily raise the intrauterine pressure to the danger point.

During birth, the uterine muscles contract, raising the intrauterine pressure to expel the child. The mother usually further increases the pressure in her abdomen by tightening her abdominal muscles and pushing the diaphragm muscle down into the abdomen. After the child leaves the uterus, the placenta still remains to be expelled. This is an especially critical time, because the placenta has large blood vessels that are directly connected to the mother's circulatory system. As long as the uterine muscles stay contracted, the pressure inside the uterus will be higher than the mother's blood pressure, and no further blood will flow into the uterus. When the flow stops, coagulation occurs quickly and the mother is out of danger. However, if for any reason the uterine muscles relax, nothing prevents the mother's blood from hemorrhaging into the uterus. Since it is very hard to stop this flow, the results are often fatal.

Pressure and Elimination

During the elimination of feces, we usually raise the pressure in the abdomen to speed up the process. The process is basically the same as that in childbirth—the abdominal muscles are tightened and the diaphragm pushes down on the intestines, raising the pressure in the abdomen. During normal elimination, this momentary pressure increase has no ill effects. However, when people strain and pressurize the abdomen for long periods of time, the pressure increase can cause several problems: The veins can become overpressurized and swell, hemorrhoids and rectal fissures can be aggravated, and the abdominal wall may tear (hernia). The increased pressure in the abdomen will also be transmitted through the blood, in accordance with Pascal's principle, causing difficulties elsewhere in the body. The heart, for instance, must work harder to pump against this increased pressure. For these reasons, no one should strain for a long time trying to eliminate feces. If healthy, the person should simply let nature take its course and, if necessary, the person should add more bulk to his or her diet. If a problem exists, it should be diagnosed and cured if possible, but in no case should a person go on straining during elimination.

Glaucoma

One bodily fluid pressure that should be measured regularly is the fluid pressure inside the eyeball. Occasionally, the drainage ducts from the eyeball become plugged. The pressure inside the eyeball then builds up, a condition known as *glaucoma.* Glaucoma can damage the optic nerve and lead to blindness. It is especially dangerous because much of the damage is done before noticeable symptoms appear.

The test for glaucoma is a simple measurement of the pressure inside the eyeball. The eye is anesthetized, and a rounded rod is used to depress the eyeball (Figure 9-16). The greater the pressure is inside the eyeball, the harder it will be to depress the eyeball. Another type of glaucoma tester uses a puff of air to do the same thing. In principle, both methods are the same as kicking a car tire to see if it is too hard or soft; however, the eye measurement is far more precise.

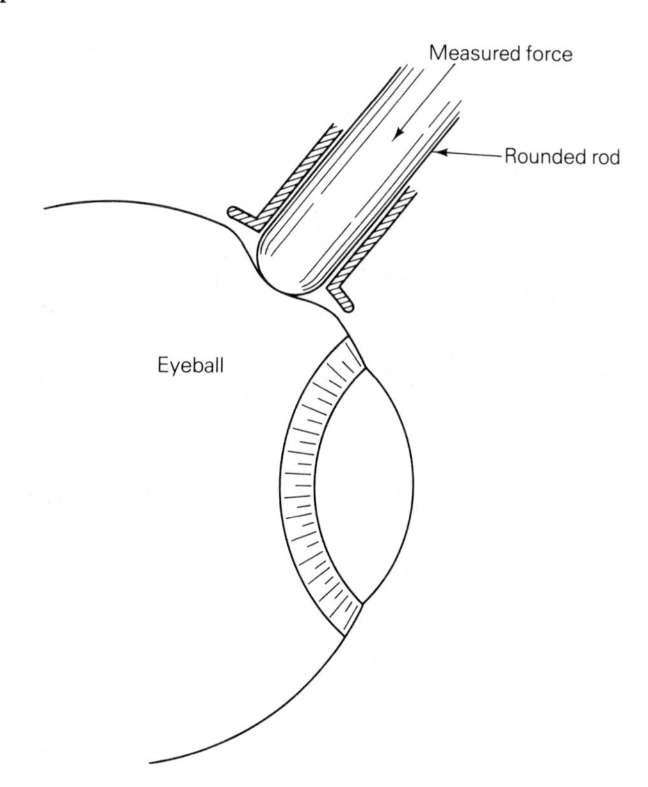

Figure 9-16 The glaucoma tester measures the fluid pressure in the eyeball by measuring the extent of the depression produced by a rounded rod when it is pressed against the eyeball with a known force.

Blood Pressure Differences Due to Gravity

The weight of the blood causes a large pressure difference between the head and the feet of a standing person. For instance, the blood pressure in the feet of a 2-meter-tall person will be greater than the pressure in the head by the following amount:

$$\begin{aligned} P &= Dhg \\ &= 1{,}050\ \text{kg/m}^3 \times 2\ \text{m} \times 9.8\ \text{m/s}^2 \\ &= 20{,}600\ \text{Pa}. \end{aligned}$$

Expressed in mm of Hg,

$$P = \frac{20{,}600\ \text{Pa}}{133\ \text{Pa/mm of Hg}} = 155\ \text{mm of Hg}.$$

This is about the same as the peak (systolic) pressure in the heart. Thus, adding the weight pressure to the pressure caused by the heart, we see that the pressure in the feet of a standing person can be twice as high as the pressure in the head.

The weight pressure is easily seen by watching the veins in the back of your hand as you raise your hand overhead from your side. When your hand is by your side, the veins are enlarged by the extra pressure, and you can easily see them under the skin. As you raise your hand overhead, the pressure drops. The veins shrink and become hard to see.

The recommended procedure for stopping bleeding makes use of the pressure difference caused by the weight of the blood: The wound should elevated as high as practical above the level of the heart. In most cases this will stop venous bleeding and reduce arterial bleeding enough so that it can be stopped with a pressure bandage. Throbbing pain, caused by the pressure of the blood acting on a nerve, can also be relieved by raising the injured part.

When we lie down, the pressure difference caused by the weight of the blood becomes very small. The heart no longer has to pump against a large pressure difference to deliver blood to the lower body, and the fluid pressures in the lower body drop. One of the main reasons why bed rest is so helpful is because the work of the heart is reduced, thus reducing the work of the intestines, lungs, and everything else that supplies food and nutrients to the heart.

The weight pressure is the direct cause of a number of afflictions, and it aggravates many others. The situation is complicated by the fact that humans have no one-way valves in the main veins (venae cavae) running along the spine. As a result, the pressure difference caused by the weight of the blood will be directly transmitted to the veins in the pelvic region. Humans don't have valves in these veins because we are descended from creatures who walked on all fours (quadrupeds). In our four-legged ancestors, the major veins ran horizontally along the spine. Thus, there was no need for any valves in those veins, because the weight of the blood didn't cause much of a pressure difference.

Hemmorhoids, the painful swollen veins in and near the rectum, are caused by the pressure difference due to the missing valves. If we still walked on all fours, our rectums would be one of the highest organs in our bodies; the pressure in the rectal veins would be very small and we would never be troubled by hemorrhoids.

Varicose veins in the legs are also aggravated by the weight pressure. Varicose veins are simply veins that have been swollen or burst by the higher pressure in the legs, and they are particularly common in people whose jobs require standing without moving. When we stand, the pressure in the legs is so large that the heart can't pump the blood back up the legs very efficiently. Proper blood flow in the legs depends on the leg muscles to squeeze the veins, which drives the blood back up. (Of course, the one-way valves in the leg veins prevent the blood from going back down again.) If a person doesn't move his or her legs much, the blood will pool in the legs, raising the pressure in the veins and causing the veins to swell. Often varicose veins will result. Any person whose job requires standing in one spot should find a way to exercise the leg muscles regularly to prevent the blood from pooling in the legs.

So much blood can pool in the legs of a person who stands motionless for a long time that the supply of blood to the head becomes insufficient, causing the person to faint. Fortunately, falling will restore the blood supply to the head, because the pressure difference will disappear when the person lies down. People who faint should not be helped up. Forcing them to sit up or stand only reduces the already insufficient blood supply to the brain. Unless some other problem exists, people who faint should remain horizontal until they recover. A person who feels faint should lie down or sit with the head between the legs to restore the blood supply to the brain.

Whenever we have any fluid buildup in our tissues (edema), we first notice it in our feet and ankles. Since the pressure is greatest there, that is where the swelling will be most noticeable. To alleviate the swelling, we should lie down and prop our feet up on a pillow. The blood pressure in the feet will drop and the swelling will usually go away rapidly.

Elastic bandages and support stockings help to counteract the effects of extra weight pressure in the leg, because they place a counterpressure on the outside of the leg, which prevents the veins from overexpanding. This is also why elastic bandages are used to prevent swelling. The extra pressure that the bandage places on the afflicted body part prevents fluids from building up there.

Pressure Changes in Diving

Because water is about 1,000 times denser than air, the pressure underwater increases with depth 1,000 times more rapidly than the pressure changes with altitude in air. For instance, let's calculate the change in depth that is needed to raise the pressure by an amount equal to atmospheric pressure, 100 kilopascals. The pressure change with depth in a liquid is

$$P = Dhg,$$

so

$$h = \frac{P}{Dg}$$

$$= \frac{100{,}000\ \text{Pa}}{1{,}000\ \text{kg/m}^3 \times 9.8\ \text{m/s}^2}$$

$$= 10.2\ \text{m}.$$

Thus, when we dive a mere 10 meters down, the pressure in our lungs goes from atmospheric pressure to twice that.

Suppose that we were using scuba (self-contained underwater breathing apparatus) tanks to breathe. The regulator automatically gives us air at the same pressure as the surrounding water, so if we breathe regularly, the volume of gas in the lungs will stay the same. Now suppose that we were 10 meters down, that our lungs were full of air at twice atmospheric pressure, and that for some reason we panicked and held our breath as we went up to the surface. Because the water pressure would drop as we went up, the air in our lungs would expand. (Remember that for a specific amount of gas, PV = constant.) When we got to atmospheric pressure at the surface, our lungs would have had to expand to *twice* their normal volume. Of course, the lungs simply couldn't expand that much—they would rupture long before we got to the surface. For this reason a scuba diver must constantly exhale when going up. A mere 2-meter rise can damage the lungs if the breath is held.

The rapid pressure increase with depth is also why goggles should never be used for scuba diving. Since the body is relatively soft, the pressure of the water is transmitted throughout the body in accordance with Pascal's principle. With a regular face mask, the eyes and nose are in the same chamber, so air can be exhaled through the nose to build up the pressure in the mask so that it

The scuba diver must breathe regularly while ascending to prevent the expanding air from damaging the lungs.

equals the pressure in the eyes (which is about the same as the water pressure). Since the pressure in the air is balanced by the pressure inside the eyes, no damage is done. However, the air pressure can't be raised in goggles, and the excess pressure in the eyeballs will rupture the blood vessels in the eye. Indeed, if the pressure gets too large, it almost blows the eyeballs out of their sockets.

Buoyancy

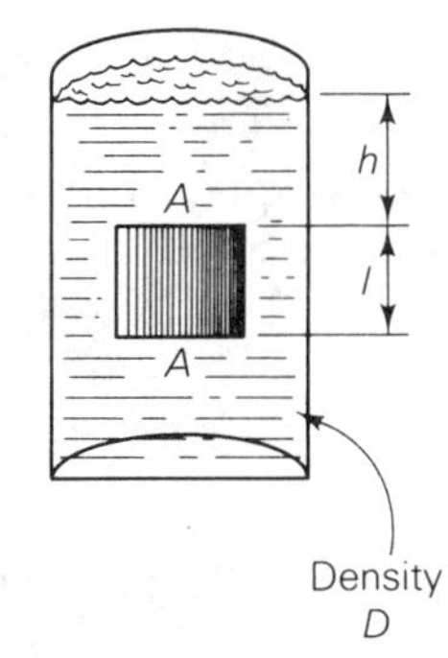

Figure 9-17

All objects seem to weigh less when they are in a fluid. For example, when we swim, we seem to weigh so little that a deep breath and a little paddling will keep us afloat indefinitely. In fact, in the case of a wood chip underwater or a helium-filled balloon in air, the buoyant force of the fluid is greater than the actual weight of the object, so the object floats upward. The apparent weight reduction—which occurs with all objects in all fluids—is a direct result of the pressure increase with increasing depth that is due to the weight of the fluid.

Let's calculate the buoyant force of a liquid on a block (Figure 9-17). Call the density of the liquid D, the vertical length of the block l, and the area of the top is A and the area of the bottom is the same. The top of the block is at a depth h below the surface of the liquid. First, we note that the horizontal force on the block must be zero, because a stationary liquid can only push perpendicularly to a surface. There is a horizontal force on each of the sides, but these four forces add up to zero because the force on one side of the block is balanced by the equal but opposite force on the opposite side. Thus, the net force of the liquid on the block must be vertical. Since there can be no vertical force on the sides, we only have to add up the forces on the top and bottom.

The upward force on the bottom can be found by multiplying the pressure at the bottom of the block by the area of the bottom. The pressure at the bottom is

$$P_b = Dg(h + l),$$

because the bottom is $(h + l)$ below the surface. Thus, the upward force on the bottom is

$$\begin{aligned} F_{\text{up}} &= P_b A \\ &= Dg(h + l)A. \end{aligned}$$

The pressure at the top of the block will be

$$P_t = Dgh,$$

so the downward force on the top will be

$$\begin{aligned} F_{\text{down}} &= P_t A \\ &= DghA. \end{aligned}$$

Thus, the net upward force of the fluid, which we shall call the buoyant force F_{buoy}, will just be the difference between F_{up} and F_{down}:

$$\begin{aligned} F_{\text{buoy}} &= F_{\text{up}} - F_{\text{down}} \\ &= DgA\,(h + l) - DghA \\ &= DgAl. \end{aligned}$$

But Al is just the volume V of the block, so

$$F_{\text{buoy}} = DgV.$$

This is exactly equal to the weight of the liquid moved aside by the block—the liquid that would occupy the volume displaced by the block. This relation was discovered by Archimedes in about 250 B.C. and is named after him.

Archimedes' Principle

The buoyant force on an object in, or floating on, a fluid is equal to the weight of the fluid displaced by the object.

For floating objects, the volume of the displaced fluid is only the volume of the object that is below the liquid level. Since a floating object is stationary, no net force acts on it. Thus, the weight of the displaced liquid is exactly equal to the weight of the floating object.

EXAMPLE An ice cube is floating in a brimful glass of water. When the ice melts, will the water level drop or stay the same, or will the water overflow (Figure 9-18)?

Answer The weight of the ice cube is exactly equal to the weight of water below the waterline of the cube. Thus, when the cube melts, the resulting water will fit into the volume displaced by the cube exactly, so the water level will stay the same.

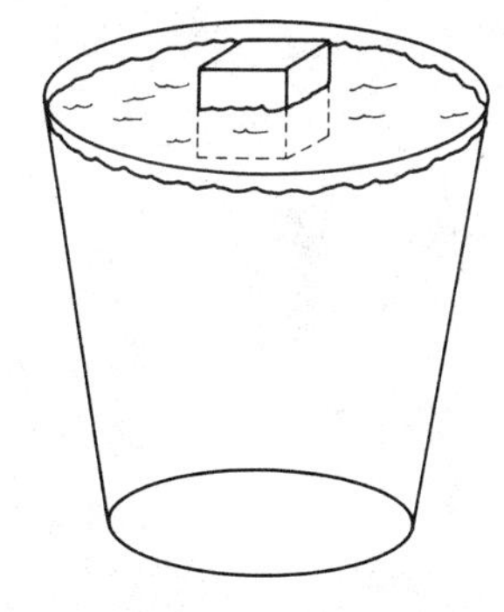

Figure 9-18 What happens to the water level in the glass when the ice cube melts?

Since the human body is more than two-thirds water, it isn't surprising that we almost float in water. That is, the buoyant force of the water is only a little smaller than our weight. Thus, humans in water can move and lift their arms and legs with very little effort. For this reason, exercise in swimming pools can be very helpful for weak patients or patients who do not have enough muscle control to exercise normally. Since there is no danger of falling and since the patients only need to lift a small part of their weight, they can achieve a much wider range of motion in the water. This increased range of motion may allow them to exercise muscles that would otherwise remain unused—muscles that would shrink (atrophy) with disuse. Since most of their weight is borne by the water, they may be able to exercise for longer periods, too.

Buoyant forces play some important roles within the body, too. For example, the brain is almost entirely supported by the watery cerebrospinal fluid that surrounds it. This buoyant support is not only very gentle but also very effective for smoothly distributing the forces of impacts. When the cerebrospinal fluid is removed to take an encephalogram (a type of brain x ray), the brain is no longer supported except where it touches the inside of the skull. The result is usually a very painful headache. Spinal taps and spinal anesthetics often cause headaches, too, because they alter the amount of cerebrospinal fluid and thus alter the buoyant support of the brain. Generally, patients who are recovering from these procedures are told to lie very still to minimize the forces on the brain.

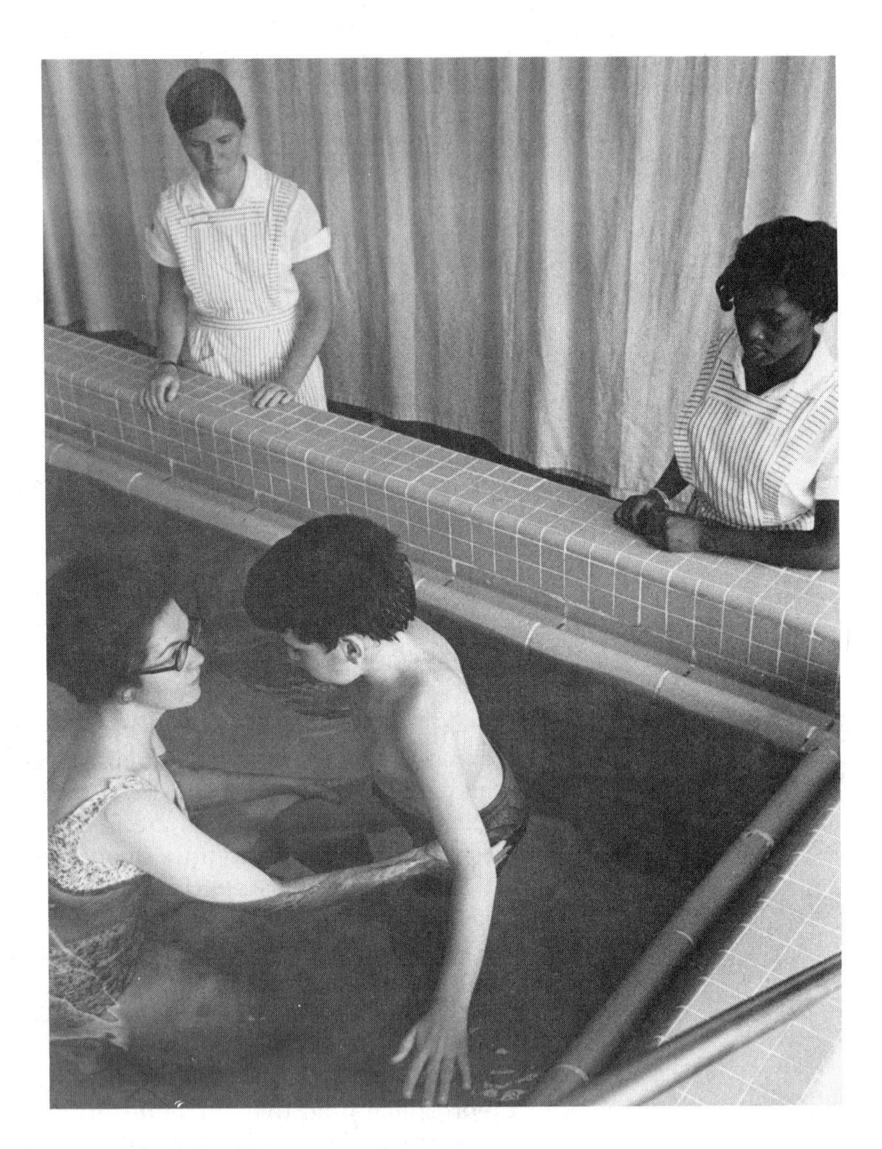

The buoyant force of the water makes exercise easier for a weak patient.

Floating and Density

What determines whether an object will float in a fluid? It clearly isn't the total weight of the object: Some very heavy things float, such as ships, but some very light things sink, such as a speck of gold dust. If an objects floats on a liquid, we know two things: One, the volume of the displaced liquid is less than the volume of the object; two, the weight of the displaced liquid is the same as the weight of the object. Let W_o, D_o, and V_0 be the weight, density, and volume of the object and W_l, D_l, and V_l be the same characteristics of the displaced liquid. We know that

$$V_o > V_l$$

and

$$W_o = W_l.$$

Since $W = mg = DVg$,

$$D_o V_o \not{g} = D_l V_l \not{g}$$

$$D_o V_o = D_l V_l$$

or

$$\frac{D_l}{D_o} = \frac{V_o}{V_l}.$$

Since V_o is greater than V_l, the relation above says that D_l must be greater than D_o. That is, in order to float in a liquid, the object must be *less dense* than the liquid. Thus, wood and oil float on water, because they are both less dense than water. Lead sinks because it is denser.

Since the specific gravity of a substance is just the ratio of its density to that of water (or air if it is a gas), it tells us immediately whether a substance will sink or float in water (or air if it is a gas). If the specific gravity of a substance is greater than 1.0, the substance is denser than water (or air for gases), and it will sink. If the specific gravity is less than 1.0, the substance will float.

EXAMPLE Does gasoline sink or float on water?

Answer Since the specific gravity of gasoline is 0.69, it floats on water.

EXAMPLE Why is carbon dioxide (CO_2) better for filling fire extinguishers than other noncombustible gases, such as nitrogen or helium?

Answer Carbon dioxide has a specific gravity of 1.5; thus, it sinks in air to the base of the fire, blocking off the oxygen and smothering the fire. Helium and nitrogen have specific gravities that are less than 1.0, so they float upward, away from most normal fires. (Helium would be great for fighting fires on the ceiling, but such fires are very rare.)

EXAMPLE Why are the electrical outlets in an operating room located much higher than normal?

Answer The specific gravities of most common anesthetic gases, such as cyclopropane, ether, and nitrous oxide, are greater than 1.0, so they tend to sink to the floor. Since many of these gases are explosive, a spark at floor level from an electrical outlet is more dangerous than a spark at about chest level. (Of course, all sparks are dangerous in an operating room.)

The density of a fluid is often a very useful thing to know. As we mentioned before, the density can be a good indicator of the condition of blood, urine, or other fluid. Also, such things as the sugar content of syrups, the alcohol content of liquors, the freezing point of antifreeze, and the condition of a storage battery can all be found quickly by measuring the density of the liquid.

Density Measurement

The density of a liquid can always be found by measuring the volume of a known mass of the liquid, but this is always time consuming and usually messy. An easier way is to float an object of known density on the liquid. The farther the object sticks out of the liquid, the denser the liquid is. It is a simple matter to put markings on the side of the object that will indicate the density or specific gravity directly. Such devices are called *hydrometers.* Hydrometers used for special purposes are sometimes given special names. For instance, the hydrometers used to measure the density of urine are called *urinometers.*

For special uses such as measuring the freezing point of antifreeze or the condition of a storage battery, the desired information is marked on the hydrometer instead of the density. This avoids the need to convert the density to the desired information.

The density of an irregular object, such as a human being, is hard to find directly because the volume can't be calculated. Archimedes discovered his principle while lying in a bathtub, trying to solve this problem. He wanted to measure the density of a gold crown, without harming it, to see if the gold had been alloyed with a cheaper metal. This was his solution: First, the weight W of the object is measured. Next, the object is immersed in a liquid of known density D_l, and its apparent weight W_a in the liquid is measured. The apparent weight is, of course, the difference between the real weight and the buoyant force. We know from Archimedes' principle that the buoyant force is just the weight of the displaced liquid, which is D_lVg, where V is the volume of the object. Thus, W_a is

$$W_a = W - D_lVg.$$

But we also know that the weight of the object is related to its density by the formula

$$\begin{aligned} W &= mg \\ &= DVg. \end{aligned}$$

Solving this relation for V,

$$V = \frac{W}{Dg}.$$

When we put this value for V back into the first equation, we find

$$\begin{aligned} W_a &= W - D_l g\left(\frac{W}{Dg}\right) \\ &= W\left(1 - \frac{D_l}{D}\right). \end{aligned}$$

The only unknown quantity in this equation is D, the density of the object. Solving for D, we get

$$D = D_l\left(\frac{W}{W - W_a}\right).$$

If the liquid is water, we can find the specific gravity by dividing by D_l:

$$\begin{aligned} \text{specific gravity} &= \frac{D}{D_l} \\ &= \left(\frac{W}{W - W_a}\right). \end{aligned}$$

This method is ideal for measuring the density of humans, because it only requires a scale tied to a diving board over a pool. Since a person's density changes as the chest volume changes, the density that is normally reported is the density after exhaling, that is, the density with the smallest chest volume.

Incidentally, the density of most people is so close to that of water that they sink when they exhale but float with no effort when they inhale deeply. This is an important fact in water safety; even very poor swimmers can usually be taught to stay afloat simply by breathing in deeply and then breathing with quick, shallow breaths. This works for most people; the exceptions are very muscular and very thin people, who sink because their densities are greater than the density of water, even when their chests are fully expanded.

Measuring a person's density in a "fat tank."

The approximate percentage of fat in the body is found from the density D (in SI units) by using the following formula:*

$$\% \text{ fat} = 100 \left(\frac{4{,}200}{D} - 3.81 \right) .$$

*This formula is what is called an "empirical relationship." This means that the formula seems to work well experimentally, although there is no theoretical reason to believe that the formula is exactly right. Thus, this formula is an educated guess based on several experiments in which the fat percentage was measured by other means.

Since few people are comfortable when they exhale underwater, the subject usually wears a snorkel or other breathing tube when the measurements are made. For precise measurements, small corrections must be made for the air left trapped in the lungs after exhaling and for the fact that the density of water at pool temperature is slightly less than 1,000 kg/m ^{3}.

EXAMPLE What is the percentage of fat of a person who weighs 734 newtons out of the water and 35 newtons in the water? Note: If the scale is marked in kilograms or pounds, the same formula can be used, because the weights appear in the density formula as a ratio. Thus, any conversion factors simply cancel.

Answer The density of the person is

$$D = D_l \left(\frac{W}{W - W_a}\right)$$

$$= 1{,}000 \text{ kg/m}^3 \times \left(\frac{734 \text{ N}}{734 \text{ N} - 35 \text{ N}}\right)$$

$$= 1{,}050 \text{ kg/m}^3.$$

When we put this density into the formula for percentage body fat, we get

$$\% \text{ fat} = 100 \left(\frac{4{,}200}{1{,}050} - 3.81\right)$$

$$= 19.$$

Surface Effects in Liquids

The forces of attraction between the molecules of a liquid, which are basically electrical in nature, are quite strong. So are the forces between the molecules of a liquid and the container walls. The attractive forces between the molecules of the liquid are called *cohesive* forces, while those between a liquid and the container walls are called *adhesive* forces. Cohesive and adhesive forces play an important part in such diverse things as the rising of sap in the trunk of a tree, cleaning things with detergents, and the elasticity of the lungs. While these forces play only a small part in the behavior of a tub full of water, they become very important in small containers. On a microscopic level, these intermolecular forces are the most important forces governing liquid behavior.

Surface Tension

The effects of the cohesive forces are generally grouped under the name *surface tension.* Since the molecules of a liquid strongly attract each other, they will pull themselves into the most compact shape possible in the absence of other forces. Of course, the most compact shape is a sphere, because a sphere has the smallest surface area for a given volume. Thus, a liquid drop will always pull itself into as spherical a shape as possible. Soap bubbles are spherical for the same reason. Raindrops are also nearly spherical (even though they are never drawn that way). The small droplets of a light drizzle are almost perfect spheres, but air currents flatten larger drops somewhat.

Because of the attractive forces between molecules, the surface of a liquid seems to be covered by a thin elastic membrane. Consider a molecule that happens to go slightly beyond the surface of a fluid (Figure 9-19). It will be attracted back into the surface by its

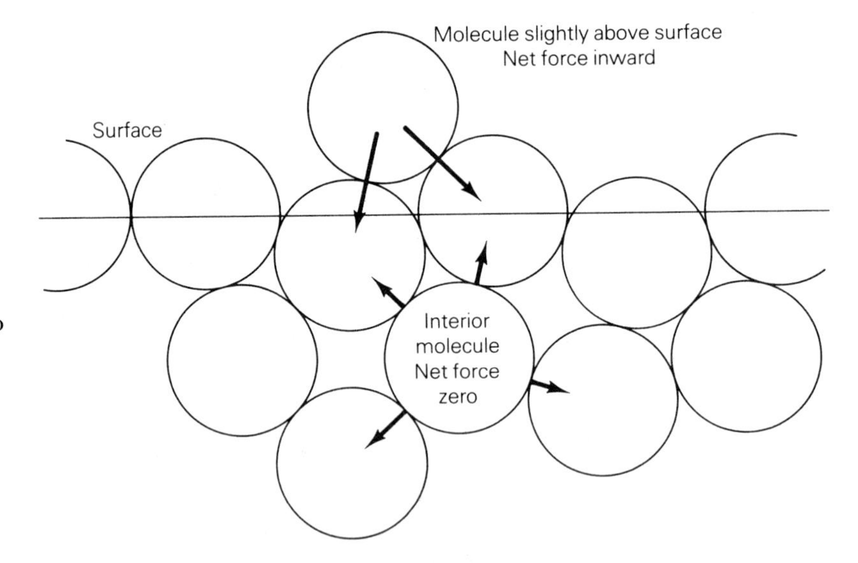

Figure 9-19 A molecule that moves outside the surface of a liquid tends to be pulled back in by the attractive forces of its neighbors. An interior molecule is pulled about the same in all directions by its neighbors and thus has no net force on it on the average.

neighbors unless it happens to be moving very rapidly. Thus, the surface acts like a membrane stretched tightly over the liquid. Although there is no actual membrane, the surface has almost all the features of one. For example, a small bug can walk across the surface of a pond if its feet don't get wet. Its feet push the surface down a bit, but the cohesive forces of the water molecules try to pull the surface flat again—thus pushing up on the bug's feet (Figure 9-20). The force isn't much, but it is big enough to support the bug. To someone watching the bug (and to the bug), a thin elastic membrane seems to cover the pond.

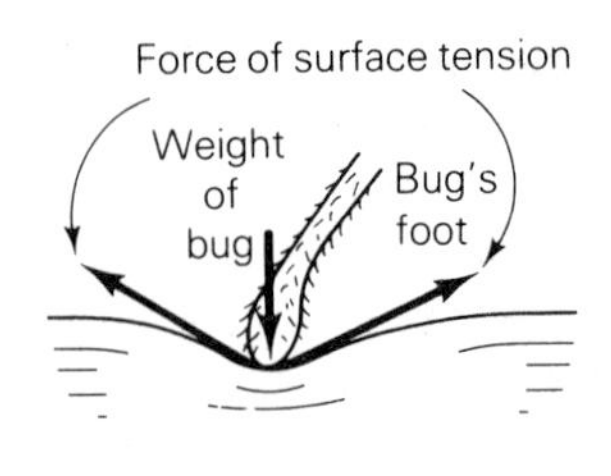

Figure 9-20

The surface tension strength s of a liquid is defined as the force per unit of length on any line lying in the surface. The line can either be the real line where the liquid surface meets a container wall or an arbitrary line on the surface. For example, the force F due to the surface tension pulling inward on a container side of length l (Figure 9-21) is

$$F = sl.$$

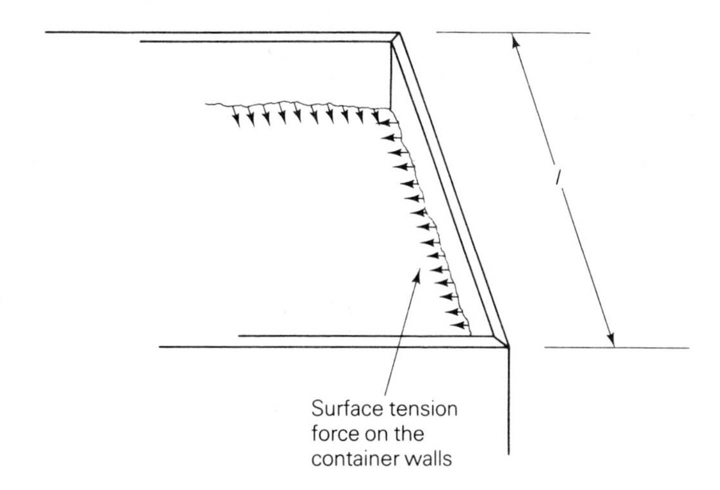

Figure 9-21

Pressure inside a Bubble

The pressure change inside a bubble as the bubble shrinks is a good example of why the intermolecular forces are very important in small objects. We will do the calculation for an imaginary cubical bubble instead of a true spherical bubble because the calculations are easier. Luckily, the answer is exactly the same for a round bubble. Look at one side of the bubble (Figure 9-22). The force of the surface tension inward must be exactly balanced by the force due to the pressure inside. The side has four edges, and there are two liquid surfaces—one on the inside of the bubble and one on the outside. Therefore, the total length of the surface along that side of the bubble is $4 \times 2 \times l = 8l$. So, the inward force to the surface tension is

$$F_{in} = 8sl,$$

where s is the strength of the surface tension. (If we were calculating the pressure inside a liquid drop instead of a bubble, F_{in} would be only half as big, $4\ sl$, because a drop has only an outside surface.) The outward force is simply the pressure in the bubble times the area of the side:

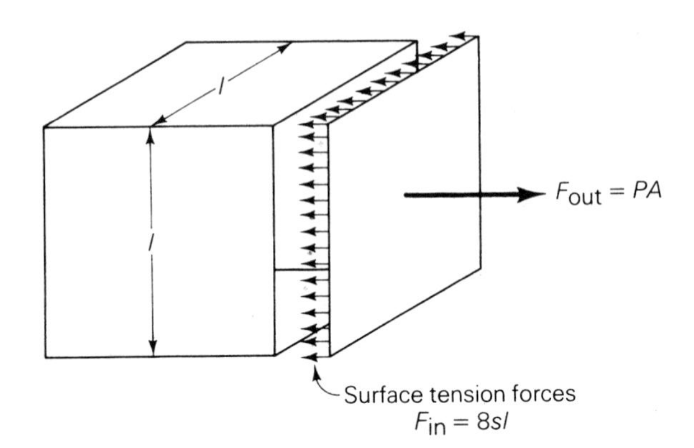

Figure 9-22 The forces on the side of an imaginary cubical bubble.

$$F_{out} = PA$$
$$= Pl^2.$$

Since the bubble is stationary, the inward and outward forces must be equal but opposite. So

$$F_{out} = F_{in},$$

or

$$Pl^2 = 8sl.$$

Thus,

$$P = 8\frac{s}{l}.$$

For a spherical bubble, the formula is exactly the same except that l is replaced by the diameter d of the bubble.

The important thing to notice is that the pressure inside a bubble is *inversely proportional* to the size of the bubble. That is, the smaller the bubble, the higher the pressure inside it. This effect is easily observable in soap bubbles. The small bubbles with high internal pressures are almost perfectly round, even when the air currents are large. But large soap bubbles are very floppy, because the internal pressure is too small to counteract the effect of air currents. This is also why balloons and bubble gum are hard to blow up at first but easy once they have been started.

Adhesive Forces and Capillary Action

The adhesive forces between the molecules of a liquid and the walls of a container (or whatever else the liquid is touching) strongly affect the behavior of the liquid. When the adhesive forces between

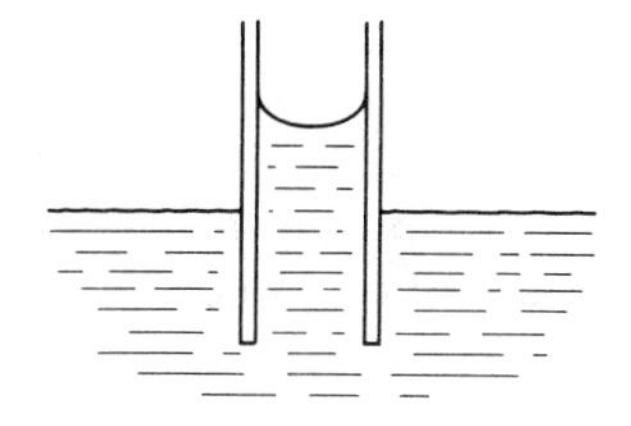

Figure 9-23 Water will be pulled up into a thin glass tube—one example of capillary action.

the liquid and the wall are greater than the cohesive forces between the molecules of the liquid, the liquid will stick more tightly to the wall than to itself. That is, the liquid wets the wall. When a liquid wets a surface, drops will spread out in a thin layeer. If the surface has any cracks, the liquid will work its way into them. This effect is easily observed by putting a thin glass tube into water. The water is drawn up into the tube by the adhesive forces (Figure 9-23); the thinner the tube is, the higher the water will rise. This effect is called *capillary action,* after the Latin name for a thin pipe.

Towels, blotters, and sponges all pick up liquids by capillary action, and a similar combination of adhesive and cohesive forces is responsible for fluid motion in plants. These forces also play a part in fluid motion in animal cells. We can see just how remarkable these forces are when we look at a giant sequoia tree standing 100 meters high. Its sap is drawn to the top entirely by its adhesive and cohesive forces.

There is another good demonstration of the strength of adhesive forces. Take two pieces of glass (microscope slides work very well). Place a drop of water between them, then squeeze out the excess water. The slides will be very difficult to pull apart, even though you can easily slide one piece over the other.

When the cohesive forces are stronger than the adhesive forces, the liquid will not wet the surface. If a liquid doesn't wet a surface, it will bead up—like the water drops on the hood of a freshly waxed car. The liquid cannot penetrate into cracks or flow down narrow channels. A towel made of such a material would be

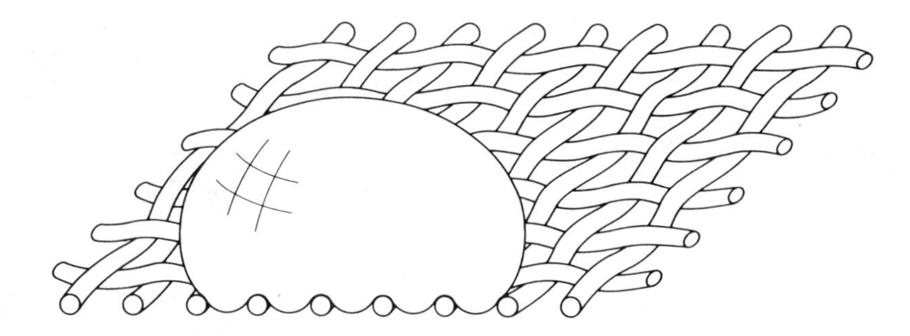

Figure 9-24 If the liquid does not wet the fibers, the surface tension will prevent the drop from flowing through the fabric.

worthless; but if we wanted to make a tent, the material would be ideal. Most waterproofing compounds work by coating the fibers of the fabric with a substance that water does not wet (Figure 9-24). These compounds are superior to waterproofing compounds that just plug up every hole in the fabric because the holes allow air to circulate on dry days.

Surfactants

Certain chemicals drastically change the adhesive and cohesive forces in a liquid. Soap is a good example. You can see the large, rapid change that soap causes by sprinkling pepper on top of water in a dish and then putting a single drop of soap in the middle of the dish. The pepper will quickly be pulled to the edge of the dish as the soap abruptly reduces the cohesive force of the water in the middle. The technical name for a material that alters the intermolecular forces, and thus the surface tension, is a *surfactant*. The name is short for "surface-active agent."

All detergents are surfactants. Detergents reduce the cohesive forces and increase the adhesive forces with most types of dirt, particularly with greases and oils. Thus, a detergent allows the water to flow into cracks more easily and to penetrate deeper into a fabric. The detergent, by acting as a kind of glue between the water and the dirt particles, also helps the water to carry the dirt away. Many disinfectants have a detergent mixed in to help them reach and destroy bacteria.

An oil slick on water is another surfactant—usually an undesirable one. It allows water to wet the feathers of water birds, thus making the birds sink. Insects that walk on the water fall through the surface and sink, too. Not only does the oil layer affect life on the surface, but it also blocks oxygen transfer at the surface, injuring all the underwater animals.

Dissolved substances, proteins in particular, alter the intermolecular forces considerably. Thus, it is not surprising that the surface tension and capillary forces of most bodily fluids are different from the forces of pure water, even though the fluids are mostly composed of water.

Surface Tension and Breathing

The cohesive and adhesive forces play a major role in respiration. Inside the lungs, the air tubes (bronchi and bronchioles) end in millions of little bubblelike sacs called *alveoli* (Figure 9-25), where oxygen and carbon dioxide are exchanged. The alveoli are coated with a layer of water mixed with a surfactant that controls the surface tension. This surface tension tends to contract the alveoli, just as the tension of a rubber balloon tends to blow the air out of the balloon. This surface tension is also one of the major sources of elasticity in the lungs. When the chest muscles are relaxed to allow the lungs to contract, this elasticity actually blows the air out the lungs.

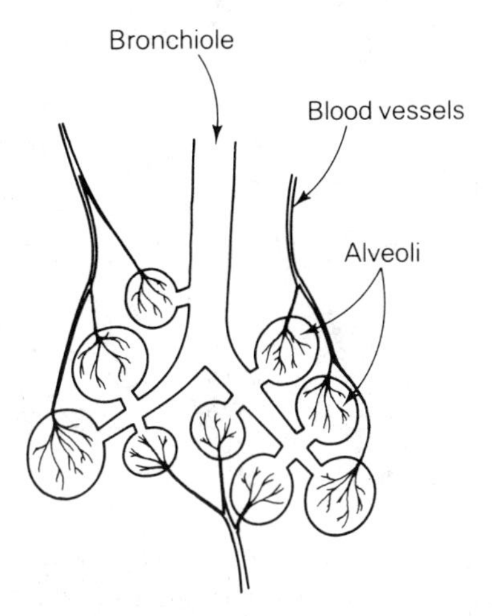

Figure 9-25 Alveoli and their air and blood supply.

The surfactant is necessary because the surface tension of pure water is too strong. If there were no surfactant, the alveoli would be very difficult to expand. Some newborns, premature ones especially, don't have the proper amount of surfactant, which causes a serious breathing problem known as hyaline membrane disease, or respiratory distress syndrome. A similar problem can be caused by too much oxygen, because pure oxygen alters the surfactant in the alveoli.

Recall how the pressure in a bubble changed with bubble size: The smaller the bubble, the higher the pressure. This is why the first breath of a newborn is so crucial. Because the newborn's alveoli are collapsed, it takes a high pressure and a great deal of effort to expand them against surface tension for the first time. This situation is exactly the same as blowing up a balloon for the first time; it takes a lot of pressure to expand a small bubble. Usually the trauma of birth is more than enough to start the infant breathing, but the delivery team must always be ready to help the occasional baby who can't quite get started breathing on his or her own.

Emphysema involves the opposite effect. The alveoli degenerate, and many small alveoli join to form a few large ones (Figure 9-26). Since the surface tension produces a much smaller pressure in a large bubble than in a small one, the enlarged alveoli have very little elasticity. Thus, they don't contract much when the chest is relaxed. When people with emphysema try to exhale, the air comes out with so little force that they sometimes can't even blow out a match. To make matters worse, a few enlarged alveoli have less surface area than many little ones do, so there is much less surface through which the gases can enter and leave. As a result, the emphysema victim has a very hard time breathing.

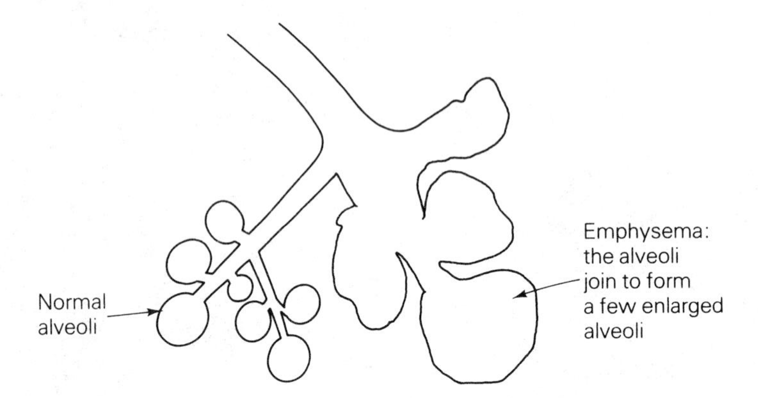

Figure 9-26

Surface tension is also important on the outside of the lungs, since the lungs are not directly attached to the chest wall. Instead, the small gap between the lungs and the chest (intrapleural region) is filled with a fluid, the intrapleural fluid, that performs two functions (Figure 9-27). The first is to lubricate the lungs so that they can slide on the chest wall without irritation. The second function isn't as obvious, but it is just as important. The intrapleural fluid holds the lungs to the chest wall in the same way that a drop of water holds two glass plates together. The fluid actually pulls the lungs outward as we inhale, thus expanding them far more effectively than the air pressure on the inside could do alone. When the intrapleural fluid fails to do this job for one of the lungs, that lung separates from the chest wall and air enters the intrapleural region. This condition is called *pneumothorax.* The surface tension in the alveoli then collapses the lung. Since the chest wall no longer pulls outward on the lung, it doesn't expand much during inhalation, and respiration becomes very inefficient.

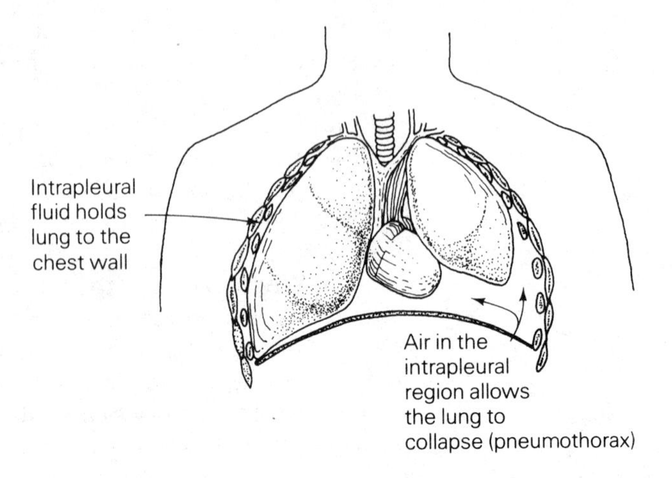

Figure 9-27

Concepts

Fluid
Solid
Liquid
Gas
Intermolecular force
Density
Specific gravity
Pressure
Hyperbaric chamber
Pascal's principle
Hydraulic jack
Manometer
Sphygmomanometer
Absolute pressure
Gauge pressure
Systolic pressure
Diastolic pressure
Aneurysm
Korotkoff sounds
Central venous pressure (CVP)
Buoyancy
Archimedes' principle
Hydrometer
Surface tension
Cohesive forces
Adhesive forces
Capillary action
Surfactant
Alveoli
Emphysema
Pneumothorax

Discussion Questions

1. What are the main features of the solid, liquid, and gaseous states of matter?
2. Explain why gases are easy to compress. Why are liquids and solids very hard to compress?
3. How is the density of an object different from its mass?
4. What is meant by the specific gravity of a substance?
5. How can you create a large pressure with a small force? Give one or two actual examples of how this can happen.
6. Explain why tracked vehicles like bulldozers and tanks go through mud so much better than ordinary wheeled vehicles.
7. Why are patients with low blood counts sometimes placed in hyperbaric chambers? How can a hyperbaric operating room reduce the surgical risk for such a patient?
8. Describe the pressure changes that occur in the lungs as we breathe.
9. Explain why it is easier to stop bleeding when the wound is positioned so that it is the highest spot on the body.
10. Explain the difference between absolute pressure and gauge pressure. Is a blood pressure measurement a measurement of absolute pressure or gauge pressure?

11. During blood pressure measurements, why are the Korotkoff sounds only heard when the pressure in the inflatable cuff is in between the systolic and the diastolic blood pressures?

12. How does the amniotic fluid in the uterus protect the fetus from injury?

13. Why are swollen ankles and feet far more common than swollen wrists and hands?

14. Why must a scuba diver not hold his or her breath when changing depth?

15. Explain why exercise in a swimming pool is often more beneficial than regular exercise for many weak patients.

16. How is the fat content of the body related to its density?

17. Explain why surfactants are often added to disinfectants.

18. Why do the lung alveoli lose their elasticity when the alveoli enlarge during emphysema?

Problems

1. A steel bar is 4 meters long, 15 centimeters wide, and 2 centimeters thick. What is its mass and how much does it weigh?

2. The average automobile gas tank holds about 60 liters. How many kilograms of gasoline does it hold?

3. If 37 cm^3 of a liquid have a mass of 46 grams, what is the density and specific gravity of that liquid?

4. If 3 liters of a liquid have a mass of 4.5 kilograms, what is the density of the liquid?

5. How many 750-cm^3 bottles would you need to hold 7 kilograms of ethyl alcohol?

6. An empty graduated cylinder has a mass of 219 grams. When 121 cm^3 of a certain liquid are poured into the cylinder, the total mass of the liquid and the cylinder is 309 grams. What is the density of the liquid?

7. Approximately how many kilograms of air does an average car push aside in an hour at 80 km/hr?

8. What is the mass of a round copper wire 2 millimeters in diameter and 18.2 meters long?

9. What is the mass of a 1-meter-long glass tube with an outside diameter of 11 millimeters and an inside diameter of 7 millimeters?

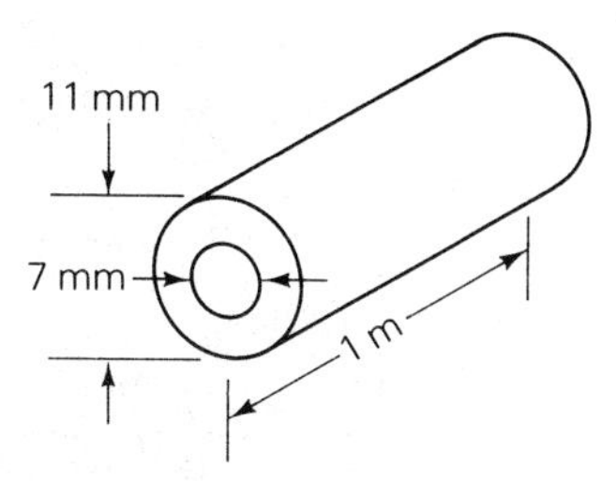

10. A rectangular bathtub 2 meters long by 72 centimeters wide is filled with seawater to a depth of 16 centimeters. Calculate the pressure at the bottom of the tub and the total force on the bottom.

11. What is the pressure under a 75-kilogram person when he is wearing shoes that are each 30 centimeters long by 6 centimeters wide? What is the pressure under that same person when he is wearing skis that are each 2 meters long by 7 centimeters wide?

12. Show that the pressure under 1 millimeter of liquid mercury is 133 pascals.

13. If a person can exert a force of up to 500 newtons on the handle of the hydraulic jack shown below, what is the largest mass that the person can lift with the jack?

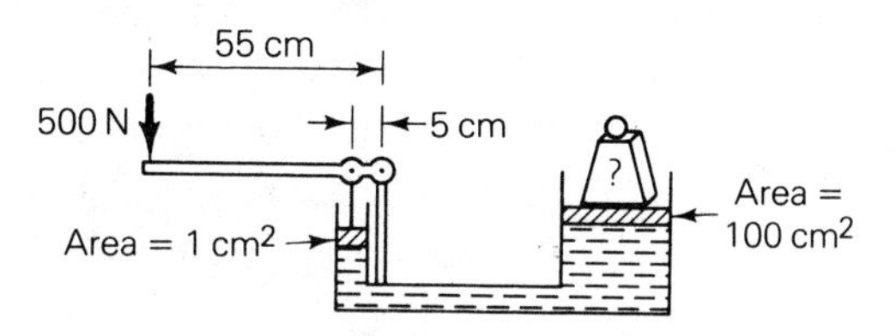

14. A sealed metal box is 51 centimeters long, 22 centimeters wide, and 9 centimeters tall. It has a mass of 10 kilograms. Does it sink or float when it is thrown into the water? If it floats, calculate the fraction of the volume underwater. If it sinks, calculate its apparent weight.

15. A person is being fed intravenously from a plastic bag of liquid through a hypodermic needle inserted into a vein. The pressure in the vein is 14 mm of Hg, and the density of the liquid is the same as the density of water. How high must the bag be raised to insure that the liquid flows in instead of the blood flowing out? If by accident the needle was inserted into an artery having a pressure of 117 mm of Hg, how high would the bag have to be raised to prevent the blood from flowing out through the hypodermic needle?

16. Surprisingly, a bed of nails made from ordinary, unsharpened construction nails is not painful to lie on if the nails are closely spaced (although no one would call such a bed comfortable). Estimate the pressure on a person's skin under a nail if the point of each nail ends

in a flat square 2 millimeters on a side and the nails are 2 centimeters apart.

17. A rock with a density of 4,000 kg/m ^{3} weighs 420 newtons in air. What would its apparent weight be underwater?

18. Scuba divers should not wear goggles because the air pressure inside cannot be equalized with the water pressure outside. (A diving mask that covers the eyes and the nose should be used. With a mask, the pressure can be easily equalized by exhaling through the nose.) Calculate the force on a round goggle lens 4 centimeters in diameter when a diver goes 50 meters underwater. Note that there is an equal but opposite force on the eye pushing it into the goggle.

19. A rectangular barge is 10 meters wide, 100 meters long, and 2 meters high, and it has a mass of 50 tonnes when empty. The barge captain wants to have at least 1 meter of the barge above the waterline when it is fully loaded. What is the maximum load that the barge can take on in fresh water? In seawater?

20. A piece of jewelry weighs 0.60 newton. When it is immersed in water, its apparent weight is 0.55 newton. What is the density of the material from which the jewelry is made? Is the jewelry made of pure gold?

21. What percentage of the volume of an iceberg is below the surface in fresh water? In seawater?

22. A certain person just barely floats in seawater after exhaling. Approximately what percentage of body fat does that person have?

23. A girl weighs 624 newtons. After exhaling underwater, she weighs 42 newtons. What is the percentage of fat in her body? Is she fat, skinny, or normal?

24. An overweight person has a mass of 98 kilograms. After exhaling underwater, a scale marked in kilograms gives the person's apparent mass as 3 kilograms. Approximately how much mass should that person lose in order to have a body fat percentage of 15%?

Experiments

1. Demonstrate that the pressure inside a small soap bubble is larger than it is inside a large one by blowing soap bubbles of various sizes. The large bubbles will be very floppy when hit by air currents because of their lower internal pressure. The very small ones will stay spherical in almost any air current.

2. The effect of soap on the surface tension of water can be seen by sprinkling pepper on the surface of a dish of water, then putting a tiny bit of any soap or detergent onto the water. Record your observations and explain why the surface behaved as it did.

3. The strength of the adhesive force of water can be demonstrated by putting a drop of water between two small pieces of glass, such as microscope slides, and then trying to pull them apart.

4. The fact that your blood pressure is lowest in the highest parts of your body and highest in the lowest parts can be observed by watching the veins in the back of your hand as you raise and lower your hand. Let your hand hang by your side for a minute or so. How do the veins look? Raise your hand above your head. How do the veins change?

CHAPTER TEN

Fluids in Motion

Educational Goals

The student's goals for Chapter 10 are to be able to:

1. Calculate the power needed to maintain a flow through a known pressure change.
2. Describe the characteristics of ideal fluid flow, laminar flow, and turbulent flow.
3. Describe the influence of the tube length, tube diameter, pressure drop, and fluid viscosity on the flow rate in laminar flow.
4. Give several examples of laminar flow and turbulent flow within the human body.
5. Explain applications of the Bernoulli effect, such as entrainment devices, curve balls, and the sails of a boat.

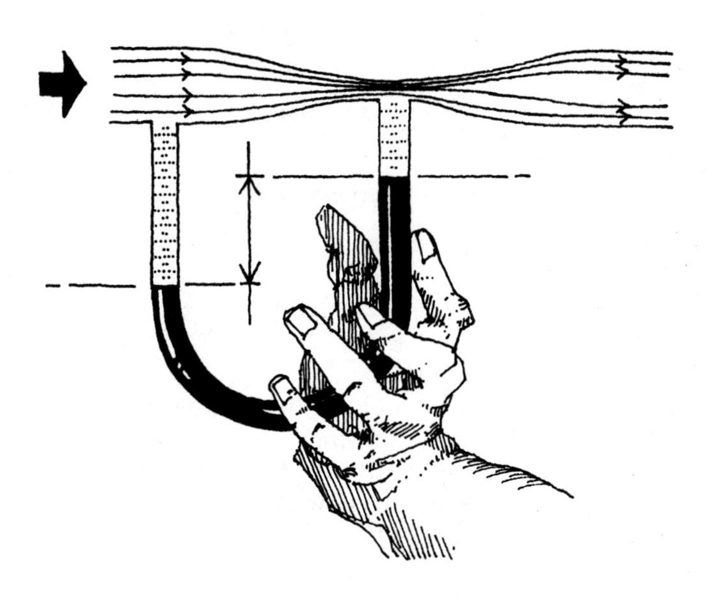

Fluid flow is essential to human life. Every part of the body, even the teeth and bones, depends on a fluid flow to deliver the food and remove the wastes. In addition, the fluids of the body also carry messages from one part of the body to another. The various glands secrete hormones into the blood, which carries the hormonal messages to the rest of the body to control a multitude of things, including growth, blood pressure, sexual drive, and sugar metabolism. Because of the many functions of fluid flow in the human body, understanding flow is a necessary part of understanding human physiology. In this chapter we will see how physics can help us in this understanding.

Flow Rate

Since fluids are normally measured in volume units, the *flow rate* of a fluid is defined as the number of volume units flowing through or out of a pipe in a unit of time. Thus, if we let the letter J stand for the flow rate, the flow rate is

$$J = \frac{V}{t},$$

where V is the volume of fluid that has passed a particular spot and t is the time it took to pass by. For example, if the flow from a pipe fills a 20-liter bucket in 50 seconds, the flow rate would be

$$\begin{aligned} J &= \frac{V}{t} \\ &= \frac{20\ \text{l}}{50\ \text{s}} \\ &= 0.4\ \text{l/s}. \end{aligned}$$

Of course, the liter is not the SI volume unit; in SI units,

$$\begin{aligned} J &= 0.4\ \text{l/s} \times \frac{\text{m}^3}{1{,}000\ \text{l}} \\ &= 0.0004\ \text{m}^3\text{/s}, \end{aligned}$$

because $1\ \text{m}^3 = 1{,}000\ \text{l}$.

The flow rate should not be confused with the velocity of the fluid. Consider water flowing through a wide hose that ends at a narrow nozzle (Figure 10-1). The flow rate through the hose is exactly the same as the flow through the nozzle. If 14 l/s pass

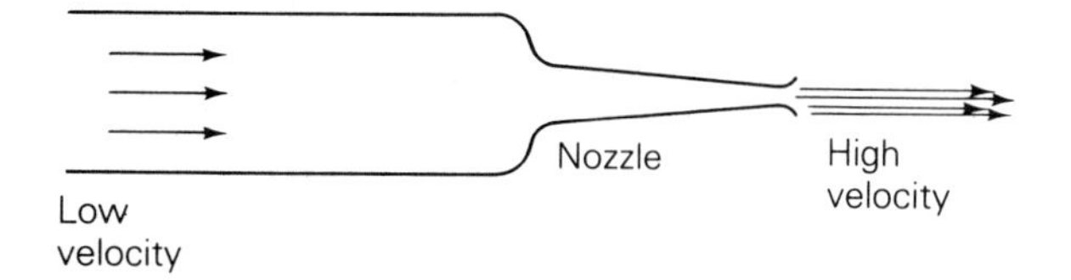

Figure 10-1 Even though the velocity of the water is higher in the narrow nozzle, the *rate* of flow is exactly the same in the hose as in the nozzle.

through the hose, 14 l/s pass through the nozzle—there is no other place for the water to go. However, the velocity of the water is higher in the narrow nozzle than it is in the wide hose.

Energy in Pressurized Fluid

When we release a pressurized fluid, the spray and noise tell us that energy has been released, too. For instance, when we open a faucet or a spray can, the fluid comes out with a high velocity and a high kinetic energy. The energy stored in the pressurized fluid is converted into the kinetic energy of the spraying fluid. Let's try to find how much energy is stored in a volume V of a fluid pressurized to a pressure P. Suppose we have a piston of area A connected to the fluid container, and we push the piston a distance d against the pressure (Figure 10-2). The energy needed to push the piston is the product of the force and the distance that the piston moves,

$$E = Fd.$$

If friction is small enough to ignore, the force on the piston is

$$F = PA,$$

so

$$E = PAd.$$

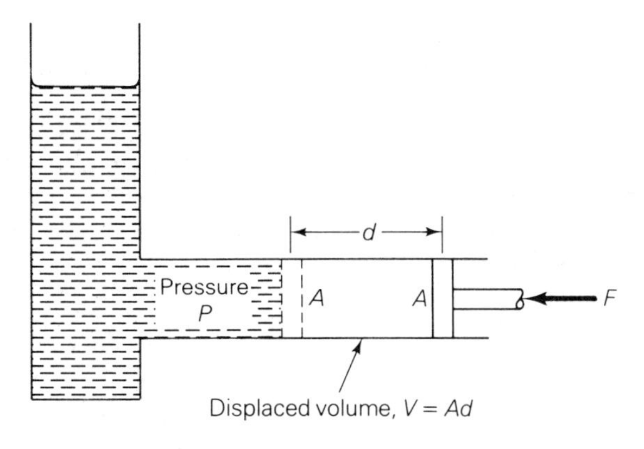

Figure 10-2

But Ad is just the volume V of the fluid displaced by the movement of the piston. Therefore, the energy required to move the piston is

$$E = PV.$$

Since energy is conserved, this energy is now stored in the pressurized fluid; that is, the energy stored in the pressurized fluid is also $E = PV$.

To find the power needed to keep a steady flow going at a rate V/t, we start from the definition of power. Power is the rate of energy production or use:

$$p = \frac{E}{t}.$$

Since $E = PV$ for the pressurized fluid,

$$\begin{aligned} p &= \frac{PV}{t} \\ &= P\left(\frac{V}{t}\right) \\ &= PJ. \end{aligned}$$

Since V/t is the flow rate J, the power needed to keep a flow going is simply the pressure times the flow rate, assuming that the pressure of the fluid drops to zero. If it doesn't, P is just the difference between the initial and final pressures.

The Power of the Heart

In a person who is lying down, the blood flow rate will be about 5 l/min, and the pressure will increase by about 100 mm of Hg as the blood is pumped through the heart. The only problem in calculating the power of the heart is converting these two numbers into SI units:

$$\begin{aligned} P &= 100 \text{ mm of Hg} \\ &= 100 \text{ mm of Hg} \times \frac{133 \text{ Pa}}{\text{mm of Hg}} \\ &= 13{,}300 \text{ Pa}. \end{aligned}$$

The flow rate in SI units will be

$$\begin{aligned} J &= 5 \text{ l/min} \\ &= 5 \times \frac{0.001 \text{ m}^3}{60 \text{ s}} \\ &= 0.000083 \text{ m}^3/\text{s}. \end{aligned}$$

Thus, the power of the heart of a resting person is about

$$\begin{aligned} p &= PJ \\ &= 13{,}300 \text{ Pa} \times 0.000083 \text{ m}^3/\text{s} \\ &= 1.1 \text{ W}. \end{aligned}$$

This doesn't sound like a lot of power until you realize that the heart must put out *at least* this much power every second of a lifetime.

Now suppose that the person stands up. This will cause a pressure increase, over the normal blood pressure, of around 150 mm of Hg due to the weight of the blood. Since the squeezing of the muscles—in the legs particularly—provides some of the extra energy, we will assume that the heart provides about two-thirds of the extra energy, about the amount associated with a pressure increase of 100 mm of Hg. In this case, the pressure increase at the heart would be about 200 mm of Hg (or 26,600 Pa). If the flow rate stays the same (5 l/min = 0.000083 m³/s), the power of the heart will be

$$\begin{aligned} p &= PJ \\ &= 26{,}600 \text{ Pa} \times 0.000083 \text{ m}^3/\text{s} \\ &= 2.2 \text{ W}. \end{aligned}$$

Thus, a simple change in position, from lying down to standing up, makes the heart double its power. As we mentioned before, this is one of the main reasons why bed rest is so important for sick patients.

The formula for finding the power of a pump, such as the heart, deserves another look, for it can tell us a lot about how we should treat our hearts. Since

$$\begin{aligned} \text{power} &= \text{pressure} \times \text{flow rate} \\ &= PJ, \end{aligned}$$

an increase in either the pressure or the flow rate will call for a proportionate increase in the power of the pump. When both factors increase, the power will increase far more rapidly than when either increases alone.

When we exercise, the body demands a greater flow of blood to deliver more food and oxygen to the muscles. But to force more blood through the body, the heart must *increase the pressure.* Since *both* the flow rate and pressure go up during exercise, the *power of the heart increases much more rapidly than the power of the other muscles.* For instance, in a poorly conditioned person, the pressure might have to be doubled in order to double the flow, which would require the heart to *quadruple* its power to double the power of the body. If the power of the body were quadrupled, the heart would have to work 16 times harder.

Fortunately, for most people the strain isn't quite so bad. Although the power of the heart does increase more rapidly than the power of the body, proper conditioning can cut down the pressure increase needed to produce a flow increase. However, people should never rapidly increase the amount of exercise that they do. All conditioning programs should stress a *gradual* buildup in the amount of exercise so that the heart and the rest of the circulatory system can have time to adjust to the change.

Fluid Flow

The rate of fluid flow in a tube is governed by several things: the pressure difference causing the flow, the length and diameter of the tube, the roughness of the walls, and the viscosity of the fluid. Many types of flow are too complicated to analyze in detail, but we can get a good idea of how these factors affect the flow rate by looking at some simple cases.

There are three basic types of fluid flow: (1) ideal fluid flow, (2) laminar flow, and (3) turbulent flow. Even though there are no perfect instances of any of the three, many common situations are close to one of the basic types.

Ideal flow occurs when the frictional losses are so small that they can be ignored. The flow is close to ideal whenever a nonviscous fluid flows out of a hole or short pipe.

In laminar flow, the friction is high enough to retard the flow, but the flow is still smooth, with no turbulence, cross-flow, or intermixing. Laminar flow occurs when there are no sharp corners or obstructions and the flow rates are not too high.

Turbulent flow has whirlpool-like vortices everywhere. The flow is not at all smooth, and the flow rate is drastically smaller than laminar flow would be for the same pressure difference. Turbulent flow occurs wherever there is a sharp bend or obstruction, but even a smooth tube can have turbulent flow at high fluid velocities.

As we will see, ideal flow and laminar flow are not too hard to analyze; turbulent flow is so hard to analyze that even the most advanced theories are not very good.

In many ways, fluid flow is similar to traffic flow. Ideal flow is like light traffic, where a few cars are traveling on a smooth road. There is little lane changing, the cars don't interact much, and the flow of traffic is at maximum speed with little energy wasted by stopping and starting. Laminar flow is similar to moderate traffic, where the cars have to slow down somewhat but there are no obstructions or drivers changing lanes unnecessarily. Even though the flow is slow, everyone gets where he or she wants to go in a reasonable time. Turbulent flow is similar to heavy rush hour traffic when it comes to a spot where four lanes suddenly narrow to two. The cars interfere with each other by making sudden lane changes, there is no smoothness to the flow, and the flow almost stops.

Ideal Fluid Flow

Since friction can be ignored in ideal flow, we can find the speed of the flow by using conservation of energy. The potential energy of pressurization will be completely converted into kinetic energy as the fluid flows out of the container.

Suppose a container has a smooth hole in it as shown in Figure 10-3. We have drawn a horizontal hole, but the direction in which the hole points doesn't make any difference. Let P be the pressure of the liquid just before it goes through the hole. Let D be the

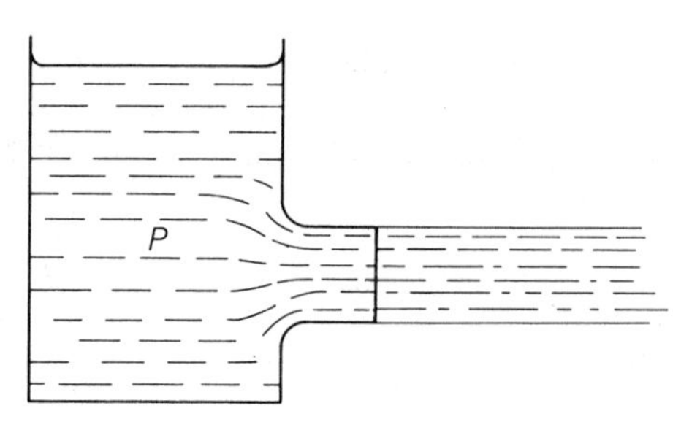

Figure 10-3

density of the liquid. Consider a drop as it passes through the hole. If its volume is V, its mass will be $m = DV$. Just before the drop goes through the hole, it has a potential energy of PV due to the pressurization of the liquid. As it passes through the hole, this potential energy is converted into kinetic energy, $\frac{1}{2}mv^2$, since there is no friction. Therefore,

$$\frac{1}{2} mv^2 = PV.$$

Substituting $m = DV$, we get

$$\frac{1}{2} D\cancel{V}v^2 = P\cancel{V}.$$

Solving for v, we find

$$v = \sqrt{\frac{2P}{D}}.$$

This relation is quite accurate for nonviscous liquids flowing through holes or very short pipes.

Once the speed of the flow is found, the flow rate is found from the following argument. Let the area of the hole be A. Imagine that we put a dye mark on the liquid as it leaves the hole (Figure 10-4). At some time t later, the dye mark would be a distance $d = vt$ away from the hole. Thus, the volume of fluid that would flow through

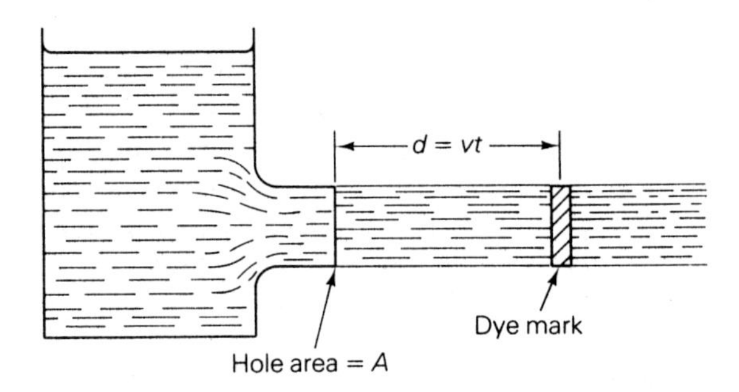

Figure 10-4

the hole after we mark the liquid would be $V = Avt$, since the volume of any uniform shape is just the length times the cross-sectional area. The rate of flow J, which is equal to V/t, would thus be

$$\begin{aligned} J &= \frac{V}{t} \\ &= \frac{Avt}{t} \\ &= Av \\ &= A\sqrt{\frac{2P}{D}}. \end{aligned}$$

EXAMPLE A person's aorta is punctured by a small hole, about 4 mm on a side (about the size of a .22-caliber bullet). Assuming that the average pressure in the aorta is about 100 mm of Hg, what is the rate of blood loss?

Answer The area of the hole will be

$$\begin{aligned} A &= 0.004 \text{ m} \times 0.004 \text{ m} \\ &= 0.000016 \text{ m}^2. \end{aligned}$$

The pressure in the aorta is

$$P = 100 \times 133 \text{ Pa}$$
$$= 13{,}300 \text{ Pa}.$$

Looking up the density of blood in Table 9-1, we find $D = 1{,}050 \text{ kg/m}^3$. Thus, the rate of flow is

$$J = 0.000016 \text{ m}^2 \times \sqrt{\frac{2 \times 13{,}300 \text{ Pa}}{1{,}050 \text{ kg/m}^3}}$$
$$= 0.000016 \times 5.0 \text{ m}^3/\text{s}$$
$$= 0.00008 \text{ m}^3/\text{s}.$$

In liters per second,

$$J = 0.08 \text{ l/s};$$

in liters per minute,

$$J = 4.8 \text{ l/m}.$$

This is a very high rate of flow, especially since the total blood volume is only about 5 liters. Of course, the actual flow would be slower, because the pressure would start to drop before all the blood poured out; nevertheless, even a small puncture of the aorta or of any other major artery is extremely dangerous and must be plugged in seconds if the victim is to survive.

The equation

$$J = Av,$$

which relates the flow rate to the fluid velocity and the cross-sectional area of the pipe, can be used to find out how liquids speed up in a narrow section of a pipe. Consider the pipe in Figure 10-5, where section 1 is wide and section 2 is narrow. Since liquids are

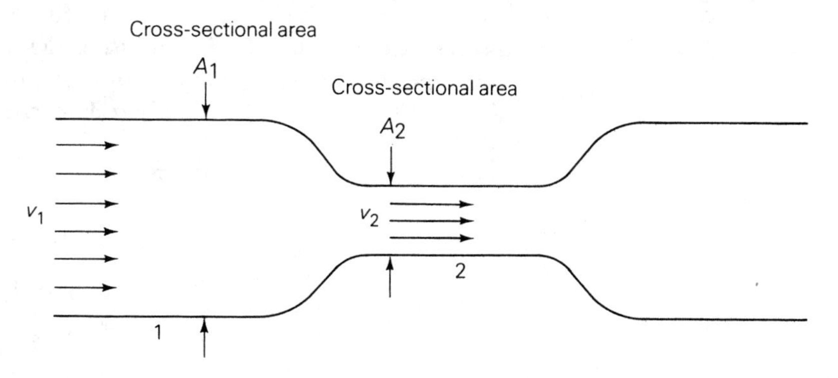

Figure 10-5

almost incompressible and there is no other place for the liquid to flow, the flow rate through both sections must be the same. Thus,

$$A_1 v_1 = A_2 v_2,$$

since

$$J = A_1 v_1 \quad \text{and} \quad J = A_2 v_2.$$

So

$$v_2 = \frac{A_1}{A_2} v_1,$$

which says that for liquids in a pipe, the velocity of the liquid is inversely proportional to the cross-sectional area. This relation doesn't work for gases, becauses gases are compressible and the volume flow rate is not the same everywhere in the pipe.

Laminar Flow

In long tubes, friction with the tube wall becomes an important factor in the flow. When friction dominates a smooth flow, the flow is said to be *laminar.* The word *laminar* means layered. We can see where the name comes from if we think of a layer of a viscous liquid, like syrup, flowing downhill. The syrup will stick to the surface of the hill, so the layer closest to the hill will hardly move at all (Figure 10-6). The next layer will stick to the first layer, so it will move only slightly faster than the first. And so it goes, with each layer moving a bit faster than the layer beneath it, and the top layer moving the fastest. To actually see laminar flow, put a spoonful of syrup on a plate and mix some pepper in it. Then tilt the plate and watch the flow.

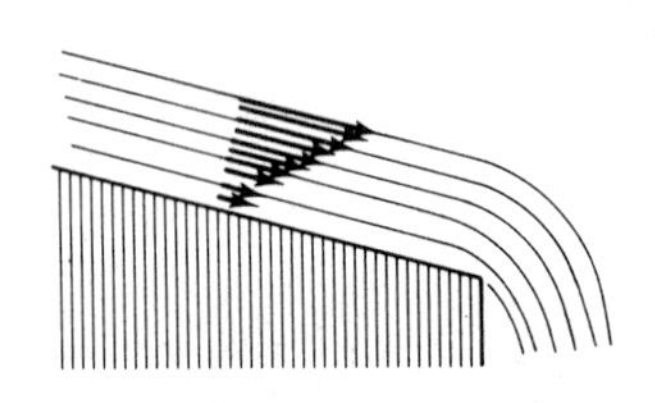

Figure 10-6 Laminar downhill flow of a viscous liquid.

Comparing different fluids, such as air, water, oil, and syrup, makes it clear that the viscosity of fluids varies widely. We won't bother with the details of how viscosities are measured; we will just give the viscosities of some common fluids to see how they vary (Table 10-1). The SI unit of viscosity turns out to be

$$\begin{aligned}\text{SI viscosity unit} &= \frac{\text{newton-second}}{\text{m}^2} \\ &= \text{N}\cdot\text{s/m}^2 \\ &= \text{pascal-second} \\ &= \text{Pa}\cdot\text{s}.\end{aligned}$$

We will use the Greek letter eta (η) to symbolize viscosity. Of course, the larger the value of the viscosity, the more viscous the fluid is and the slower it will flow. Conversely, a fluid with a lower viscosity value flows more rapidly under the same conditions.

Table 10-1 Viscosities of Some Common Fluids

Fluid	*Viscosity (in Pa · s)*
Liquids	
Acetone	0.00032
Alcohol (ethyl)	0.0012
Blood (whole)	0.004
Blood plasma	0.0015
Gasoline	0.0006
Glycerine	1.49
Mercury	0.0016
Oil (light)	0.11
Oil (heavy)	0.66
Water	0.001
Gases	
Air	0.000018
Helium	0.000019
Methane	0.000020
Nitrogen	0.000018
Oxygen	0.000020
Water vapor (steam)	0.000013

The viscosities in Table 10-1 are for room temperature (20°C) except the viscosities for blood and blood plasma, which are for normal body temperature (37°C), and the viscosity for steam, which is given at 100°C. The viscosity of most fluids changes quite rapidly with temperature, usually becoming less viscous as the temperature goes up.

Temperature changes the viscosity of blood in the normal way. This is an important factor in the treatment of a person who is in shock. Since the temperature of the extremities drops rapidly because of the reduced blood flow, the blood becomes more viscous. This reduces the flow even further, complicating the victim's condition. This is one of the reasons why accident victims should usually be kept warm (but not hot which appears to be harmful, too).

Calculating laminar flow rates is not easy because the various layers move at different speeds. For this reason, we will only look at the case of laminar flow in a circular tube. The fluid next to the wall

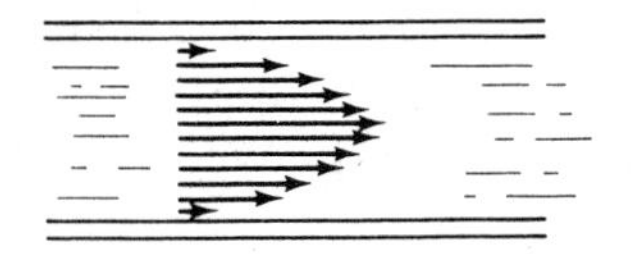

Figure 10-7 Laminar flow in a tube. The fluid at the center moves fastest.

of the tube will be nearly stationary. The fluid speed will increase as we go toward the center, where it will be fastest. The velocities at various points are shown in Figure 10-7. The edge of the flame in a Bunsen burner has exactly the same shape because the flow of the gas up the burner tube is laminar.

The calculation of the flow rate is too complicated to go through, so we will just give the result:

$$J = \frac{2\pi d^4 P}{\eta l},$$

where d is the diameter of the tube, l is the length of the tube, P is the pressure difference between the two ends of the tube, and η is the viscosity of the fluid. The formula looks complicated, but if we look at the factors separately, they affect the flow rate roughly as we would expect—with one important exception. We expect the flow to increase with pressure and to decrease with viscosity and the length of the tube, so P, η, and l appear in the formula as expected. We expect the flow to increase as the tube gets bigger, but the fact that it increases so rapidly with diameter—d to the fourth power—is surprising. This means that doubling the diameter of the tube increases the flow rate *16 times.*

Upon reflection, the d^4 isn't so surprising after all. In a very thin tube, all the fluid would be very close to the wall and would therefore flow very slowly. In a large tube, the percentage of fluid close to the wall would be much smaller, so a higher percentage of the fluid would go at a higher speed.

Turbulent Flow

Turbulent flow is caused by the momentum of the fluid. A slowly flowing fluid will flow over an obstruction, but at high speeds the fluid will "jump" off the top of the obstruction. The situation is similar to a motorcycle going over a hump in the road as shown in Figure 10-8. At low speeds the motorcycle won't leave the road—it just goes over the hump smoothly. But at high speeds the momentum of the motorcycle will carry it into the air, off the top of the hump.

In turbulent flow, the fluid just behind the obstruction tends to be nearly stagnant. Where the rapid flow meets the stagnant fluid, there will be the typical eddies of turbulent flow. In a sense, these whirlpools act like ball bearings to permit a rapid transition from the stagnant flow of the "backwater" region to the rapid flow of the fluid that jumps the obstruction. Unlike ball bearings, however, turbulent flow is very inefficient.

Because turbulent flow is so chaotic, it is difficult to analyze.

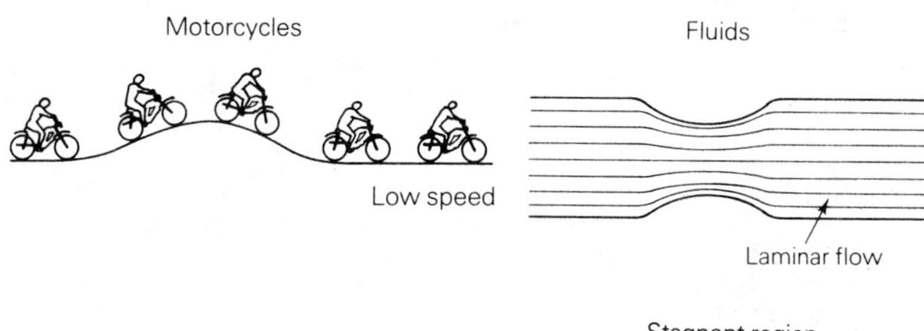

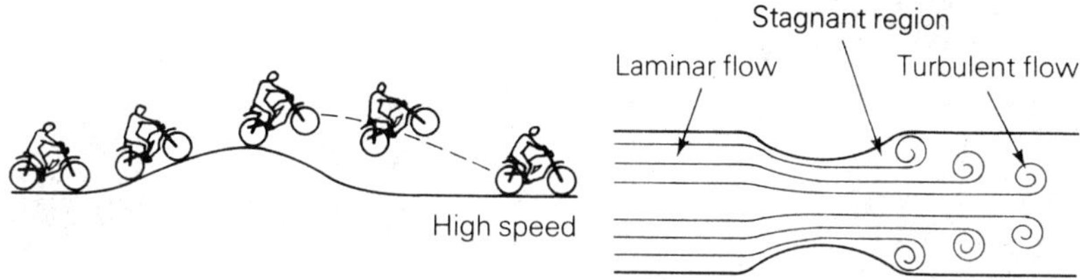

Figure 10-8 At high speeds, the momentum of fluids (and motorcycles) causes them to "jump" as they flow past an obstruction.

But some results of the turbulence are worth noting: (1) The flow rate is much lower than that for a comparable laminar flow; this is not surprising, since turbulent flow has almost as much backward motion as forward motion. (2) The fluid becomes well mixed. In fact, you must have turbulent flow to mix fluids efficiently; there is almost no mixing in either ideal or laminar flow. (3) A great deal of noise will accompany most turbulent flows. In fact, the simplest way to detect turbulent flow is to listen for noise, as we do when we use a stethoscope to hear the blood spurt past the inflated cuff of the sphygmomanometer.

Fluid Flow in the Body

All three flow types occur in the body. For instance, the flow is nearly ideal when we blow air through our lips, and turbulent flow occurs around the valves of the heart as they open and close. However, laminar flows are the most common. In fact, the formula for laminar flow rates was developed by a French physician, Poiseuille, in an effort to understand blood circulation better.

Since blood flow is nearly laminar in most of the vessels, small changes in vessel diameter can change the flow rate very effectively. In fact, changes in vessel diameter are the principal way that blood flow is controlled in the body.

EXAMPLE Blood flow is controlled mainly by the arterioles, the last small arteries before the capillaries. The arterioles are also where the major drop in blood pressure occurs. If all the arterioles in the hand contract by 10%, by what percentage will the flow of blood decrease?

Answer Before the arterioles contract, the flow through them will be

$$J_{\text{initial}} = \frac{2\pi d^4 P}{\eta l}.$$

If everything else stays the same while the diameter drops to 90% of its initial value, the new flow rate will be

$$J_{\text{new}} = \frac{2\pi (d')^4 P}{\eta l}$$

where d' is the new diameter. Since $d' = 0.90d$, the new flow rate will be

$$\begin{aligned} J_{\text{new}} &= \frac{2\pi (0.90d)^4 P}{\eta l} \\ &= (0.90)^4 \frac{2\pi d^4 P}{\eta l} \\ &= (0.90)^4 J_{\text{initial}} \\ &= 0.90 \times 0.90 \times 0.90 \times 0.90 \times J_{\text{initial}} \\ &= 0.66 J_{\text{initial}}. \end{aligned}$$

Thus, a mere 10% reduction in arteriole diameter will reduce the flow to 66% of its initial value; that is, the flow will be reduced by 34% .

Because minor reductions in vessel diameter cause major drops in the flow capacity, even small deposits of cholesterol inside the arteries are dangerous. The reduction in blood carrying capacity is far greater than the decrease in the arterial diameter. To partially restore the flow, the heart must pump at a higher pressure. Thus, high blood pressure usually goes hand in hand with arterial disease. It also makes the heart work harder, and it increases the danger of strokes from ruptured blood vessels in the brain.

The blood flow rate can also be controlled by such things as changes in the pulse rate, but these things are not as versatile in controlling flow as changes in vessel diameter are. For instance, a change in pulse rate changes the flow to all parts of the body by the same percentage, with no regard for the different needs of the different parts. A reduction in the pulse rate can impair the functioning of the brain, which must have a certain minimum blood supply to function properly; a large drop in the blood supply of the brain will cause permanent damage. Conversely, changes in vessel diameter can be done selectively, reducing the blood supply to sections that don't need a large flow and increasing the flow to the parts that need it most. For example, blood flow is diverted from the skin when we are cold, and it is sent to the intestines after a meal and to the muscles when we work hard.

About half of the blood volume of a normal person is made up of cells, which are so closely packed that they usually touch each other. Since an individual cell must move as a whole—the part of a cell closest to the wall cannot move more slowly than the part near the center—blood flow cannot be truly laminar. Nevertheless, the assumption that blood flow is laminar doesn't cause any serious errors except in evaluating the flow in the tiny capillaries. The capillaries are so small that the blood cells must pass through them single file and folded in half.

The flow sounds in the body are caused by turbulent flow. The sounds of the heart are due to the turbulent flow of blood around the valves. The Korotkoff sounds heard during blood pressure measurements are caused by the turbulent flow of blood spurting past the inflated pressure cuff. The sounds of breathing are also caused by turbulent flow. A healthy person who has no serious obstructions in the air tubes breathes quietly, but breathing can be quite noisy when the air breaks into turbulent flow around infections that obstruct the air tubes. The sounds of the voice also start with the turbulent flow of air around the vocal cords.

The turbulence around obstructions in the lungs sharply reduces the flow of oxygen and greatly increases the effort needed to breathe. This turbulence can be reduced if the patient breathes a helium-oxygen mixture instead of air, which is a nitrogen-oxygen mixture, because the lighter helium flows around obstructions with less turbulence. The use of helium-oxygen to reduce the effort of breathing is discussed in Chapter 18.

The Bernoulli Effect

The Bernoulli effect explains why the wings hold an airplane up and why baseballs, tennis balls, and golf balls curve in flight. It also explains the workings of some sprayers and suction devices. The effect is that the pressure in a moving fluid drops when the fluid moves faster. That pressure changes as the fluid picks up speed isn't surprising; what surprises most people is that the pressure drops.

The effect is easy to demonstrate. Take a piece of paper and hold it by one edge with the other edge hanging down (Figure 10-9). Now blow across the top of the paper. The paper will lift *up* because the pressure of the moving air on top is *less* than the pressure underneath the paper, where the air is stationary.

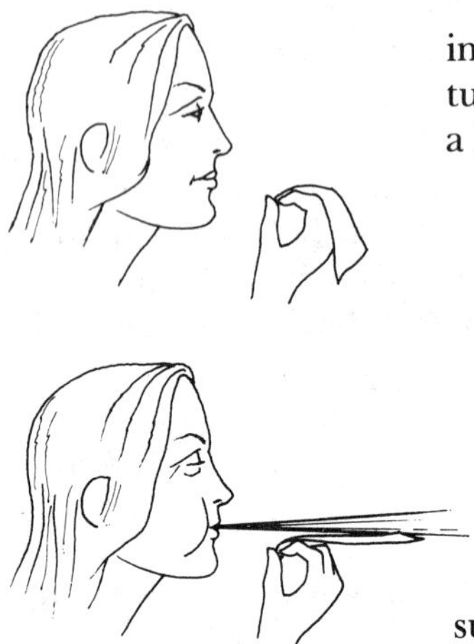

Figure 10-9 The Bernoulli effect. The pressure in the moving air above the paper is lower than it is in the stationary air below the paper. Thus, the paper rises.

This pressure drop makes sense when we look at the energy involved in the flow. Consider a liquid flowing without friction in a tube that suddenly narrows down (Figure 10-10). (We will consider a liquid because the volume of a liquid doesn't change with pres-

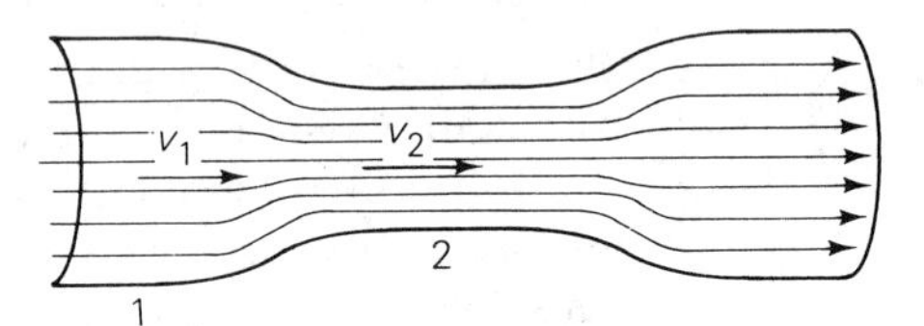

Figure 10-10

sure, unlike the volume of a gas. The basic idea still applies to a gas, but the details are more complicated.) The energy in any small part of the liquid is partly potential energy from pressurization and partly kinetic energy. At any given time, the total energy is

$$E = PV + \frac{1}{2}mv^2.$$

Since $m = DV$, we can also write the energy as

$$E = PV + \frac{1}{2}DVv^2.$$

As the liquid flows through the narrow section, the energy stays the same, because there is no friction with the walls and no energy can be supplied from the outside. We showed earlier that the velocity of a liquid in a pipe is inversely proportional to the cross-sectional area of the pipe; thus, the liquid must be moving faster in the narrow section. If so, its kinetic energy must be larger. Since the total energy stays the same, the potential energy must drop. Because the volume of the liquid doesn't change, the pressure must be less in the narrow section, and the pressure must be smallest where the liquid moves fastest. In summary, the Bernoulli effect is that as a fluid speeds up, its pressure drops.

Entrainment Devices

We often want to mix a gas or liquid spray into the flow of another gas. For instance, in oxygen therapy and anesthesia, the gas must be mixed into the air flow. In respiration therapy, liquid medications are delivered deep into the lungs by mixing a very fine spray of the medication into the air that the patient breathes. The finer the spray is, the better; because the finer the drops are, the slower they fall and the deeper they penetrate into the lungs before they settle out of the air.

The process of mixing a gas or liquid spray into the flow of another gas is called *entrainment,* and a device that breaks a liquid up into a fine spray is called a *nebulizer* or *atomizer.* In many devices, the entrainment and the nebulization take place at the same spot.

Most gas-mixing devices make use of the Bernoulli effect in the following way. The main flow tube has a narrow spot where the gas speeds up. A side tube carrying the gas or liquid to be entrained enters at the narrow spot (Figure 10-11). As the gas flows through the main tube, the pressure drops in the narrow spot due to the Bernoulli effect. The gas or liquid in the side tube is pushed into

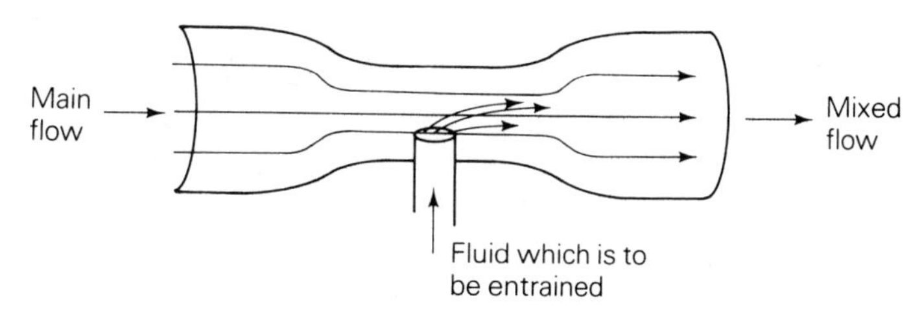

Figure 10-11 Bernoulli effect entrainment device.

the main tube by the pressure difference and mixes into the main flow. When a liquid is entrained, it breaks up into a fine spray as the main flow hits it. Some other designs for entrainment devices that use the Bernoulli effect are shown in Figure 10-12. A common nonmedical entrainment device that uses the Bernoulli effect is the automobile carburetor, which entrains the gasoline in the air going to the engine.

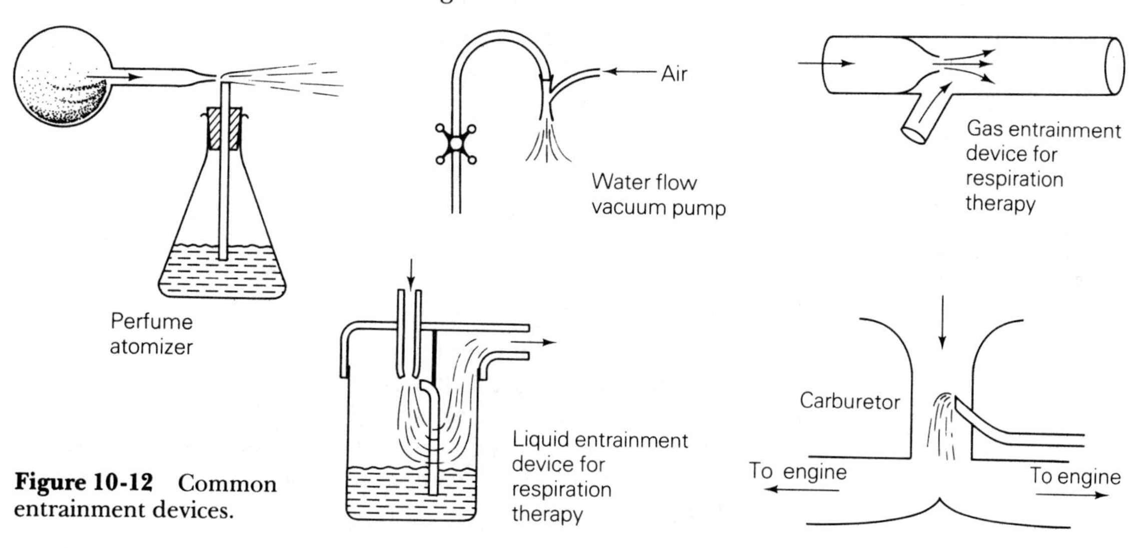

Figure 10-12 Common entrainment devices.

Airplane Wings and Boat Sails

Airplane wings, ski jumpers, and sailboats all use the Bernoulli effect. Take an airplane wing, for example. The top of the wing is curved, while the bottom is nearly flat. As shown in Figure 10-13, the air going over the top has farther to go, it must move faster than the air going under the wing. Thus, the pressure on top of the wing is less than the pressure below it, producing a net upward force on the wing that holds up the airplane. The ski jumper takes on a very similar shape to generate lift for increasing the length of the jump.

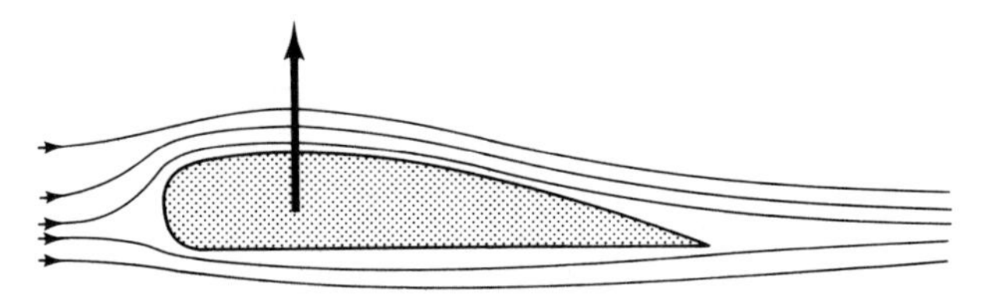

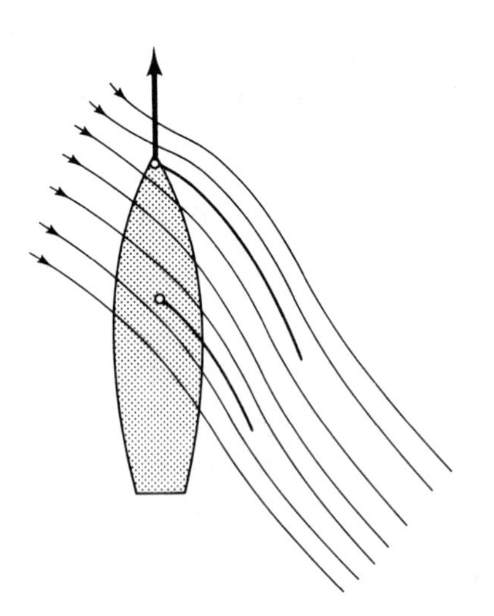

Figure 10-13 Airplane wings and sails both use the Bernoulli effect to generate the desired force.

Sails work in the same manner with one added feature to improve the efficiency. Most sailboats have two sails, which are adjusted to create a narrow channel or "slot" between them (Figure 10-13). The air flows through this slot very rapidly, thus creating a large pressure drop. This creates a larger net forward force than an single sail could generate.

The curve of the ski jumper's body helps generate lift to extend the jump.

Hooks, Slices, and Curve Balls

A spinning ball drags a thin layer of air along with it; thus, when a spinning ball is thrown through the air, the air will flow faster past one side of the ball than the other. Consider the ball in Figure 10-14, which is moving to the right and spinning counterclockwise. The thin air layer next to the bottom of the ball is moving opposite to the general air flow past the ball. Thus, the air at the bottom is moving more slowly than the air at the top, where the general flow and the thin layer move in the same direction to produce a higher net velocity. The Bernoulli effect produces a higher pressure on

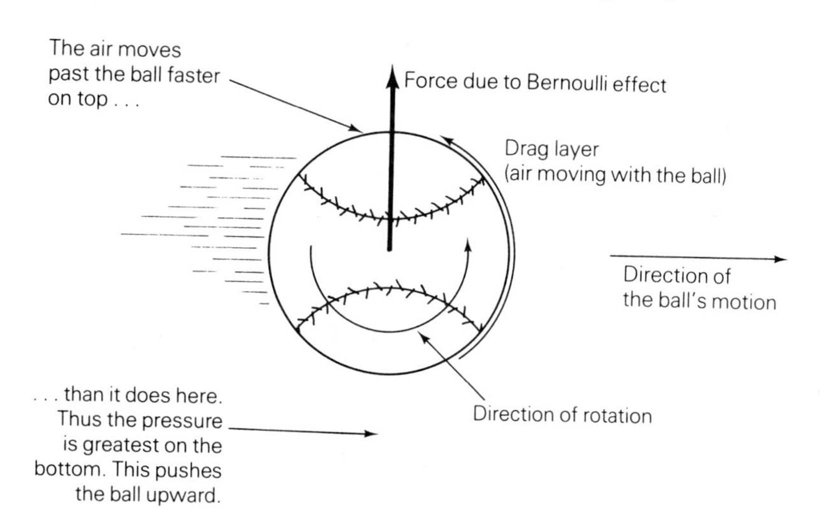

Figure 10-14 The force on a spinning ball.

the bottom, thus creating a net upward force on the ball. If the ball was spinning about a vertical axis, it would curve to one side. Naturally, the faster the ball spins, the more the ball curves.

The dimples on a golf ball increase the amount of air that the ball drags along with it, thus increasing the Bernoulli effect—not to increase a hook or slice, but ideally to give the ball more lift and extra distance (a properly hit golf ball has backspin, which creates lift). The seams on a baseball and the fuzz on a tennis ball have the same effect—as do cuts, dirt, and spit.

Concepts

Flow rate
Flow velocity
Energy of pressurization
Ideal flow
Laminar flow
Turbulent flow
Viscosity
Bernoulli effect
Entrainment
Nebulizer
Atomizer

Discussion Questions

1. Why does the heart have to work harder when you are standing up than when you are lying down?
2. When you exercise, why does the power of the heart have to increase more than the muscle power?
3. Give the main characteristics of ideal flow, laminar flow, and turbulent flow.
4. Why do laminar flows usually change to turbulent flows when the fluid velocity becomes large?
5. In which type of flow does the most mixing occur? Explain why.
6. Give examples of laminar flow and turbulent flow that can be found within the body.
7. Why are slight contractions and expansions of blood vessel size very effective in controlling the flow of blood within the body?
8. How do entrainment devices use the Bernoulli effect? Give an example of an entrainment device used in your field.
9. Why does a spinning baseball curve?

Problems

1. A person lives near a 10-meter-high waterfall. The average flow over the falls is 2 m^3/s. If the water was piped to the bottom of the falls to a turbine generator instead of going over the falls, what would the power output of the generator be? Ignore any frictional losses in the pipes or the turbine.

2. The cardiac output (rate of blood flow through the heart) during violent exercise is about 25 l/min. The blood pressure increases by about 150 mm of Hg as it is pumped through the heart of a trained athlete during exercise, but by as much as 250 mm of Hg in an untrained person. Calculate the power of the heart during violent exercise for a trained athlete and an untrained person.

3. If your bathtub is filled 18 centimeters deep, how fast will the water start to flow out when the plug is pulled if the drain has a diameter of 3 centimeters? Assume that the flow through the drain hole is nearly ideal.

4. An astronaut is orbiting the earth in a spacecraft that is pressurized with air at atmospheric pressure (100 kilopascals). The craft is hit by a meteor, which punctures the cabin with a square hole 1 centimeter on a side. At what rate does air start to leak out of the cabin?

5. Water flows into a tank at a rate of 1 l/s. The tank has a round hole in the bottom 2 centimeters in diameter. To what height will the water rise in the tank? If the fluid was changed from water to a nonviscous oil but everything else stayed the same, would the oil rise to the same height?

6. A tank is filled to a height H with water. The water flows out at the bottom through a short pipe pointed straight up. If the flow through the pipe is ideal, how high will the stream of water go?

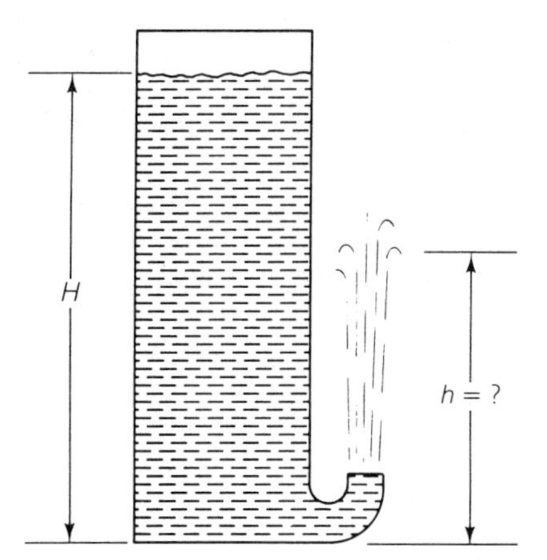

7. If oxygen flows laminarly out of a long tube at a rate of 50 ml/s, at what rate will water flow through the same tube? The pressure is assumed to be the same in both cases.

8. If a blood vessel increases in diameter by 25% and the pressures at the ends stay the same, by what factor will the blood flow increase?

9. A baseball pitcher throws a ball spinning clockwise (as viewed from above) toward the batter. Which way will the ball break, toward the pitcher's right or left?

Experiments

1. To observe laminar flow, mix some pepper with a spoonful of syrup on a plate. Then tilt the plate. The layers next to the plate will hardly move, while the outer layers will move much faster.

2. Demonstrate the Bernoulli effect by blowing over a sheet of paper as shown in Figure 10-9.

3. To demonstrate entrainment, cut a soda straw in half and put one half in a glass of water. Then blow over the half in the water with the other half as shown in the figure.

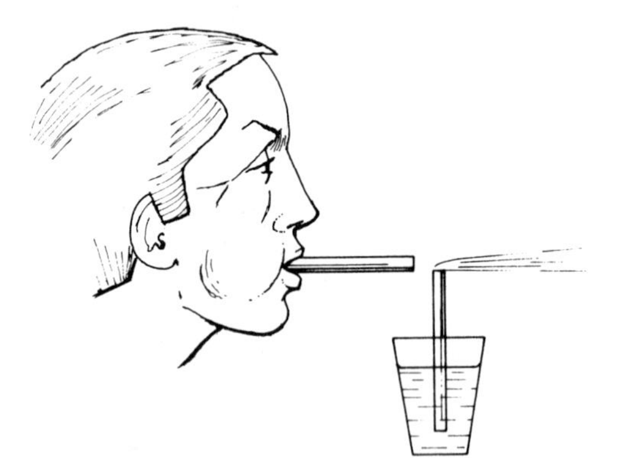

4. Stick a common pin through a piece of paper, then put the pin in the center hole of a spool of thread. Next, blow through the other end of the spool and try to blow the paper off the spool. Why can't you blow the paper off? Explain in terms of the Bernoulli effect.

CHAPTER ELEVEN

Heat

Educational Goals

The student's goals for Chapter 11 are to be able to:

1. Explain what heat is and explain the difference between heat and temperature.
2. Convert Celsius temperatures to Kelvin temperatures, and vice versa.
3. Discuss the effect of heat on the solid, liquid, and gaseous states of matter.
4. Describe how conduction, convection, and radiation transfer heat and what the roles of these three methods of heat transfer in the human body are.

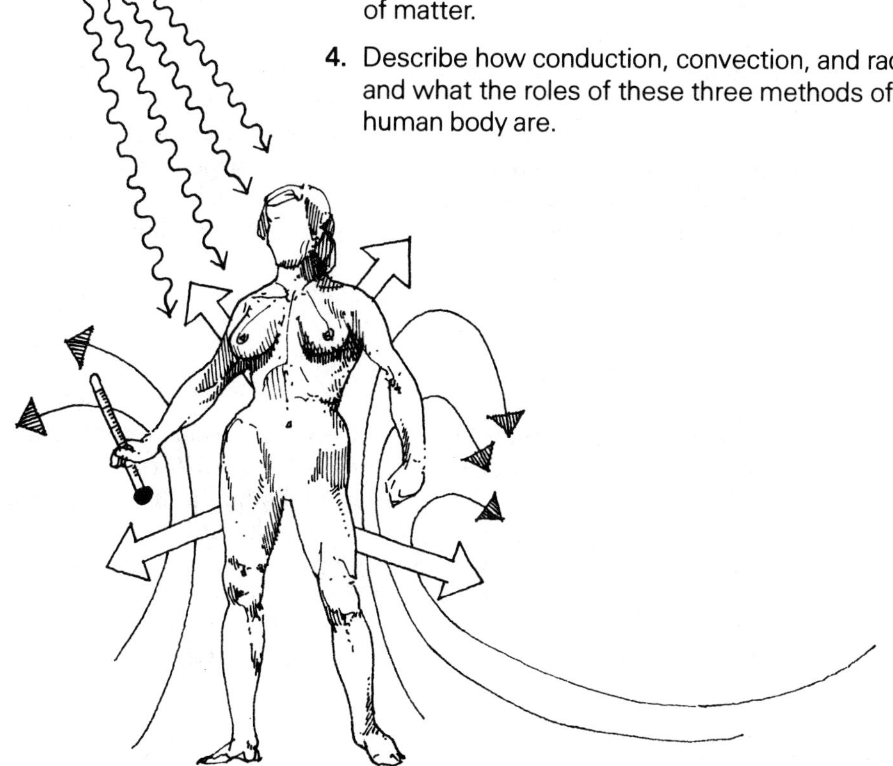

The true nature of heat was not discovered until the nineteenth century. At about the same time that people realized that matter was made up of atoms, they figured out that heat was just the energy of random atomic motion. Fundamentally, heat is no different from any other form of mechanical energy except that it is always random and disorganized, while the other forms of mechanical energy may or may not be disorganized. Since all energy is the same regardless of the form or organization, other forms of energy, such as mechanical, electrical, and chemical energy, can be transformed into heat energy. Conversely, heat can be transformed into the other forms of energy, too, but not at 100% efficiency, because changing disorder to order is difficult.

It took a long time to discover the true nature of heat because our senses are not able to see things at an atomic level. Heat was thought to be some strange sort of substance that flowed from object to object. Indeed, our senses are of little help in correcting this false assumption. The sensation we feel when we touch a hot object is not that different from the purely chemical sensation we feel when our tongue touches a "hot" pepper.

Temperature

The temperature of an object is closely related to its heat energy, but it isn't the same thing. Generally speaking, temperature can be thought of as an indicator of the average energy of atomic motion. Precisely, temperature defines the direction in which heat will flow: Heat always flows from a higher temperature to a lower one. When two objects at the same temperature touch, no heat flows between them.

Celsius Temperature Scale

Temperatures are usually measured on the Celsius temperature scale (which used to be called the centigrade scale). Zero (0°C) is the freezing point of pure water, 100°C is the boiling point of water at sea level atmospheric pressure, and there are 100 evenly spaced degrees in between. Normal body temperature is 37.0°C, and the temperature of a cool but comfortable room is 20°C.

Despite the vast improvement of the Celsius scale over such scales as the Fahrenheit scale, where zero and 100 don't correspond to anything in particular, the Celsius scale also has some disadvantages. The biggest one is that zero Celsius does not correspond to zero energy of atomic motion. Zero Celsius is not the lowest possible temperature, and it isn't even very cold. The normal winter temperatures in the northern United States are far below zero Celsius. Although the temperature may be uncomfortable on a winter day, the atomic motions certainly haven't stopped. For this reason, the Celsius scale is *not* the base scale of the SI system. For the same reason, using the Celsius scale can cause trouble when a calculation involves atomic motion. For instance, problems concerning the speed of chemical reactions, expansion of gases with temperature, and the energy of atomic motions are difficult to work out if the Celsius scale is used.

The Kelvin Temperature Scale

The base temperature scale of the SI system is the Kelvin scale. Zero on the Kelvin scale is the lowest possible temperature.* This temperature is called *absolute zero,* and it is equal to −273°C. At absolute zero, the atomic motions have been reduced to their lowest possible values.

The Kelvin scale uses the same size unit as the Celsius scale, so conversions between the two are simple. This unit is called the **kelvin** (K). To convert a Celsius temperature to a Kelvin temperature, just add 273. To go from Kelvin to Celsius, subtract 273. Expressed as formulas, the conversions are

$$K = C + 273 \quad \text{and} \quad C = K - 273,$$

where K stands for the Kelvin temperature and C for the Celsius temperature. Some temperatures of interest are shown on both scales in Figure 11-1.

Incidentally, the word *degrees* is not used for a temperature on the Kelvin scale. Since the Kelvin scale has its zero at the true zero of temperature, it was felt that the unit of temperature should be treated like all the other units. Thus "37 degrees Celsius" is the same temperature as "310 kelvin."

---

*Absolute zero cannot actually be reached. To do so would require an object colder than absolute zero to which the last little bit of heat energy could flow. However, you can theoretically get as close to absolute zero as you want. In practice, the lowest temperature attained is less than 0.001 kelvin, and temperatures of only 1 kelvin are routinely produced in many laboratories.

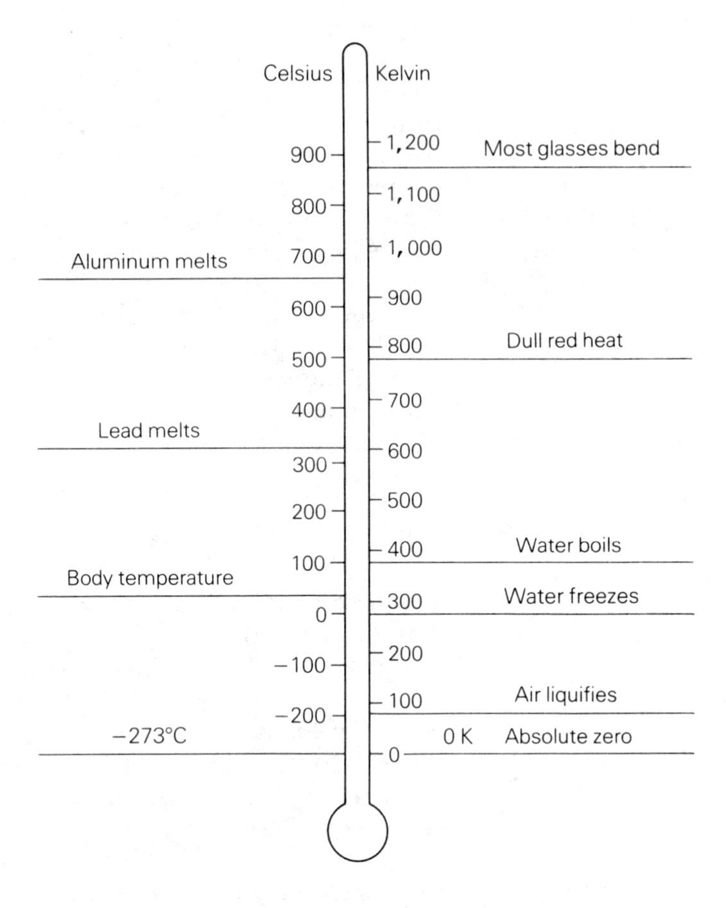

Figure 11-1

EXAMPLE A human can live reasonably comfortably, with the proper clothes, at temperatures from as high as 50°C to as low as −50°C. What are these temperatures on the Kelvin scale?

Answer On the Kelvin scale, 50°C is

$$\begin{aligned} K &= C + 273 \\ &= 50 + 273 \\ &= 323 \text{ K}. \end{aligned}$$

And −50°C is

$$\begin{aligned} K &= -50 + 273 \\ &= 223 \text{ K}. \end{aligned}$$

So humans can exist comfortably in the range from around 223 kelvin to 323 kelvin.

The skin is very temperature sensitive. The most sensitive regions, such as the upper lip, can detect changes of a fraction of a degree. However, the skin is more sensitive to changes in temperature than to the actual temperature. For instance, the skin will detect a sudden, small temperature change easily, but a gradual change of many degrees may pass unnoticed. Also, the reaction of the skin to extremely high or low temperatures can be misleading. Very cold water can easily be mistaken for very hot water, for example; and the pain of a "burn" caused by a supercold liquid, such as liquid air, is very similar to the pain of an ordinary burn.

Heat and the States of Matter

Let's watch the behavior of water as the temperature changes. Most other substances behave in a similar way except that changes such as melting and boiling occur at different temperatures. Suppose we start out with water in the form of ice at a very low temperature, close to absolute zero (0 kelvin). The molecules will be in a regular, ordered arrangement, and they will almost be motionless (Figure 11-2). As we add heat energy, the molecules will start to vibrate slightly, but they won't move far because of the strong intermolecular forces. Occasionally, a molecule might jump one position, but this is very rare. As more heat is added and the temperature climbs, the size of the vibrations will increase. Naturally, the larger the vibrations, the more room a molecule takes up. Thus, the entire piece of ice expands as the temperature goes up. With a few rare exceptions, all substances expand with heat.

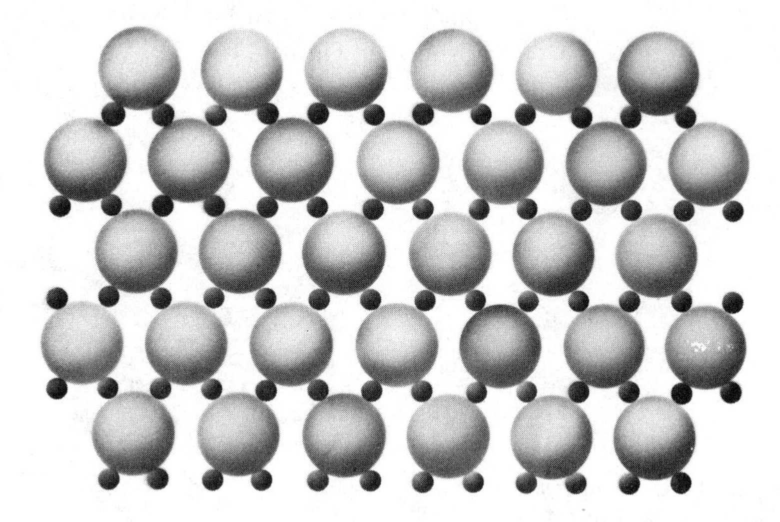

Figure 11-2 Ice (H_2O) near absolute zero (0 K). The molecules are nearly motionless. (The crystal structure is shown in a two-dimensional sketch, but the actual structure is three-dimensional.)

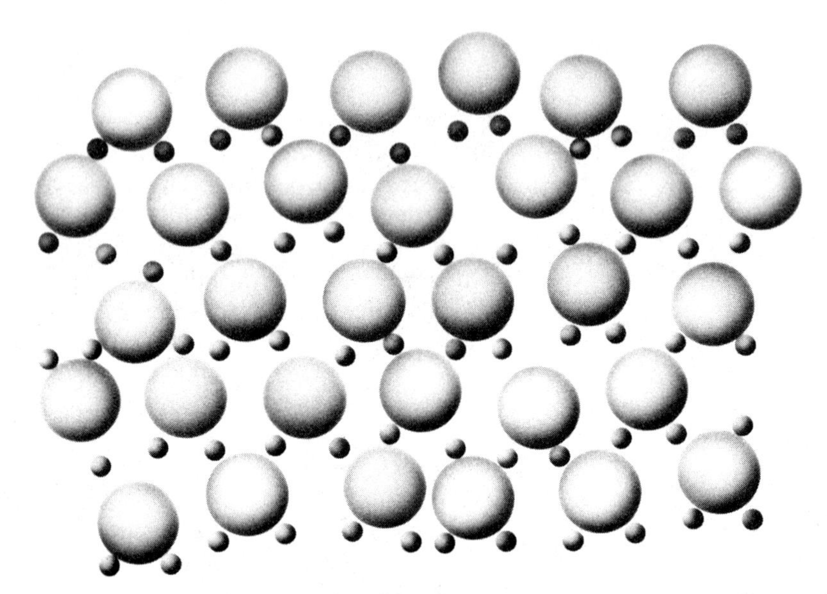

Figure 11-3 Ice (H_2O) just below the melting point.

The molecular structure of water just before it melts is shown in Figure 11-3. As the temperature goes up, the vibrations will eventually get so large that the molecules break away from their neighbors. At this point the intermolecular forces can no longer maintain order, and the solid ice turns into liquid water at 273 kelvin (0°C). Because it takes a great deal of energy to break the intermolecular bonds, a great deal of energy is needed to melt the ice. That is, a substantial amount of energy is required to convert the ice at 0°C into water at 0°C with no change in temperature. This is why ice at 0°C is far more effective for cooling a drink than cold water at 0°C is. The amount of energy needed to melt ice without changing the temperature is 335,000 J/kg (80 cal/kg). This is called the latent heat of fusion of water. The molecular structure of water is shown in Figure 11-4.

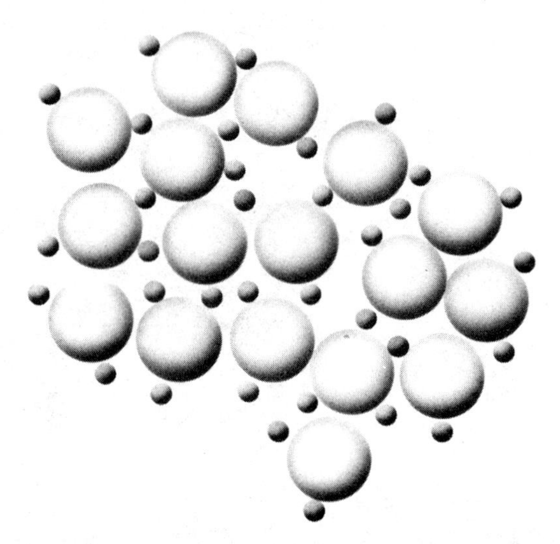

Figure 11-4 Liquid water (H_2O) just above melting.

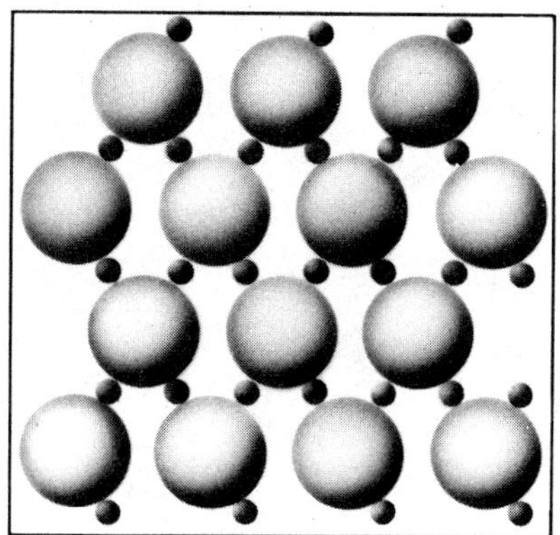

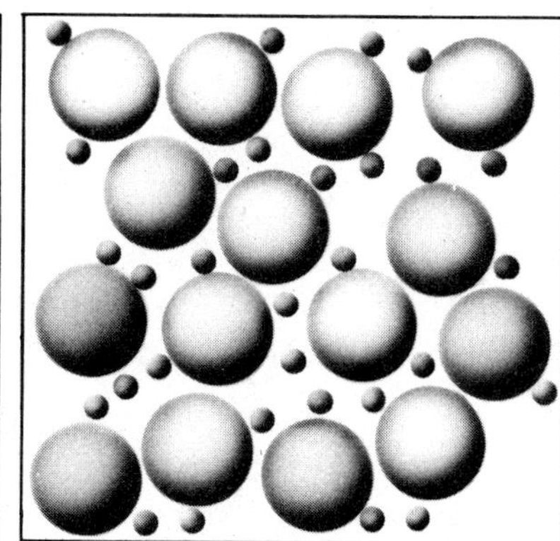

Figure 11-5 Because of the unusual shape of an H_2O molecule, more can be packed into a random pattern (water) than can be packed into an ordered arrangement (ice). This is not true for most substances; generally, the solid is denser than the liquid.

Most substances expand when they melt because an ordered arrangement generally takes up less space than a disordered one. Water is one of the few exceptions. Since water molecules have a complex shape that can be packed more closely in a disordered arrangement than in an ordered one, ice is less dense than water (Figure 11-5). This irregular behavior also affects liquid water. As the temperature increases from 0°C to 4°C, the liquid water shrinks. As the temperature increases above 4°C, water expands in the same way that most liquids do.

As the liquid water is heated, the molecular motions become more rapid and more disordered. Eventually, the molecular motions become so violent that the intermolecular forces cannot even hold the molecules close to each other. They then fly off in totally random motion, well separated from each other (Figure 11-6). That is, the liquid water becomes a gas. At sea level, this occurs at 100°C. The gas is called steam; however, the word *steam* is also used for the fog of water droplets that forms when the gas turns back into liquid water. True steam is colorless and odorless, just like air.

As the water molecules fly apart to form steam, every intermolecular bond is broken. Breaking these bonds takes a large amount of energy, even more than is required to turn ice into water. To turn water into steam without changing the temperature takes 2,260,000 J/kg (540 cal/kg), more than 6 times the energy required to turn ice into water. This is called the latent heat of vaporization of water.

Cooling the body by the evaporation of perspiration is very efficient because of the large amount of energy carried away. Even below 100°C, some liquid water evaporates and becomes water vapor. Every time a water molecule evaporates from the skin, the heat required to break the bonds is carried off by the departing molecule, thus cooling the skin. Normally we sweat about ½ liter of

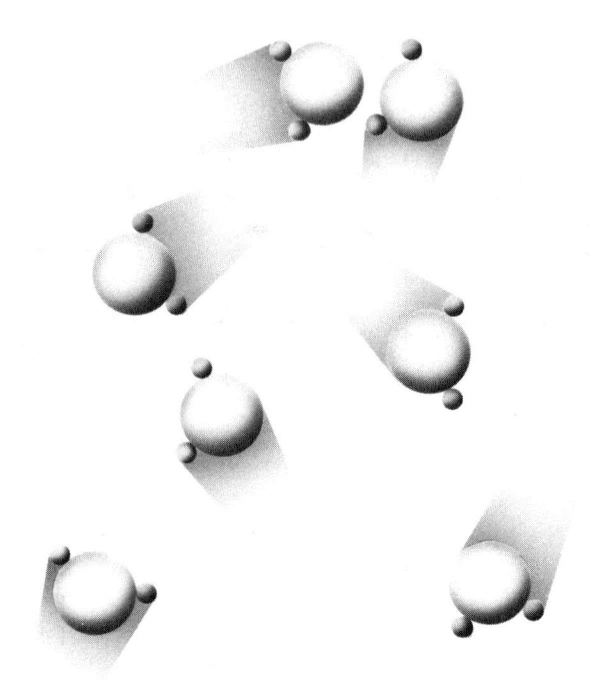

Figure 11-6 Water vapor (H_2O), which is also called steam.

water per day, so around 1 megajoule of energy is carried off daily by evaporating sweat (approximately half the energy in a pizza). On a hot day we sweat far more than that. When we exercise, we also perspire more to remove waste heat. An athlete can easily lose several liters of water during a hard workout.

When the air has a lot of water vapor, cooling by evaporation doesn't work; nearly as much water returns from the air as evapo-

The white steam is actually a cloud of fine drops of liquid water.

rates. Since evaporation is normally our most effective way of cooling, we become very hot and uncomfortable on a humid day.

Let's look at steam again. The molecules fly around at random. Usually they are fairly far apart, but they do collide occasionally. As more heat is added, the molecules move more and more rapidly. Eventually, the collisions become so violent that the molecules are torn apart into hydrogen and oxygen atoms. Of course, the temperature can still go higher, but we will no longer have water as we know it. As far as we know, there is no upper limit to temperature, but no ordinary matter stays together above a few thousand kelvin.

Gases expand with increasing temperature in a simple way. When the pressure stays constant, the volume of the gas is directly proportional to the Kelvin temperature. Conversely, when the container is sealed to keep the volume constant, the pressure is proportional to the Kelvin temperature. These two ideas can be combined into what is called the ideal gas law. Expressed as a formula,

$$PV = \text{constant} \times T.$$

When using the ideal gas law, be sure the temperature is given on the Kelvin scale, and be sure the pressure is the absolute pressure and not the gauge pressure. As long as the mass of gas stays the same, the constant stays the same.

Chemical Reaction Rates

The speed of a chemical reaction depends on two things: how often the proper molecules collide and how hard they collide. Both factors depend on how rapidly the molecules are moving, so both increase as the temperature goes up. This fact, combined with the fact that many reactions have a threshold—a minimum impact speed that must be exceeded for the reaction to occur—means that chemical reaction rates increase very rapidly as the temperature goes up. A rule of thumb is that a reaction rate will be 10 times faster for every ten-degree temperature rise. This is just a rough rule; the exact temperature dependence of the reaction rate varies among reactions.

Thousands of different chemical reactions take place in the body, and they must all proceed at their proper rates if the body is to function properly. A mere 1°C increase in the temperature of the body throws this delicate balance off and we feel out of sorts. A 3°C rise is a raging fever. Conversely, if the body temperature is lowered, the reactions proceed much more slowly. This is dangerous when uncontrolled, but the effect can be useful in many types of surgery, particularly heart surgery. The temperature of the body is carefully lowered, slowing down all bodily processes. This decreases the need for oxygen—especially important for the brain—so the surgeon has much more time to perform the operation.

Because the body's processes slow down with temperature, people who drown in cold water have a better chance of being revived after a long time underwater than people who drown in warm water. A person who drowns in water water cannot be revived after more than a few minutes underwater. However, some people have been resuscitated after being submerged in cold water for more than a half hour, with no noticeable brain damage. Rescue workers should not stop the cardiopulmonary resuscitation of a drowning victim too quickly, no matter how lifeless the person looks—particularly when the person was submerged in cold water.

Heat Transfer

There are three main ways of transferring heat: *conduction, convection,* and *radiation.* All three are important in keeping the body at its proper temperature. When we are too hot, the body must get rid of the excess heat by these three means plus evaporation; when we are too cold, the loss of heat must be reduced.

Conduction

Conduction is simply the transfer of energy resulting from one atom bumping into another. Consider a long rod that is being heated in a flame at one end. The flame raises the energy of the atoms at that end, making them vibrate more rapidly. These atoms collide with their neighbors, starting them in motion. Then the next layer of atoms is set into vibration, and so on, until the heat energy is transferred down the entire length of the rod.

The rate of heat transfer by conduction through an object depends on the material from which the object is made. Materials vary widely in their ability to conduct heat. Metals are generally good conductors, silver and copper being among the best. Most nonmetals are poor conductors; this includes all the materials from which the body is made.

A quick way to see the large differences in conductivity is to take a sterling silver spoon, a stainless steel spoon, and a plastic spoon and put them into a cup of hot coffee or water. The handle of the silver spoon gets uncomfortably hot almost immediately, the stainless steel spoon only gets warm, and the plastic spoon never warms up. Even though sterling silver tableware is hard to come by, the effect is so striking that it is worth the trouble of borrowing a silver spoon.*

In order for heat to be transferred by conduction, the objects must actually touch so the atoms can bump into each other. To insulate against conduction, just pull the objects apart.

Convection

The transfer of heat through the physical motion of a heated object is called *convection.* Whenever a hot object moves from one place to another, heat energy is transferred by convection. There are some examples of convection by solids, but most often convection occurs in fluids.

In many types of convection, the fluid is simply pumped from one place to another. This is the case in the body, where one of the blood's jobs is to carry heat from inside the body to the skin. Most home heating systems use either water or air to circulate the heat. Water-cooled car engines move heat away from the engine to the radiator by circulating water. The radiator then transfers the heat to the air primarily by conduction. In spite of its name, a car radiator doesn't radiate much heat.

One very important type of convection is due to the earth's gravity. As most fluids are heated, they expand. (Water below 4°C is the most important exception.) Thus, the heated fluid has a lower density than the surrounding cooler fluid, and in accordance with Archimedes' principle, it starts to float upward, also moving the heat upward. These currents are called convection currents, and they can be as gentle as the currents rising from a hot asphalt road or as violent as a tornado.

Gravity convection is an important factor in most fluid heating. Consider what happens when you heat a pot of water on the stove (Figure 11-7). As the cool water is heated at the bottom, it starts to rise. This brings more cool water to the bottom, resulting in a strong convective flow. This flow is very efficient in that it always brings the coldest water to the bottom to be heated. If this convective flow didn't occur, heating water would be very hard. If we try to

---

*You can tell if you are being served with genuine sterling silver just by putting the spoon into hot coffee. If the handle gets hot very quickly, it is made of real silver.

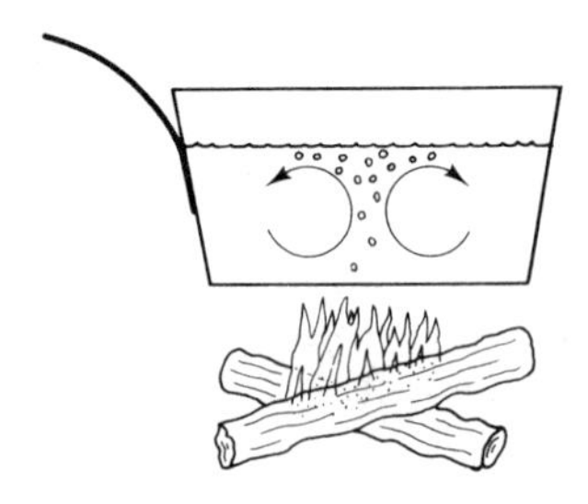

Figure 11-7

heat water from the top—with an electric heating element, for instance—there will be no convective current, and only the water at the top will get hot. In fact, if the pot isn't stirred, the water on the top will boil before the water at the bottom gets warm.

The cooling coils inside a refrigerator are usually near the top to make the best use of convective air currents. As the air is cooled by the coils, it becomes denser and sinks to the bottom. Convective currents then bring warm air up to the coils to be chilled. The fact that ice floats makes it ideal for cooling warm water. The chilled water sinks, bringing more warm water up to the ice. These currents continue until the water reaches 4°C. The convection currents then stop because water no longer contracts as the temperature falls below that temperature.

Convective flow is also responsible for the earth's weather. The air that is heated at a warm spot rises. This causes the surrounding cooler air to flow toward the warm spot, producing a wind. All winds, from gentle breezes to hurricanes, result from convective air currents.

Water's strange behavior below 4°C has a major effect on the earth's climate. In the fall, the water at the top of lakes contracts and sinks as the colder air cools it. The resulting convection currents bring the warmest water to the top, so the lake cools quickly—until it gets to 4°C. At that point, the convection currents stop; the cold water stays on top and the warm water (if 4°C water can be called warm) stays at the bottom. Gradually the water at the top freezes, but it doesn't freeze very far down, because the only way that the heat can be transferred from the water is by conduction through the ice—which is very slow. If water behaved like most other liquids—always contracting as it cools and contracting even more as it freezes—lakes would freeze solid in the winter and melt only near the surface in the summer. Fish could never live in the lakes, and the climate would be much colder.

The best way to insulate against convection is to remove the fluid—no fluid, no convection. The common thermos bottle (or vacuum bottle), used to keep drinks hot or cold, uses this principle. The thermos bottle is really two bottles, one inside the other (Figure 11-8). Nearly all the air is pumped out from the space in between, and this resulting vacuum stops convection (and conduction).

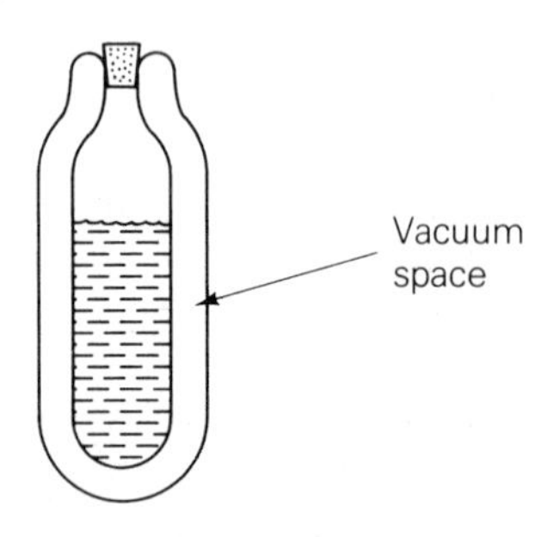

Figure 11-8 The thermos bottle stops convection by removing the air around the inner bottle.

The next best way to insulate against convection is to use something that stops the convection currents. Almost anything that has lots of small air spaces will work—ground-up newspapers, fiberglass, plastic foam, or cloth. Generally, the smaller the individual spaces are, the better the insulation will be (Figure 11-9).

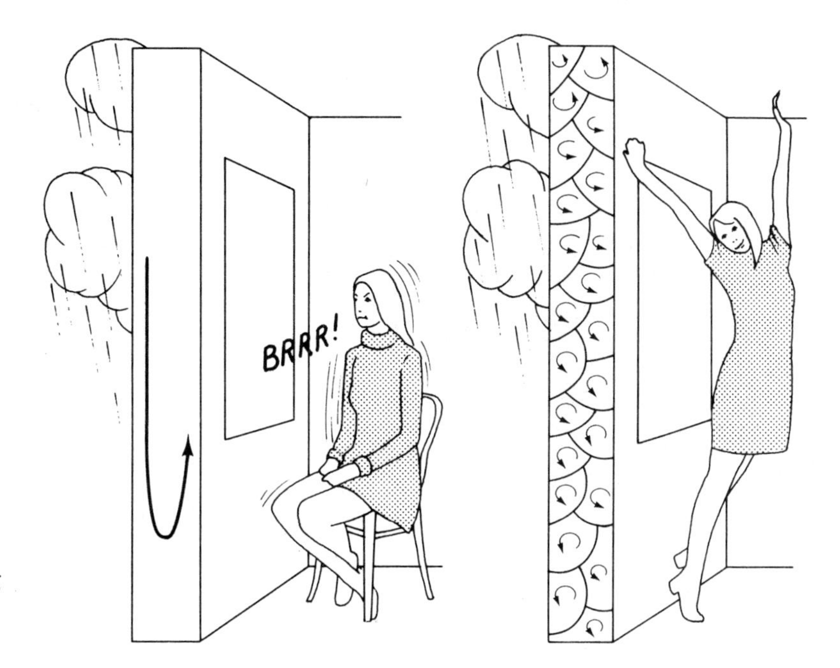

Figure 11-9 Convection in the air space in a wall can be reduced by dividing the space into many small spaces.

Radiation

When atoms in a heated object vibrate, they give off electromagnetic radiation. Light is the electromagnetic radiation that we are most aware of, but radio waves, microwaves, infrared rays, ultraviolet rays, and x rays are also forms of electromagnetic radiation. We will go into the details of electromagnetic radiation later; for now we will just point out two important features: First, the rays carry energy, so they can transfer heat; second, they can travel through a vacuum (in fact, they move fastest in a vacuum). Thus, radiation can transfer heat between objects that are far apart from one another, such as the sun and the earth.

At high temperatures, the radiation is clearly visible. A hot object can glow red or even white. At lower temperatures, heated objects don't give off as much radiation; this radiation, called infrared radiation, is not visible.

The earth is heated almost entirely by radiation from the sun, which gives off enormous amounts of radiant energy. This radiation warms the earth and evaporates water from the oceans, and a small part is converted into chemical energy by plants. Some of this energy was converted into fossil fuels eons ago—fuels that we are running out of rapidly.

The amount and type of electromagnetic radiation that an object gives off depend on the temperature of the object and the color of its surface. At temperatures below 0°C, objects give off only a tiny amount of radiant energy, which is in the form of radio waves and microwaves. As the temperature of the object goes up, more and more energy is given off. At room temperature the object starts to give off infrared radiation. The eyes can't see these rays, but the skin can feel them. The warmth we feel from a fire is mainly the sensation of the fire's infrared radiation striking the skin. Many animals that live in caves or that hunt in the dark have very sensitive infrared detectors to track their prey. The pits on the heads of pit vipers are such detectors. Even though humans can't detect infrared radiation particularly well, we can sometimes detect other people in a dark room by their infrared radiation, particularly if they are not too far away.

When an object is heated to around 600°C, it starts to give off visible light and to glow with a red color, in addition to giving off large amounts of infrared radiation. The hotter the object becomes, the whiter the light gets. The filament in an electric light bulb is at about 3,000°C, and it radiates large amounts of white light and infrared radiation.

The color of an object's surface is an important factor in determining how much radiation the object will give off. The darker the color is, the more it will radiate. Thus, at a given temperature, black objects radiate the most heat, white objects radiate very little, and shiny, silvery objects radiate least of all. Color is equally important in determining how much radiation an object will absorb. The best absorbers are the same colors that radiate energy well. For instance, a black object left out in the sun will heat up very rapidly, while a white or shiny object will not absorb much of the sun's heat.

In very hot or cold climates, the right clothing color can make a big difference in comfort. When working under the hot sun, we don't want to absorb more of its heat than we must. Thus, white is a good color to wear because it doesn't absorb much radiant energy. When working in the sun in the winter, black is the best color, because it absorbs the sun's rays best; however, if there was no sun, white would be better, because white clothing would not radiate the body's heat away as quickly.

Since radiation can travel through a total vacuum, it is difficult to insulate an object to prevent radiative heat losses. The first step is to paint the object a light color or, better yet, to cover it with a shiny metal foil. The next step is to surround the object with mirrors to reflect the radiation back to the object. These are the reasons why a thermos bottle is silvered. The vacuum can't stop the radiation losses, but the silver coating on the inner bottle reduces the amount that is lost. The silver coating on the outer bottle then reflects much of the radiation that does escape back toward the inner bottle. For the same reason, many types of insulation have a metal foil layer to cut radiation losses. The very light, silvery emergency blankets—called space blankets because they were developed for the space program—also insulate by cutting radiation losses. They don't work as well as a goose down sleeping bag, but they make a good emergency substitute, especially since they can be folded up and carried in a pocket.

In industries such as steel making, where the workers must be insulated from the heat of red-hot objects, blocking radiation is of critical importance. Some methods for shielding workers from radiated heat are explained in Chapter 18.

Heat Transfer in the Body

All three types of heat transfer are important in the body. The large amounts of heat produced inside the body are brought to the skin both by conduction and by convection of the blood. Of course, the convection of the blood is forced by the pumping of the heart. This convection is far more efficient than conduction because the body parts are poor conductors. Without the blood's convection, the center of the body would be at too high a temperature for the body to work properly. So not only does the blood deliver food to the cells and remove the waste products, but it also serves as the body's main coolant.

The body regulates its temperature mainly by controlling the flow of blood to the skin. When the body is overheated, the blood vessels in the skin expand, increasing the flow of blood and thus the flow of heat to the skin. When the body is cold, these blood vessels shrink to reduce the flow of blood and of heat to the skin.

At the skin, heat is removed by conduction, convection, and radiation, plus evaporative cooling: Heat is conducted to everything that we touch; the heated air next to our bodies rises, starting convection currents that carry away heat; exposed parts radiate heat away; and our sweat cools us by evaporating. Evaporative cooling is not only very efficient but also very controllable, because the body can sweat more than a liter per hour or almost nothing at all, depending on how much heat it has to get rid of.

The main purpose of clothes is to insulate us from the environment, which for the most part is too cold for naked humans. Even though the human race lives in all parts of the globe, we are basically tropical animals. The naked human cannot live at temperatures much below 20°C. Clothes insulate against all three types of heat transfer: They retard conduction by placing a very poor conductor between us and the environment, they slow down the convective currents that flow along the body, and they reflect some of the radiated heat back to us.

Sometimes people put on extra clothes or even rubber insulating suits to "sweat away some fat." Some mass will be lost, but it will only be the water that the body puts out in its effort to cool down. There is no extra fat loss, and the weight reduction doesn't last far beyond the first water fountain. Besides failing to reduce excess fat, the method is also dangerous because the increase in body temperature caused by the extra insulation can be harmful. Many people have died running around in rubber exercise suits because their bodies could not handle the temperature increase or because they lost so much water that they became dehydrated. For this reason, such practices as having athletes run in rubber suits or holding back on water to "toughen them up" have no value and can be dangerous. Careful trainers and coaches take immediate action at the first sign of overheating or dehydration to prevent serious problems.

Hypothermia

Hypothermia, the lowering of the body temperature, is one of the big dangers of many cold weather outdoor activities. Falling into cold water is especially dangerous, because water and wet clothes rapidly absorb body heat. Thus, anyone who falls into cold water must be taken out fairly quickly to survive. In 10°C water, the survival time is only an hour or two. The water temperature in most of the United States is usually below 10°C in the spring and fall.

When the body is losing too much heat, it first cuts the heat loss by reducing the blood flow to the skin. If this isn't enough, the body increases heat production by shivering, which will raise heat production to 4 to 6 times normal. However, shivering can only last as long as the carbohydrate reserves of the body do. When they run out, the body temperature starts to drop. Blood pressure and flow drop, and the victim becomes very listless and confused, and usually starts to mumble. After the body temperature slips below 33°C, the victim will not be able to produce enough heat to restore proper body function; heat must be applied from the outside to revive the person.

The listlessness and mumbling of early hypothermia are easy to confuse with fatigue. Because of this, hypothermia can become seriously advanced before it is recognized. For this reason, all members of a group should be alert for the early signs of hypothermia, particularly if anyone has gotten wet. When the signs appear, the safest action is to seek shelter and warmth immediately. Hypothermia in its early stages is easy to stop, even in a fairly primitive shelter. The advanced stages can only be helped with proper medical facilities, which are much harder to find than simple shelter in the woods, lakes, and mountains.

Local Treatments with Heat and Cold

Both heat and cold can be very useful for treatment of pain and injury. Since heat speeds up molecular motions, it isn't surprising that heat generally speeds up bodily processes while cold slows them down. (There are some exceptions to this rule, especially when the treatments last a long time.)

When heat is applied, the blood vessels expand, circulation speeds up, and healing is usually faster. Muscles relax faster when they are warm, so hot packs can be useful for treating muscle spasms. In contrast, cold shrinks the blood vessels, reducing circulation. Because cold also slows down nerve response, cold packs can act as a local anesthetic. When a minor injury is treated with heat and cold, the usual method is to apply cold packs immediately after the injury. This slows down the swelling by slowing down the buildup of intercellular fluid, and it also cuts down on the pain. Later on, usually the next day, heat might be applied to speed up the circulation to the injured area, which will remove the excess intercellular fluid and speed up healing.

Because the body is not a very good heat conductor, local hot and cold packs cannot do much to change the temperature of the deeper-lying tissues and joints. In order to heat these areas, a diathermy machine is used. Diathermy machines work on the same principle as microwave ovens. Radiowaves or microwaves are beamed at the injured body part, and the radiation energy is converted to heat uniformly throughout the radiated region.

Burns

The severity of a burn is determined more by the amount of heat transferred than by the temperature. For instance, you can pass your hand through a candle flame (around 1,000°C) quickly with no damage. But if you put your hand into 100°C boiling water for the same time, it would be badly burned, because water conducts heat to the hand much more rapidly than the hot gas in a flame does. For this reason, contact with hot solids and liquids is often

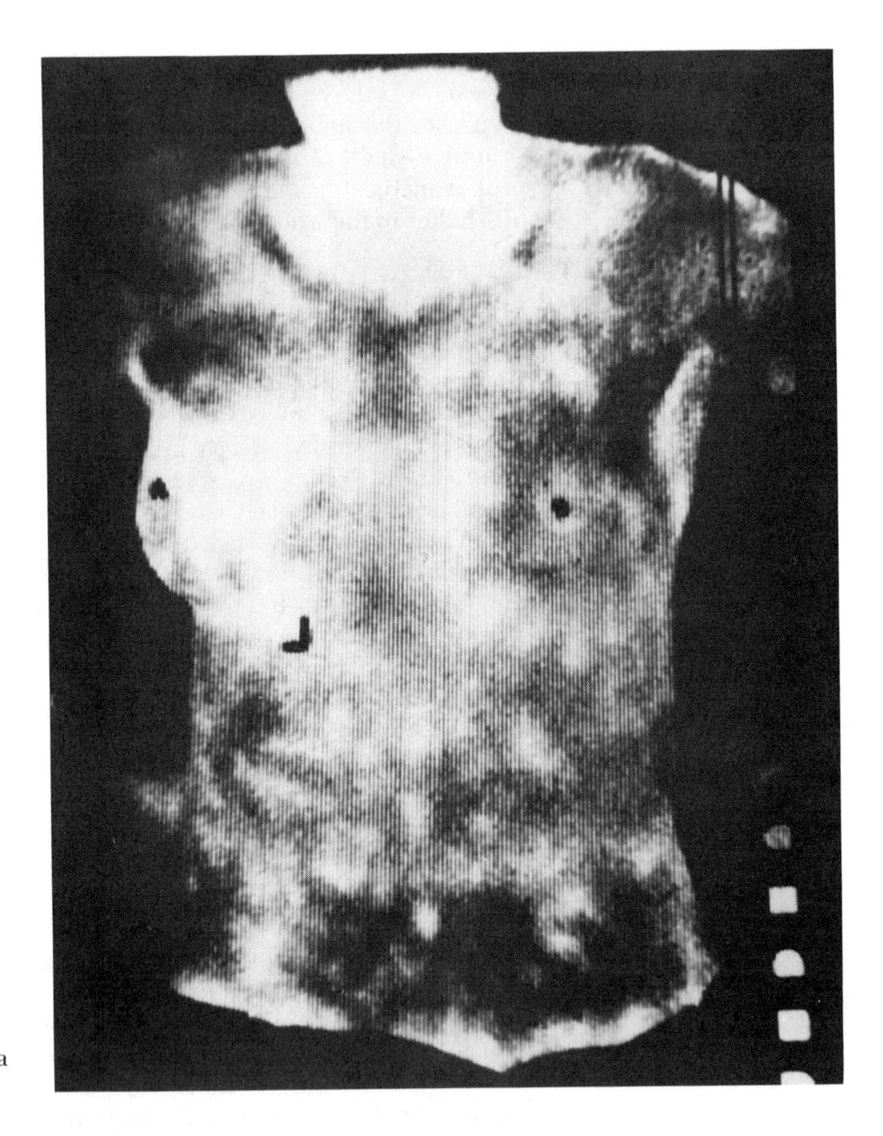

The warm area on the left part of the thermograph indicates the possible presence of a tumor.

more damaging than contact with hot gas, even if it is flaming. Some synthetic fabrics are quite dangerous in this respect; the burning fabric melts into a hot, sticky liquid that sticks to the skin and transfers heat into the skin very rapidly.

One of the best first aid treatments for burns is to reduce the temperature of the burned region quickly with cold water from the faucet or with ice water if it is available. If the burn is large, prompt medical attention is necessary because a burn is one of the hardest injuries to treat.

Diagnostic Thermography

Very sensitive infrared cameras can detect temperature differences of far less than 1 degree. They can be used to spot certain diseases because many diseased regions are at a different temperature than normal. An inflamed region will be hotter than normal because of the increased circulation there. Tumors and cancers grow faster than normal tissue, so they will also be hotter than normal because of their more rapid metabolism. Unfortunately, many normal things cause small temperature differences, too. So although diagnostic thermography is useful, it is not foolproof for screening patients.

Thermography has recently been used to solve the common problem of finding a suitable vein for an IV needle by using small skin-temperature differences. Certain solutions called liquid crystals change color very rapidly with temperature. Such a solution is painted on the skin; then the warmer area above the veins changes color, making the veins easy to see.

Concepts

Heat
Temperature
Celsius scale
Kelvin scale
Absolute zero
Latent heat of fusion
Latent heat of vaporization
Evaporation
Chemical reaction rate
Conduction
Convection
Radiation
Insulation
Hypothermia
Diagnostic thermography

Discussion Questions

1. Why is ice far more effective for cooling than an equal mass of cold water?

2. Why are we much more uncomfortable on a hot, humid day than on a hot, dry day?

3. Describe how heat is transferred by conduction, convection, and radiation.

4. How is heat different from temperature?
5. Explain why lowering the body temperature will allow a surgeon more time to perform an operation.
6. Why does an aluminum softball bat feel colder than a wooden bat, even though both bats are at the same temperature?
7. Explain why a goose down jacket is superb insulation when it is dry but not when it is wet.
8. Why are cloudy winter nights usually warmer than clear ones?
9. Why does a person wet his or her finger before holding it up to determine which way the wind is blowing?
10. Discuss how heat is transferred within and away from the body. Explain the roles of conduction, convection, radiation, and evaporative cooling in the transfer of body heat.
11. Explain how diagnostic thermography can be used to locate diseases.
12. Why do hot liquids generally burn a person worse than hot gases?

Problems

1. Oxygen liquifies at 90 kelvin. What temperature is this on the Celsius scale?
2. At 600°C an object will be glowing a dull red. What is this temperature on the Kelvin scale?
3. What is the temperature of a comfortable room on the Kelvin scale?
4. How would you feel on a day when the temperature was 237 kelvin?
5. Explain why most, but not all, things expand when they are heated.
6. If a tire is inflated to a pressure of 400 kilopascals at 20°C, what will the pressure in the tire be when it warms up to 50°C? The cords and belts inside the tire will keep the tire from expanding, so the volume of the tire is nearly the same at both temperatures.
7. If a rigid can is filled with air and sealed at atmospheric pressure (100 kilopascals) at −40°C, what will the pressure in the can be when it is heated to 100°C?
8. By what percentage is hot, 40°C summer air less dense than cold, −20°C winter air, assuming that the barometric pressure is the same on both days?
9. During vigorous exercise the body produces heat at a rate of about 400 watts. The amount carried away by each of the four different types of heat transfer depends on clothing, temperature, and humid-

ity. If evaporation is removing body heat at a rate of 200 watts, how much sweat must a person produce during one hour of vigorous exercise?

Experiments

1. Build a thermometer from a pop bottle, a thin glass or plastic tube, and a cork. Cut a hole in the cork for the tube and insert the tube into the cork. Next fill the bottle to the top with water. A drop or two of food coloring will make the water easier to see. Then put the cork and the tube into the bottle without trapping any air bubbles. Adjust the height of the water in the tube so that it comes up about halfway. Place your thermometer in a refrigerator and watch how the water level changes. Bring it back to a warm room and watch as it warms up. In what ways are regular thermometers better than your simple one?
2. Wrap a small bit of wet cloth around the bulb of a thermometer, then wave it around in the air. Why does the thermometer indicate a lower temperature?
3. Put a sterling silver spoon, a stainless steel spoon, and a plastic spoon into a cup of hot water. The differences in the heat conductivity can be observed by feeling the handles. If you can't borrow a sterling silver spoon, you can use a piece of copper pipe as a substitute.

CHAPTER TWELVE

Membranes and Molecular Motion

Educational Goals

The student's goals for Chapter 12 are to be able to:

1. Explain how molecules move through fluids by random diffusion.
2. Explain how osmosis through a cell membrane leads to a pressure difference across the membrane.
3. Explain what hypotonic, isotonic, and hypertonic solutions are and what their effects on blood cells are.
4. Explain why a steady energy flow is necessary for life.

Life cannot exist without continuous molecular motion. While no one knows exactly what life is, everyone agrees that it is a highly ordered, mobile state of matter. Certainly one of the key differences between living matter and inert matter is that living matter has a pattern that is passed along from generation to generation. As we will see in this chapter, this fact alone demands that the molecules always stay in motion. For it is a physical law that a pattern—a specific ordering of matter—can only be maintained when energy flows into the system continuously. Thus, life cannot exist without an energy source or without continuous molecular motion.

The Cell Membrane

The cell membrane acts mainly as a selective barrier allowing food, oxygen, and other nutrients into the cell and keeping most wastes and harmful molecules out. It must also keep molecules that the cell needs inside, but all waste products must be discharged. In many cases, the membrane does its job in a simple way. It allows certain molecules through and blocks others with no regard to whether they are going into or out of the cell. This process is called *osmosis,* and the net flow can be either inward or outward, depending on whether the concentration of that type of molecule is higher inside or outside. In some other cases, the membrane seems to select out a particular molecule and actually carry it through the membrane in the proper direction. This process is called *active membrane transport.* The big difference between these two types of transport is that active membrane transport takes energy, while osmosis does not.

Most plant and animal cell membranes are constructed like a sandwich, with a protein layer on the top and bottom and a fat (lipid) layer in between (Figure 12-1). The proteins are amino acid polymers, which are quite kinky in most cell membranes. The proteins are therefore quite flexible and elastic, a property that allows the membrane to move, stretch, and grow. The total thickness of the membrane is about 8 nanometers—about 50 atoms thick. Each protein layer is about 2 nanometers thick, and the lipid layer about 4 nanometers thick. Of course, the cell membrane is difficult to study because of its incredible thinness. It is invisible under an ordinary microscope, and even an electron microscope reveals only the rough structure; the details remain invisible. The membranes don't seem to have any actual holes or pores—at least not any large enough to see under an electron microscope. If any pores are pres-

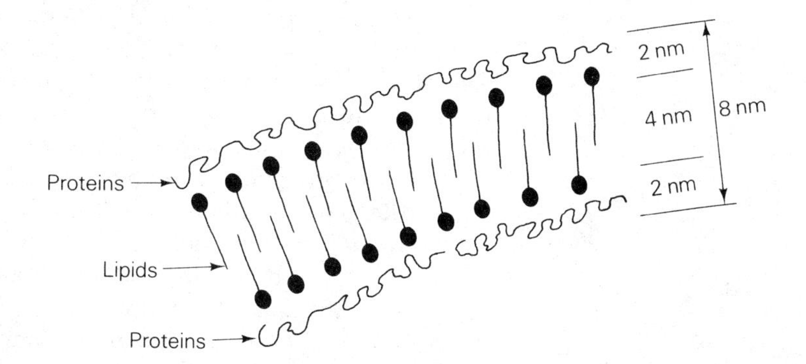

Figure 12-1 Structure of the cell membrane.

ent, they must be very small—more like a gap between molecules than an actual hole.

Because the membrane center is a fat, fat-soluble compounds pass through cell membranes more rapidly than water and other fat-insoluble compounds, such as sodium and potassium salts. Many drugs are fat soluble so they can penetrate the cell membrane quickly. Many poisons and insecticides are fat soluble for the same reason.

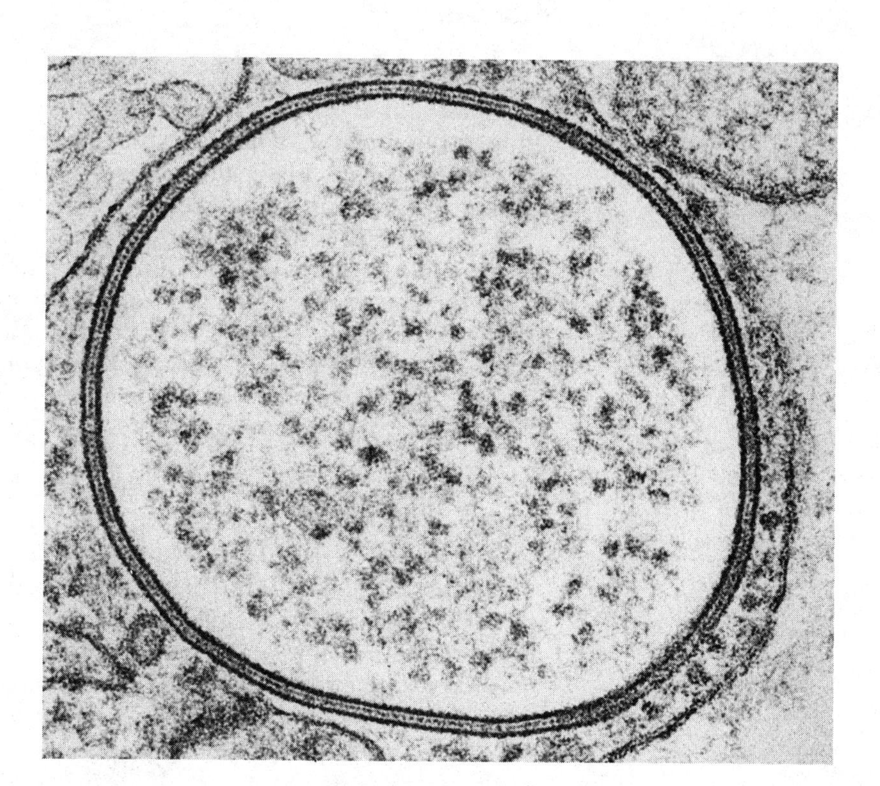

A cell membrane. This electron micrograph shows the protein-lipid-protein sandwich structure of a cell membrane.

The kinky proteins at the membrane surface resemble the edge of an unfinished jigsaw puzzle, and they function in a similar way. A molecule floating by in the intercellular fluid will not attach to the membrane unless it fits; that is, it must match the edge of the puzzle properly. This property leads to a high degree of accuracy in picking out the right pieces and rejecting the wrong ones. It also allows the cells to adhere closely to the proper neighboring cells. The cell's ability to identify other cells by the surface structure of the proteins is quite remarkable. When chick embryo heart and kidney cells are mixed in a fluid, they will gradually segregate and the cells of each type will clump together.

At one time, the cell membrane was considered to be like a plastic bag that separated a molecular soup on the inside from a different molecular soup on the outside. This isn't the case at all. The material inside the cell isn't solid, but it isn't a true liquid, either. Roughly speaking, the cell interior is a gel. Even this is a simplification, because under some conditions the material behaves like a viscous liquid while under others it is very elastic.

Diffusion

When we put a drop of food coloring into a glass of still water, the color gradually spreads throughout the water because of the random motion of the molecules. This slow mixing is known as *diffusion,* and it occurs in all substances—solid, liquid, and gaseous. Of course, diffusion in solids is much slower than in liquids and gases.

The diffusion rate depends on differences in the concentration of the molecules and the speed of molecular motion. The faster the molecules are moving, the more rapid the mixing will be. The two biggest influences on molecular speed are the temperature and the mass of the molecules. Of course, the molecular speed increases with temperature; and at any given temperature, light molecules move faster than heavy ones.

Rate of Diffusion

Suppose you dumped a basket of advertising handbills in your backyard. Soon your neighbors would get the message; a week or so later a few people in the next block might get a copy, but almost no one in the next state would ever see the handbill. If you put a drop of food coloring in still water, you will see that the mixing takes place rapidly near the drop but much more slowly far from the coloring. Because the molecular motions are completely random, a molecule of coloring is unlikely to go very far in a short time. For this reason, diffusion is very efficient for moving molecules at the cellular level but extremely inefficient for moving molecules a long distance.

When we compare the anatomy of various-sized animals, one of the striking things is how much more "plumbing" large animals have than small ones. The one-celled amoeba has no tubes or vessels because it obtains nourishment by diffusion from the pond water surrounding it. Diffusion also carries the nutrients within the amoeba, and it carries the amoeba's wastes back to the water. Small worms have a simple gut and a few rudimentary circulation vessels, but diffusion brings most food in and takes most wastes out.

In larger animals, diffusion doesn't carry the nutrients where they are needed or remove the wastes fast enough. The larger the animal is, the more tubes and vessels are needed to supplement diffusion. Diffusion still transports things at the cellular level, but the main transport from the outside is by a pumped flow. A human has tubes everywhere—arteries to bring fresh blood from the heart, veins to take it back, lymph ducts to remove excess intercellular fluid, tubes to bring air in and out, tear ducts, gland ducts, urinary tubes, bile ducts, and so on. All of these tubes are necessary because diffusion cannot move molecules over long distances.

Direction of Diffusion

In diffusion, the net movement of a substance is from where it is most concentrated to where it is least concentrated. This might sound as if the individual molecules know which way to go, but that isn't the case at all. The individual molecules move completely at random, and any one molecule has as much chance of moving in one way as in another. But because the high-concentration regions have more molecules, more molecules leave these regions than enter them from the low-concentration regions. As the difference in the concentrations of a substance becomes smaller, the net flow gradually decreases. Diffusion continues until the substance becomes completely mixed and a uniform concentration exists. Of course, the molecular motions don't stop when a uniform concentration exists. The individual molecules still move as they did before, but there are just as many molecules moving in one direction as in any other, so there is no *net* movement of the substance.

Since diffusion always moves a substance toward lower-concentration regions, a cell can only eliminate wastes by diffusion when the concentration of wastes is lower in the surrounding water. Thus, some means must exist of either taking the wastes out of the surrounding water or providing fresh water. Otherwise, the cell would be killed by its own waste products. One example of this cellular suicide occurs in fermentation, where yeast cells live on sugar and give off ethyl alcohol as a waste product. In the beginning, the yeast cells easily transfer the ethyl alcohol out to the water because the water has no alcohol in it. As the alcohol concentration builds up in the water, it builds up even higher in the cells. Finally, when the alcohol concentration in the water reaches 12%, the cells can no longer survive and fermentation stops. This is why the ethyl alcohol concentration doesn't exceed 12% in naturally fermented alcoholic beverages.

In the lungs, gas is exchanged by diffusion. The air in the lung alveoli has a high concentration of oxygen and a very low concentration of carbon dioxide. On the other side of the alveolar membrane, the blood returning from the cells contains a lot of carbon dioxide and very little oxygen. So the oxygen diffuses into the blood and the carbon dioxide diffuses out of the blood into the air until the concentrations on both sides of the membrane become roughly equal; at this point we exhale the old air and bring in a fresh breath.

When we are exercising, we want to take in oxygen as fast as possible. So we breathe rapidly, keeping the oxygen concentration in the alveoli as high as possible and the carbon dioxide concentration as low as possible. If, for whatever reason, a sick person isn't getting enough oxygen into his or her blood, the amount can be increased by increasing the oxygen concentration in the air, thus increasing the rate of oxygen diffusion into the blood.

Diffusion in the Eye

The eye cornea has no blood vessels, since blood vessels would block the light. Rather than getting its oxygen from the blood, the cornea gets oxygen by diffusion through the tear fluid. This creates a potential problem with contact lenses. There is no problem when a person is awake, because the constant blinking movements rock the lenses and pump tear fluid underneath to bring oxygen to the corneas. But when the person sleeps, eye motions stop and the lenses block the tear fluid from the corneas, thus depriving the corneas of oxygen. For this reason, contact lenses should not be worn when sleeping.

Osmosis

The process of diffusing through a membrane is called *osmosis.* Even though osmosis causes some remarkable effects in cells, it is not fundamentally different from diffusion—it is just the random motion of molecules through a membrane. The remarkable effects occur because not all molecules can go through cell membranes. That is, cell membranes are *semipermeable;* some molecules go through easily, some only slowly, and others not at all.

Several factors control membrane permeability. Small molecules such as water usually go through faster than large ones, and fat-soluble molecules go through faster than ionic salts. The adhesive forces between molecules and the membrane also affect permeability, as does the specific chemistry of the molecules. The cell membrane can also change its permeability somewhat; it may let certain molecules pass at some times and block them at others.

Because cell membranes are semipermeable, there is a pressure difference when the concentrations of dissolved substances on the two sides of the membrane vary. A pressure of this kind is called an *osmotic pressure,* and it is an important factor in maintaining proper water content in the cells.

Let's look into the cause of osmotic pressures on a molecular level. Consider a semipermeable membrane that lets water molecules pass through but stops salt (NaCl) molecules and ions. (Most cell membranes behave this way, incidentally.) Now assume that a strong salt solution is on one side of the membrane and that fresh water is on the other. Assume that the initial pressures are the same on both sides. On the saltwater side, the pressure is due to both water molecules and salt molecules; however, on the water side, the pressure is due to water molecules alone. Thus, water molecules must be striking the membrane more often on the water side than on the saltwater side. Therefore, the chances of a water molecule passing through the membrane from the water side to the saltwater side are far greater than they are for a water molecule going in the other direction. As a result, there will be a net flow of water by osmosis to the saltwater side. This flow will continue until the pressure on the saltwater side becomes so high that the chances of a water molecule going into the fresh water are the same as the chances of one going to the salt water.

The more concentrated the salt water is originally, the greater the pressure must become to make the backward and forward diffusion probabilities equal. That is, the greater the concentration difference is, the greater the osmotic pressure will be.

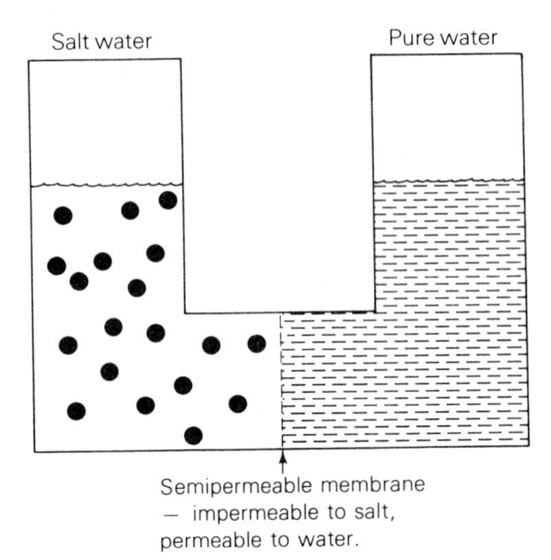

Figure 12-2

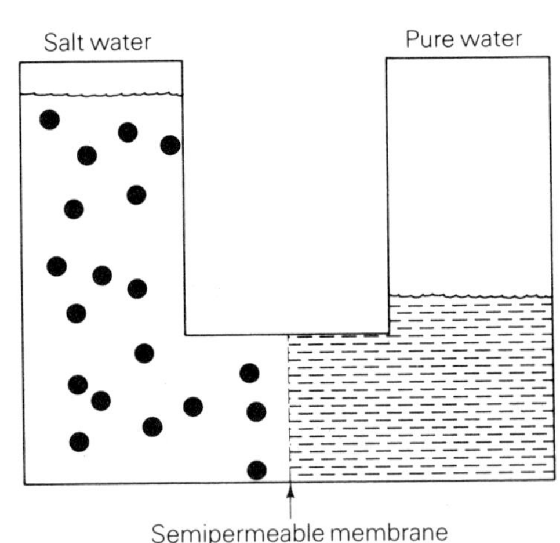

Figure 12-3 The water will flow through the membrane into the saltwater side until the chances of a water molecule going in either direction are equal.

Diffusion and Osmosis in Cells

The fluid around a cell must have the right concentration of salt to maintain the correct osmotic pressure in the cell. If the fluid outside has too much salt, the water will flow out and dehydrate the cell. If the fluid has less salt than is in the cell, the water will flow into the cell, and the cell will swell and burst.

The behavior of the red blood cell (erythrocyte) is a good example of osmotic pressure. When the erythrocyte is in normal plasma with the right amount of salt, the cell will be just filled with water, and the membrane will be tight but not stretched. The salt concentration in the cell will be almost the same as it is in the plasma. If the cell is put into fresh water, the water rapidly diffuses into the cell, swelling and bursting the cell. But if the salt concentration in the plasma increases, as it does when a person becomes dehydrated, the water leaves the cell for the saltier plasma, causing the cell to shrink. If the salt concentration in the plasma gets too high, the cell will eventually shrivel up and die.

In hundreds of bad detective stories (and one or two good ones), the detective is confronted with the question "Did the victim drown in the ocean or was she drowned in her bathtub and then thrown into the ocean?" A knowledge of osmosis makes it easy to answer this question. When a person drowns in fresh water, water flows into the blood through the lung membranes; the blood thins out and the red blood cells in the lungs swell and burst. When a person drowns in salt water, just the opposite happens. Water leaves the blood plasma, so the blood becomes thicker and saltier; the red blood cells in the lungs become dehydrated and shriveled. So examining the victim's lungs quickly answers the detective's question.

Some technical terms are used to describe the fluid balance. When the salt concentration is just right, the fluid is said to be *isotonic*. When there is not enough salt, the fluid is *hypotonic*. If there is too much salt, the fluid is *hypertonic*. The effect of these conditions on red blood cells is shown in Figure 12-4.

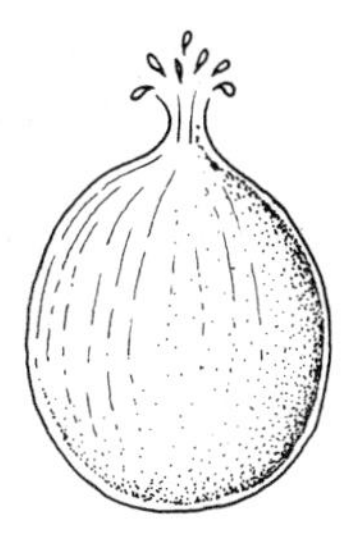

Figure 12-4 The amount of water inside a red blood cell (erythrocyte) depends on the salt concentration in the water outside the cell.

Of course, the salt concentration of intravenous fluids must be nearly isotonic to prevent damaging the cells by either swelling or dehydration. The various salts of sodium, potassium, and the other elements dissolved in the bodily fluids are called *electrolytes* because such solutions can conduct electricity. Thus, when a patient's electrolyte balance is off, it doesn't mean that the person has too much or too little electricity; it means that the salt balance of the fluids is off.

Careful control of a patient's electrolyte balance is essential in the treatment of many diseases, especially those that lead to excessive fluid loss by vomiting and diarrhea. The importance of careful fluid management was first discovered during the cholera epidemics that swept the world in the nineteenth century. Cholera victims suffer from vomiting and diarrhea, their skin loses its elasticity, their blood becomes very thick, urination stops, and without treatment the mortality rate is very high. Eventually physicians realized that most cholera symptoms resulted from dehydration due to the water loss. They then started treating the disease by intraveneously injecting water and electrolytes, which drastically cut the mortality rate. The physicians who first tried this treatment on cholera patients described the results as almost miraculous.

They told of almost dead patients who were brought back to consciousness and a healthy appearance a half hour after the fluids were given.

Fluid loss from sweating produces the same sort of symptoms as a dehydrating disease, although the symptoms are usually not as severe. However, if the dehydration is long-term or chronic, it can be dangerous, particularly if the person must work or exercise when it is hot. Athletes should always be encouraged to drink water whenever they want it during a workout. Fruit juices and the commercial electrolyte drinks can help to replace lost electrolytes along with the water. Some old-time coaches used to withhold water during practices to "toughen 'em up," but this technique only makes the players more susceptible to heat exhaustion.

People sometimes wonder why shipwrecked mariners cannot quench their thirst with seawater. The reason is that seawater has much more salt than cell plasma; thus, drinking seawater draws water *out* of the body into the intestines, thus dehydrating the person even more.

The amount of salt (sodium chloride, or NaCl) that is consumed must be watched fairly closely, because the body tends to hoard sodium whether it needs it or not. When a person takes in too much sodium, water is drawn from the cells into the fluid between the cells. This makes it harder for the blood to deliver oxygen and nutrients to the cells, so the blood pressure goes up to compensate for this deficiency. If the blood pressure is already high, more salt just makes it worse. For this reason, patients with high blood pressure are often encouraged to substitute potassium chloride (KCl) for ordinary salt (NaCl). Since most normal foods contain plenty of salt, people should be encouraged not to use a lot of salt. Most people should only eat salty foods occasionally. The only exception is people who work or exercise in hot weather, who might need a little extra salt sometimes. However, the amount of extra salt should be watched closely, since it is easy to overuse salt tablets.

Fluid Balance in Burn Victims

Burns cause particularly severe fluid and electrolyte shifts. The burned capillaries become permeable to both water and electrolytes, so the blood plasma leaks out quickly. Even in a minor burn, the leaking fluid quickly causes large blisters. When the burn covers a large area, the plasma losses are severe. The fluid leakage also causes extreme swelling in the burned area. Thus, careful restoration and maintenance of the blood plasma is one of the major tasks during the first few days of burn treatment. Because of the fluid loss, burn patients become very thirsty. At one time they were allowed to drink whatever they wanted, whenever they wanted it, until it was observed that patients who drank mostly water went into convulsions after a few days. However, patients who drank a lot of chicken soup did much better. Of course, it wasn't anything magical in the chicken soup; it was the salt. Because both water and salt were lost through the damaged capillaries, replenishing the water alone resulted in a severe sodium deficiency. Without the salt in the chicken soup, the plasma became far too dilute and the osmotic pressure in the cells was thrown off.

The Kidneys

The kidneys have two important functions: They remove any waste products from the blood, and they keep the body's electrolytes and water in balance. When the blood has too much water or too much salt, the kidneys step up the removal of whichever substance is in excess. We won't go into detail on kidney function, since the kidneys clean the blood using every type of membrane transport that we have talked about, as well as some that we haven't. Instead, we will just look at the relationship between kidney function and blood pressure—a relationship that is responsible for a good deal of high blood pressure (hypertension).

The blood-cleansing process starts in a little bundle of capillaries (the glomerulus) enclosed in a small sac (Bowman's capsule), which is drained by a small tube (proximal convoluted tubule) (Figure 12-5). The capillaries of the glomerulus have very fine pores that pass water, electrolytes, and other small molecules into Bowman's capsule. Blood cells, protein molecules, and other large molecules are kept in the blood. It is here that pressure becomes a crucial factor. If the pressure was the same on both sides of the capillary wall, the osmotic pressure would take the water and other small molecules in the wrong direction, from Bowman's capsule into the blood. In order to force the water out of the capillaries, the water molecules in the capillaries must hit the capillary walls more often than the water molecules on the outside do. That is, the blood pressure inside the capillaries must be far greater than the pressure outside. The artery that delivers blood to the kidneys (renal artery) connects directly to the aorta, the large artery coming right from the heart, so the blood pressure at the kidneys is basically the full pressure of the heart. If for some reason this pressure isn't high enough to keep the flow in the right direction, the kidneys secrete a powerful hormone (renin) that shrinks all the other blood vessels in the body. Because this tends to reduce the blood flow, the blood pressure goes up to compensate. If the kidneys can function correctly with only a small increase in pressure, no harm is done. However, if a large pressure increase is needed, all the problems of high blood pressure come with it, including the very serious possibility of capillary damage in the kidneys themselves.

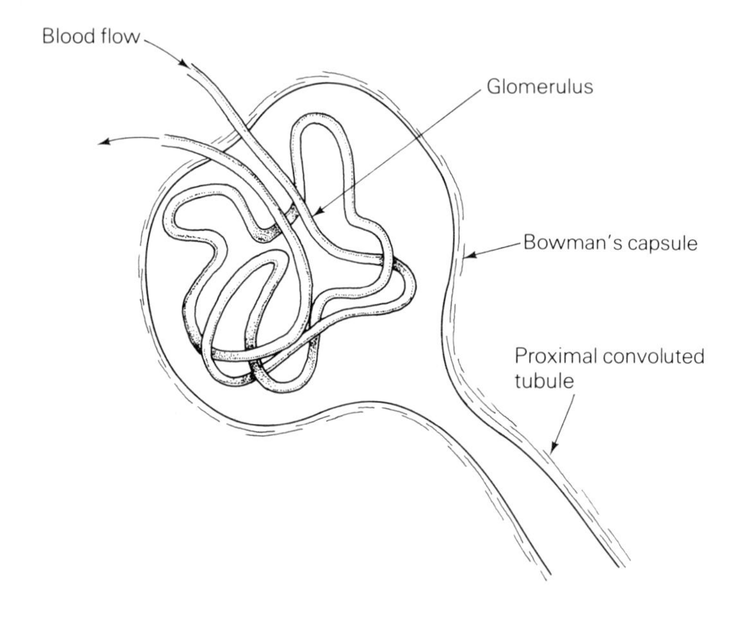

Figure 12-5 The blood-cleansing process in the kidneys starts with blood entering the glomerulus at high pressure. Water and wastes pass by osmosis into Bowman's capsule.

Active Membrane Transport

Diffusion and osmosis are not the only forms of membrane transport; they are just the easiest to explain. Every cell has substances that must be moved to regions where they are more concentrated; that is, they must be moved opposite to the direction of diffusion. This is done by a process called *active membrane transport.* It is "active" because energy is required to move a substance from a low-concentration region to a high-concentration region; in active membrane transport, the substance must be pumped.

In most books, including this one, the discussions of active membrane transport are short. They are short not because the phenomenon is unimportant but because it is poorly understood. Somehow molecules are picked up at one side of the membrane and combined into the structure of the membrane. Next, the molecule is moved through the membrane to the other side, the high-concentration side, where it is released. There are some theories on how energy is delivered to the molecule to move it, but no one really knows how active transport works in detail. However, don't let the crudeness of the explanation make you think that active transport can be ignored. The details of active membrane transport are one of the key missing pieces in the puzzle of cell metabolism.

Cellular Thermodynamics

Thermodynamics started out as the study of steam engine efficiency. It still applies to that subject today, but it has also become an essential part of the study of chemical reactions, including those that happen in living things. Thermodynamics even sheds a bit of light on the question "What is life?"—the central question of biology. This is the aspect of thermodynamics that we will consider.

Thermodynamics starts with two simple laws. The first law of thermodynamics is just conservation of energy. That is, in any process, the sum of the heat and other forms of energy coming in equals the sum of the heat and other forms of energy going out. Of course, this law applies to cells and steam engines alike. In a muscle cell, for instance, the sum of the mechanical energy and heat that the cell produces must be equal to the chemical energy that the cell takes in.

The second law of thermodynamics is also simple: Heat flows from the hot areas to the cold areas in a stationary system. But what does this simple law have to do with life? The connection turns out to be as simple as the connection between order and disorder. Heat is the disordered energy of atomic motion, so heat and disorder generally go together. All other things being equal, the hotter something is, the more disordered its atomic motions will be. In contrast, life implies atomic order. For life to exist, there must be a pattern to the atoms, and they must be capable of organizing other atoms into the same pattern. In other words, life requires that atoms be continually organized into states of greater and greater order.

Suppose we have two identical blocks, one of which is very cold—near absolute zero—while the other is hot. When we place them in contact, both common sense and the second law of thermodynamics tell us that the cold block will warm up and the hot block will cool until they are both at the same temperature, which will be about halfway between the two original temperatures. What happens to the order when we put the blocks together? Does it increase or decrease, or does it stay the same? At the beginning, the cold block is very highly ordered, and its atoms are nearly stationary. The hot block is very disordered, and its atoms are moving randomly. As the cold block warms up, it goes from almost perfect order to a much more disordered state. As the hot block cools, it becomes a bit more ordered, but the slight increase in its order in no way makes up for the enormous loss of order in the cold block. The situation resembles what happens when you open the door between a very clean room and a very dusty room: The dust spreads evenly throughout the two rooms. That the dirty room has only half as much dust in it as it did originally does not make up for the loss of cleanliness and order in the clean room. One clean room and one dirty room is a far more ordered and organized state than two half-dirty rooms. Similarly, one hot block and one cold block is a more ordered state than two blocks with the heat energy evenly distributed between them.

Since the second law says that heat flows from hot to cold, it also says that disorder increases during any process that involves heat flow and is unchanged during any process that does not. In fact, that is another way of saying the second law.

Second Law of Thermodynamics (Alternate Statement)

The amount of disorder in an isolated system cannot decrease. If there is any heat flow, the amount of disorder must increase.

The technical word for molecular disorder is *entropy;* the technical statement of the second law is that the entropy of an isolated system never decreases.

What the second law says about life is that the ordering of matter to form new cells can *only* take place if an even greater disordering occurs someplace else. Thus, as a cell organizes matter to grow and maintain life, it must disorder its surroundings even more. That is, heat must be given off by the cell while it is alive; therefore, in accordance with the first law of thermodynamics, cells must take in energy to live.

Thermodynamics tells us that life can only exist if there is a continuous flow of energy through the cell to its surroundings. When we talked about energy before, we made the point that watching the energy flow, the metabolism, of a cell was very important in understanding the life of a cell. Thermodynamics now shows us why that flow is so important. Without that flow, life would be impossible.

Concepts

Membrane
Diffusion
Osmosis
Semipermeable membrane
Osmotic pressure
Isotonic
Hypotonic
Hypertonic
Electrolytes
Hypertension
Active membrane transport
Thermodynamics
First law of thermodynamics
Second law of thermodynamics
Entropy

Discussion Questions

1. Describe the structure of a typical cell membrane.
2. Why is the amount of "plumbing" in a large animal much greater than in a small animal?
3. What are hypotonic, isotonic, and hypertonic solutions?
4. Explain what causes the osmotic pressure difference across a semipermeable membrane.
5. Explain why red blood cells shrivel up in very salty water. Why do they burst in fresh water?
6. Why can't a shipwrecked sailor live on seawater?
7. Explain why careful control of bodily fluids and electrolytes is an essential part of the treatment of burns and dehydrating diseases.
8. Why does an excess of salt in the diet often lead to high blood pressure (hypertension)?
9. Explain why athletes should be encouraged to drink plenty of water or juice during and after vigorous exercise.
10. What is the difference between osmosis and active membrane transport?
11. Explain why life cannot exist without a continuous flow of energy through the organism.

Experiments

1. Diffusion can be observed by taking a glass of water and placing a drop of food coloring in it. The water should be very still, so it should

be allowed to sit undisturbed overnight before the food coloring is added. Notice that the water close to the coloring becomes colored rapidly, but the coloring takes a long time to spread throughout the water.

2. Random diffusion is efficient for moving things short distances but very inefficient for moving things long distances. This can be seen in the following way. Place a marker on the middle line of a lined sheet of paper. Now start flipping a coin; move the marker up one line for every head and down one line for every tail. Keep track of where the marker goes by putting a pencil mark on the line every time the marker lands on it. Do you ever get very far from the center?

3. The different effects of salt water and fresh water on cells can be seen by soaking one hand in fresh water and the other in very salty water. The salt water should have about 0.25 kilogram of salt per liter of water. The experiment takes about one hour of soaking. Describe and explain what happens to your hands. Don't go near any electrical devices, such as a TV, without first drying your hands.

CHAPTER THIRTEEN

Electricity

Educational Goals

The student's goals for Chapter 13 are to be able to:

1. Describe the atomic origins of electricity.
2. Explain the differences between conductors, insulators, and semiconductors and give examples of each.
3. Give definitions of current and voltage.
4. Calculate the power of an electrical device from the current and voltage.
5. Explain the difference between AC and DC electricity.
6. Calculate the resistance of objects connected in series and in parallel.
7. Describe how electrical heaters, lights, meters, and motors work.
8. Give examples of how electricity is generated by static electricity, magnetic induction, and chemical batteries.
9. Explain how osmosis leads to a resting electrical potential in a cell and what changes happen in the cell membrane when the action potential occurs.
10. Give the five main rules for electrical safety and explain the reasons behind them.

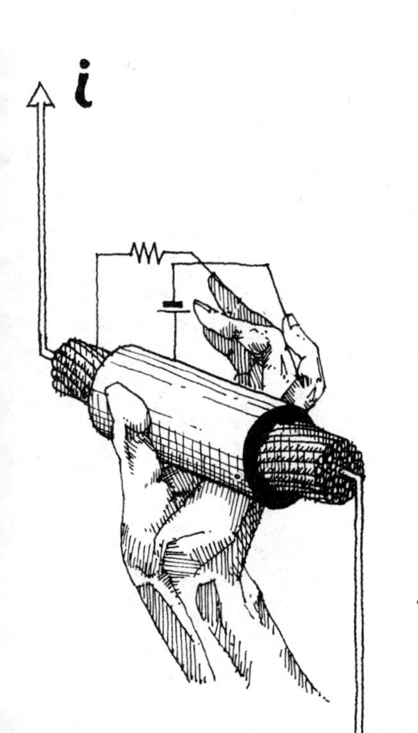

The most remarkable thing about electrical energy is how easily it can be controlled. By just flipping a switch, we can turn on a light or start a motor—whether it is across the room or across the state. By pushing a few buttons on a phone, we can ring a bell in any home or office in the country. No physical law stops us from doing these things mechanically, but you can imagine how complex and expensive a mechanical system would be that could ring a bell wherever we wanted.

Because electricity is so easy to control, electrical impulses are at the heart of every modern information processing system. Telephones, radios, televisions, computers, and pocket calculators all use electricity. We can see how fast electricity has changed things by comparing a $20 pocket calculator with the mechanical calculator that it replaced. Twenty five years ago, the mechanical calculator sold for $1,000—a cheap car wasn't much more at the time—and it broke down constantly. Now $1,000 will buy a minicomputer (but not a car).

Electricity's controllability also makes it well suited for transmitting energy over long distances. Electricity is produced at generating plants and hydroelectric dams, then sent to wherever it is needed over high-voltage lines. At the end of each line, the electricity is transformed to a lower, safer voltage and used to run such things as motors, lights, TVs, and computers and to generate heat.

The only serious drawbacks to electricity are that it is more expensive than most other forms of energy and that it cannot be stored easily. At present, over 99% of the electrical energy in the world is consumed a few milliseconds after it is produced. Tiny amounts of electricity can be stored in batteries, but no one knows how to store large amounts. Consequently, generating stations must have very large reserve capacities that are big enough to power all the air conditioners on the hottest day. If electrical energy could be stored cheaply, we could get along with far fewer generating plants. Very little can be done to cut the cost of electricity. Since electricity must be made from some other form of energy, it has to cost more than the energy from which it is made.

As important as electricity is for our civilization, it is even more important to us for the roles it plays within our bodies. In humans, all information gathering and processing is done with electrical impulses. Brain activity, sight, hearing, touch, taste, smell, pain, and muscle control directly involve electricity. Even tissue and bone healing seem to be controlled to some extent by electricity.

The impulses in nerves and muscles are not very large, but they are easily detected with modern electronics. The signals from the brain, heart, and muscles have proven very useful in diagnosing and treating diseases in these organs, but there is still much to learn about what these signals tell us about the body.

Atomic Origins of Electricity

We don't know exactly what electricity is any more than we know what mass is or what life is. However, we do know that electricity has its origins in the structure of the atom. An atom consists of a heavy central core, called the nucleus, surrounded by a cloud of light electrons. The nucleus is made up of two different subatomic particles: neutrons and protons. Both types have about the same mass, and they are heavy as subatomic particles go.* The neutrons have no electric charge, the protons are positively charged, and the electrons are negatively charged. The electrical attraction between the negative electrons and the positive protons keeps the electrons orbiting around the nucleus.

In many ways, an atom resembles a miniature solar system (Figure 13-1). The light electrons orbit the heavy nucleus in much the same way that the planets orbit the sun. The proportions of atoms and the solar system are even similar: The nucleus occupies about the same percentage of the atom as the sun does of the solar system. Of course, this analogy shouldn't be carried too far; subatomic particles have a wave character that is not noticeable in large chunks of matter.

The atom is electrically neutral; that is, the positive charge of the protons is exactly balanced by the negative charge of the electrons. However, the electrons are not attached to the atom very

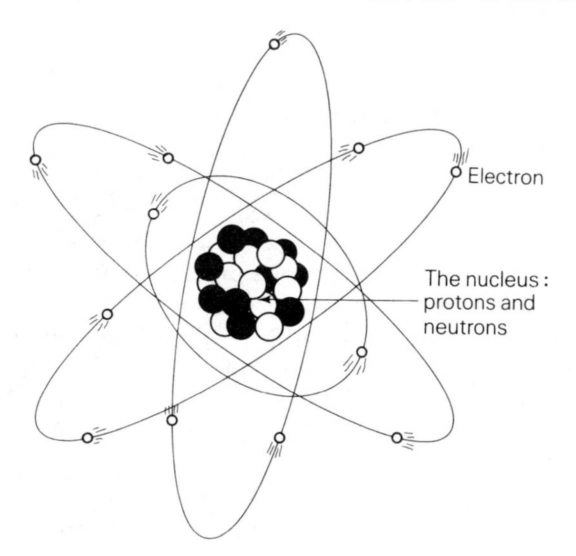

Figure 13-1 An atom. The nucleus is greatly enlarged. If the nucleus were drawn to the same scale as the rest of the atom, it would be the size of a pinpoint.

*There are many other types of subatomic particles, but electrons, protons, and neutrons are the primary ones in ordinary matter.

tightly; the outside electrons in particular can be detached fairly easily. When an electron is removed, the atom has a net positive charge; the atom to which the detached electron goes has a net negative charge. Incidentally, atoms with an electric charge are called *ions.* If they have excess electrons and thus a negative charge, they are called *anions;* if they have missing electrons and thus a positive charge, they are called *cations.*

There are many ways to remove electrons from atoms. Simply rubbing two different materials together will transfer electrons from one material to the other because different materials attract electrons differently. This is what causes ordinary static electricity. When you rub your shoes on a carpet, electrons from the rug stick to your soles and then move to you. The electrons stay on your body until you touch something that conducts electricity, like a metal doorknob. Then the electrons jump from you to the doorknob, producing a spark and a slight shock.

Electrons can also be transferred from atom to atom by chemical means. Atoms of different elements have different affinities for electrons. When an atom with one or more loosely bound electrons meets a second atom to which electrons cling strongly, the loose electrons will migrate from the original atom, leaving it positively charged (cation) and making the second atom negative (anion). The science of chemistry is basically the study of how electrons move between atoms and how this movement influences the arrangement of atoms in molecules.

Conductors, Insulators, and Semiconductors

The rate of electron flow in a substance depends on how tightly atoms hold their electrons. When the atoms of a substance hold their electrons very tightly, there will be no free electrons and no electron current; thus, electricity will not move through the material easily. Such materials are called *insulators.* Materials whose atoms hold their electrons loosely will carry an electric current readily. These materials are called *conductors.*

Metals are the best conductors. Silver is the best of all, and copper is a close second. Another important class of conductors are liquids containing dissolved ions, such as salt water. In these liquids, the electric current is carried by ions instead of electrons. Because ions are much larger and heavier than electrons, the speed of electrical impulses is much slower in ionic liquids than in metals. This is why metals are shiny and ionic liquids are not. In order to reflect light, the electric charges in a material must wiggle back and forth very rapidly. Ions don't wiggle fast enough because they are too big and heavy.

Ionic conduction is the most important kind of conduction in animals. Nerve signals, for instance, travel by ionic conduction. Because ions can't move as fast as electrons, nerve signals travel much slower than signals in, say, a metal telephone wire. In a millisecond, nerve signals travel about 1 meter, while a telephone message can go 100 kilometers. As we mentioned before, the salts of sodium, potassium, and other compounds which dissociate to form ions, when dissolved in the bodily fluids, are called electrolytes because the ions can carry an electric current.

As long as they aren't wet, most nonmetallic solids are insulators. Most plastics, glass, and ceramics are very good insulators, as are most organic chemicals. Static cling is a problem with most modern synthetic fabrics because they are extremely good insulators; once they become electrically charged, the charge takes a long time to flow away. Since cotton is not as good an insulator, it is used when static electricity could be dangerous—when working with explosive chemicals, oxygen, or explosive anesthetic gases, for instance.

A small class of materials conducts electricity weakly but very controllably. These are *semiconductors,* and they are the key materials in solid-state electronics. Every modern electronic device uses semiconductors. Silicon and germanium are the two most important semiconducting elements.

Electrical Forces

There will always be a force between two electrically charged objects. This is the force that causes static cling, and it is also the main force that holds atoms and molecules together. The direction of the force depends on the charges. The basic rule is that unlike charges attract and like charges repel (Figure 13-2). Thus, two negative charges or two positive charges will tend to move away from each other, while a positive and a negative charge will tend to move toward each other.

The size of the force depends on two things: the size of the electric charges and the distance between the charges. As we would guess, the force increases as the charges get bigger and it decreases as the charges move away from each other. The electric force between two charges is inversely proportional to the square of the distance between the charges. That is, doubling the separation cuts the force down to one-fourth its original value, tripling to one-

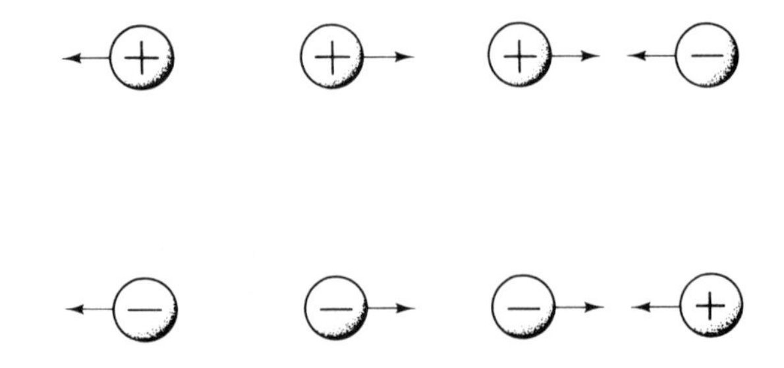

Figure 13-2 (a) Like electrical charges repel each other. (b) Unlike electrical charges attract each other.

ninth, and so on. As you move away from a light, its apparent brightness decreases in exactly the same way.

The SI unit of electric charge is the **coulomb** (C). The size of a coulomb cannot be described easily, but the fact that a static electricity spark between you and a doorknob involves about a microcoulomb gives you some idea of its size. A coulomb is also equal to the charge on 6 billion billion protons (6×10^{18}). We will use the letter Q to denote the charge in formulas.

Current, Voltage, and Power

Current

We will usually be more concerned with the rate of charge flow than the amount of charge present. This rate is called the electric current, and it is measured in coulombs per second (C/s). Since current is used so often, the C/s is given the name **ampere** (A), or amp for short:

$$\text{SI unit of electric current} = \frac{\text{couloumb}}{\text{second}} = \text{C/s} = \text{A}.$$

The letter I is used for current. Algebraically,

$$I = \frac{Q}{t}.$$

The ampere is mentioned most often on fuses and circuit breakers, devices that cut off the current if it gets dangerously high. For instance, a 15-amp fuse will carry any current up to 15 amps. When more than 15 amps flow through, the thin metal strip in the fuse melts and stops the flow.

Voltage

The voltage—or electrical potential difference, as it is sometimes called—is closely related to the energy required to move a charge from one place to another. Suppose we wanted to move a small positive charge Q from point A to point B, which is close to a large positive charge (Figure 13-3). Since the two charges repel each

Figure 13-3

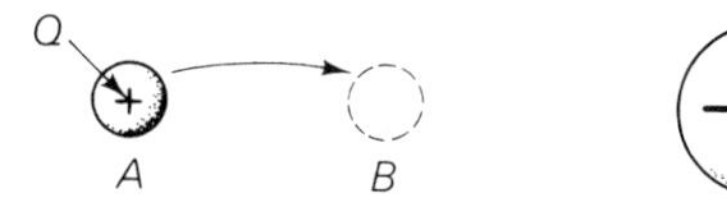

other, we must furnish energy. The voltage V between A and B is defined as the energy E per unit of charge needed to move a charge from A to B. So,

$$V = \frac{E}{Q}.$$

If the charge and voltage were known and we wanted to find the energy,

$$E = QV.$$

The SI voltage unit is the **volt** (V). It takes 1 joule of energy to move 1 coulomb of charge through a voltage of 1 volt.

The voltage in an electrical problem is very similar to height in a mechanical problem. The energy needed to lift something is proportional to the height to which the object is raised; the energy needed to move a charge is proportional to the voltage between the two points. Thus, a person at a high voltage is like a person standing on the edge of a high cliff. If the latter person drops over the cliff, a great deal of energy will be released. If a charge flows through the former person—and it will if the person touches anything—a lot of energy will be released and the person will receive a dangerous shock. However, being at a high voltage is not dangerous as long as no charge moves through the person.

The voltage is sometimes compared with the pressure in a fluid flow problem. The analogy between voltage and pressure isn't quite as good as the analogy to height, but it is useful because a large voltage difference across a conductor produces a large current, just

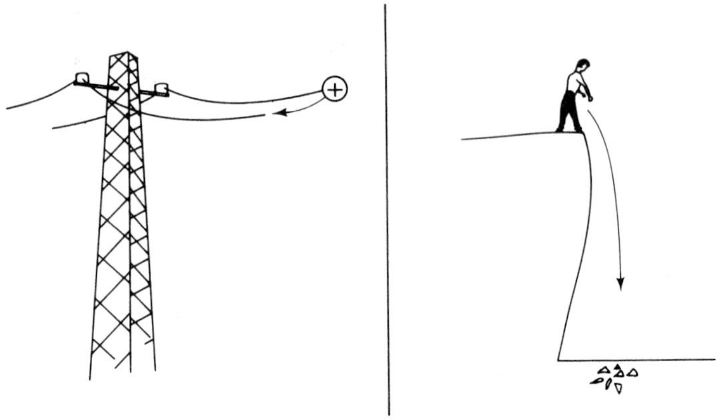

Figure 13-4 A charge at a high voltage is very similar to an object on top of a cliff—energy will be released when the charge drops to a lower voltage and when the object drops over the cliff.

as a large pressure difference across a pipe produces large fluid flow.

EXAMPLE A heavy-duty, 12-volt car battery has a rating of 90 ampere-hours, which means that we can draw a current of 90 amps for 1 hour, 1 amp for 90 hours, 2 amps for 45 hours, and so. How much energy is stored in this battery?

Answer An ampere-hour rating is really a rating of how much charge is stored in the battery. We can take the charge out fast or slowly, but the total amount of charge will be about the same. The total amount of charge in a 90-ampere-hour battery can be found by rearranging the equation for current

$$I = \frac{Q}{t}.$$

Rearranging to find the charge, we get

$$Q = It.$$

In the case of a 90-ampere-hour battery, we know

$$\begin{aligned} Q &= 90 \text{ A} \times 3{,}600 \text{ s} \\ &= 324{,}000 \text{ C}. \end{aligned}$$

(Time, as always, must be expressed in the SI time unit, the second, when we are doing problems.)

The total amount of energy stored in the battery is

$$\begin{aligned} E &= QV \\ &= 324{,}000 \text{ C} \times 12 \text{ V} \\ &= 3{,}900{,}000 \text{ J}. \end{aligned}$$

At residental rates of around 2¢ per megajoule, this is less than 10¢ worth of electrical energy. Batteries cannot store much electrical energy; their main advantages are their convenience and portability.

Power

Since the energy needed for an electrical device is $E = QV$, finding the power is simple: We divide the energy by the time, so

$$\frac{E}{t} = \frac{QV}{t},$$

which can also be arranged as

$$\frac{E}{t} = \left(\frac{Q}{t}\right)V.$$

But E/t is the power p, and Q/t is the current I; thus, the equation above is

$$p = IV.$$

Thus, the power of any electrical device is just the product of the voltage and the current. Of course, the power will be in the SI unit of power, the watt, as long as the voltage is in volts and the current is in amperes.

EXAMPLE Suppose a 1,200-watt electric heater and a large, 1,000-watt light are plugged into the same 110-volt circuit. Will the 15-amp fuse in the circuit blow if they are both turned on at the same time?

Answer Since

$$p = IV,$$

the current can be found from

$$I = \frac{p}{V}.$$

Of course, the total power is the sum of the power of the heater and the lamp. So p = 1,200 watts + 1,000 watts = 2,200 watts, and the current is

$$I = \frac{2{,}200 \text{ W}}{110 \text{ V}} = 20 \text{ A}.$$

The fuse will blow.

EXAMPLE Electrical energy is sold by the kilowatt-hour, although when we become fully metric it will probably be sold by the megajoule. If the price of electricity is 10¢ per kilowatt-hour, how much does it cost to leave a 100-watt light bulb burning for 24 hours?

Answer The first step is to convert kilowatt-hours into the SI energy unit, the joule.

$$1\ \text{kW}\cdot\text{hr} = 1{,}000\ \text{W} \times 3{,}600\ \text{s}$$
$$= 3{,}600{,}000\ \text{J}.$$

In 24 hours, a 100-watt bulb consumes

$$E = pt$$
$$= 100\ \text{W} \times 24 \times 3{,}600\ \text{s}$$
$$= 8{,}640{,}000\ \text{J}.$$

Thus, the cost would be

$$10¢ \times \frac{8{,}640{,}000\ \text{J}}{3{,}600{,}000\ \text{J}} = 24¢.$$

EXAMPLE The better exercise bicycles work against an electrical generator instead of a friction brake. The energy that the person produces by turning the pedals is changed to electrical energy by the generator. Then the energy goes to a small heater, where it is turned into heat. The two big advantages of this kind of exercise bicycle are that the force needed to turn the pedals can be controlled by turning a knob and the power output can be measured easily by measuring the current and voltage. Such an exercise bicycle is called an *ergometer.* Ergometers are widely used for testing athletes and by people recovering from heart disease.

If a bicycle ergometer is putting out a current of 3 amps at 100 volts, what is the power of the person turning the pedals? If a person keeps pedaling at this rate for 15 minutes, how much energy will he or she produce?

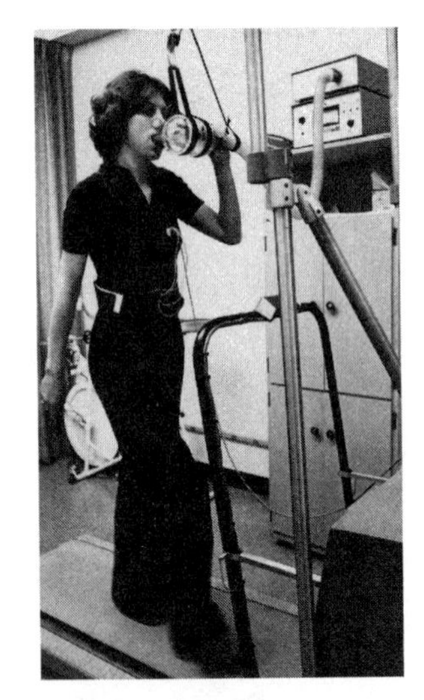

In this treadmill ergometer, the work done by the subject is converted into electrical energy and then recorded.

Answer The power will be

$$p = IV$$
$$= 3\ \text{A} \times 100\ \text{V}$$
$$= 300\ \text{W}.$$

If the person puts out this power for 15 minutes, he or she will produce

$$E = pt$$
$$= 300\ \text{W} \times 15 \times 60\ \text{s}$$

$$= 270{,}000 \text{ J}.$$

This is less than 1¢ worth of energy.

Electric Circuits

Surprisingly, not much happens when a well-insulated object touches a high-voltage terminal. A small spark jumps and a little bit of charge flows to the object, but that's about all. The charge stops flowing because like charges repel. The charge transferred by the spark repels the charges still on the terminal, stopping further charge transfer. Only a small amount of energy is given to the object, since the amount of charge transferred is tiny. This is what happens if we build up a static charge while walking on a rug and a spark jumps when we touch a doorknob. Although a voltage difference of 10,000 volts or more may exist between us and the doorknob, very little charge actually jumps.

If we want to transfer large amounts of energy, charge cannot build up in any spot, because it would repel further flow. Excess charge must be drained off and returned to its source. That is, a complete circuit must be set up so that the charge flows in a continuous cycle.

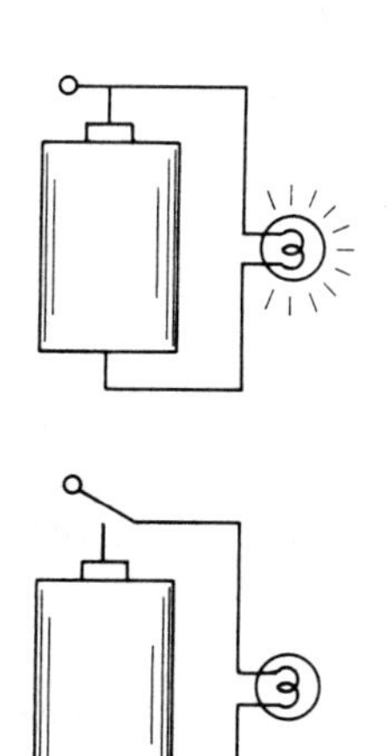

Figure 13-5 A complete circuit is necessary for a current to flow.

For instance, if we want to light a flashlight bulb, we must connect a path so that the electrons leaving the negative terminal of the battery can return to the positive terminal. Inside the battery, chemical energy pumps the electrons back to the negative terminal. To turn the bulb off, we just break the circuit at any point so that the electrons cannot go around the full circuit (Figure 13-5). Switches usually stop current in this way. Usually a switch is merely a mechanical device that physically pulls two wires apart to open the circuit. You can learn many of the important ideas about electric circuits by taking a flashlight bulb, a battery, and a couple of pieces of wire and hooking them up to light the bulb.

A bird can safely land on a bare high-voltage wire because it doesn't form a complete circuit with the wire. The most that the bird feels is a small static spark. Of course, if the bird touched another wire at a different voltage, it would be immediately electrocuted. Fortunately, most birds are too small to touch two wires at the same time.

It is well worth remembering that you must make simultaneous electrical contact with *two* objects that have a large voltage difference in order for a dangerous shock to occur. It isn't dangerous for one part of the body to touch a high voltage if the rest of the body is insulated; thus, if you avoid touching high voltages and keep your body insulated from other conductors, you have very little danger of being electrocuted. It is important to avoid wet floors, water pipes, and the ground when working with electricity because these objects provide a return path for electricity—for reasons that we shall discuss when we study the U.S. electrical system.

Currents in Circuits

A current flows whenever the two terminals of a voltage-generating device are hooked together with a conductor. However, there is a bit of confusion over the direction of flow because the terms *positive* and *negative* were chosen—by Benjamin Franklin—long before electric currents were understood. We therefore talk about currents going from positive to negative, even though the current is usually carried by electrons flowing from negative to positive. This mix-up doesn't cause much of a problem because the results remain the same.

AC and DC Electricity

Two types of current are used in electric circuits: direct current (DC) and alternating current (AC). Direct current always flows in the same direction—the electrons flow out at the negative terminal and back in at the positive terminal; the negative terminal stays negative and the positive terminal stays positive. Batteries are the DC sources that we use most often. The direction of alternating current changes rapidly; first it goes one way, then the other. This change is caused by rapidly switching the terminal voltages. First one terminal is positive and the other is negative, and then, a moment later, the voltages are reversed. Graphs of DC and AC voltages versus time are shown in Figure 13-6.

The rate at which an AC current is switched is called the frequency, which is measured in **hertz** (Hz). The number of hertz is the number of complete cycles per second. For instance, normal residential and industrial electricity makes 60 full cycles—from positive to negative and back again—every second, so it is called 60-hertz AC electricity.

You might wonder why the current direction is changed every few milliseconds. The reason is that changing the voltage of AC electricity is very easy with a transformer; changing the voltage of DC electricity is very difficult and inefficient. Electrical energy can be shipped efficiently over long distances only when it is at a high voltage, but high-voltage electricity is unsafe to use in the home or

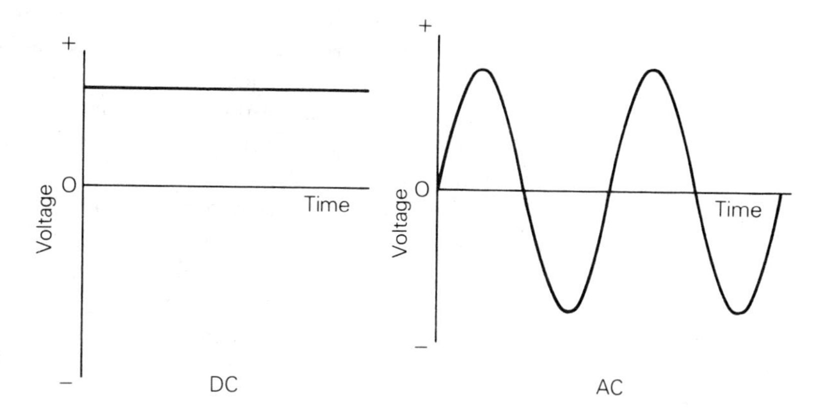

Figure 13-6 Graphs of DC and AC voltages versus time.

office. Therefore, to have a safe *and* efficient electrical system, the voltage must be increased for long-distance transmission and decreased for safe usage. These changes can only be made with AC. It is also convenient to be able to produce special voltages for specific uses, which transformers can do with AC.

Ohm's Law

The current in a conductor depends on two things: the voltage difference between the two ends of the conductor and the quality of the conductor itself. Since the voltage is like the electrical pressure, the greater the voltage is, the greater the current will generally be. The quality of the conductor depends on several things. The most important factors are the size and shape of the conductor and its composition. Fortunately, the conduction quality of most substances does not depend too much on the size of the current. As long as the current is not so large that it overheats the material, the conduction properties of most materials are nearly independent of the current. (Transistors and similar devices are the most important exceptions to this rule.)

The conduction quality of a particular object can be characterized by its resistance to the flow of electricity. The resistance of an object is measured in **ohms** (Ω). (Ω is the Greek letter omega.) As you would expect, an object with a large resistance conducts electricity poorly, while an object with a small resistance conducts electricity easily.

The resistance of an object, a wire for example, depends on its size and shape. A long, thin wire will have a much greater resistance than a short, thick one, just as a long, thin hose will have a greater resistance to water flow than a short, thick hose. Extension cords should always have the smallest possible resistance, especially when they are connected to high-power (high-wattage-rating) appliances. For instance, when you use an extension cord to hook up an electric drill, you should use a short, thick cord. Using a thin cord, like the ones used for lamps, is asking for fire, shock, and a burned-out drill.

The substance from which a wire is made is also a major factor in determining its resistance. Wires made from good conductors, such as copper, have a resistance several times smaller than wires made of poorly conducting metals, such as nickel, and billions of times smaller than wires made of insulators.

We won't bother with the details of calculating the resistance of an object. In most cases, either the resistance is given, it can be easily calculated from the voltage and wattage ratings, as we will learn later, or it can be easily measured with an *ohmmeter,* which is a meter designed to measure resistance. Table 13-1 will give you a rough idea of how the resistances of objects vary.

Table 13-1 Resistances of Some Objects

Object	*Resistance*
Copper rod	0.2 mΩ
Iron rod	1.3 mΩ
Electric coffee pot	15 Ω
100-watt, 110-volt light bulb	120 Ω
Human	
With wet skin	1 kΩ
With dry skin	100 kΩ
Wood rod	100 GΩ
Glass rod	100 TΩ

Note: All rods are 1 centimeter in diameter and 1 meter long.

The relationship between current, applied voltage, and resistance is called Ohm's laws. "Law" is somewhat of a misnomer, because the word is normally reserved for rules that *all* things obey, such as Newton's laws of motion or the second law of thermodynamics. Some substances do not obey Ohm's law, so it would be better to call it Ohm's relation—but nobody does.

The current I through an object with a resistance R is directly proportional to the voltage V applied across the object and inversely proportional to the resistance. That is,

$$I = \frac{V}{R}.$$

This is Ohm's law.

EXAMPLE How much current flows through a 180-ohm heating pad when it is plugged into a 110-volt socket? At what rate (in watts) does the pad produce heat?

Answer The current is found from Ohm's law:

$$\begin{aligned} I &= \frac{V}{R} \\ &= \frac{110\ \text{V}}{180\ \Omega} \\ &= 0.61\ \text{A}. \end{aligned}$$

The power of any electrical device is

$$\begin{aligned} p &= IV \\ &= 0.61\ \text{A} \times 110\ \text{V} \\ &= 67\ \text{W}. \end{aligned}$$

Of course, all the electrical energy ends up as heat in the pad.

EXAMPLE How much current would flow through a person with dry skin if he or she had a 1,000-volt difference between the hands and feet? How much current would flow through a person with wet skin when the voltage was 110 volts?

Answer In the first case, the resistance for dry skin is around 100,000 ohms, so the current would be

$$\begin{aligned} I &= \frac{V}{R} \\ &= \frac{1{,}000\ \text{V}}{100{,}000\ \Omega} \\ &= 0.01\ \text{A}. \end{aligned}$$

In the second case, the resistance for wet skin is only about 1,000 ohms, so the current would be

$$I = \frac{V}{R}$$

$$= \frac{110 \text{ V}}{1{,}000\ \Omega}$$

$$= 0.11 \text{ A},$$

which is 11 times more than the current through the person touching 1,000 volts with dry skin. While both persons are in severe danger, the second person is far more likely to be electrocuted.

Series and Parallel Circuits

When many objects are connected to the same voltage terminals, the current naturally depends on the resistances of all the objects, but it also depends on how the objects are connected. Objects can be connected in endless ways. We will only consider the two most useful methods: the series connection and the parallel connection (Figure 13-7).

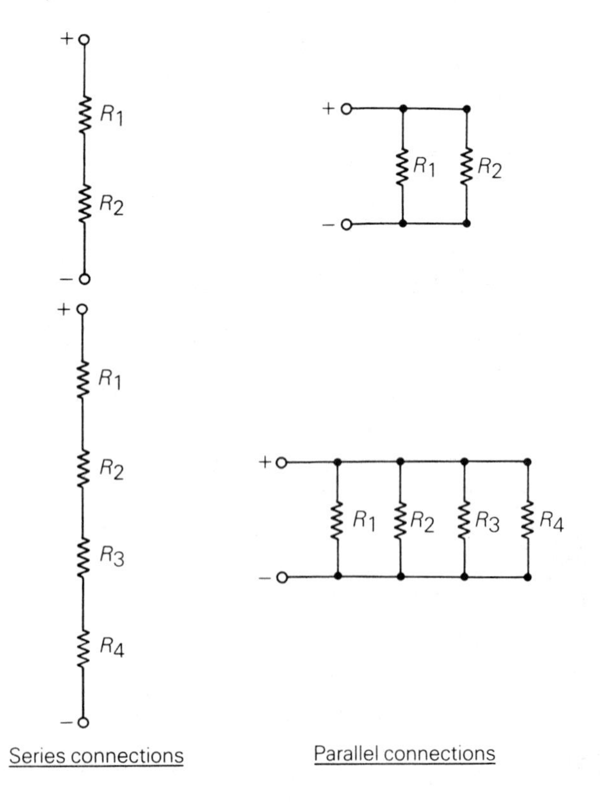

Figure 13-7 The symbol —ʍʍ— stands for a resistance.

In the series connection, the current flows through each object, one after the other. Since the current cannot flow off to the side, exactly the same current flows through each object. Since each additional object tends to retard the flow, the resistance of a set of objects hooked in series is simply the sum of all the individual resistances:

$$R_{\text{total}} = R_1 + R_2 + R_3 + \ldots.$$

The series connection isn't used often for electrical appliances because if one unit fails, the current stops. The only common household use of the series connection is for small Christmas tree lights. There are many uses for the series connection in electronics, however.

You will encounter the series connection most often when you hook up two or three extension cords to make one long cord. Since you hook up the cords in series, the resistance goes up as you add more cords. Because of the increased resistance, the applicance at the other end might not get enough current, especially if it is a high-power appliance.

In the parallel connection, the current divides every time it comes to a junction, so different currents flow through the different objects. Every object that is added in parallel adds another path through which the current flows. Thus, adding objects in parallel makes it easier for the current to flow; that is, it lowers the resistance. The situation is similar to traffic flow. When there is only one road between two towns, the resistance to traffic flow will be very large because everyone must use the road. If new roads are added, travel between the two towns becomes easier; the total resistance to traffic drops as new roads are added.

The lights and electrical sockets in all buildings are connected in parallel. Therefore, when we plug more appliances into the same circuit, we add them in parallel. Thus, whenever another appliance is plugged in, the total resistance drops and the total current flow increases—at least until the fuse blows.

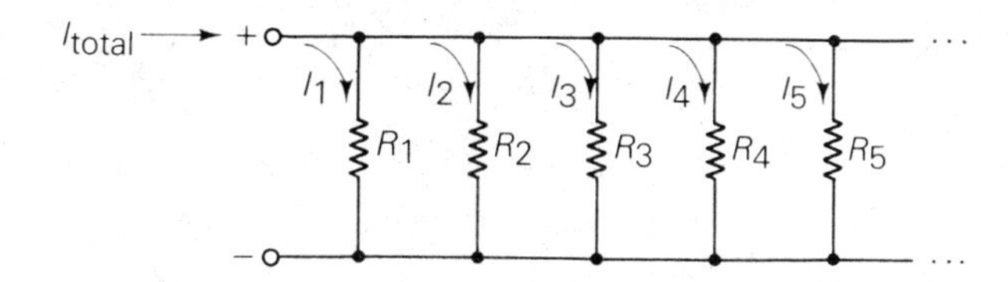

Figure 13-8

In the parallel connection, shown in Figure 13-8, the total current through the circuit is just the sum of the currents through the individual objects:

$$I_{total} = I_1 + I_2 + I_3 + \ldots .$$

Because all the objects are connected to the same two terminals, they all have the same voltage V applied to them. Thus, if we apply Ohm's law to the entire system, we know

$$I_{total} = \frac{V}{R_{total}}.$$

Ohm's law also holds for each of the individual objects;

$$I_1 = \frac{V}{R_1}, \quad I_2 = \frac{V}{R_2}, \quad I_3 = \frac{V}{R_3},$$

and so on. When we substitute these relationships for the currents in the first relation, we find that

$$\frac{V}{R_{total}} = \frac{V}{R_1} + \frac{V}{R_2} + \frac{V}{R_3} + \ldots .$$

Since V is the same for the individual objects as it is for the system, it can be canceled. The remaining equation is the relation for finding the total resistance of objects connected in parallel.

$$\frac{1}{R_{total}} = \frac{1}{R_1} + \frac{1}{R_2} + \frac{1}{R_3} + \ldots .$$

If we try this relation on a few examples, it will be clear that the total resistance of objects connected in parallel is always smaller than any of the individual resistances.

If only two resistances are in parallel, the relation above can be written as

$$R_{total} = \frac{R_1 R_2}{R_1 + R_2}.$$

This form is sometimes easier to use.

EXAMPLE A 10-ohm, a 40-ohm, and a 90-ohm light bulb are connected in series across a 14-volt generator (Figure 13-9). What is the current through the bulbs?

Answer The total resistance will be

$$\begin{aligned} R_{total} &= R_1 + R_2 + R_3 \\ &= 10\ \Omega + 40\ \Omega + 90\ \Omega \\ &= 140\ \Omega. \end{aligned}$$

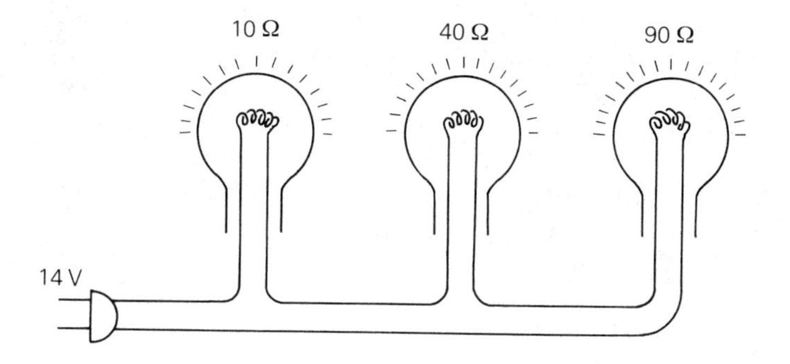

Figure 13-9

The current will be

$$I = \frac{V}{R}$$

$$= \frac{14\ \text{V}}{140\ \Omega}$$

$$= 0.1\ \text{A}.$$

EXAMPLE A 5-ohm heater and a 10-ohm light bulb are connected in parallel. What is their total resistance? What would the total current be if they were connected to a 12-volt battery?

Answer For two resistances connected in parallel,

$$R_{\text{total}} = \frac{R_1 R_2}{R_1 + R_2}$$

$$= \frac{5\Omega \times 10\Omega}{5\Omega + 10\Omega}$$

$$= \frac{50\Omega}{15}$$

$$= 3.3\ \Omega.$$

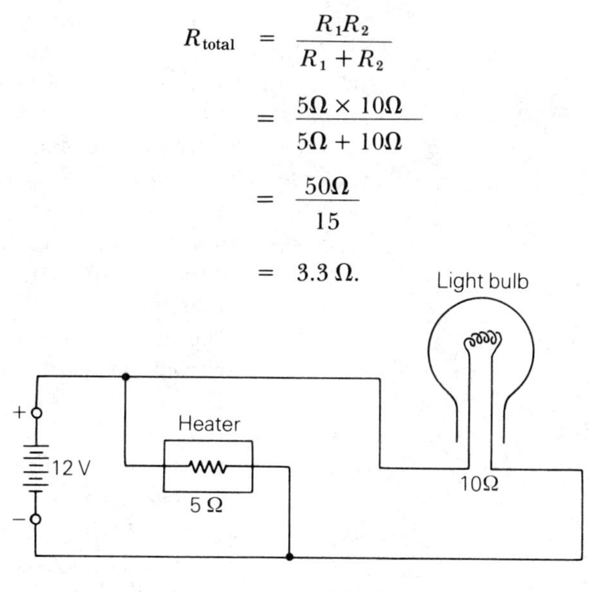

Figure 13-10

$\overset{+}{-}\!\!|\!|\!|\!|\!|\!\!\overset{-}{-}$ is the standard symbol for a battery.

As expected, this is less than either of the two individual resistances. The total current is found from Ohm's law,

$$
\begin{aligned}
I &= \frac{V}{R} \\
&= \frac{12\ \text{V}}{3.3\ \Omega} \\
&= 3.6\ \text{A}.
\end{aligned}
$$

Magnetism

The ancient Greeks knew that certain iron-rich rocks, called magnets, picked up bits of iron and either attracted or repelled each other, depending on their placement. Except for the compass, which was discovered about 1300 A.D., there were no important applications of magnetism until the early 1800s. In 1820, a discovery was made that changed the course of history: It was found that all magnetism was produced by moving electric charges. Electricity and magnetism were just two aspects of the same phenomenon. This discovery made the practical use of electricity possible.

Any moving electric charge produces a magnetic field; that is, a moving electric charge will exert a magnetic force on a magnet or a piece of iron. This is easily demonstrated by putting a compass near a wire and turning a direct current on and off in the wire. The compass needle will move back and forth. (This is exactly how the connection between electricity and magnetism was discovered by H. C. Oersted, a Danish physics teacher.) The reverse is also true; a magnetic field produces a force on a moving electric charge. To see this effect, put a magnet by a TV screen and watch the effect on the electron beam that "paints" the picture on the screen.

Since all materials contain moving electrons, we might wonder why all materials are not magnetic. The reason is that nonmagnetic materials have as many electrons spinning clockwise as counterclockwise. The equal but opposite currents thus cancel, and no magnetic field is produced. Only a few materials, such as iron, cobalt, and nickel, when under the right conditions (including not too high a temperature), have significantly more electrons going one way than the other.

The simple trick of picking up iron with a magnet has been put to many uses. It is used to locate iron ore deposits and to sort out steel cans from trash. Steel splinters are removed from the eyes with a small magnet, and steel splinters that are too small to show up on x-ray photographs can be located with a special probe that senses the charged magnetic field around the splinter. One especially useful application of magnetism is for guiding catheter tubes through the blood vessels. A iron tip is placed on the catheter tube, then magnets are used to guide the tip through the right branches of the vessels.

Recently, the attraction between magnets and iron has been put to use to treat babies with incomplete esophagi. In this birth defect, the tube between the throat and the stomach has a missing section, which prevents food from reaching the stomach. The surgical problem is that the esophagus cannot be stretched and sewn together, because rapid stretching cuts off the blood circulation and destroys the tissue. The tissue must be stretched slowly over a period of months by alternating the stretching with periods of rest every minute or so to allow the blood to circulate properly. This slow stretching is done by placing a plastic-coated piece of iron at each end of the missing section of esophagus; the baby is then put in an electromagnet, which pulls the two pieces of iron toward each other, thus stretching the esophagus. The magnet is turned off every other minute to relax the stretching and allow proper circulation. When the two sections finally meet, they are sewn together. This technique does not seem to cause the infant any pain at all.

The two ends (or poles) of a magnet are not the same. They are usually called the *north* pole and the *south* pole after the directions they point to when the magnet is free to turn. The forces between

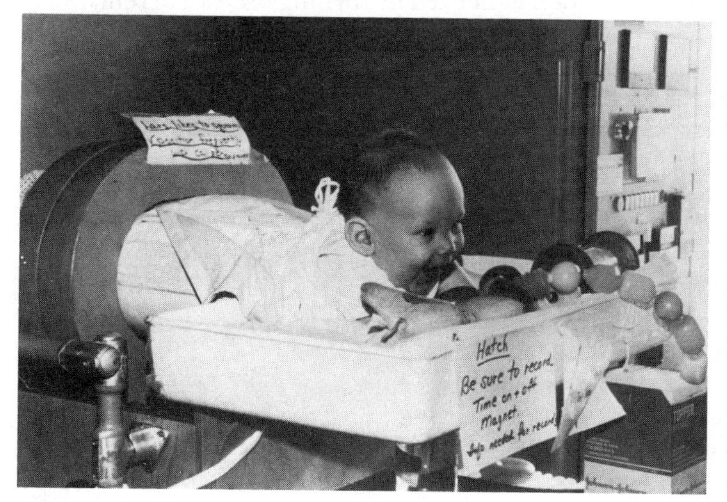

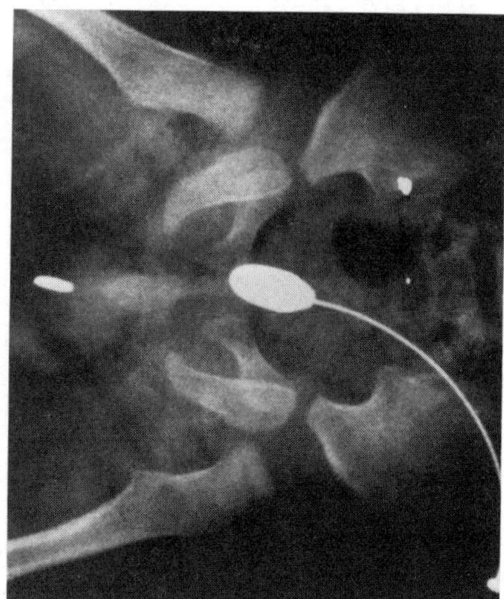

The baby, whose esophagus is being stretched magnetically, is obviously quite comfortable.

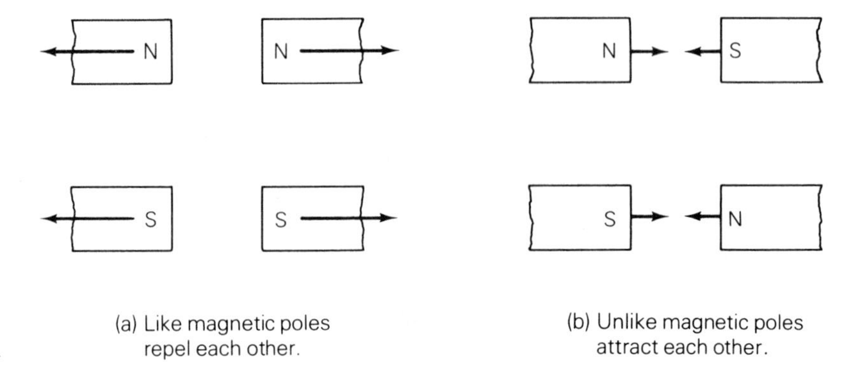

Figure 13-11

two poles are repulsive when the poles are alike and attractive when unlike. Thus, two north poles (or two south poles) repel each other, while a north and a south pole attract (Figure 13-11). In this sense, magnetic forces are similar to electric forces. However, unlike the case of electric charges, there is no known way to separate north poles and south poles. If you chop a magnet in half, you don't get one with just a north pole and another with just a south pole; you get two weaker magnets, each with its own north and south poles.

All current-carrying wires are surrounded by a magnetic field; however, the most efficient way to make an electromagnet is to concentrate the magnetic field by winding the wire in a coil (Figure 13-12). Such coils are commonly called *solenoids.* Placing an iron core within the coil concentrates the magnetic field even more. Such an electromagnet behaves exactly like a permanent magnet, except that it can be demagnetized by turning off the current.

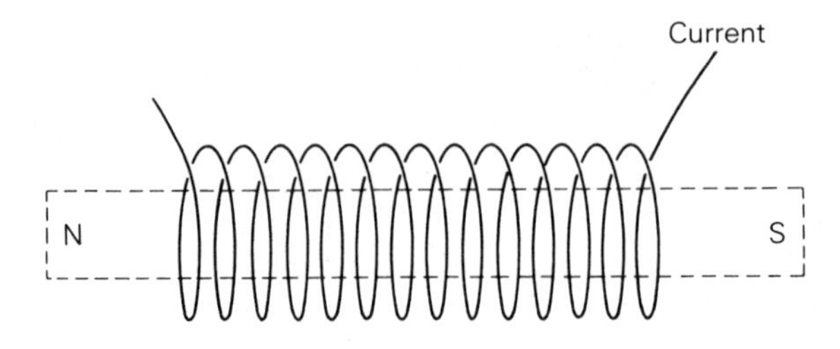

Figure 13-12 A current-carrying coil has the same magnetic properties as a bar magnet.

Electromagnets are often used to open such things as valves and doors. The valve or door latch is connected to an iron bar, which is near an electromagnet. When the current is turned on, the

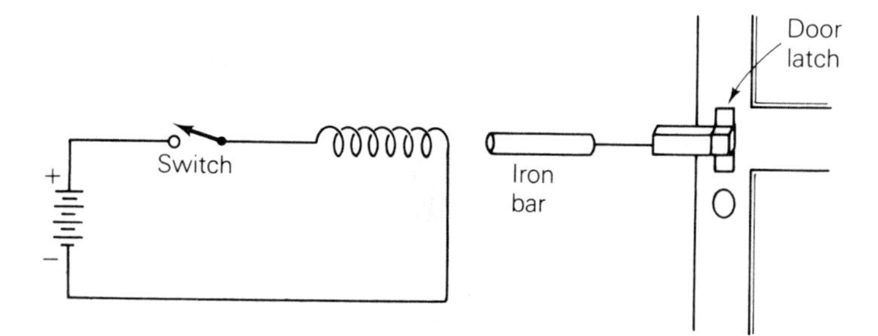

Figure 13-13 Electromagnets are often used to open doors by remote control.

electromagnet pulls on the bar, thus opening the valve or door (Figure 13-13).

Magnetic Induction of Electricity

By placing a magnet next to a TV screen, we saw that a magnetic field produces a force on a moving electric charge. Can a magnetic field produce a force on a stationary electric charge? It really doesn't matter if we move a charge through a magnetic field or vice versa; the same force will be on the charge either way.* If we move a magnet (either a permanent magnet or an electromagnet) near a stationary charge in a wire, we will produce a force on the charge, which will move down the wire if the circuit is complete. This effect goes by the rather complicated name of *magnetic induction of an electric current.*

We can induce a current by moving an electromagnet around a wire, but we can move the field of an electromagnet in other ways, too. Turning off the current removes the magnetic field, just as removing the magnet would. Thus, changing the current in an electromagnet causes its magnetic field to move either inward or outward, depending on whether the current was decreased or increased. For a charge in a nearby wire, the movement of the field has the same effect as moving the magnet. The moving field creates a force on the charge, and a current will be induced by the changing magnetic field if a complete circuit exists.

*This simple statement turns out to be one of the most important scientific statements of the twentieth century: It is the cornerstone of Einstein's theory of relativity. Once you accept this statement, all the results of relativity follow from reasonably simple algebra. Einstein's genius was not in inventing complicated mathematical theories but in pointing out how seemingly difficult things can be easily understood when they are approached in the right way.

Heaters, Lights, Meters, and Motors

The bulk of the electrical energy produced is used for one of three simple tasks: producing heat, light, or mechanical motion. Let's look at how these jobs are done.

Heaters

As electrons move down a wire, they bump into the atoms. This causes the atoms to vibrate randomly; that is, the electric current heats the wire. Thus, electrical heaters are made by simply winding wire into whatever shape is needed. Heaters work best when the resistance is high enough that the electrons bump into lots of atoms but not so high that little or no current flows.

In addition to moderately high resistance, two other physical factors must be considered. The heater wire should have a high melting point, and it should not oxidize rapidly when it is hot. Most heating elements are made from a nickel-chrome alloy, which meets all three criteria fairly well.

Lights

The incandescent light bulb works in the same way that heaters do except that enough current is passed through the wire to heat it white hot. Most metals melt long before white heat is reached, but tungsten and tantalum remain solid; they also have a high resistance. However, since both metals oxidize quickly at high temperatures, the wires must be enclosed in a glass bulb that is filled with an inert gas, such as argon, which contains no oxygen (Figure 13-14). Unfortunately, incandescent bulbs are not very efficient—far more energy leaves the bulb as heat than as light.

The fluorescent light works by continuously passing a spark through mercury vapor at low pressure. The spark knocks electrons off the mercury atoms, and as these electrons fall back into place, they emit light. (We'll discuss the details of the process later on). The big advantage of the fluorescent light is its efficiency: It puts out twice as much light per watt as an incandescent light. However, fluorescent lights require some complicated circuitry to get the spark started and to keep it going. Even so, fluorescent light bulbs are usually cheaper to use in the long run.

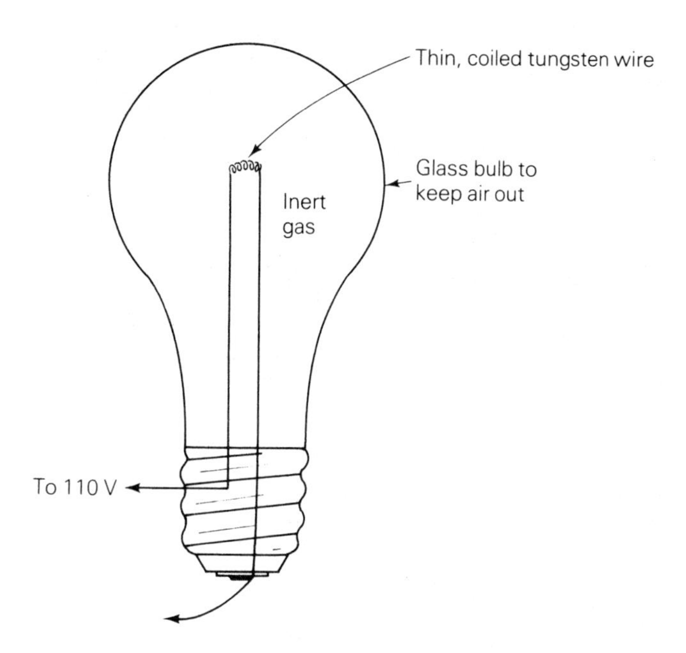

Figure 13-14 An incandescent light bulb.

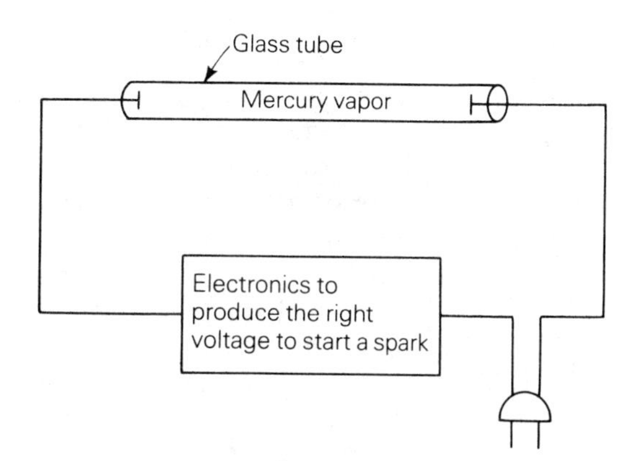

Figure 13-15 A fluorescent light.

Meters

In a pointer-type meter, the current to be measured passes through an electromagnet coil, which is mounted on a pivot and is free to turn. A light pointer attached to the coil indicates how far the coil turns, and a small spring attached to the pointer pulls it back to the zero position. A permanent magnet is placed on each side of the coil (Figure 13-16).

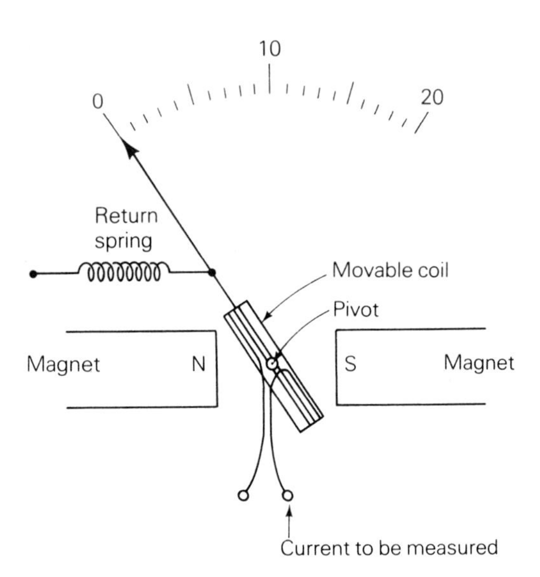

Figure 13-16 The basic construction of a pointer-type meter.

When no current passes through the coil, the magnets have no effect on the coil and the spring pulls the pointer needle to the zero position. When a current passes through the coil, the coil becomes an electromagnet, and the pull of the permanent magnets tends to twist the coil farther from the zero position. The more current there is passing through the coil, the farther the coil and the pointer move from the zero position. A scale is placed behind the pointer so that the current can be read on the scale.

The basic pointer meter measures current, but with slight modifications it can measure voltage, resistance, and other things. The main drawback of pointer meters is that they are delicate. In general, they will not withstand abuse, either physical or electrical.

Nowadays more and more pointer meters are being replaced with digital meters. The digital meter uses a built-in microcomputer to analyze the current or voltage, then it displays the result on a numerical display. Sometimes the meter even sends the result to a larger computer for analysis. Digital meters are very accurate and fairly rugged, so they are almost always used where high accuracy is required. The old-fashioned pointer meter does have one advantage, however. Just as graphs convey information to the mind better than tables, the movement of the pointer is easier to follow than a display of flashing numbers. The pointer meter is still superior for monitoring trends if high accuracy is not required.

Motors

There are many types of electric motors, but they all use electromagnetism to convert electrical energy into mechanical energy. The basic motor, shown in Figure 13-17, operates in the following way. An electromagnet coil is mounted on the motor shaft, and a permanent magnet is placed on each side of the coil. When a current magnetizes the coil, the coil and the shaft turn so that the south pole of the coil points toward the north permanent magnet pole. The direction of the current in the coil is then reversed; the south pole of the coil becomes the north pole and is repelled from the north permanent magnet pole, causing the shaft to make another half turn. A switching mechanism attached to the shaft reverses the current automatically as the motor turns, enabling the motor to go around continuously.

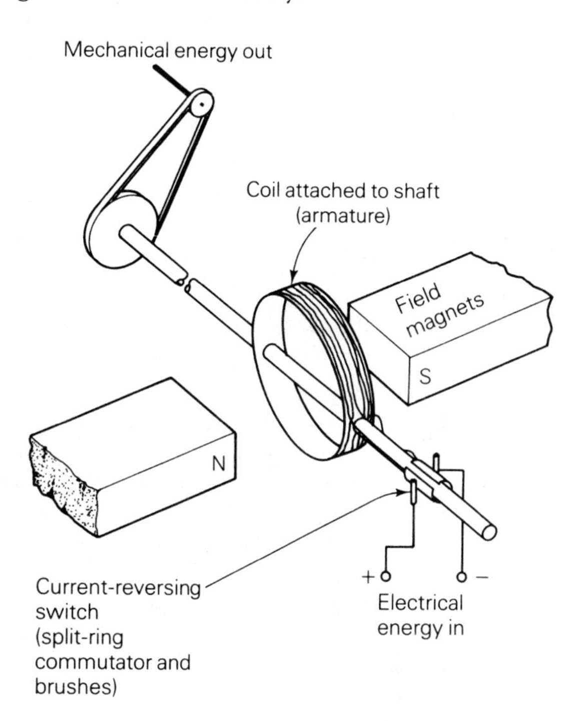

Figure 13-17 The basic electric motor.

Generating Voltages

All methods of voltage generation employ the same fundamental principle: Electric charges are physically moved from one place to another. Only the details of the forces and motion vary. The motion may be large, as in the case of static electricity, where the charge from a rug is carried to a doorknob some distance away; or the motion may be small, as in a living cell, where a voltage is generated by moving ions from one side of the cellular membrane to the other. Sometimes the motion is even smaller: Some voltages are produced by moving an electron from one side of an atom to the other.

Static Electricity

Static electricity isn't an ideal name, because charges must be moved to produce a voltage in this and in every other type of electricity. But static electricity is the common name, so we will use it, too.

Static electricity is produced by physically rubbing electrons from one material to another and then moving the two materials apart. Very high voltages can be produced this way—doorknob sparks often exceed 10,000 volts. However, it is difficult to separate very much charge in this way, so the total amount of energy stored by static electricity is usually not very high.

Static electricity isn't a very efficient way to produce electricity because the mechanical losses are great. However, it is a good way to generate ultrahigh voltages when only a small current is needed. The largest static electricity generators, which are used to make high-energy beams of subatomic particles, can produce over 10 megavolts at currents of up to a milliamp.

Lightning, the greatest display of static electricity, is simple static electricity generated by air currents in a storm. As the water droplets in the air swirl and collide, electrons move from one drop to another; winds then separate the drops until the voltage is so large that lightning discharges to the ground or to another cloud. The voltage of a lightning bolt is very high, about 100 megavolts. The energy in lightning is also very large; large trees have been blown into splinters by a single lightning bolt.

Common static sparks rarely damage the body because the energy in a spark is quite small, even though the voltage is high. However, sparks can be dangerous in some cases. Since a person usually reflexively jerks when shocked, a brain surgeon or a diamond cutter would not want to get a static shock at a crucial moment. A spark to the eye can cause permanent damage, and static sparks can set explosives off, especially explosive gases. Some anesthetic gases, such as ether and cyclopropane, are explosive; when these gases are used, extremely careful antistatic measures must be taken to prevent static sparks.

AC and DC Generators

Generators are the most efficient way to turn mechanical energy into electrical energy. Over 99% of the electrical energy in the world is produced in generators.

The basic construction of a generator (Figure 13-18) is very similar to that of a motor. In fact, turning the shaft of most motors will generate electricity by mechanically moving the electrons in the coil wire through the field of the permanent magnets. This causes a force on the electrons that makes them flow around the coil. By choosing the proper switching arrangements on the shaft, either an AC or a DC voltage can be produced.

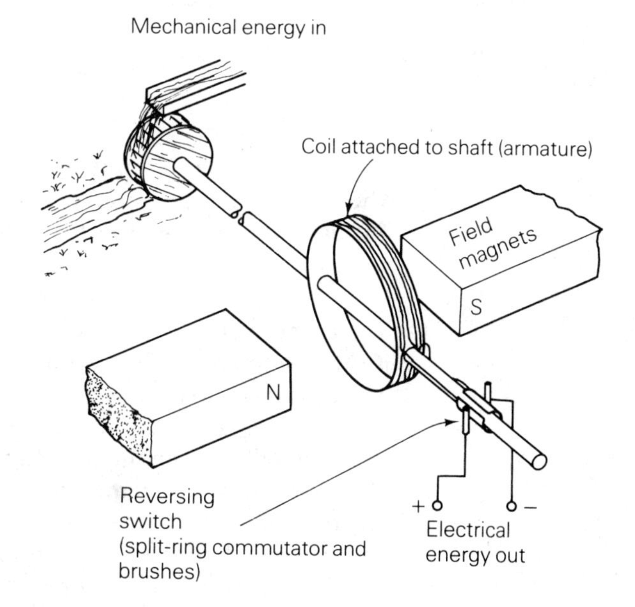

Figure 13-18 The basic DC electrical generator.

Transformers

Transformers increase or decrease AC voltages as needed. A transformer consists of two coils placed so that their magnetic fields overlap as shown in Figure 13-19. An iron core is usually added to increase the overlap of the magnetic fields.

An AC current is sent through the input coil, which creates a changing magnetic field. This field then produces an electrical voltage in the output coil by magnetic induction. The ratio of the output voltage to the input voltage is the same as the ratio of the number of turns in the output coil to that in the input coil. For instance, when the output coil has twice as many turns, the output

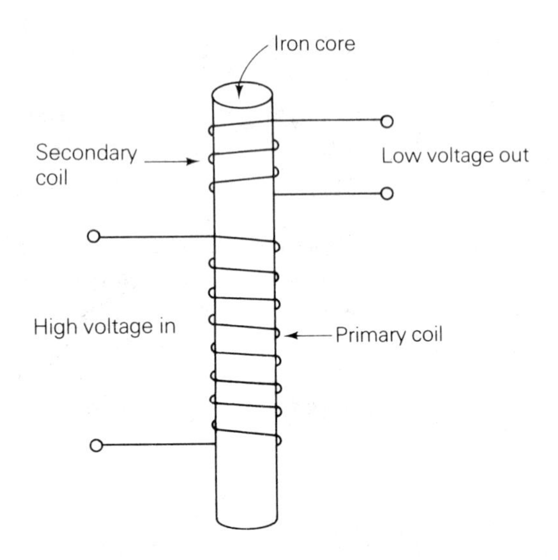

Figure 13-19 The transformer. A transformer can change an AC voltage but not a DC voltage. Note that there is no direct electrical connection between the two coils. The wire is insulated from the iron core. This transformer is a *step-down* transformer because the output voltage is lower than the input voltage. A *step-up* transformer would have more turns in the secondary coil than in the primary coil.

voltage is twice the input voltage. Of course, the electrical power at the output of the transformer must be less than or equal to the input power because of the conservation of energy. Because

$$p = IV,$$

the current is decreased when a transformer steps up the voltage. Conversely, a step-down transformer, which decreases the voltage, increases the current.

The transformer will not work on DC electricity, because direct current will not change the magnetic field around the coil. Because the field does not change or move, there is no induced voltage at the output coil.

Notice that the two coils have *no direct electrical connection.* Therefore, no charge ever flows from the input to the output, or vice versa. For safety purposes, this feature can be very useful for preventing shock currents from completing the circuit by going to the ground. The transformers used to isolate patients and apparatus from connections with the ground are called *isolation transformers.*

Photoelectric Effect

When light strikes the nearly free electrons in a metal, some of the electrons may be knocked out of the metal. Basically what happens is that the light collides with the electrons in the metal, tossing the electrons in all directions, including away from the surface of the

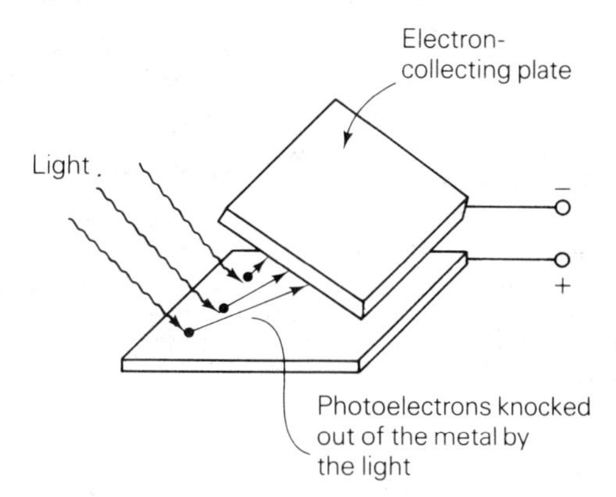

Figure 13-20 A photoelectric cell.

metal. Some of these electrons escape from the metal and fly off into space. If a second metal plate is placed near the first surface, it will gather the electrons and a voltage will be produced between the two surfaces (Figure 13-20). This is the photoelectric effect, which converts light energy directly into electrical energy.

Unfortunately, photoelectric conversion of light into electrical energy is not very efficient, and photoelectric cells are expensive to produce. Therefore, photoelectric production of electricity from the sun is not the answer to the energy crisis at present. However, solar cells are very useful in areas where conventional electricity isn't available, for instance, in communications satellites and remote lighthouses. The cost is fairly high, however. A solar cell big enough to run a 100-watt light bulb now costs around $1,000, but the costs are coming down; someday solar cells may produce large amounts of electrical energy.

The photoelectric effect is most often used for detecting and measuring light levels. TV cameras convert light to an electrical signal by the photoelectric effect. The light meters in cameras use the photoelectric effect to measure the amount of light available for a photograph.

Piezoelectricity

Certain insulating crystals, quartz in particular, generate a voltage when they are bent. This effect is known as *piezoelectricity.* When a crystal is squeezed, every atom is distorted slightly. In piezoelectric materials, the distortion moves slightly more electrons to one side of the atom than the other, producing a very small voltage across the atom (Figure 13-21). The individual atomic voltages add up to

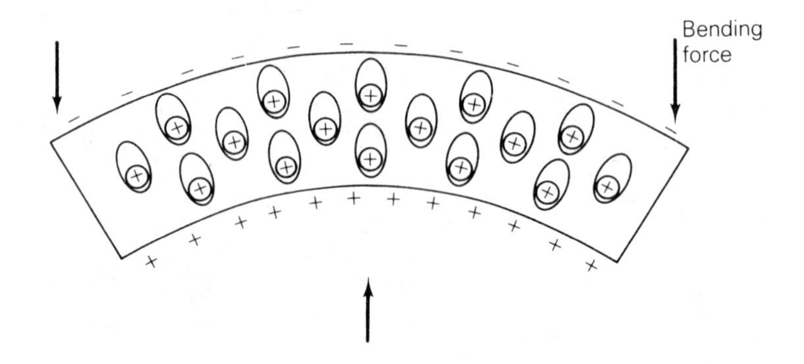

Figure 13-21 Piezoelectricity. Bending certain insulating crystals distorts the electron orbits, causing one side of the crystal to become negatively charged and the other to be positively charged.

Piezoelectric crystals are too costly to use for the routine production of electricity. They are used most frequently for *transducers,* devices that convert mechanical motion, acceleration, sound, pressure, and other physical quantities into electrical signals. Microphones and phonograph pickup cartridges, both transducers, convert mechanical vibrations into electrical signals, and both can be made with piezoelectric crystals.

The piezoelectric effect can also be turned around to convert an electrical signal into a mechanical motion: A voltage applied to a quartz crystal will bend it slightly. Quartz clocks and watches use tiny quartz crystals, which vibrate when an electrical signal is applied. This oscillation performs the same role as the swinging of a pendulum, only it is much more rapid and much more accurate. Ultrasonic (ultrahigh frequency) sound generators and radio transmitters also use the piezoelectric effect.

Ultrasound is used in a number of clinical situations. It is used in special cleaning machines and in respiration therapy to humidify oxygen by breaking up water droplets. It is also used to measure blood flow rates, and it can be used to form images of internal organs. Piezoelectric crystals are used to generate the sound waves in many of the devices used for these purposes. We shall look at the details of some of these applications when we study sound.

Batteries

Batteries produce electricity by chemical means. That is, batteries convert the chemical energy stored in certain materials into electrical energy. As in the case of static electricity, chemically produced electricity results from the different affinities that different substances have for electrons. In static electricity, the electrons move to

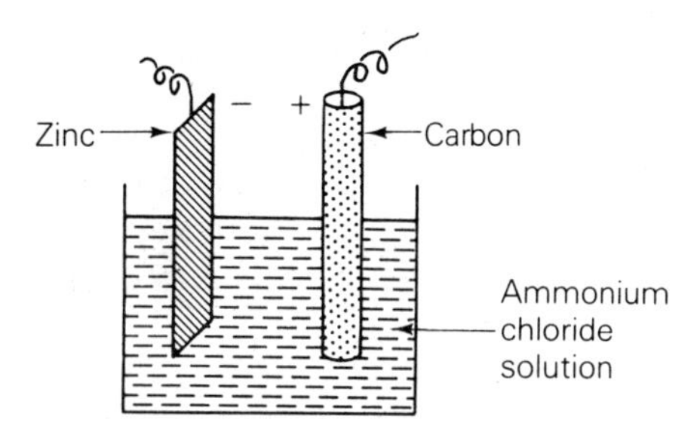

Figure 13-22 A simple electric battery. Most flashlight batteries use these materials to produce electricity.

the most attractive of the two surfaces being rubbed together and then the surfaces are mechanically separated. In chemical production, the energy needed to separate the charges comes from the chemical energy of the materials.

The conversion of chemical energy into electrical energy is accomplished in the following way. Suppose we have two substances that have different attractions for electrons. We will consider zinc, which has a great affinity for electrons, and carbon. If zinc and carbon are put into an ionic salt solution as shown in Figure 13-22 (ammonium chloride in water is commonly used), the greater electron affinity of zinc causes a chemical reaction that leaves the zinc negatively charged with electrons and the carbon positively charged. When the two materials of the battery are connected through an outside circuit, the electrons flow from the zinc through the circuit to the carbon. The current can light a flashlight, run a radio, or do any of the hundreds of things that batteries are used for.

Batteries do not produce energy for free; in a zinc-carbon battery, the zinc is chemically consumed, and the battery dies when the zinc is used up. A similar process occurs in all batteries, which makes them rather expensive sources of electrical energy, since all the practical battery materials are expensive. Typically, energy from batteries is about 1,000 times more expensive than energy from the wall socket. However, the safety and convenience of batteries often make up for the higher cost.

All batteries can also be used to "store" electrical energy. That is, electrical energy can be reconverted to chemical energy when the current is reversed. However, most batteries don't reconvert the energy very effectively. Batteries that are not easily recharged are called primary batteries, while those that are easily recharged are called storage batteries. The lead-acid battery used in cars and the nickel-cadmium battery used in many rechargeable appliances are two common types of storage batteries. All storage batteries have a rather high initial cost, and they are sensitive to electrical abuse.

Thus, the manufacturer's recommendations for charging and discharging should be followed.

Unless an unforeseen breakthrough happens, batteries will continue to be one of the most expensive ways to produce electricity. But since they are so convenient, they will continue to be an important source for small amounts of electricity.

Thermoelectricity

When two different metals are joined together and the junction is heated, a small voltage is generated between the two metals (Figure 13-23). Such a junction is called a *thermocouple,* and the effect is known as the *thermoelectric* effect. The flow of heat in a metal, for some very complicated reasons, tends to carry some electricity with it. The size of the effect varies among metals, but the voltage is always small, usually in the millivolt range.

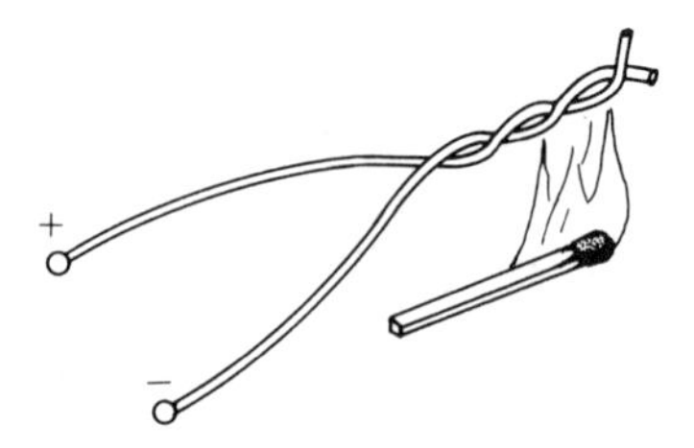

Figure 13-23 Thermoelectricity. If two different metals are joined together and the junction is heated, a thermoelectric voltage will be produced.

Thermoelectric voltages are too small to produce electricity commercially. Thermocouples are, however, very useful for thermometers. The thermoelectric voltage goes up with increasing temperature, so, in principle, a thermoelectric thermometer could consist of just a thermocouple hooked up to a sensitive voltmeter. In practice, a microcomputer circuit is added to convert the voltage reading directly into a temperature reading. Thermocouples can be used from near absolute zero to white heat with an accuracy of a small fraction of a degree.

The thermoelectric effect can also be turned around to cool the junction if a current is sent through a thermocouple in the right direction. Most minirefrigerators and laboratory cold plates use this principle, because they are so simple to build, requiring only the proper junction material and a supply of electricity. No ice or complicated refrigeration systems are needed. Such thermoelectric freezing systems are also used in neurosurgery to freeze small groups of cells.

Bioelectricity

Nearly every method of voltage generation plays some part in plant and animal life. The retina of the eye, for instance, converts light into an electrical signal through a photoelectric process. Piezoelectricity is involved in the inner ear, where sound vibrations produce electrical signals in the auditory nerve. Smell and taste involve electrochemical voltages, and there is clear evidence that homing pigeons use voltages generated electromagnetically by the earth's magnetic field to help them navigate. All these voltages are very complex and only partially understood. It is likely that even more electrical interactions in living things have yet to be discovered.

Membrane Potentials

The most common type of biologically generated voltages arises from the diffusion of ions through the semipermeable cell membranes. Suppose a semipermeable membrane separates water with potassium chloride salt dissolved in it from fresh water. The salt water would initially contain large but equal numbers of positive potassium ions and negative chlorine ions. Thus, both the fresh water and salt water would be uncharged initially, and no voltage difference would exist across the membrane. Now suppose the membrane is permeable to potassium ions but not to chlorine ions. When a positive potassium ion diffused through the membrane into the fresh water, it would carry a positive electric charge. As the

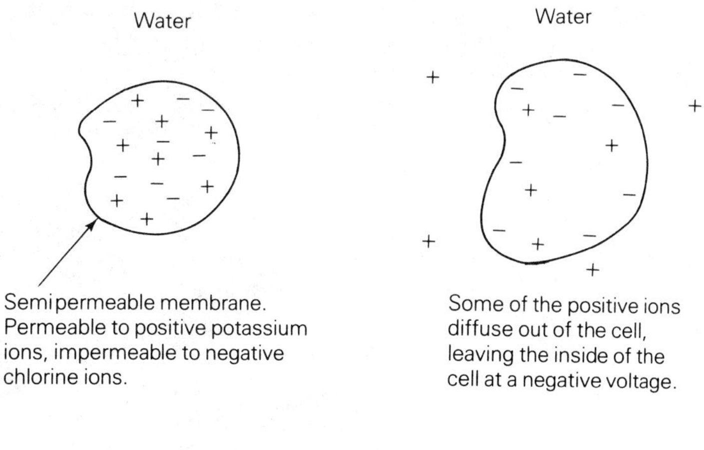

Figure 13-24 Semipermeable membranes often produce bioelectric voltages.

diffusion continued, the fresh water would become more and more positively charged, leaving the salt water side negatively charged.

Eventually the diffusion of potassium would stop because the positively charged fresh water would repel the potassium ions back toward the saltwater side, but by this point a rather large voltage difference would exist across the membrane.

The situation in a cell is very similar. The fluid inside the cell contains potassium chloride, while the fluid outside contains mostly sodium chloride. The membrane of a normal resting cell is far more permeable to the potassium ions than to the sodium ions. Potassium ions diffuse out of the cell, leaving the cell interior negatively charged. The outward diffusion of potassium continues until the inward electrical force on the potassium ions is big enough to make the chances of a potassium ion leaving equal to the chances of a potassium ion returning. This occurs when the voltage in the cell is −85 millivolts. This voltage is called the *resting potential* of the cell.

Even though −85 millivolts is a relatively large voltage—almost 0.1 volt—the fraction of the potassium ions that leave the cell is quite small. Since cells are so small, a slight charge will change their voltage by a large amount.

Membrane Structure and Electric Forces

Because the role of electric forces in membrane structure is very poorly understood, many discussions of cell structure just ignore it. However, if we calculate the voltage change per meter across the cell membrane, it becomes clear that these electric forces must be considered. The cell membrane is about 8 nanometers thick, so the voltage change per meter is

$$\frac{0.085 \text{ V}}{0.000{,}000{,}008 \text{ m}} = 11{,}000{,}000 \text{ V/m}$$

$$= 11 \text{ MV/m}.$$

This is equivalent to 20 megavolts applied between our head and feet. Such a large voltage change in such a small distance must play an important role in membrane structure, even if we don't know or understand the details.

Many of the molecules in membranes, proteins and water in particular, have electrically charged regions. For example, the electrons of water molecules are more likely to be near the oxygen atom; thus, the oxygen part of the molecule will be negatively charged and the hydrogen part positively charged. A molecule with

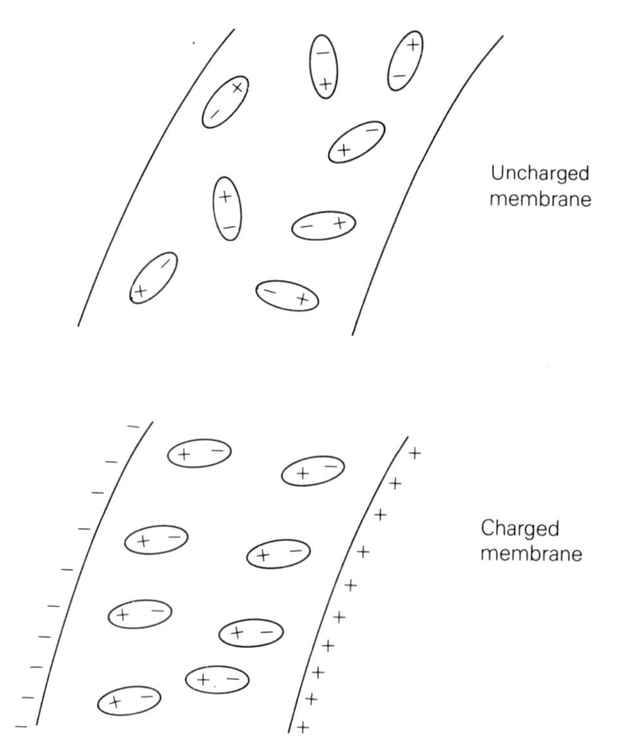

Figure 13-25 An electrically charged membrane will tend to line up the polar molecules within it. Thus, the electrical forces play a role in membrane structure.

such charged regions is called a *polar* molecule. Because the large electric forces caused by the resting potential tend to line up polar molecules, they partially determine the arrangement of the molecules in the membrane (Figure 13-25).

Action Potential

The cell membrane is not always impermeable to sodium; a number of mechanical, chemical, and electrical stimuli can trigger changes that make the membrane highly permeable to sodium. When this happens, the positive sodium ions rapidly diffuse into the cell, quickly increasing the interior voltage of the cell to about +60 millivolts. This potential is called the *action potential* of the cell. Since the change is from −85 millivolts to +60 millivolts, the net change is 145 millivolts—quite a change, if you consider that the membrane is only 50 atoms thick. When the cell reaches its action potential, the membrane again becomes impermeable to sodium but remains permeable to potassium. So the potassium rushes out, bringing the cell back to its resting potential. The entire process

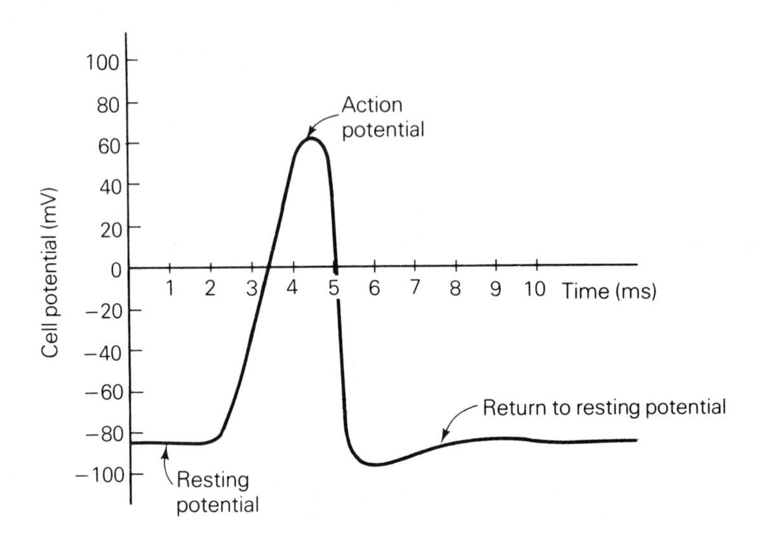

Figure 13-26 The potential inside a cell when it is electrically activated.

takes only a few milliseconds (Figure 13-26). The amount of sodium that diffuses into the cell is quite small, since only a little bit of charge is needed to produce the voltage change. This excess sodium is gradually pumped from the cell by active membrane transport.

One cell's switch to the action potential can trigger neighboring cells to switch, too. The positive charge between the two cells is drawn into the first cell when it switches to its action potential. Therefore, where the two cells touch, the voltage difference across the membrane of the second cell is somewhat less than the normal −85 millivolts. This voltage change is thought to cause changes in the membrane structure that allow sodium to penetrate the second cell. This switches the second cell to its action potential, and the signal spreads in a dominolike reaction through the entire group of cells (Figure 13-27).

The nerve cells (neurons) that control the voluntary muscles are especially well suited for intercellular communication (Figure 13-28). The neuron has a long tail (axon) covered with a layer of myelin, a very good electrical insulator. The construction is much like that of a telephone wire consisting of a good conductor covered by a good insulator. When one end of the neuron is activated, the electrical impulse travels down the axon quite rapidly. The cells near the side of the axon are not activated because they are insulated by the myelin sheath. Only the cell in the next neuron in the chain going to the muscle cell is activated. In multiple sclerosis, the nerve sheath breaks down, interfering with the proper transmission of nerve signals.

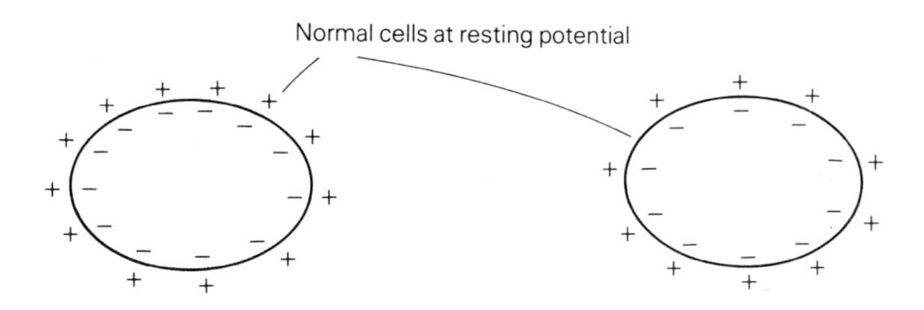

A cell switching to its action potential

Figure 13-27 A cell switching to its action potential will pull in positive charges from a neighboring cell. This tends to make the neighboring cell switch to its action potential also.

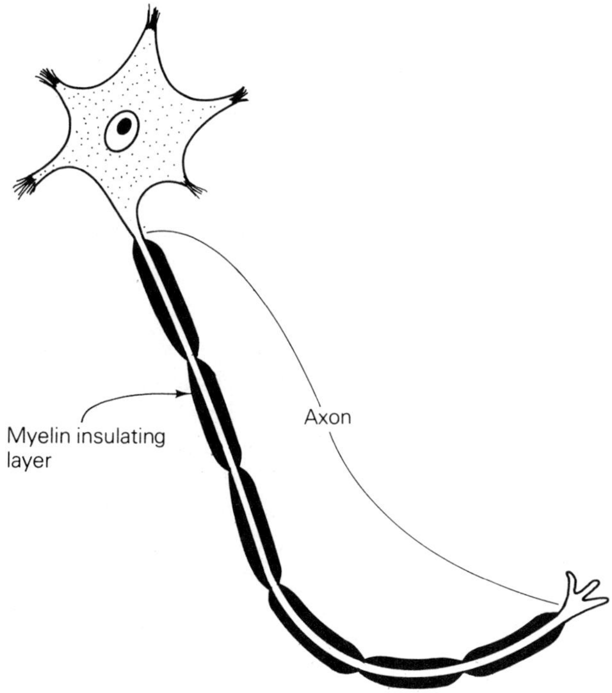

Myelin insulating layer

Axon

Figure 13-28 A nerve cell (neuron). The long axon (as long as 1 meter) is constructed much like a telephone wire with a conductor on the inside covered with an insulator.

Electricity in the Body

Certain disorders change the electrical signals produced by the muscles, heart, and brain. Thus, recordings of electrical activity can be very helpful in accurately diagnosing the condition of these organs.

The signals from the organs are not very big, and they are even smaller by the time they reach the skin. However, if they are enhanced with a modern amplifier, they can be detected if the interference from things like microwave ovens, TV stations, and electric motors is filtered out. In addition to being very informative, these measurements are close to ideal diagnostically because they are safe, painless, and nearly free of unwanted side effects. Generally, the most serious effect is the apprehension that patients suffer when they see the electrodes and wires that are going to be hooked up to them; however, their fears can usually be overcome if the procedure is explained.

Electricity is also widely used in the treatment of diseases. Outside electrical signals can be used to stimulate the response of several organs, electric currents can combat pain, drugs can be driven under the skin with an electric current, and growth patterns can be altered with electricity to speed up healing.

The future is likely to bring even more advances in the diagnostic and therapeutic uses of electricity. Thought-controlled artificial arms and legs, TV cameras that send signals directly to the brain of a blind person, and brain wave analyzers that pinpoint the cause of mental disorders seem like science fiction ideas today, but they are all likely possibilities for the future. Promising starts have already been made on these ideas.

Muscles

The electrical activity of muscles is one of the most sensitive indicators of muscle activity; it is far more accurate than sight or touch. In addition, the electrical signals can be timed to millisecond accuracy. The technique of measuring electrical signals from muscles—which is called *electromyography* (EMG)—has lead to great improvements in the understanding of human motion.

Electromyography has been especially useful in the kinesiology of athletics. Before electromyography, muscle function was analyzed by visual and tactile observation. While such "look and touch" methods worked for slow motions of major muscles, they weren't too helpful for analyzing the activities of muscles deep below the surface or for the multitude of rapid motions of athletic activity. Electromyography, on the other hand, shows exactly which muscles are used, when they are contracted and released, and to what extent they contract. With this information, a coach can often find ways to make important improvements in an athlete's performance.

Physical therapists often use electromyography in the rehabilitation of patients with injured or diseased muscles. In order to learn to reuse a muscle or to use a muscle in a new way, a patient needs some indication that the muscle is being activated. The idea behind this is the same as the idea behind learning to wiggle your ears. To learn to wiggle the ears, a person usually stands in front of a mirror and tries. Eventually the person will hit on the right combination and an ear will twitch. After that, all that it takes to become proficient is further practice. Electromyography is far more sensitive than watching a mirror, because it can detect the first, feeble contractions long before they can be seen. Thus, if the muscle is hooked up to an electromyograph, the patient will detect the slightest success in using the muscle. After the first twitches are observed, the process of training the muscle usually goes smoothly.

Electric currents from outside the body will activate muscle contractions, as we have all experienced when our hand jerks away from a doorknob after a static spark. Such electrical stimulation is used to exercise paralyzed muscles and to test muscle response. At present, electrostimulation of muscles is too crude to restore any practical use to a paralyzed muscle, but in the future electrical stimulation may be used to bypass the broken nerves.

The Heart

An electrocardiogram (ECG)—the recording of the heart's electrical activity—is a very good indicator of the heart's condition. A typical one is shown in Figure 13-29. A signal from the heart can be picked up at every point of the trunk and arms, but it is not the same at all points. At any particular spot, the signal is a mixture of the signals from the various parts of the heart, and the proportions of the mixture change from place to place. For instance, the signal at one spot might consist of a strong signal from the left ventricle and weak signals from the rest of the heart; the situation might be just the opposite in another spot. Therefore, electrocardiograms usually measure the electrical signals at several different spots to get a complete picture.

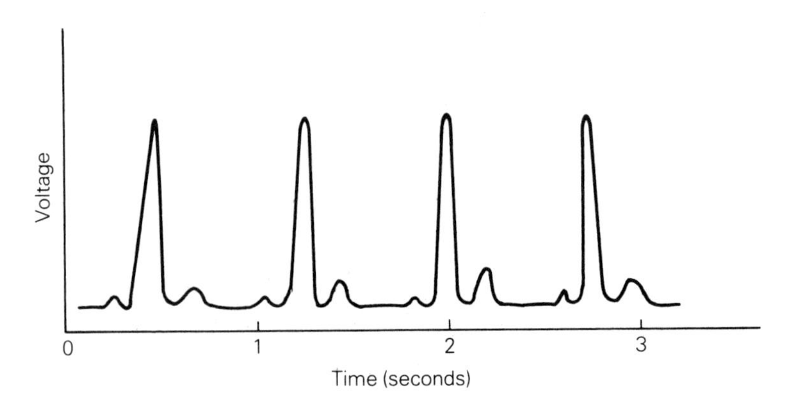

Figure 13-29 A typical electrocardiogram. The various peaks correspond to various parts of the heart contracting. The large spike corresponds to the main ventricular contraction.

The large amount of information contained in a multielectrode electrocardiogram takes a great deal of training and experience to interpret; thus, their interpretation has been somewhat expensive. Recently, computers have been used for more and more of the analysis, since they sort large amounts of data much faster than humans do. No one expects computers to replace humans for making medical judgments—at least not for some time to come—but more and more of the routine analyses of ECGs and other data will be done electronically.

An electrocardiogram usually doesn't show any abnormalities until *after* the heart muscle is damaged. Because several heart diseases do not damage the heart muscle until the first heart attack, ECGs will not find every type of heart disease. They are most useful for assessing heart damage after a heart attack and for determining whether chest pains are due to mild heart attacks or other causes.

In a normal heart, contractions are started by one signal that rapidly spreads over the entire heart. Thus, the heart contracts in a coordinated, wavelike motion. Of course, the heart pumps blood most efficiently this way. Certain heart disorders break down the communication paths between the parts of the heart, causing the different sections to pump with different rhythms. Because of this loss of cooperation between the different parts of the heart, the amount of blood pumped is greatly reduced.

Such disorders can be treated by using an electrical signal to stimulate the heart so that it beats all at once. The devices that generate such signals are called *pacemakers.* They consist of electrodes attached to the heart muscle and the electronics that generate the electrical pulses. The entire package is usually implanted in the body surgically so that the patient can resume a near normal life. Batteries are the biggest problem with implanted pacemakers.

The best conventional batteries last only a few years and then must be replaced by an operation. Some pacemakers use a small amount of radioactive plutonium that heats a thermocouple to generate the electricity. Such nuclear batteries will last for a decade or more, but they demand extra care to prevent the spread of radioactivity if the patient should be involved in an accident.

Another device is used to stimulate the heart electrically is the *defibrillator.* During heart attacks and electrical shocks, the heart starts to flutter in an uncoordinated way. This fluttering is called *fibrillation,* and it doesn't pump blood through the heart. The defibrillator sends a massive, quick jolt of electricity to the heart that contracts it completely. This sometimes stops fibrillation and restores normal pumping.

The Brain

The study of the electrical activity of the brain—electro-encephalography (EEG)—is still in a very early stage. The brain is so complex that not much progress has been made in interpreting the fine points of the patterns.

When a person is relaxed and not concentrating on anything, the brain pattern is dominated by the alpha rhythm—a slow, steady pattern with a frequency of around 10 hertz. When a person concentrates, the frequency of the signal increases and the so-called beta rhythm dominates, which has a frequency of about 20 hertz. During sleep, the pattern is less regular, and the frequency drops to around 4 hertz (Figure 13-30).

At present, the diagnostic applications of electroencephalography are limited by the difficulty in interpreting the patterns. Currently, diagnosis is limited to epilepsy, brain tumors, and a few other diseases that change brain wave patterns significantly. However, as interpretation of brain wave patterns improves, EEGs will be used more frequently in the treatment of brain disorders.

Recently, the old indications of death—absence of breath and absence of heartbeat—have been totally outmoded by medical advances. This has again raised the question, When is a person dead? There isn't any hard and fast answer yet, but generally a person is considered alive as long as the brain shows detectable electrical activity.

There is evidence that the brain can be stimulated by an outside electric current, but there hasn't been much progress in controlling the results of the stimulation. However, one encouraging bit of progress has been made in the treatment of some severe mental problems such as schizophrenia and uncontrollable, violent

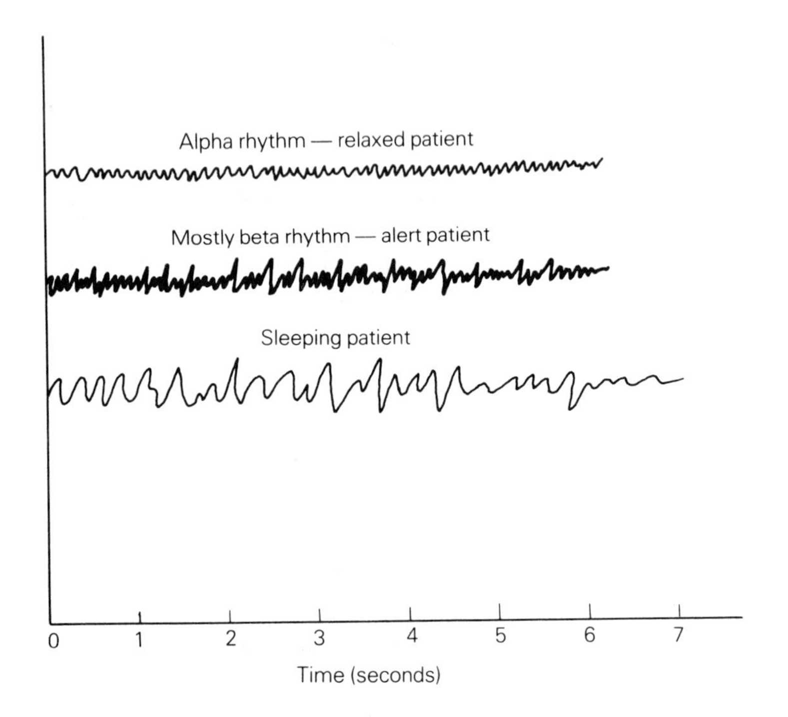

Figure 13-30 Common brain wave patterns.

behavior. Certain spots on the brain are known to be associated with pleasurable feelings. When these spots are stimulated electrically by implanted electrodes, some patients improve enough to lead near normal lives outside of institutions.

Electric currents are also used to block nerve signals from reaching the brain. A small current is sent through skin electrodes to the nerve, which can reduce pain in a number of cases. This method is now used routinely for certain types of back pain. Further advances in the electrical control of pain are likely.

The Bones

One of the surprising recent discoveries about electricity is its role in bone growth and healing. A bone grows fastest where the stress is largest, which results in a shape that is well suited to carrying the load normally placed on the bone. For instance, the spongy center of a load-carrying bone, such as the femur, has a web-like growth pattern that follows the stress pattern almost perfectly. What makes bone grow fastest where the stress is greatest?

The answer has two parts, and both involve electricity. First, the bones are piezoelectric—they produce a weak electric current when they are bent. Naturally, the current is greatest where the bending is greatest. Second, electric currents stimulate bone growth. Thus, the two effects together are perfect for guiding bone growth. Since the bone bends the most where it is weakest, electric currents there are the strongest. This in turn stimulates the bone to grow most rapidly where it is weakest. This effect also seems to be important in fracture healing. Once the bone has healed enough, mild exercise may speed up healing by producing piezoelectric currents that stimulate growth around the weaker fractured region.

External currents can also be used to speed fracture healing. A current is passed through electrodes that are attached to the bone on each side of the fracture, healing the fracture much faster than it would heal otherwise. The method isn't used very often for treating uncomplicated fractures, because natural regrowth is usually fast enough. However, the method has shown great promise in the treatment of fractures that do not respond well to normal methods.

Iontophoresis

Electric currents can carry medications underneath the skin if the molecules are ionized. This method is known as *iontophoresis*. Suppose that the medication was positively ionized. The medication is placed on the affected area, and a positive electrode is then placed over this area and a negative electrode is placed over another part of the body. A current is applied, which goes from the positive electrode, through the body, and then out of the body at the nega-

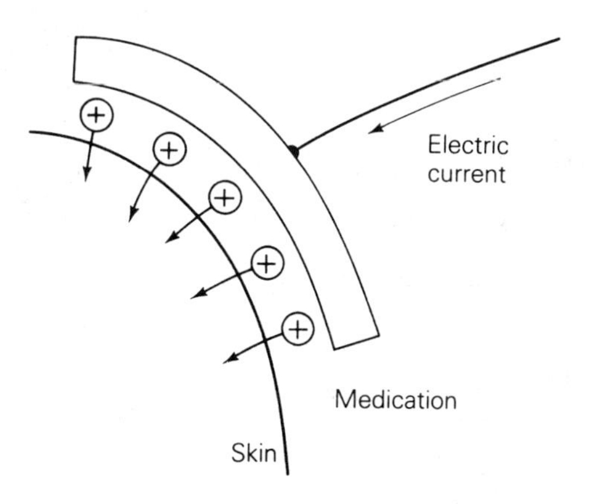

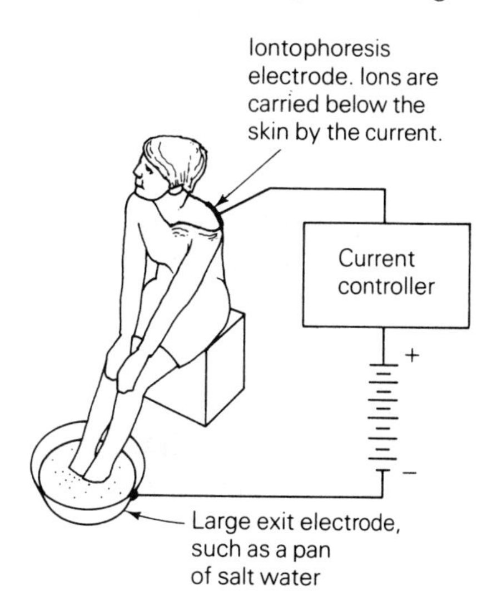

Figure 13-31 Iontophoresis.

tive electrode. The current carries the positive ions of the medication through the skin. However, once the current passes through the skin, so many other ions are available that the medicated ions only carry a small fraction of the current. Thus, the current will only draw the medication just under the skin surface. Consequently, iontophoresis is not very useful for delivering medications to deep tissues.

If the medication is negatively charged, the current is simply reversed: The negative electrode is placed over the affected region and the positive exit electrode is placed somewhere else.

Transducers

We often want to change a nonelectrical quantity into an electrical signal so that we can measure it more conveniently, send the information by a radio signal, analyze it with a computer, or do any of the many other things that can be done so conveniently with electricity. Devices that convert nonelectrical quantities, such as force, pressure, and temperature, into electrical signals are called *transducers.*

An example of a transducer is the IV drip rate monitor (Figure 13-32). This transducer consists of a light and a photocell placed on opposite sides of the IV drip chamber. Every drop that passes through the chamber interrupts the light beam, shutting off the photocell. Thus, each drop produces an electrical signal that can be counted and analyzed electronically.

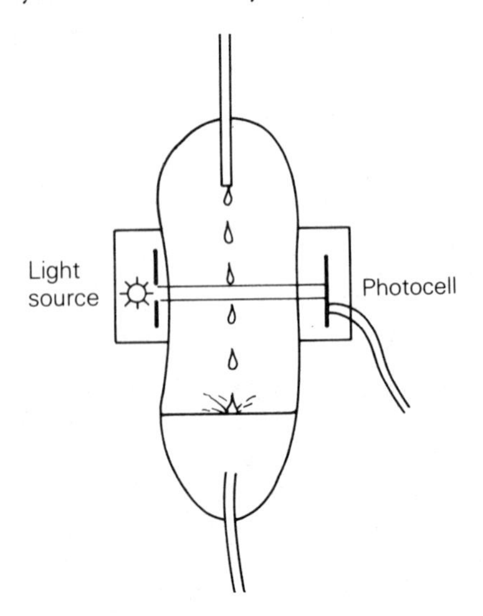

Figure 13-32 An IV drip rate monitor. The drops break the light beam, and the interruption is detected by the photocell.

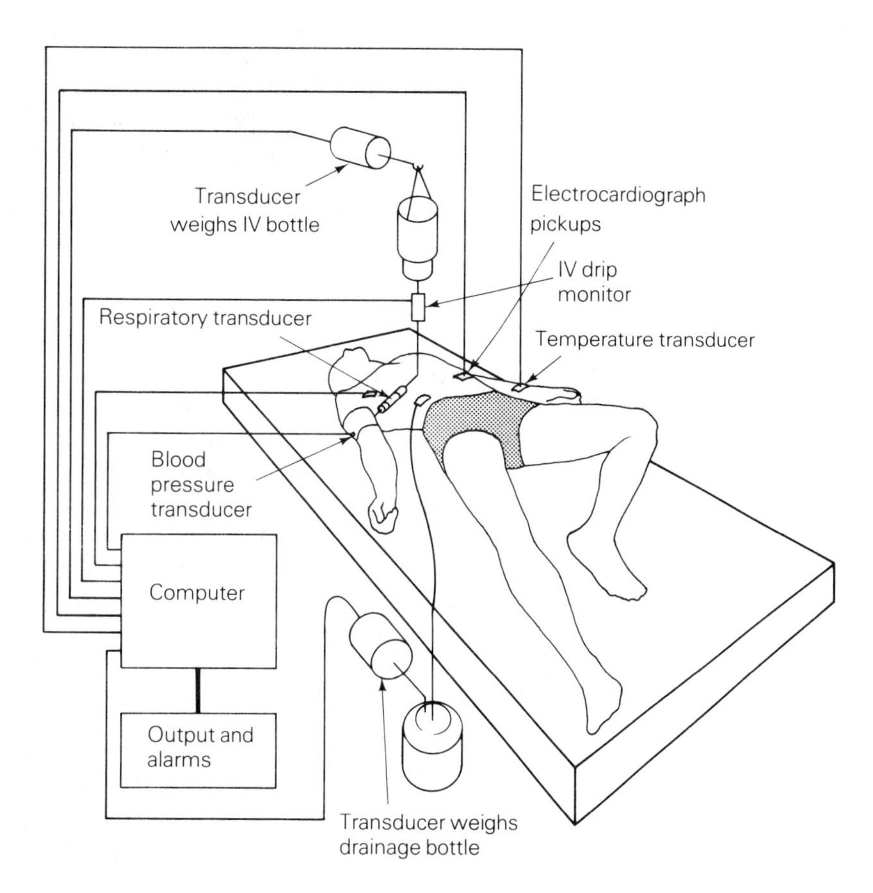

Figure 13-33 In intensive care units, a patient's vital signs are often measured with transducers, and computers analyze the measurements.

In intensive care units and open-heart surgery recovery rooms, it is desirable to watch every bodily function possible for signs of change in the patient. No human can possibly make all these measurements, record them, cross-relate them to other measurements, and still see the first critical signs of change. Computers are far better at this kind of data analysis, but they can't listen to stethoscopes or take temperatures. A nurse could take the measurements and enter them into the computer with a typewriter, but such a process isn't very efficient, and it takes the nurse away from the patient. Thus, devices are needed to convert blood pressures, temperatures, fluid outputs, and other measurements into electrical signals. Transducers can perform these tasks. Once the computer knows the values, it can watch the patient for a critical change far more efficiently than a human can. If any crucial change occurs, the computer can signal the doctor and nurses. Some of the transducers that are frequently used in intensive care units are shown in Figure 13-33.

Of course, transducers are also used in other fields. They are widely used in industry, where transducers are available for converting almost every physical quantity. Some examples are microphones, which convert sound waves into electrical signals, the force transducer in an electronic scale, which converts weight into an electrical signal, and the radar guns that police use to measure your speed.

Electrical Safety

Our ancestors lived with dangers such as falls and fires for millions of years, so we instinctively avoid many of these hazards. We have no such protection from electricity. No organs warn us of electrical dangers, nor do we have any instinctive fear of electricity. We must therefore compensate for our lack of natural defenses by treating electricity with special care.

The important point about electrical safety is that the *normal line voltage of 110 volts can be deadly!* Any electrical applicance showing the slightest sign of an electrical problem should not be used until it has been checked and repaired. Even appliances in perfectly good condition are not safe when they are used under conditions for which they were not designed. For instance, using a hair dryer in the bathtub is the electrical equivalent of smoking in a gunpowder factory. Even an indoor table lamp at an outdoor party is hazardous because it does not have the proper insulation; it is doubly dangerous if the party is a pool party.

Effect of Electric Shocks

Very high voltage shocks sometimes kill by overheating the cells, but such shocks are rare. Most shocks damage the cells very little; they kill by disrupting nerve and muscle functions. The electric current activates the cells, contracting the muscles and sending false nerve signals. After the shock, the heart and the respiratory muscles usually do not resume their normal functions, causing death.

Shocks often cause the heart to flutter, that is, to fibrillate. The heart rarely returns to its normal beat spontaneously. Surprisingly, high-voltage shocks sometimes do less damage to the heart than low-voltage shocks, because high-voltage shocks will cause the heart to contract totally. The heart will often restart by itself after complete contraction but only rarely after fibrillation.

Since most shocks don't damage the cells very much, a shock victim has a good chance of survival if resuscitation is started *immediately.* The first step is to disconnect the victim from the electricity with a good insulator, such as a dry plastic rod. Then mouth-to-mouth resuscitation and closed chest-heart massage (cardiopulmonary resuscitation, CPR) are applied until help arrives. The brain will be damaged very quickly by lack of oxygen, so cardiopulmonary resuscitation must be started within minutes. Cardiopulmonary resuscitation should also be used on people hit by lightning. In spite of the awesome appearance of a lightning bolt, lightning victims usually have a good chance of being revived if resuscitation is started quickly.

The seriousness of a shock depends mainly on the current, although people vary slightly in their sensitivity to electricity. Currents smaller than a milliamp cannot be felt by most people; currents of 1 or 2 milliamps can be felt but generally don't do any harm; 100-milliamp currents will cause heart fibrillation and death; and currents of more than 1 ampere may cause burns in addition to disrupting nerve and muscle functions.

The voltages needed to produce these currents vary because the resistance of the body varies. Under most circumstances, voltages under 50 volts will not produce any sensation. At around 60 volts a shock can be felt; at 90 volts the shocks are quite painful and dangerous. Of course, caution should be used below 50 volts. For one reason, what you *think* is less than 50 volts might be more. For another, severe burns can result from a short circuit even when the voltage is only a volt or two.

The resistance of the body is mainly in the skin, because the ionic fluids inside the body are good conductors. Dry skin is a poor conductor, but just a little bit of perspiration lowers the skin resistance dramatically because the salt makes sweat a very good conductor. Thus, the severity of a shock depends greatly on the wetness of the skin. A shock through wet skin might be 100 times worse than a similar shock through dry skin.

Since the amount of perspiration changes rapidly with emotional change, skin resistance is a good indicator of emotional changes; thus, it is one of the physiological variables measured with lie detectors. Although a sudden drop in skin resistance doesn't necessarily mean the person has lied, it is a strong indication that the person is under emotional stress. Measuring skin resistance is also a favorite trick of many phony doctors. A skin resistance meter is easy to build, and since it does show changes in a person's emotional state, it can convince a gullible person that a quack has something to offer.

A current cannot flow through the body unless a complete

circuit exists. That is, the current must have a place to exit as well as enter the body. The severity of a shock also depends on where the entry and exit points are. When the current flows into one finger and out the next, the shock will usually not be severe. But when the current enters through the right hand and exits through the left, it passes through the entire trunk of the body, including the heart and lungs; such a shock is extremely dangerous. An electronics technician who works around live high-voltage wires reduces the danger both by standing on insulating pads and by working with one hand in a back pocket. Thus, a current has no place to leave the body, so no current can flow if the technician accidentally touches a high-voltage wire.

One of the important side benefits of modern solid-state electronics is that the voltages in most solid-state devices are low and seldom dangerous. Old-fashioned tube-type equipment has lethal voltages at almost every spot.

Electrical Safety Precautions

A shock can't pass through your body unless a connection with a high voltage and a complete circuit through the body happen at the same time. Therefore, the basic principle of electrical safety is to avoid both whenever possible.

Electrical equipment should be as safe as possible. When you purchase new equipment, check to see that it has been approved by Underwriters Laboratory (UL) or another appropriate safety agency. Of course, such approval does not mean that the device is safe under all possible conditions. If a device is to be used outside or in a wet environment, it should be certified as suitable for those conditions.

Because all devices, no matter how well designed, wear out eventually, you must be alert for the first signs of an electrical problem. At the slightest hint, send the item in for repair or replace it. Worn power cords, missing safety covers, sparks, smoke, or slight shocks are the most common signs of electrical problems, but even a hunch that something is wrong is enough to have the device checked. Of course, unplug the device before you try to repair it. Most simple appliances are safe to work on once they have been unplugged; however, some electronic devices, such as TVs, ECG units, and electrotherapy devices, have parts that store dangerous amounts of charge long after they have been unplugged.

Avoiding a complete circuit is tougher than it seems because touching anything that is connected to the ground also connects you to the electrical system. Let's look at the commercial electrical system in detail to see why so many unsuspected things provide a return path for an electric current. The electrical line voltage in the United States is 110 volts, 60 hertz AC. It is brought to the wall sockets on two wires, *one of which is permanently connected to the ground.* (This connection provides a safe path along which lightning can flow to the ground.) Thus, the ground and everything connected with the ground, such as water pipes, most wet floors, bathtubs, sinks, and heating ducts, provide a *direct return path for shock currents.* Of course, you cannot completely avoid using these things, so when you do you must avoid touching a high voltage. When you must use electricity near things that are grounded, use only those devices that are *specifically made for use under such conditions.*

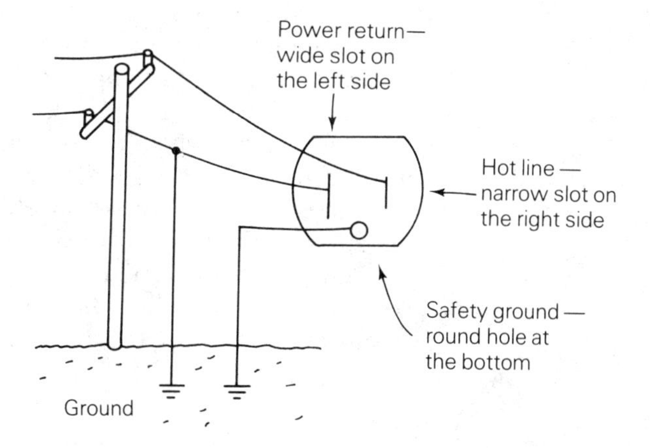

Figure 13-34 The electrical socket connections in the United States and Canada.

Three wires go to each modern wall socket, and they are connected as shown in Figure 13-34. The narrow slot on the right is connected to the hot wire, and it is always at 110 volts AC. The wide slot on the left is connected to the so-called power return wire; it returns to the power station, but it is also permanently connected to the ground. Normally, current returns through the power return wire, but since everything connected to the ground also connects to the power return wire, touching the hot wire and the ground can complete the circuit, just as touching the power return wire does. The round hole at the bottom of the socket is the safety ground, which is connected to the ground with a separate wire from the power return. The safety ground prong on a plug is always the longest so that it is the first to connect when the plug is inserted into the socket. Not all sockets installed before about 1960 have safety ground connections. These sockets should be replaced with properly grounded three-pronged sockets wherever appliances with three-pronged cords are used. The use of three-to-two prong adapters is not safe, particularly since few people ever ground the safety ground wire.

The purpose of the safety ground wire (which is normally green) is to provide a safe path for the current to return to ground whenever a short circuit develops in the appliance. To see how the safety ground works, consider an old-fashioned electric drill. The cord has only two wires, and the drill case is made of metal. When the drill is new, the metal case does not touch any of the wires; but with use (and abuse), the hot wire can contact the case. (This is commonly called a short circuit.) The danger is that the drill will continue to work without showing any signs of the short circuit. As long as the operator stands on a perfectly dry floor and does not touch anything connected to the ground, the drill will work normally. (Sometimes a very slight tingling can be felt when short circuits occur; if a tingling is felt, the device should be checked.) However, if the person should stand on a wet floor and use the drill, the person will be electrocuted. The safety ground wire on a modern drill (Figure 13-35) prevents this. It connects the metal case directly to the ground, so when the hot wire touches the case, the current flows back through the safety ground instead of through the operator. This flow through the safety ground usually blows the fuse in the hot line, disconnecting the tool from the high voltage and indicating that the tool needs to be fixed.

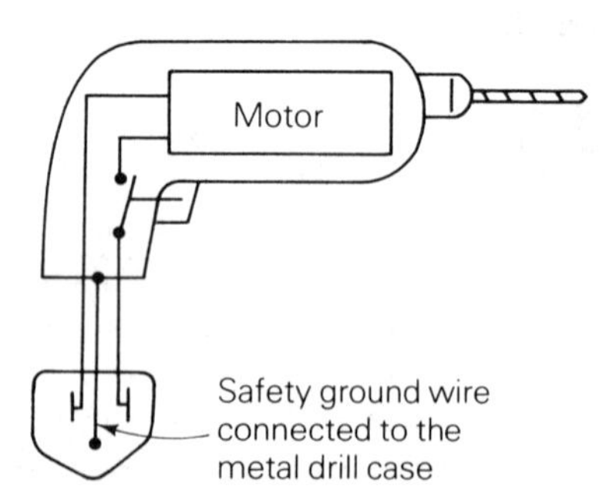

Figure 13-35

It is extremely important that electrical devices with three-pronged plugs be used correctly. The ground prong should never be cut off, nor should the safety ground be bypassed in any way. The device might work fine without the safety ground, but it will provide no protection whatever from dangerous shocks.

The current that flows through the safety ground due to a short is called a *fault current*. A device that is working properly will have no fault current (or a very tiny one at most). Since a fault current of any size indicates that the device is developing a short circuit and becoming dangerous, many hospitals and other institutions have fault current meters, which monitor the safety ground and switch the electricity off if a fault current is detected. Most electronic repair shops have fault current meters to check individual devices. Checking the fault current in all devices brought to the shop should be a standard practice.

Ordinary two-pronged appliances with no safety ground should only be used where there is no danger of touching a ground. Many institutions are phasing out two-pronged appliances, and everyone should be careful not to use two-pronged appliances under abnormal conditions.

Some devices avoid the need for the safety ground by using insulation and plastic handles to reduce the danger of shock. These devices will be labeled with the term *double insulated* to distinguish them from ordinary two-pronged devices. Double-insulated devices meet the safety standards of most institutions.

Another device that can prevent dangerous fault currents is the *isolation transformer*. Remember that there is no actual electrical connection between the two coils of a transformer. The electrical energy going to the input coil is transformed into magnetic energy; then it is retransformed into electrical energy in the output coil. No electrical charge actually flows between the coils. When an isolation transformer is connected between the device and the electrical supply wires, the power return of the device is no longer connected to the ground (Figure 13-36). Thus, an isolation transformer prevents shocks through water pipes and wet floors. A shock can only occur if the hot line and the power return line are touched at the same time. This might happen, but is is much less likely than touching a wet floor.

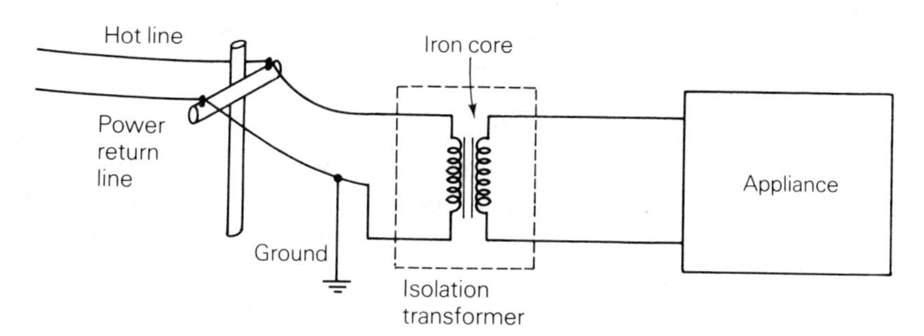

Figure 13-36 The isolation transformer has no direct electrical connection between the input and output coils. Thus, the power return of the output is *not* connected to ground.

Many electronic devices have built-in isolation transformers, and some hospitals have gone to the expense of connecting all wiring to an isolation transformer. However, it is usually difficult to find out whether a device has such protection. The safest policy is to insist on isolation transformers when they are available on new equipment, then treat every device as if it didn't have one.

No one list of dos and don'ts will prevent all electrical accidents; however, the following list covers the major points of electrical safety practice:

1. Remember that 110 volts can easily cause fatal shocks. Treat it with the respect that it deserves.
2. Do not use any electrical device that shows signs of electrical defects. A cord with even one small crack should be replaced.
3. Unplug an appliance before trying to repair it.
4. When working outdoors, in wet areas, or at other hazardous locations, use only electrical equipment that has been specifically designed for use in those areas.
5. Do not tamper with safety devices such as safety ground wires, fuses, isolation transformers, and safety interlock switches. Blown fuses should not be replaced with higher amperage fuses.

When proper care is taken, electricity is quite safe. But you should never forget that it can shock, kill, and start fires without warning when it is handled improperly.

Special Precautions for Patient Care

The high resistance of the skin is bypassed during most electrical diagnostic procedures and treatments. The electrodes are either attached directly to the skin with a highly conductive paste or needle electrodes are inserted beneath the skin. The patient then has none of the normal protection of the skin; thus, even small fault currents can be dangerous. Whenever electrical apparatus is attached to patients, special precautions must be taken to prevent electrocution.

The first rule is *never* to ground the patient. This prevents fault currents from flowing through the patient to the ground. The patient should be kept away from water pipes, metal instrument cases, floor drains, and all other possible grounds. When older instruments without built-in isolation transformers are used, external isolation transformers should be used.

You must look closely to get rid of all possible grounds. For example, consider electrotherapy, where the patient is often seated in a tub of water. An ordinary bathtub would be totally unsuitable, since it is connected directly to ground by both the water and sewer pipes. The tub must stand by itself on a wooden or plastic table. The hoses used to fill the tub must be removed, the drain hose must be disconnected from the drain, spilled water must be mopped up, and no electrical wires can be near the tub except for those used during the treatment. You must also allow for the possibility that the patient will do something such as reaching for the water faucet to let in more hot water. All grounds must be so far away that the patient cannot reach them without getting completely out of the tub.

All IV, drainage, and catheter tubes should be thought of as wires, because they always contain ionic fluids, which are very good conductors. Since they always lead under the skin, such tubes provide very good paths for fault currents when they become grounded. For instance, if a patient's intravenous fluid bag is hung on a water pipe, the patient may be grounded; if a drainage bottle is placed over a floor drain, the patient may be grounded; and so on.

Pacemaker leads and catheter tubes leading to the heart require special care. Since they can conduct a current directly to the heart, a small current can cause serious problems. Only a few microamps can send the heart into fibrillation. Very small voltages can cause such currents when there is a direct path to the heart.

Concepts

Electric charge
Atom
Nucleus
Electron
Proton
Neutron
Ion
Anion
Cation
Insulator
Conductor
Semiconductor
Ionic conduction
Current
Series and parallel circuits
Magnets
Electromagnetism
Magnetic induction
Transformer
Transducer
Resting potential
Action potential
Electromyography
Electrocardiogram
Electroencephalography
Iontophoresis
Electrical safety practice
Ground

Voltage
Electrical energy
Electrical power
Electrical circuit
AC and DC electricity
Ohm's law
Resistance
Power return
Hot line
Safety ground
Short circuit
Fault current
Isolation transformer

Discussion Questions

1. Describe the structure of an atom. What role does electricity play in holding the atom together?

2. Why do some materials conduct electricity very well, while others are insulators that do not conduct electricity at all? Name some materials that are conductors and some that are insulators.

3. Why is static cling a bigger problem with synthetic fabrics than with cotton?

4. Give the definition of electrical voltage. How is voltage similar to height in mechanical problems and pressure in fluid flow problems?

5. Why is it a poor practice to hook several long, thin extension cords together to reach an appliance, particularly when the appliance requires a high current?

5. Explain the difference between AC and DC electricity. What are some of the advantages and disadvantages of each?

7. Explain how motors turn electrical energy into mechanical energy and how generators do the opposite.

8. Explain how electrical heaters and light bulbs work.

9. Describe several important roles of electricity in the body.

10. Explain how osmosis often leads to a voltage across a semipermeable membrane.

11. Explain how electric currents are used in iontophoresis to drive medications under the skin. Why is the direction of current flow so important? That is, why would reversing the plus and minus terminals make the treatment worthless?

12. Explain why a bird can land on a bare power line without being electrocuted.

13. What big safety advantage do transformers have?

14. List some of the advantages and disadvantages of battery-produced electricity.

15. What are the proper steps for rescuing a person from electrocution or lightning?

16. Explain why wet floors, water pipes, and the ground drastically increase the risk of shock and electrocution when you are working with electricity.

17. What are the five main rules of electrical safety? What are some of the reasons for these rules?

18. Explain the function of the safety ground on three-pronged appliance plugs.

19. What is a fault current?

20. Why must such things as IV tubes, catheters, and pacemaker lead wires be watched carefully when working with electricity around a patient?

Problems

1. Consider the three electrically charged objects shown below. In what direction will the middle object move when it is released?

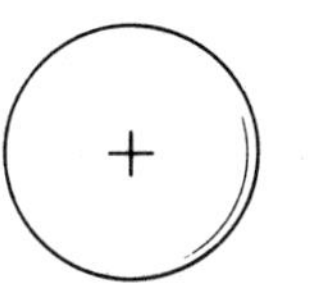

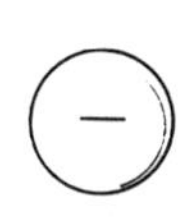

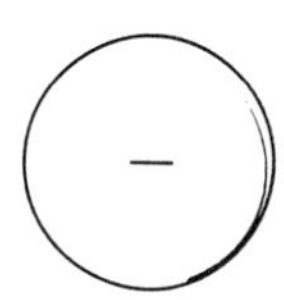

2. The two positively charged objects at the base of the triangle each have the same charge. In what direction will the top, negatively charged object move when it is released?

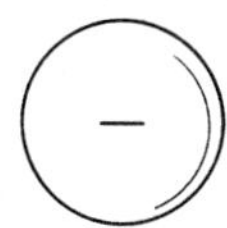

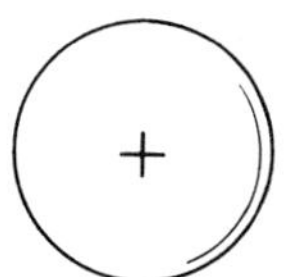

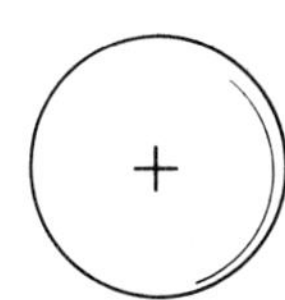

3. What is the maximum number of coulombs of electric charge that can flow through an 8-amp fuse in 1 minute?

4. How much energy does it take to move 6 coulombs of electric charge through a voltage of 220 volts?

5. How much does it cost to run a 1,500-watt electrical heater for 8 hours if electrical energy costs 9 ¢ per kilowatt-hour?

6. What is the largest wattage appliance that can be hooked up to a 110-volt circuit with a 20-amp fuse?

7. A large electrical heater is rated at 5.5 kilowatts at 110 volts. What should the current carrying capacity of the cord be?

8. How many 100-watt light bulbs can be safely hooked up on the same 110-volt circuit if the circuit is protected by a 15-amp fuse?

9. A certain 6-volt battery will put out a current of 3 amps for 1½ hours before it dies. How much energy is stored in the battery? If the battery costs $5, compare the cost of the energy in the battery to electrical energy from the wall socket, which costs about 2¢ per megajoule.

10. How much current does a 700-watt, 110-volt hair dryer require and what is its resistance?

11. If a light bulb has a resistance of 4.5 ohms, how much current will it draw when it is connected to a 12-volt battery?

12. A 110-volt light bulb has a resistance of 161 ohms. What is the wattage rating of the bulb?

13. How much current would each of the circuits draw if they were plugged into a 110-volt socket?

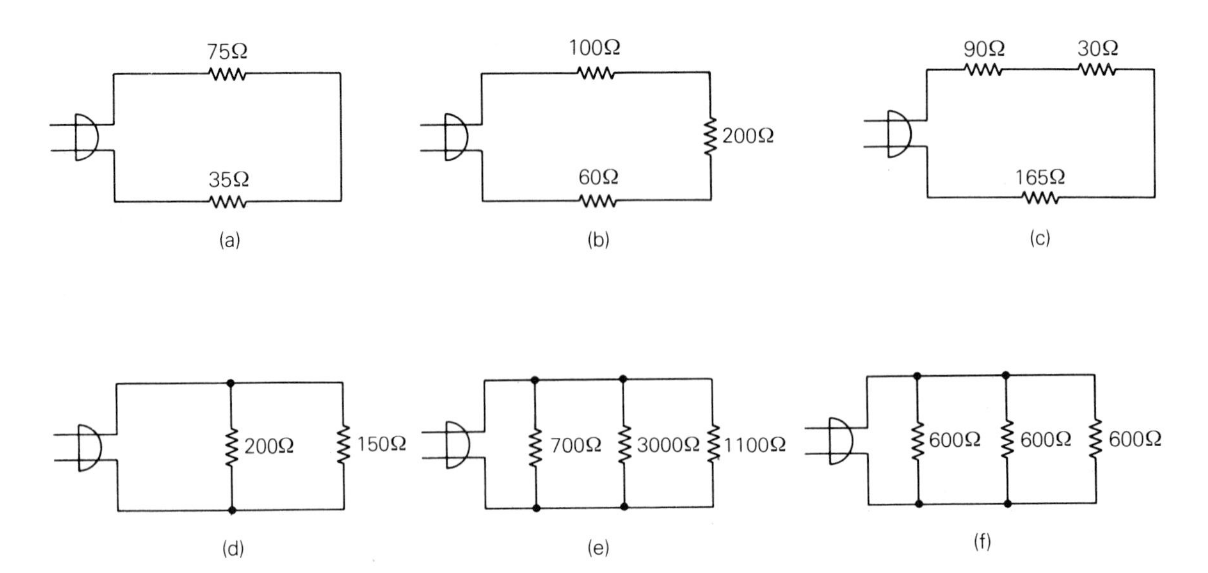

14. What is the total resistance of the four resistors in the circuit shown below?

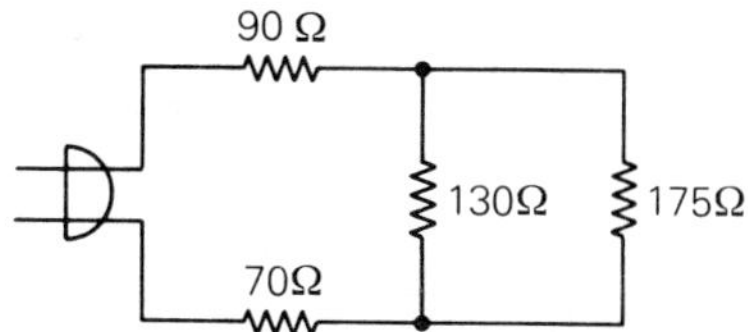

Experiments

1. If you haven't had any experience in wiring circuits, a great deal can be learned by taking a flashlight battery, a flashlight bulb, and a piece or two of wire, and then hooking them up to light the bulb. Observe that opening the circuit at any of the connection points will turn off the light.

2. Demonstrate the force of a magnet on a moving electric charge by placing a magnet near a TV screen. The magnet will change the path of the electrons in the tube, causing the picture to distort. Don't do this experiment on a color TV, because the magnetic field can distort the color permanently.

3. Show that electric currents create magnetic fields by placing a wire above a compass parallel to the compass needle, then passing a current from a flashlight battery through the wire. What does the compass needle do when the current is turned on? What happens when the direction of the current is switched?

4. Build an electromagnet by wrapping about 50 turns of insulated wire around an iron nail. Connect the two ends of the wire to a flashlight battery, then pick up paper clips and other iron objects with the magnet.

5. The shape of the magnetic field around a magnet can be seen by laying the magnet under a sheet of cardboard (or better yet, glass) and then sprinkling iron filings on the cardboard. Give the cardboard a few light taps, and the iron filings will line up along the magnetic field lines.

6. An interesting experiment shows a curious physiological effect of electricity on worms. Take a common earthworm and lay it on a paper towel wetted with a weak salt solution. Then take two wires from a battery with a voltage between 6 volts and 12 volts, and place one wire on the paper close to one end of the worm and the other wire near the other end of the worm. Then switch the wires. With the current one way, the worm will try to stretch out; with the current the other way, the worm will contract. No one knows of any simple explanation for this; but it does demonstrate that there is still a lot to learn about the physiological effects of electricity.

CHAPTER FOURTEEN

Waves

Educational Goals

The student's goals for Chapter 14 are to be able to:

1. Identify the frequency, period, amplitude, and wavelength of a wave.
2. Give several examples of how waves are important to human life.
3. Explain the difference between a longitudinal and a transverse wave.
4. Describe what wave polarization is and explain why a longitudinal wave cannot be polarized.
5. Give examples of wave diffraction and interference.

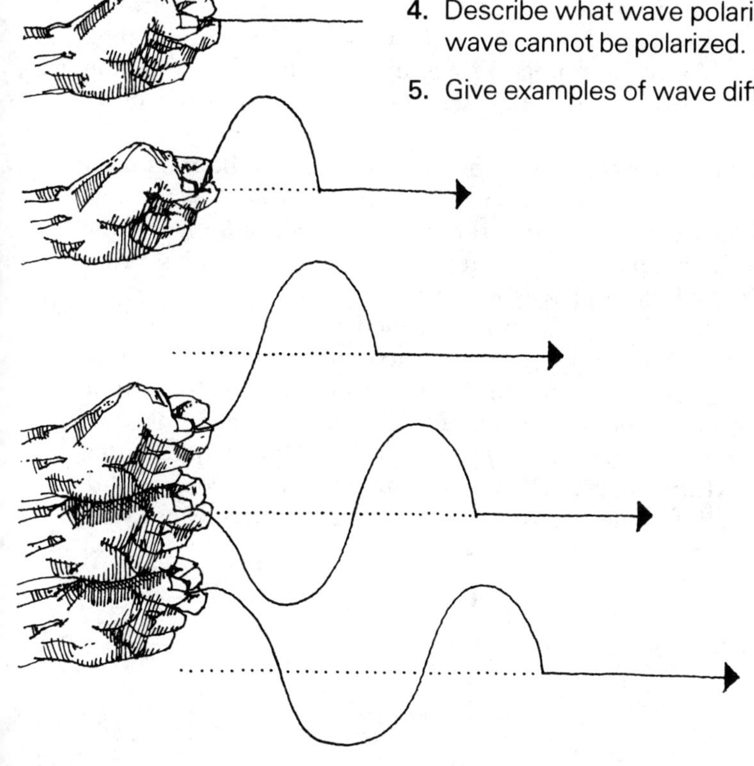

Waves affect our lives in many different ways. Most of the information that we receive comes to us as sound waves and light waves. Our food is moved through our esophagus and intestines by peristaltic waves of contraction and relaxation. We use x-ray waves and ultrasound waves to image the inside of the body. Microwaves are used to cook food and to heat sore joints by diathermy. From radio and TV waves to the waves on a beach, waves influence us in countless ways; indeed, matter itself has many wavelike characteristics. Studying waves in general is helpful before seeing how specific types of waves are important to us.

A wave is any disturbance that essentially keeps the same shape as it moves. We usually think of water waves first, and they are well worth watching if we want to understand waves. Since motion is a fundamental feature of waves, a drawing of a wave is of limited value. Filling a bathtub and sprinkling pepper on the surface of the water is a good way to study wave motion. The pepper will help you to distinguish the motion of the individual water drops from the wave motion.

Periodic Motions

The individual particles do not move along with a wave as it travels through a substance; they only move back and forth and then return to their starting points. Prove this to yourself by watching pepper on the surface of bathwater: The pepper does not move along with the wave.

Water waves are a good example of wave motion.

Motions that are repetitive and regular, such as the motion of an individual particle in a wave, are called *periodic motions.* The beating of the heart, breathing, the swing of a pendulum, and the vibrations of a drum are some examples. There are countless other examples in the body and in everyday life.

Two pieces of information are needed to describe a periodic motion: how fast it swings back and forth and how far it swings. The speed of the swing can be described in two ways—the time needed for one complete swing (the *period*) or the number of swings per unit of time (the *frequency*). For example, suppose that the alpha rhythm of a brain made 10 complete cycles every second. The frequency f would be 10 hertz (recall that *hertz* means cycles per second). The period t would be

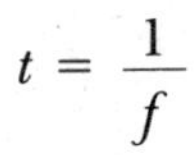

$$t = \frac{1}{f}$$

$$= \frac{1}{10 \text{ Hz}}$$

$$= 0.1 \text{ s.}$$

This relation between t and f holds for all periodic motions.

The size of the swing is described by the *amplitude* of the swing, which is the distance from the particle's rest position to the point farthest from the rest position (Figure 14-1). When the particle moves the same distance from the rest position on each side of the swing, the amplitude is one-half the total swing distance. (The amplitude is commonly given instead of the total swing distance because the amplitude makes most formulas simpler.)

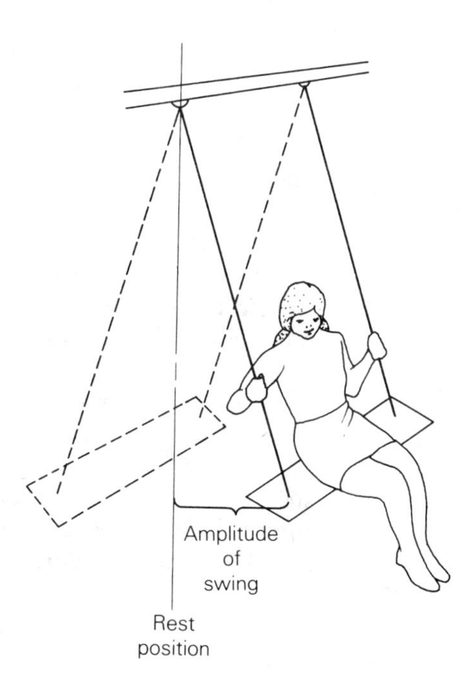

Figure 14-1 The amplitude is defined as the distance from the rest position to the position farthest from the rest position.

The simplest periodic motion is when the particle moves back and forth without jerking, like a child swinging gently on a swing. This type of periodic motion is called *simple harmonic motion*, and it is the easiest type of periodic motion to study (Figure 14-2). It is important because any periodic motion, no matter how complicated, can be thought of as a combination of simple harmonic motions.

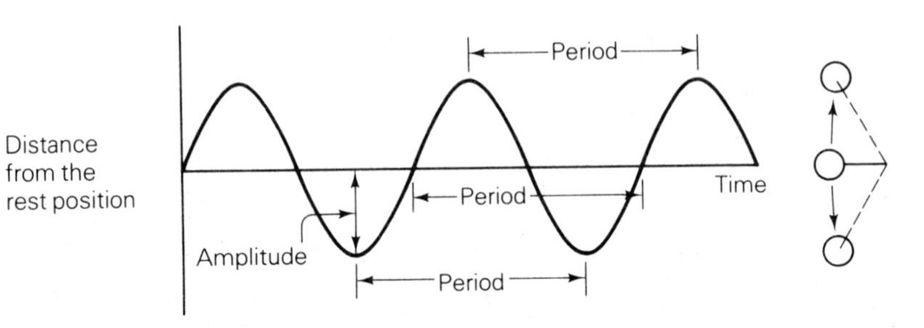

Figure 14-2 The distance from rest versus time for an object in simple harmonic motion, such as a swinging pendulum.

Waves

Waves are very common because wave motion will occur whenever two conditions are satisfied. First, the disturbance must affect neighboring particles more than ones far away. Second, the disturbance must take some time to move from place to place. The time can be long or short, as long as it is roughly the same for one pair of neighboring particles as for another. If these two conditions are satisfied, the shape of the disturbance will not change much as it moves.

Epidemic diseases have both characteristics; thus, they spread in waves, or at least they did before the days of rapid long-distance travel. The disease would break out in one place, then gradually move in an ever widening circle. When allowances were made for geography, the pattern of an epidemic wave would be much the same as the wave pattern of a stone thrown in a pond.

The simplest type of wave is called a *sine wave* (from the word *sinuous,* meaning having many curves). Small ripples on water are sine waves, and snakes move along by generating sine waves with their bodies. In a sine wave, the motion of the individual particles is simple harmonic (Figure 14-3).

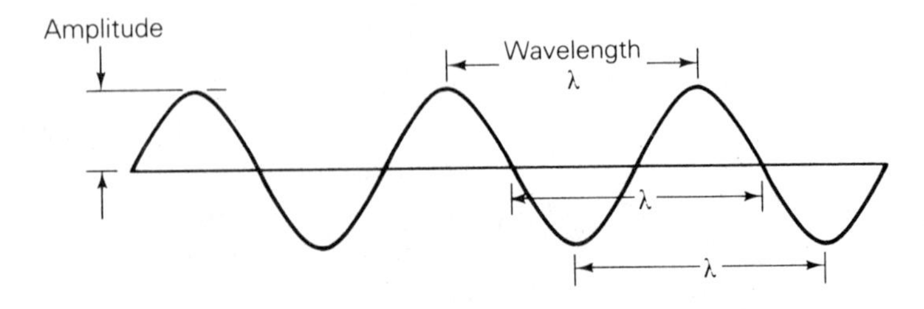

Figure 14-3 A sine wave.

Of course, waves can have almost any shape at all, but most can be considered as combinations of sine waves. An electronic organ uses this principle. Various sine waves are generated by the organ;

then the various keys and stops are used to combine the sine waves into a multitude of notes and sounds.

Several things are needed to describe a sine wave. The frequency, period, and amplitude of a sine wave are the same as they are for the individual particles in the wave. That is, the frequency is the number of complete swings back and forth per unit of time, the period is the time for one complete swing, and the amplitude is the distance from the rest position to the maximum position. Two more things are needed to describe a wave: the speed of the wave v and the wavelength λ (lambda). The wave speed is just the speed of the *disturbance*—it is *not* the speed of the individual particles. The wavelength is the distance between two neighboring crests (or two neighboring troughs) of the wave.

The wavelength and the period of a wave are closely related to the wave speed. Imagine a cork bobbing on the surface of the ocean. The time for the cork to go down from the top of one wave and come back up to the top of the next would be the period of the cork (and the wave) t. In that time, the original crest would have moved one wavelength (Figure 14-4). Thus, the speed of the wave is the wavelength divided by the period,

$$v = \frac{\lambda}{t}.$$

This relation holds for all sine waves, including sound, light, and radio waves. Since the frequency is

$$f = \frac{1}{t},$$

the wave speed can also be expressed as

$$v = f\lambda.$$

This relation also applies to all sine waves.

The above relations explain why big musical instruments have lower pitches than small ones and big people tend to have deeper voices than small people. The pitch of a sound wave is determined by its frequency; the higher the frequency, the higher the pitch. The speed of sound waves in air is the same for all waves, regardless of pitch. Therefore, the relation

$$v = f\lambda$$

says that when the frequency goes up, the wavelength must decrease if v is to stay the same. The relationship between wavelength and the size of the instrument is complex, but roughly speaking the size of an instrument must be about the same as the wavelength of

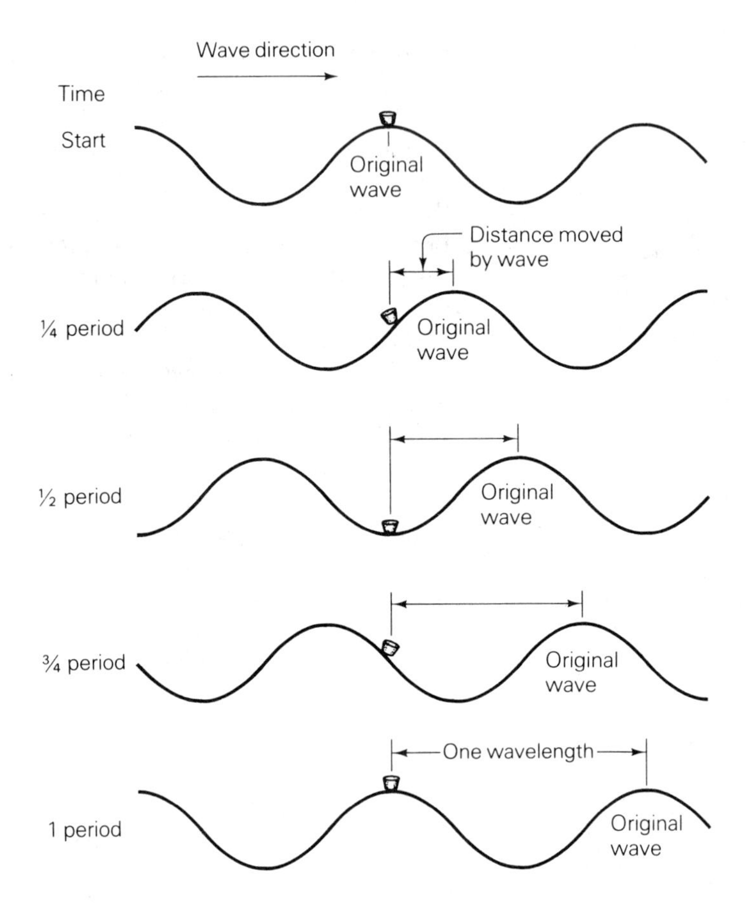

Figure 14-4 In one period of oscillation, a wave always moves one wavelength. Thus, $v = \lambda/t$ and $v = f\lambda$ for all waves. Note that the cork only bobs vertically; it does not move horizontally with the wave. Likewise, the individual water drops only move up and down.

its sound. Thus, large instruments must produce low-frequency sounds and small instruments high-frequency sounds. Because the relation holds for all sine waves, it doesn't matter how the instrument makes the sound. Thus, the bigger a stringed instrument, horn, drum, or human is, the lower the pitch will generally be.

EXAMPLE A person with good hearing can hear sounds in the frequency range between about 20 and 20,000 hertz. What wavelength range does this correspond to? The speed of sound in air is 344 m/s.

Answer Solving the relation $v = f\lambda$ for the wavelength gives

$$\lambda = \frac{v}{f}.$$

So when $f = 20{,}000$ hertz, the wavelength of sound will be

$$\lambda = \frac{344 \text{ m/s}}{20{,}000 \text{ Hz}}$$
$$= 0.017 \text{ m}$$
$$= 1.7 \text{ cm.}$$

When $f = 20$ hertz, the wavelength will be

$$\lambda = \frac{344 \text{ m/s}}{20 \text{ Hz}}$$
$$= 17 \text{ m.}$$

The Energy of a Wave

Only the pattern of the disturbance and the energy contained in the wave actually move. As the individual particles move back and forth, the energy is passed from particle to particle.

Since no matter moves along with the wave, waves can be ideal for transferring energy and information. For instance, the same microwaves that transfer microwave energy into a Thanksgiving turkey also carry most of the long-distance phone calls in the United States.

The energy of a wave goes up with the square of the amplitude. Thus, if a wave is doubled in size, it carries 4 times as much energy. This relation seems to correspond with the kinetic energy relation

$$E = \frac{1}{2} mv^2,$$

and it does. If the amplitude of a wave (and of a particle in the wave) is doubled without changing the period, an individual particle must move twice as fast to get back and forth in the same time. Since the individual particles are moving twice as fast, their kinetic energy and the energy of the wave are quadrupled. As you would guess, the amplitude of a sound wave is closely related to its loudness and the amplitude of a light wave is closely related to its brightness.

Longitudinal and Transverse Waves

The direction of the vibration of the individual particles doesn't have to be related to the direction of the wave, but in most waves it

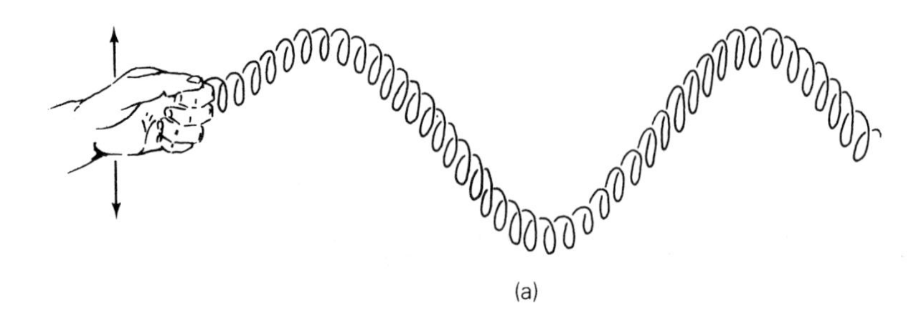

(b)

Figure 14-5 (a) Transverse and (b) longitudinal waves in a Slinky toy.

is. Most waves are either *longitudinal,* with the particles vibrating parallel to the direction of wave movement, or *transverse,* with the particles moving perpendicularly to the wave.

One of the best devices for demonstrating both types of waves is the Slinky toy, the long, coiled spring that children and physics teachers play with. If you stretch a Slinky out on a smooth floor, you will get a longitudinal wave when you shake it along its length and a transverse wave when you shake it perpendicularly to its length (Figure 14-5).

Water waves, the waves in a plucked guitar string, radio waves, and light waves are transverse waves. Sound waves are the most important form of longitudinal waves.

Polarization

Knowing that a wave is transverse does not indicate in exactly what direction the particles are moving. Many different directions are perpendicular to the direction of wave motion. For instance, suppose a girl tied one end of a rope to a wall, pulled it tight, and shook the other end. She could shake her end of the rope up and down or horizontally. Or she could combine the two motions in some way (Figure 14-6). Each motion would send transverse waves down the rope. The first type is said to be *vertically polarized,* because the particles move vertically. The second type is said to be *horizontally polarized.* The third type, where the particles move in

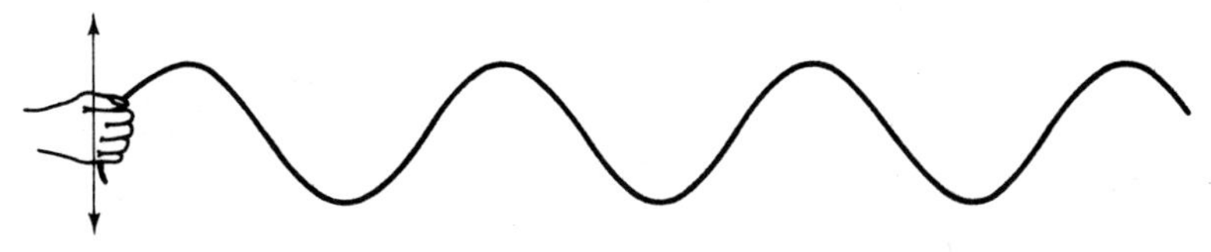

(a) Vertically polarized wave on a rope

(b) Horizontally polarized wave on a rope

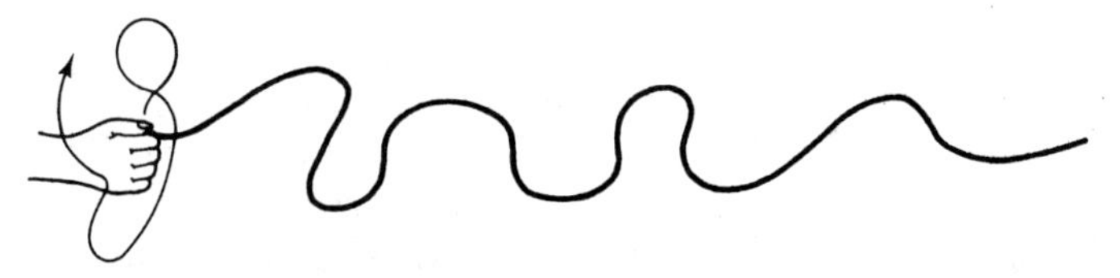

Figure 14-6 (c) Unpolarized wave on a rope

various directions perpendicular to the wave motion, is said to be *unpolarized.* Longitudinal waves cannot be polarized, since only one direction is parallel to the direction of the wave.

Diffraction and Interference

Waves travel in straight lines when their paths are not obstructed. All waves will bend around an obstruction, a characteristic known as *diffraction.* Generally, the amount of diffraction tends to be larger with longer wavelengths.

Sound waves illustrate diffraction well. When you are outside a house with an open window, you can hear much of what goes on in the house whether or not you can see the sound source. This is because sound waves diffract around the edges of the window as they pass through it (Figure 14-7). Water waves also diffract easily. Watch the waves in a bathtub move around your foot; they will diffract around your foot with only minor changes in their shape.

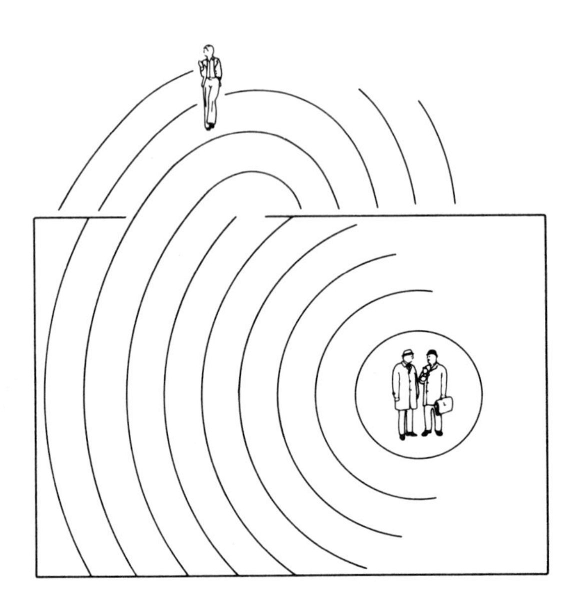

Figure 14-7 Sound waves diffract around corners very easily.

Even light diffracts a little bit. An optical microscope or telescope cannot be perfect, because the diffraction of light in the instrument limits the ultimate resolution.

When two waves arrive at the same point, the combined wave is usually just the sum of the two individual waves. (However, explosion shock waves and the almost explosive light pulses from a high-power laser don't follow this rule.) One crucial factor must be taken into account when two waves are added: the time of arrival of the wave. Suppose the peaks of two equal waves arrive at the same time, as shown in part (a) of Figure 14-8; such waves are said to be *in phase.* The combined displacement of the two waves at any instant is just the sum of the individual displacements, so the wave $A + B$ is simply twice as big as one of the original waves. When two waves combine to produce a bigger wave, the phenomenon is called *constructive interference.*

Can two waves combine to produce a smaller wave or even to destroy each other? The answer to both questions is yes, even though it seems paradoxical that two somethings can combine to produce a nothing. Suppose two equal waves A and B arrive with exactly one-half period time difference between them, as shown in part (b) of Figure 14-8. Thus, a crest in A arrives at the same time as a trough in B, and vice versa. Adding the two displacements, we find that the troughs in A are filled in by the crests of B, and vice versa. The final result is no wave at all. Two waves that have a one-half period time difference are said to be completely *out of*

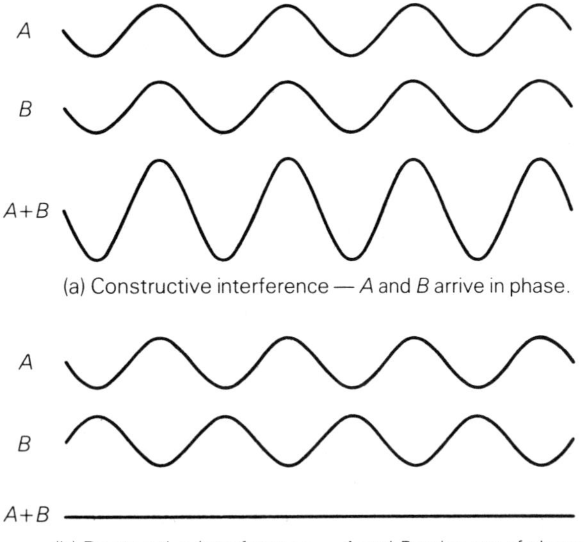

(a) Constructive interference — *A* and *B* arrive in phase.

Figure 14-8 (b) Destructive interference — *A* and *B* arrive out of phase.

phase. When two waves combine to form a smaller wave, the effect is called *destructive interference*.

Because of destructive interference, you might think that you could shine two flashlights on a wall to form a dark spot. You can't with ordinary flashlights, because they don't give off one light wave—they give off many. Thus, the destructive interference is balanced by constructive interference. But you can shine two laser beams on a spot and have them "destroy" each other.

The energy of waves is not destroyed by destructive interference. Destructive interference at one spot is always balanced by an equal amount of constructive interference someplace else. Thus, the total energy of the waves remains the same, as conservation of energy predicts.

You don't need fancy equipment to see interference; you need only your fingers and a light bulb. Form a narrow slit between two fingers and look at the bulb through the slit. If you look closely, you will see dark fringes, which are where the light has destructively interfered. Diffraction is also visible in this experiment, causing the edges of your fingers to appear somewhat fuzzy.

Reflection

When a wave bounces off a barrier, it might lose its pattern or stay basically the same. When the pattern is not changed much, the bounce is called a *reflection*. The smoother the barrier is, the more likely the bounce will be a reflection. Of course, smoothness is a relative thing: A barrier that seems smooth to a long-wavelength wave might be rough to a short-wavelength wave. Generally, a barrier will reflect the wave when the bumps on the barrier are smaller than one wavelength of the wave. For instance, an average sound wave has a wavelength of about a meter, while light waves average a wavelength of less than a micrometer. Thus, sound waves might reflect off the face of a rock cliff to form an echo, but light would not reflect because the bumps would be far bigger than a micrometer.

Almost any sudden change in material will form a partially reflecting barrier. For example, both air and glass transmit light very well, but the sudden change from air to glass is enough to reflect part of the light, as you can see in any window. The change need not even be a barrier—any sudden change can produce a reflection. An example is sound going past the open end of a pipe. Even though the end is completely open, most of the sound is reflected back down the pipe. The change from a small pipe to the open air is just too sudden. Trumpets gradually open up at the end (Figure 14-9) because this cuts down on the sound reflected back into the instrument, thus increasing the amount of sound getting into the open air.

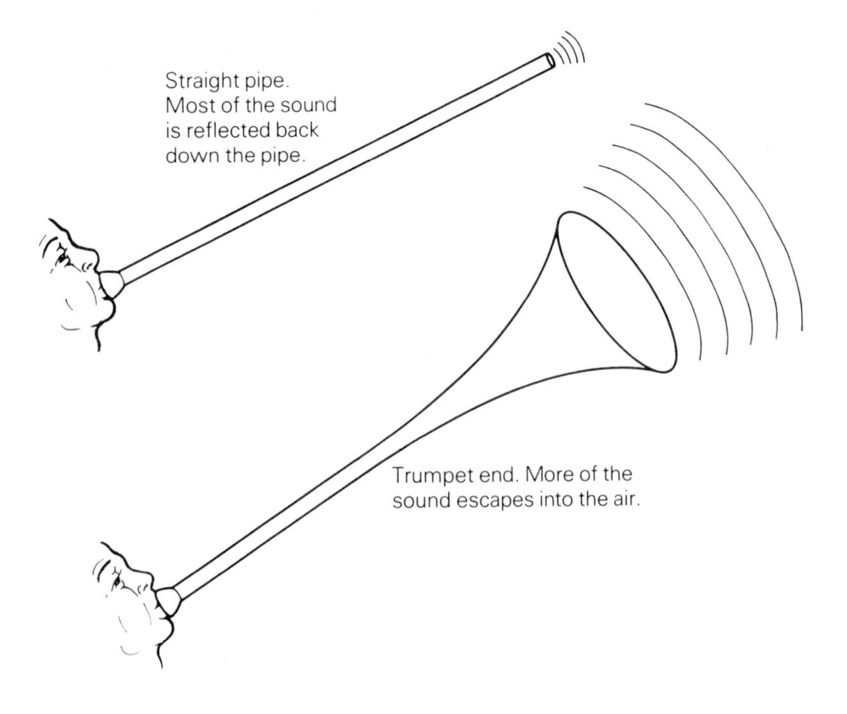

Figure 14-9

Concepts

Wave
Periodic motion
Period
Frequency
Amplitude
Simple harmonic motion
Sine wave
Wave speed
Wavelength
Longitudinal wave
Transverse wave
Polarization
Diffraction
Interference
Constructive interference
Destructive interference
Reflection

Discussion Questions

1. Give three or four examples of periodic motion that were not mentioned in the text or in class.
2. Why does a trumpet have a higher pitch than a tuba?
3. Does the water move along with the waves in deep water?
4. How are longitudinal waves different from transverse waves?
5. Explain why transverse waves can be polarized but longitudinal waves can't.
6. The most powerful microscopes cannot see the vibrations in a light wave, yet we know that light is a transverse wave. What fact proves this?
7. Is the sum of two equal waves always twice as big as one of the original waves? Explain why or why not.

Problems

1. If the frequency of a child on a swing is 0.25 hertz, what is the period of the swing?
2. If the period of a certain musical note is 0.00125 second, what is its frequency?

3. A water wave is going past a seawall painted in a checkerboard pattern, as shown in the figure. The squares of the checkerboard are 1 meter on a side. What is the wavelength and amplitude of the wave?

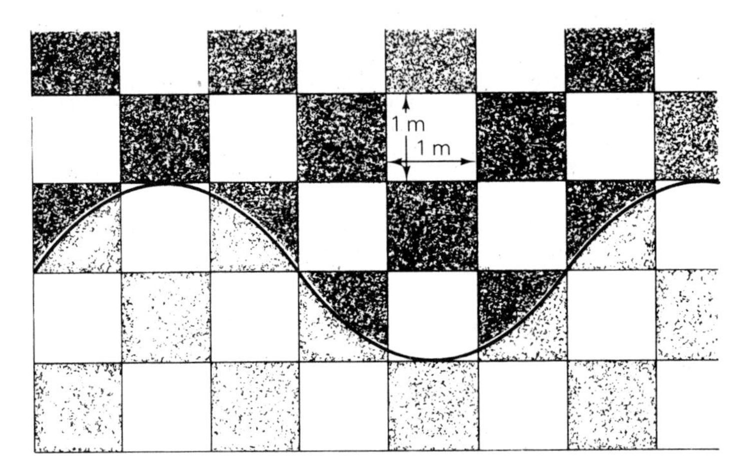

4. The lookout on a sailing ship is in a crow's nest on the mast 30 meters above the waterline. In a severe storm, the lookout notes that when the ship is at the bottom of a trough between two waves, he is just level with the tops of the waves. What is the amplitude of the waves?

5. The musical note treble C has a frequency of 523 hertz. What is its wavelength in air? What is the wavelength underwater, where the speed of sound is 1,500 m/s?

6. Radiowaves travel at the speed of light: 300,000,000 m/s. What is the wavelength of an AM radio signal from a station whose frequency is 710 kilohertz?

Experiments

1. Make a pendulum by tying a piece of string to any small object. Adjust the pendulum so that it is 1 meter long. Show that the period of this pendulum is 2 seconds. Try different objects and swing amplitudes, and show that the period is always 2 seconds as long as the pendulum is 1 meter long. Try some other lengths. How does the period change with the length of the pendulum?

2. Show that individual water drops do not move along with a wave by sprinkling pepper on the surface of the water and then disturbing the surface to create a wave.

3. Stretch a coiled Slinky on a smooth floor between you and another person. Demonstrate longitudinal waves by moving your end of the Slinky toward and then away from the other person. Next demonstrate transverse waves by moving your end back and forth perpendicularly to the Slinky.

4. Diffraction and interference of light can be easily seen by forming a thin slit between two fingers and looking at a light. The fuzziness of the edges is due to diffraction, and the dark bands in the slit are caused by destructive interference. Two razor blades taped together with a very narrow slit in between will show the effect quite strikingly.

5. Diffraction of water waves can be observed by placing two bricks in your bathtub with a 2-centimeter slot between them. The water should not quite come to the top of the bricks. Now make some waves on one side of the bricks and watch them come through the slot. Do they leave the slot in a narrow beam, or do they form circular waves?

6. The pulse travels as a pressure wave in the arteries at a fairly slow speed. The speed of the wave depends mainly on the elasticity of the arterial walls; the harder and stiffer the arteries are, the faster the pulse travels. In fact, measurements of pulse wave speed are now being used to detect hardening of the arteries and other circulatory problems. You can see that the pulse travels as a wave by comparing the arrival time of your pulse at your neck with the time at your wrist. The pulse arrives about 0.1 second sooner at the neck—a small but noticeable time difference. The pulse in the neck is in the carotid artery, which is under and slightly ahead of the bend in the jaw bone. Use your right hand to feel the neck pulse while you feel the pulse in your right hand with your left. Your hands are positioned correctly if it looks like your right hand is trying to strangle you while your left tries to rescue you.

CHAPTER FIFTEEN

Hearing and Sound

Educational Goals

The student's goals for Chapter 15 are to be able to:

1. Describe the physical characteristics of sound waves in air.
2. Relate the psychological characteristics of sound, such as pitch, loudness, and quality, to the physical properties of sound waves.
3. Use the decibel scale to describe the loudness of sounds.
4. Describe the principal features of sound production and hearing in the body.
5. Explain what can be learned from audiograms about hearing loss.
6. Give three or more common uses of ultrasound in the health professions.

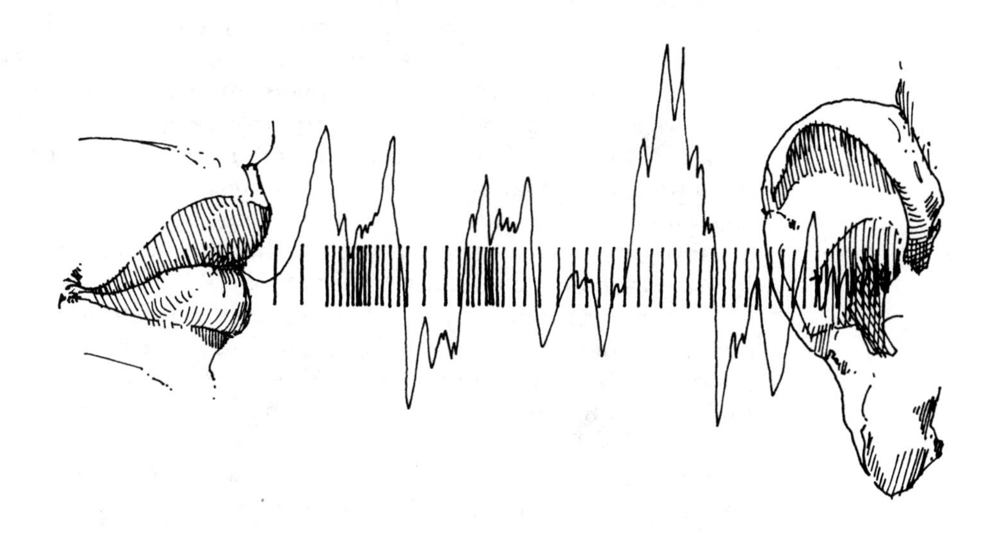

Sound is a highly efficient way of transmitting information. The percentage of sound information actually used by the brain is much higher than the percentage of visual information. For instance, when you listen to someone talking on the radio, nearly every word you hear makes an impression on your brain. But when you watch TV, the brain ignores everything outside the screen and much of what is on it. Only a small fraction of what you see makes an impression on the brain.

The efficiency of sound is of utmost importance in human communication. In order to communicate, we must be able to send information as well as to receive it. Because visual images require so much information, the body cannot send an effective visual image other than the rudimentary images formed by gestures. But sound is so simple and efficient that an organ that probably began as a simple valve to keep us from choking on our food is now able to produce a wide range of sounds that enable us to communicate with others very effectively.

The Nature of Sound

When discussing sound, we must remember that the physical characteristics of sound waves and the sensations that those sounds produce in the brain are two different things. The physical characteristics are easily measured; the sensations produced are different in different people and cannot be measured with instruments—they can only be described by the person who felt them. For example, the sensation of loudness and the power of a sound wave are related. In general, the more powerful sound will sound louder, but this is only approximately true. The loudness of a sound depends on many other factors, many of which are entirely psychological. If you ask a large group of people which is the noisier of two sounds of identical power, a well-played piano or someone banging garbage cans together, most would say the garbage cans and a few with different tastes would say the piano. They are all right. Since loudness refers to what their brains perceive, there are as many correct answers as people.

Let's look into the physical characteristics of sound first. In air and other fluids, sound waves are longitudinal waves of molecular vibration. The waves are longitudinal because the molecules of a fluid are free to slide sideways by each other. A significant force can be transmitted between molecules in a fluid only when the molecules collide almost head-on (Figure 15-1). Because sound waves in fluids are longitudinal, they cannot be polarized. In solids, sound waves

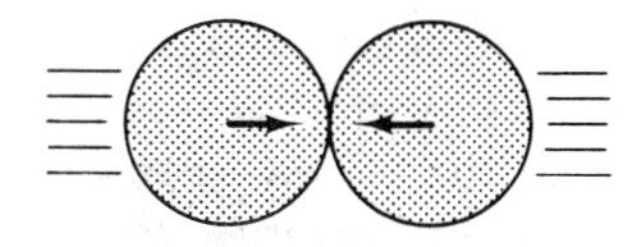

(a) Large collision force

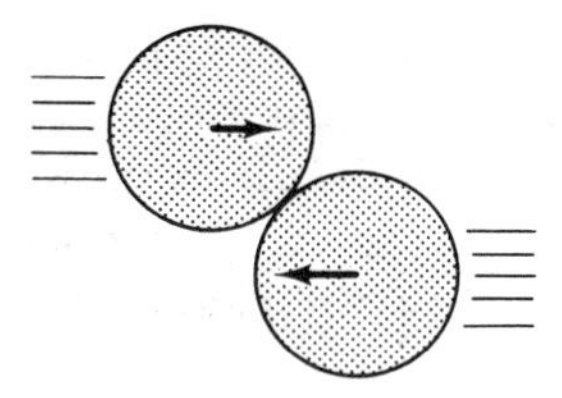

(b) Small collision force

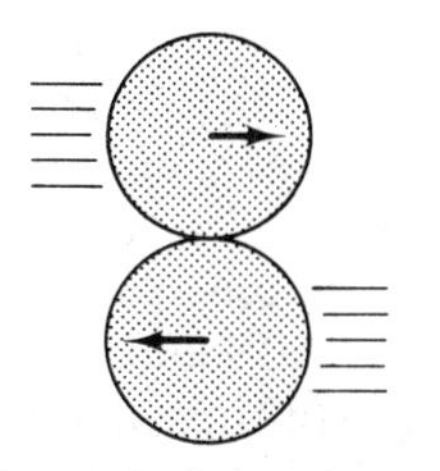

(c) Almost no collision force

Figure 15-1 The forces in a molecular collision are large only when the molecules collide nearly head-on.

can be either transverse or longitudinal; the transverse waves, of course, can be polarized.

How do the molecules move in a sound wave in air? Consider the molecules as sound is produced by a loudspeaker. A loudspeaker has a paper cone that is pushed back and forth magnetically by an electric current. As the cone moves forward, the air molecules ahead of it are bunched together. When the cone moves backward a moment later, the molecules are pulled apart. The places where the molecules are bunched together are called *condensations,* and the places with fewer molecules are called *rarefactions* (Figure 15-2). Of course, the pressure in a condensation is slightly higher than normal and the pressure in a rarefaction slightly lower than normal. Thus, the speaker sets up a chain of condensations and rarefactions that moves away from the speaker as a sound wave. The wavelength is the distance between two adjacent condensations (or rarefactions). Note that the air doesn't move off as a wave; the pattern of condensations and rarefactions is what moves. There is no net movement of air.

The speed of sound in air is 344 m/s at 20°C over a wide pressure range, and it is basically the same for all frequencies of sound. Temperature, however, does affect the speed of sound, which increases by about 2% for every 10°C increase.

The speed of sound is fast enough that we don't notice any time lag when the sound is produced a few meters away. Over longer distances we start to notice that the sound takes time to reach us. This is why marching bands sound so bad when they spread out over a football field. The sounds from the two sides of the band can reach the grandstand with a difference of up to a third of a second. This is more than enough to spoil the music. Sound takes almost 3 seconds to travel a kilometer, a worthwhile fact to remember for estimating the distance of a thunderstorm.

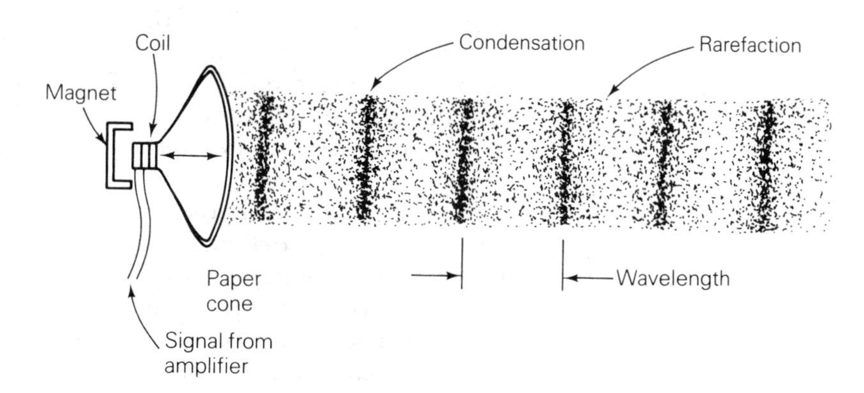

Figure 15-2 Sound waves are a series of molecular condensations and rarefactions. The amount of condensation and rarefaction shown is highly exaggerated.

The speed of sound in a gas drops as the molecular weight of the gas goes up. However, the molecular weight of most common gases is not very different from that of air, so the speed in most common gases is nearly the same as in air. The important exceptions are the ultralight gases, such as helium and hydrogen, which have sound speeds of over 1,000 m/s.

The speed of sound in liquids and solids is higher than in gases because the molecules in these substances are much closer together. As a result, the disturbance will travel between molecules rapidly. The speed in most liquids is similar to that of water, 1,500 m/s. Solids have the highest sound velocities; the velocity in glass is over 5,000 m/s.

Pitch and Frequency

The sensation of pitch is closely related to the frequency of a sound wave. Generally speaking, the higher the frequency of a wave, the higher the pitch will be. However, the relation is not exact. Two waves of identical frequency but different power will seem to have slightly different pitches. Also, other notes played at the same time can affect the pitch perceived by the brain. The differences are not large, however.

The only musical "instrument" that produces a pure sine wave is a tuning fork. All other instruments produce waves that are more complicated to varying degrees. Flutes and well-trained voices can produce nearly sinusoidal waves, while cymbals produce sounds that are too complicated to even call notes (Figure 15-3). The brain is very sensitive to the difference in sensation caused by waveform differences, which is called the *quality* of the tone.

Different people have different qualities to their voices. That is, they produce different-shaped waves when they speak. Because the brain detects the quality differences quite well, we can identify people very accurately from the sound of their voices. Computer analysis of voice wave forms has advanced to the point where a person's voiceprint, the analysis of the voice wave shape, is a fairly accurate method of identification. Soon locks and keys may be replaced with voice-activated locks that respond only to authorized persons.

Noise is another term that pertains to the waveform of a sound. Noise has no exact definition because what is music to one person may be noise to another. One of the few waves that almost everyone agrees is noise is a waveform in which the condensations and rarefactions occur randomly. This is called white noise, because all frequencies are present in about equal amounts, just as white light

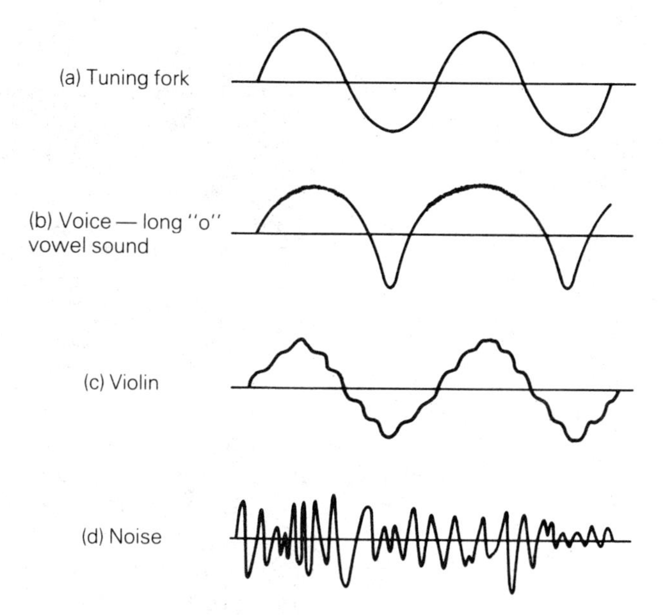

Figure 15-3 The waveforms of various sounds. The waves are drawn as transverse waves to show the pattern clearly. Sound waves are, of course, longitudinal waves.

contains equal amounts of all the light frequencies. Examples of white noise are the noise of air rushing through an air conditioner and the noise of a CB radio when no one is talking.

Actually, white noise is not unpleasant unless it is loud. Many people sleep better with a little bit of white noise than they do in total quiet, and sleeping aids are on the market that merely put out white noise.

Loudness and Intensity

The intensity of a wave is the power per unit of area struck by the wave. The SI unit of intensity would therefore be the watt per meter squared (W/m^2). Loudness is the psychological sensation related to intensity, but the relation is complex. One complication is that people find disagreeable sounds louder than pleasant sounds of equal intensity. This is the cause of many family arguments over the loudness of the TV. Frequency is another complicating factor. The ear is most sensitive in the frequency range between 500 and 5,000 hertz. Sounds outside that range of equal intensity are not as loud. Sounds below 20 hertz and above 20,000 hertz cannot be heard by the human ear at all. Thus, a sound outside that range is not loud even though it could be intense enough to damage a person's hearing.

The above complications are not that troublesome, and proper allowances can easily be made for them. The main complication in the relation between loudness and intensity is that there is no simple, direct proportion between them. That is, if you double the intensity of the sound, the loudness doesn't double. Surprisingly, most people just barely notice a doubling of sound intensity, probably because our ancestors needed both to hear small sounds to survive and to tolerate occasional loud sounds. For instance, the sound of someone screaming in your face is at least a billion times more intense than the smallest audible sound, but the brain would have difficulty handling a stimulus that was a billion times larger.

The evolutionary solution to this problem was to keep the ear very sensitive to small noises and to gradually discount larger sounds. That is, as the sounds become larger, the ear and the brain gradually reduce the percentage of the sound that produces a sensation. Thus, the sensation produced by a sound will not increase as fast as the sound intensity.

No single mathematical relation can fully describe the relationship between intensity and loudness. The relations that seem to be most accurate are too complicated for everyday use. An approximate relation that works well enough for most purposes is to assume that the ear has a *logarithmic* response that progressively discounts the effect of large stimuli in the following way: When the sound intensity increases by a factor of 10, the sensation of loudness only increases by 1; when the sound intensity increases by a factor of 100, the loudness goes up by 2; and so on. Thus, a trillion-fold increase in sound intensity only results in an increase in the loudness of 12.

Sound wave intensities are generally not reported in W/m^2. Since the response of the ear is roughly logarithmic, the intensity in W/m^2 does not really describe how loud the sound will seem. Instead, the intensities are given on the **bel** scale, which is logarithmic. The name *bel* honors Alexander Graham Bell, more for his pioneering work with the hard-of-hearing than for his invention of the telephone. The bel scale follows the same pattern we described above. If a sound is 10 times more intense, it is 1 bel (B) louder; if it is 100 times more intense, it is 2 bels louder, if it is 1,000 times more intense, it is 3 bels louder; and so on. (Note that the bel rating of loudness is not the true psychological loudness, but it is close enough to justify calling it the loudness.) The zero of the bel scale is set at the threshold of hearing; that is, a zero-bel sound can just barely be heard by a person with good ears (at the frequency where the ear's sensitivity is greatest). In W/m^2, zero bel is 1 pW/m^2.

An easy way to remember the bel scale is to recall that 10 is equal to 10^1, so a sound that is 10 times more intense is 1 bel louder.

Similarly, $100 = 10^2$, so a sound 100 times more intense is 2 bels louder, and so on.

The size of 1 bel is slightly large for reporting changes in loudness, so the decibel (dB) (0.1 bel) is commonly used. Thus, in decibels, a sound that is 10 times more intense than threshold sound has a loudness of 10 decibels, a sound 100 times more intense has a loudness of 20 decibels, and so on. If the increase in intensity is not an even multiple of 10, the decibel rating must be figured out on a calculator with a logarithm button; however, just knowing how to calculate the decibel rating for the multiples of 10 is sufficient for routine work on loudness.

EXAMPLE Roughly how loud, in decibels, is a sound that is 7,000 times more intense than threshold sound?

Answer First, let's find the decibel rating for sounds 1,000 times and 10,000 times more intense, the two powers of 10 on each side of 7,000.

$1{,}000 = 10^3$ is equivalent to $3\text{ B} = 30\text{ dB}$ louder.

$10{,}000 = 10^4$ is equivalent to $4\text{ B} = 40\text{ dB}$ louder.

Thus, the sound is somewhere between 30 and 40 decibels louder than threshold sound. Since 7,000 is closer to 10,000, a rough estimate would be 37 decibels. (The exact answer is 38.5 decibels.)

The Doppler Effect

Suppose you got sick of singing in the shower and decided to share your talent with the world by running down the street singing. The sound would leave your mouth at the same speed, 344 m/s, in all directions. But as you ran, you would tend to catch up with the sound waves ahead of you and to run away from the waves behind you. Now consider what a person sitting on the sidewalk would hear as you ran by. As you approached, the waves would be more closely spaced—by the distance you ran in one period of the wave —than they would be if you were stationary. The person would therefore hear waves with a shorter wavelength and thus a higher frequency. After you passed the person, the waves would be further apart than normal; they would have a longer wavelength and a lower frequency. So the person would hear a sudden drop in pitch. This shift in frequency is called the Doppler effect, and it occurs whenever the sound source, the listener, or the air is in motion.

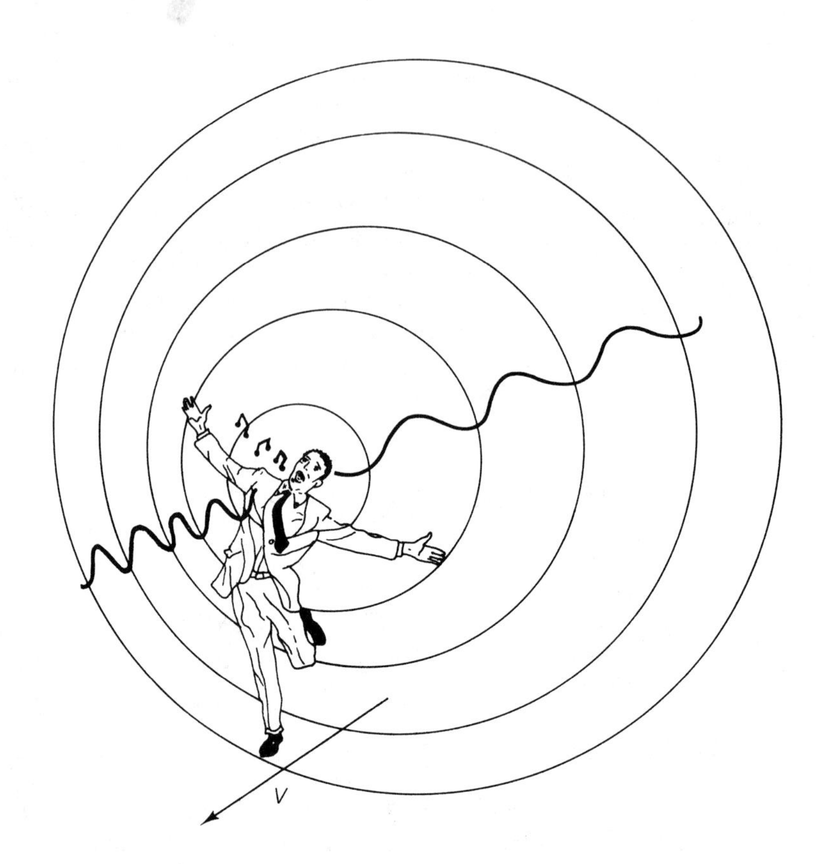

Figure 15-4 The Doppler effect. Because a moving sound source catches up with the sound waves ahead of it and runs away from the waves behind it, stationary listeners hear a higher-frequency sound when the source is approaching them and a lower-frequency sound when it moves away.

The Doppler effect is often heard on the highway. When a car blowing its horn goes by, the pitch of the horn drops sharply. The Doppler effect occurs for all types of waves, a fact that is applied also on the highways by police radar units. When a radar beam bounces off a moving car, the beam is shifted to a higher frequency by the Doppler effect. The speed of the car is easily determined by the size of this change.

The Doppler effect is often used in hospitals. For instance, when the heart of an unborn baby starts to pump blood, the sound cannot be heard over the sounds of blood movement in the mother. However, Doppler effect instruments can easily detect the motion of the fetal heart. In another example, the Doppler effect is used to detect blood clots in the legs. Such blood clots are extremely dangerous if they move out of the legs, because they can lodge in the arteries of the heart or brain and quickly kill the patient. Since a clot in a leg vein stops the blood flow there, sound waves that are bounced off the blood cells in that vein will have no Doppler frequency shift.

Sound Production in the Body

The human voice is an extremely versatile sound-producing instrument. It has a frequency range from a few hertz to several thousand hertz, and its intensity ranges from a mere whisper to an ear-shattering bellow—a range of more than 1 million (60 dB). It can change from one sound to another in a split second as it produces different words. The voice is by far the fastest and most efficient way that we have of communicating.

The voice starts at the vocal cords, two fibrous bands in the larnyx—the bony air passageway in the throat. When we speak, the vocal cords form a narrow slot as shown in Figure 15-5. When air is blown through the slot, the vocal cords vibrate, causing the air to vibrate. The pitch of the voice is controlled mainly by varying the tension in the vocal cords. The more tightly the vocal cords are stretched, the higher their vibration frequency.

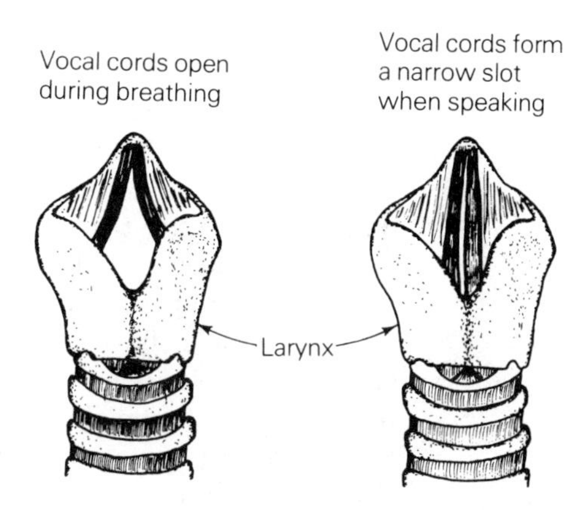

Figure 15-5

The vocal cords are not the only place where sounds are made. The lips, teeth, and tongue produce parts of the sounds of many consonants. Some sounds, like "sh," don't involve the vocal cords at all. Neither do whispers.

The vocal cords are important in that they start the sound production process, but the sound that finally leaves the mouth is vastly different from the sound of a pair of vocal cords alone, just as the sound of a bow on a lone string is vastly different from the sound of a violin.

When the sound leaves the vocal cords, it doesn't go only out of the mouth—it goes in every direction, including back into the lungs. Every passage that the sound wave penetrates, including some of the nasal sinuses, plays a role in forming the final sound

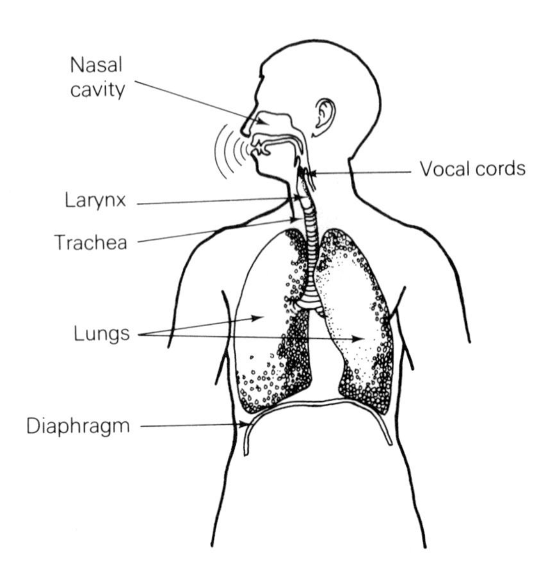

Figure 15-6 The entire respiratory system plays a role in forming sound.

(Figure 15-6). Nor does the sound pass through these passages just once. It reflects off all the passage walls, the teeth, and everything else that it hits. These reflected sound waves interfere with the others, increasing some frequencies and reducing others. Thus, the shape of the entire respiratory system determines the sound, and as we change the shape of this system, we change the tone and quality of the sound to form different words and sounds.

People who have lost their vocal cords can still produce very understandable speech. Some use an electric vibrator placed on the throat to vibrate the air, which produces monotonic speech because the vibrator produces a single frequency. Other people learn to swallow air into their stomachs and set the esophagus into vibration, expelling the air in a controlled burping process. This method is called esophageal speech, and it does an amazingly good job of replacing the function of lost vocal cords.

Hearing

The entire ear is shown in Figure 15-7. Hearing starts at the pinna, the skin-covered cartilage we commonly call the ear. The pinna reflects part of the sound down into the ear canal, thus amplifying the sound slightly. The amount of amplication isn't very great, because the pinna isn't a very good reflector. Something like the

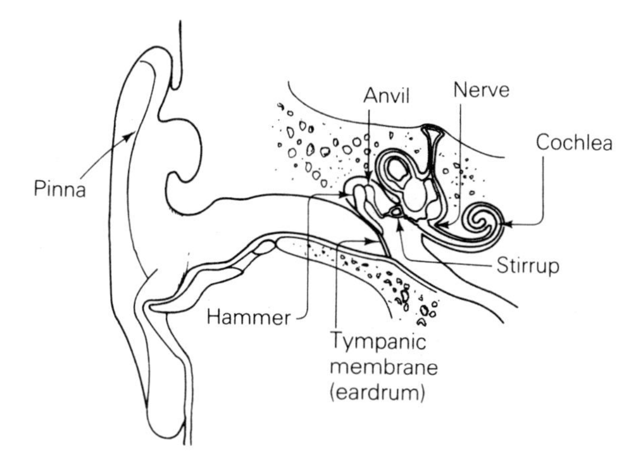

Figure 15-7

ears of a German shepherd or, better yet, trumpet-shaped ears would gather sound much better. Before electronic hearing aids were available, the hard-of-hearing used ear trumpets. The size of the pinna was most likely determined by two conflicting evolutionary forces: the need to hear well and the need to not have something that enemies could grab or bite off. Evidently, the present size of the pinna is a good compromise—it is about as large as it can be and still be hard to grab.

The sound travels from the pinna down the short ear canal, which ends at the eardrum (tympanic membrane). The term *eardrum* is a good one because it does exactly the same thing that a drum does except in reverse. Instead of spreading the beat of a drumstick over a wide surface to send out sound waves efficiently, the eardrum concentrates sound and sends it into the first of three small bones, the hammer. The hammer, anvil, and stirrup (malleus, incus, and stapes, respectively) act as levers to transmit the sound to the liquid-filled cochlea of the inner ear. The pivot points and lever arms are placed to cut down the sound energy reflected backward as the air wave is converted to a liquid wave. If the sound wave in air just hit the liquid-filled cochlea, the sound energy wasted by backward reflection would be much larger.

The bones can also adjust to provide some protection against loud sounds. When the sound becomes too loud, small muscles pull the stapes away from the cochlea to reduce the sound transmitted. However, because these muscles take some time to act, this mechanism cannot protect against unexpected loud sounds. For this reason, sudden sharp noises (such as gunshots) are more damaging to the ears than steady loud noises. Both, however, are quite harmful.

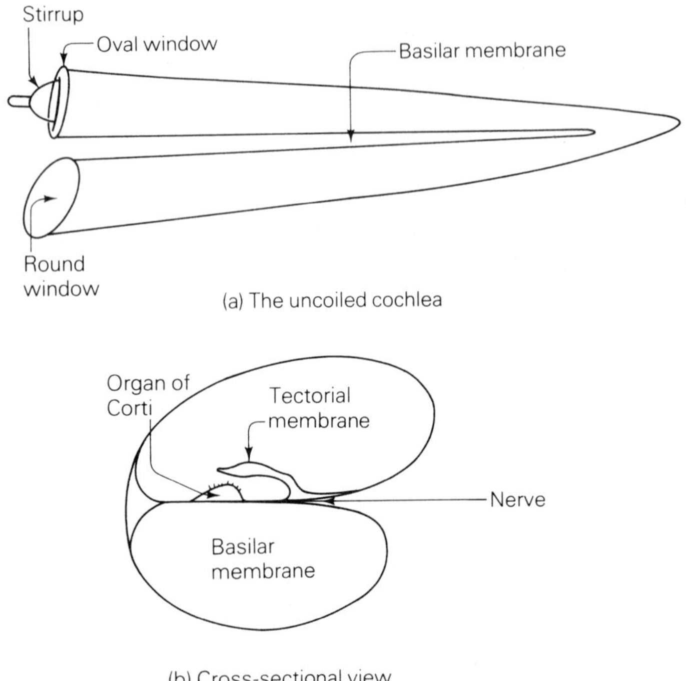

Figure 15-8

The sound wave is converted into a nerve signal in the snaillike cochlea. The cochlea is shown uncoiled in Figure 15-8. The cochlea is divided in two by the basilar membrane, which is where the sound is converted to a nerve impulse. Located on the basilar membrane is the organ of Corti, which has hairlike projections connected to nerve endings. The hairs are covered by the tectorial membrane. The sound enters the cochlea through the oval window and then travels through the fluid. Vibrations in the fluid wiggle the tectorial membrane, which moves the hairlike projections. Then, by a poorly understood piezoelectric process, the hair movements trigger the nerve cells, sending an impulse to the brain.

Different sounds move the basilar membrane differently because every frequency diffracts, interferes, and reflects differently inside the cochlea. This is apparently how different frequencies are distinguished.

Stereo

Sounds arrive at one ear before the other except when the sound source is either directly ahead or directly behind. These small time differences—never greater than 0.5 millisecond—along with the difference in loudness between the two ears, enable us to locate the direction of sound fairly accurately.

Monoaural recordings have only one sound track. They cannot duplicate these small differences in arrival time at the ears. To capture the time differences, two separate microphones must be used to make two separate recordings. Of course, the two recordings are on the same record or tape for convenience, but electronically they are completely separate. When the two recordings are played through two different speakers, we get the stereo effect, which comes close to duplicating the original time differences. Quadraphonic systems, which use four microphones and four speakers come even closer.

The Sensitivity of the Ear

We know our hearing is extremely sensitive, since we can hear pins drop, bubbles burst, and insects chewing on leaves. The ear is so sensitive that it is hard to believe just how tiny a sound it can detect. At the threshold of hearing (zero decibel), the amplitude of the vibration of the basilar membrane is less than 1 picometer—one-millionth of one-millionth of a meter. This is less than *1/100th of an atomic diameter.* Although some animals can hear slightly better than we can, human hearing is close to the limit. If our hearing was much more sensitive, we could hear the random thermal motions of air molecules.

As stated earlier, the normal ear can detect sounds in the frequency range between 20 and 20,000 hertz. Of course, the sensitivity for various frequencies is not the same. A graph of a person's response to various frequencies is called an *audiogram;* a typical normal one is shown in Figure 15-9. The lower part of the curve is the threshold of hearing for the various frequencies. The upper line is the intensity at which hearing changes to pain. Note that pain doesn't depend much on the frequency; It starts at about 120 decibels for all frequencies.

Our hearing is most acute in the range between 1,000 and 4,000 hertz, partly because the length and shape of the ear canal are best suited to transmitting sounds of 3,000 hertz. Zero-decibel sounds in this range can be heard by some people. Outside of this range, the sensitivity falls rapidly. At 100 hertz, the hearing threshold is about 30 decibels; that is, a 100-hertz sound must be 1,000 times more intense than a 1,000-hertz sound to be heard.

The rapid drop in hearing sensitivity at frequencies below 1,000 hertz points up one of the problems in reporting loudness on the decibel intensity scale. As long as the frequency of the sound is

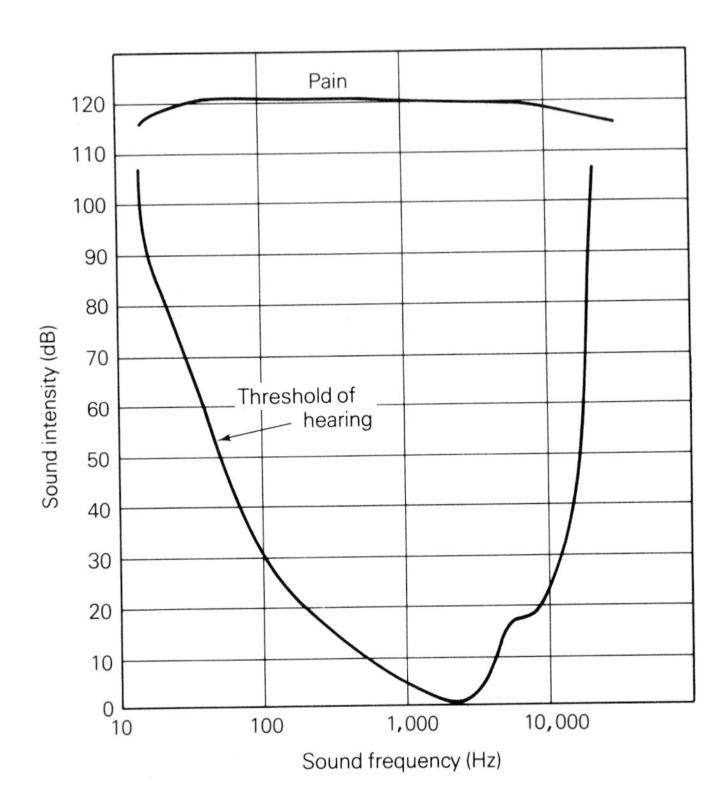

Figure 15-9 An audiogram for a normal person with good hearing.

in the midrange, between 500 to 5,000 hertz, the intensity in decibels is a good indication of loudness. Outside this range, however, corrections must be made to compensate for the changes in hearing sensitivity. Architects, automobile designers, and others often use this principle to reduce the psychological effects of vibrations. If they cannot eliminate a vibration from a machine, they try to make its frequency as low as possible. People often will not object to a fairly intense 50-hertz sound. However, if the frequency is raised to 500 hertz, the sound might be extremely aggravating.

The Dangers of Noise

The few loud sounds that our ancestors heard didn't last very long. The evolutionary result is that we are not well protected against long-term exposure to loud sounds. Although occasional loud sounds won't harm our hearing much, continuous loud sounds, such as those of a noisy factory or a loud band, inevitably lead to hearing loss. Unfortunately, the sensation of pain in the ears doesn't warn us soon enough that sounds are too loud. Steady,

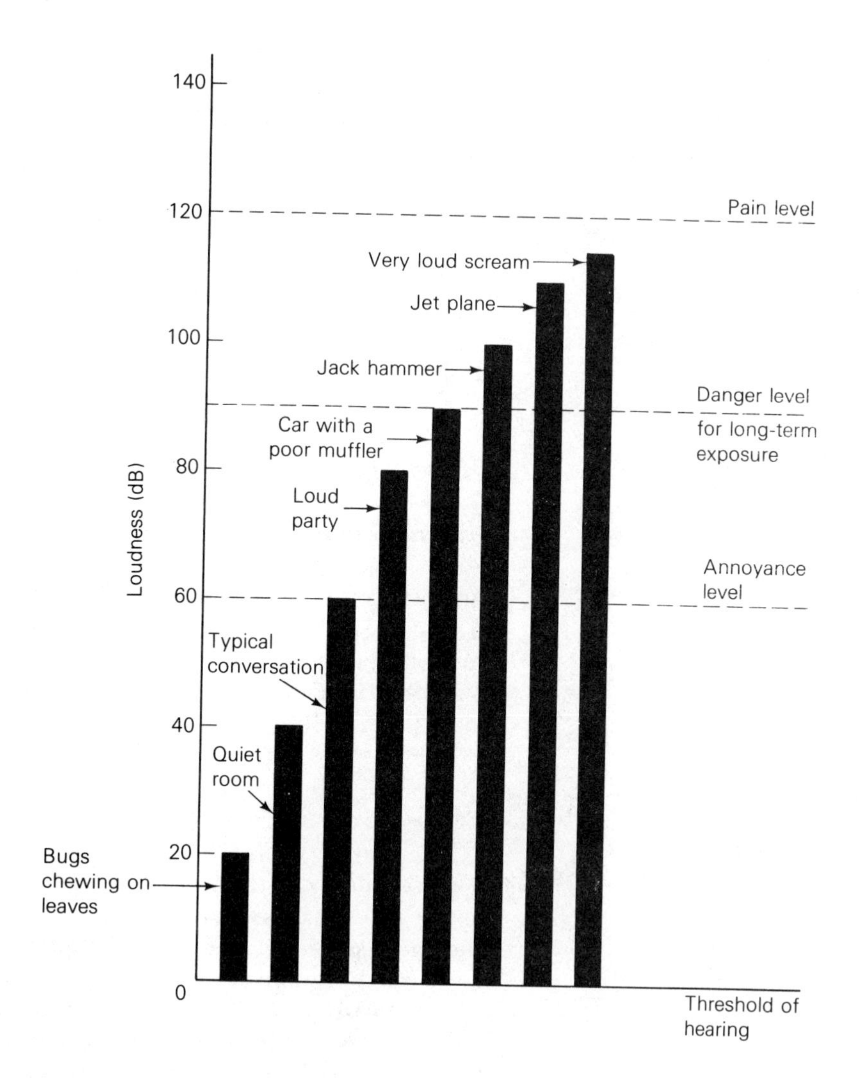

Figure 15-10 Intensities of various sounds.

eight-hour-per-day exposure to 90-decibel noise will cause hearing loss, but pain isn't felt until 120 decibels—1,000 times higher in intensity. Figure 15-10 shows that many common sounds exceed the 90-decibel level. Anyone who works or plays where the sound levels exceed 85 to 90 decibels should either find a way to reduce the loudness or wear ear protection. At present, people are allowed to work an eight-hour day in a 90-decibel environment, but most hearing experts feel that this level should be cut to no more than 85 decibels as soon as possible.

Getting people to protect their hearing is a hard job, because dangerous loud noises usually produce no pain and no noticeable day-to-day change. By the time the loss is noticeable, it is impossible to restore the hearing except by wearing a hearing aid. While a hearing aid can help, it cannot restore the full, rich range of sounds that the normal person can hear. As in other cases where danger is not immediate and obvious, people tend to ignore the serious, long-term hazards of overexposure to loud noise. For this reason, special efforts must be made to educate people about the dangers.

The logarithmic nature of the decibel scale creates another problem in dealing with noise pollution. For instance, a factory owner whose plant is too noisy by a few decibels might ask, "What difference does a few decibels make? Can't we just forget about it?" In terms of perceived loudness, a few decibels don't mean much—the average person can just barely notice them. But a few decibels is a *doubling of sound intensity*. Therefore, an increase of a few decibels in the sound level can result in doubling the hearing damage.

Incidentally, people seem to be annoyed by any steady noise above 60 decibels. This 60-decibel annoyance level is surprisingly uniform among people. In a test community near a noisy highway, traffic engineers found that nearly everyone who heard a constant highway noise level of at least 60 decibels was willing to pay to build a noise barrier. No one living beyond where the noise was this intense was willing to pay.

Hearing Losses

The sensitivity of normal hearing varies over a wide range. We have called zero decibel the threshold of hearing, but actually only the most sensitive 10% of the population can hear a sound that weak. The average person's threshold is about 10 decibels; anyone whose threshold of hearing is below 20 decibels is still in the normal hearing range. Fortunately, people can function quite well when they can only hear a 40-decibel sound, even though the weakest sound that they can hear is 10,000 more intense than zero-decibel sound. Not until a person has difficulty hearing a normal conversation does the loss become a serious problem. This happens when a person can't hear below 50 or 60 decibels.

Aging gradually deteriorates our hearing ability. Little by little, we must move closer to the sound or turn up the volume. The effects of age are most noticeable at high frequencies—it becomes harder to hear sounds above 8,000 hertz. Fortunately, the loss is generally much less severe at lower frequencies. In the range from 100 to 1,000 hertz, which is the range of most speech sounds, the hearing loss can be as small as 10 decibels in a 70-year-old who has taken care of his or her hearing and had the good fortune to stay healthy. A 10-decibel loss only reduces the hearing of a person who had good normal hearing to average normal hearing; this is a loss, of course, but one that is too small to cause any serious problems.

Hearing Tests

Hearing is tested with an *audiometer,* which is basically a tone generator hooked up to a set of earphones. Both the tone frequency and loudness can be controlled and measured. The operator varies the loudness and frequency to map out the subject's hearing threshold as a function of frequency. The results are graphed on an audiogram. Since locating the pain line of an audiogram is seldom necessary, most audiometers only go up to about 100 decibels for safety. The audiograms of a hard-of-hearing person, a person with noise damage, a normal 60-year-old, and a young person with good normal hearing are shown in the sample audiogram (Figure 15-11).

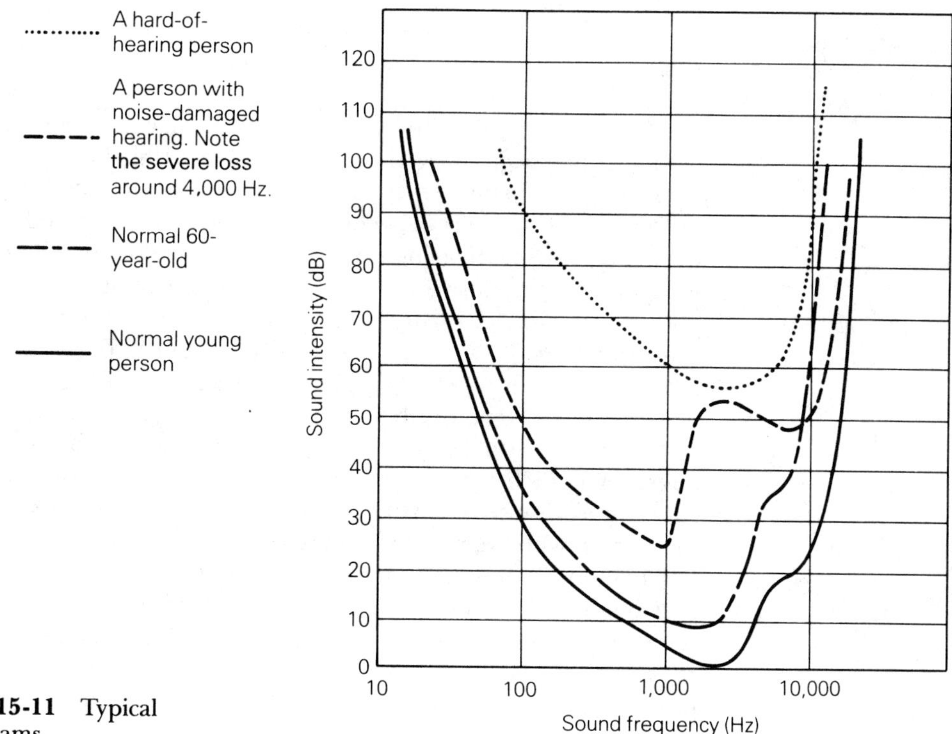

Figure 15-11 Typical audiograms.

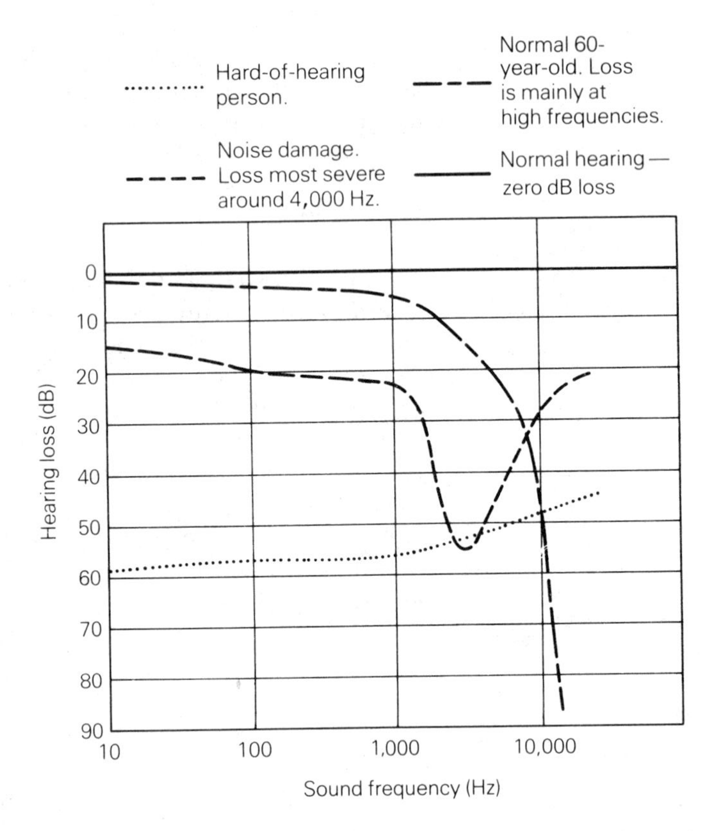

Figure 15-12 Typical hearing loss graphs.

Because the sensitivity of normal hearing varies with frequency, quickly judging the amount of hearing loss from an audiogram is difficult. To make the loss easier to understand, the difference in decibels between the subject's hearing and a person with good normal hearing is plotted at each frequency. The decibel scale in a hearing loss graph is reversed from an audiogram; so the poorer a person's hearing, the lower the line on the loss graph. A loss graph for the same four people used in the previous figure is shown in Figure 15-12. The person with good normal hearing is a straight line at zero decibel. The normal 60-year-old shows only a small loss at lower frequencies but a severe loss at the higher frequencies (which, as we have discussed, doesn't cause any severe communication problems). The sharp dip around 4,000 hertz for the person with hearing loss caused by loud noises is typical of noise impairment. The inability to hear 4,000-hertz sounds is often the first sign of a hearing loss due to noise, and it can show up even in the first year of a noisy job. Any person with a hearing loss in this range should be questioned about their exposure to noise. If they work or play in a noisy environment, they should either stop or wear ear protectors.

Hearing Aids

Hearing aids are electronic amplifiers connected to a microphone and an earphone. The advances in electronic technology have made it possible to make hearing aids small enough to fit in the pinna of the ear—microphone, amplifier, batteries, earphone, and all. Even very powerful models can be hidden in eyeglass frames. Modern hearing aids are very helpful in many cases of hearing loss; however, just like eyeglasses, they must be fitted to an individual's particular problem. For instance, an elderly person with a high-frequency loss but no serious low-frequency impairment would not want or need an amplifier for all frequencies, because the lows would be too loud. Someone who is hard-of-hearing from exposure to loud noises needs a hearing aid that amplifies frequencies around 4,000 hertz more than other frequencies. Others whose loss is fairly uniform would need equal amplification at all frequencies. Thus, no one hearing aid will work on all people. Of course, a hearing aid cannot help every type of hearing loss. In some cases of nerve damage, no amount of amplification will do any good.

Hi-Fi

The purpose of hi-fi (high fidelity) sound equipment is to reproduce the sound of performers and instruments as faithfully as possible (not as loud as possible, as faithfully as possible). To achieve this, the amount of extra noise from the system must be minimized and the system must reproduce all the frequencies of the original sound. Most of the extra noise comes from the mechanical parts of the system, such as the turntable and tape drives. Other noise sources are record scratches and the recording tape itself, which has a hissing noise due to magnetic effects. The main areas of loss in fidelity are the recording process, the electronic amplifiers, and the loudspeakers. Of course, the average listener can't do much about the recording process, but most record companies are fairly careful to preserve sound fidelity.

The two frequency areas most difficult to reproduce accurately are the extreme highs and lows. Almost any reasonable amplifier and loudspeaker will do a good job on midrange frequencies. It is no surprise that improving the high- and low-frequency response is expensive. Consider the lows, for example. In order to maintain the sound quality and loudness, the amplifiers must be more powerful and the speakers must be bigger than those used for the midrange frequencies. You can see why the amplifiers must be more powerful by looking at the audiogram of a normal person. At

100 hertz, the hearing threshold is around 30 decibels; at 20 hertz, the threshold is around 80 decibels—50 decibels higher. Thus a 20-hertz sound must be 50 decibels (100,000 times) more intense than a 100-hertz sound in order to sound as loud; and the amplifier must be more powerful to produce the additional intensity.

The speakers will have to be larger because the wavelengths of low-frequency sound waves are longer than those of midrange frequencies. Remember that most sound sources are rather inefficient unless their size is close to the wavelength of the sound. We know that

$$v = f\lambda$$

for all waves. Since the speed of sound is the same at all frequencies, this means that when the frequency is, for example, 5 times smaller, the wavelength is 5 times larger. Thus, the speakers would have to be larger, perhaps as much as 5 times larger.

The fact that *both* amplifiers and speakers must be changed to reproduce low-frequency sound brings up a point about buying hi-fi equipment. It generally doesn't pay to buy a system having components of widely different quality. In the case above, changing only the speakers or the amplifier would do very little good; the system would still be limited by the unchanged component.

Another point about hi-fi equipment: There is no point in buying a system that is better than your ears are. Who needs a sound system that produces beautiful sound from 20 hertz to 20,000 hertz if their ears can only hear sounds between 50 and 12,000 hertz? Many aging hi-fi buffs have thought their sound systems were going bad on the highs and lows, and so they have bought new equipment only to find out that their ears had gradually deteriorated, not their hi-fi equipment.

Ultrasound

Sound with a frequency greater than 20,000 hertz is called *ultrasound* or *ultrasonic sound.* There isn't any fundamental difference between ordinary sound and ultrasound; the only difference is that humans can't hear the latter.

Ultrasound is often used for cleaning. The extremely rapid vibrations can pull water molecules apart for an instant to form a little cavity (a process called *cavitation*). Then the cavity snaps closed just as quickly. If the cavity was on a hard surface, the impact of the cavity closing against the surface removes dirt very efficiently. Ultrasonic cleaning is ideal for areas that ordinary methods can't

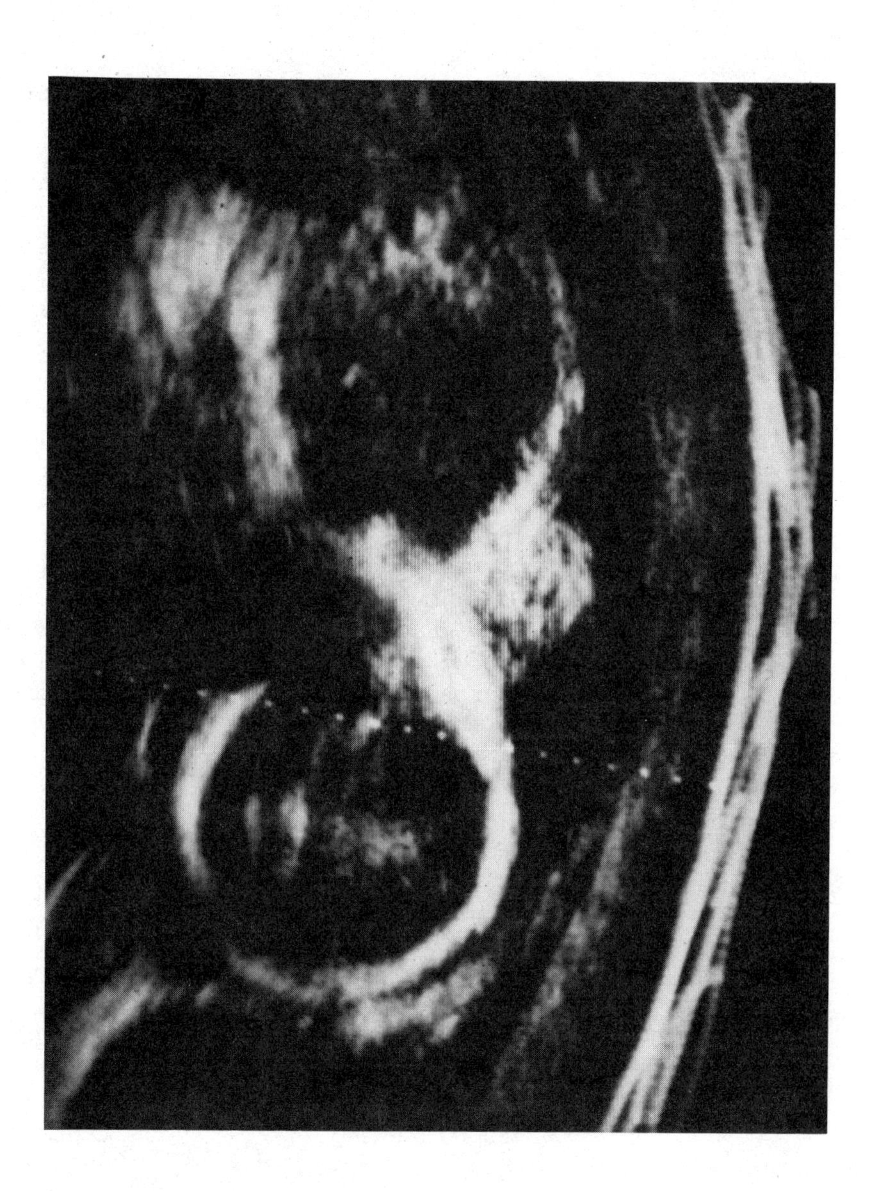

An ultrasonic image of a fetus inside the uterus.

reach, such as the inside of hypodermic syringes and needles. It has also been used successfully for cleaning teeth.

Ultrasound is widely used to study many internal organs of the body. Ultrasound waves reflect off the organs inside the body, and with the help of a computer, the reflected waves can produce an image of the organs. Ultrasound is better than ordinary sound for this purpose because short-wavelength waves can resolve finer details. Ultrasonic images have two big advantages over x-ray images: Unlike x rays, low-intensity ultrasound has no known harmful effects. Second, ultrasound waves can detect differences in soft tissue without the injection of special dyes, as must often be done for x rays. Ultrasonic imaging has proven very useful for measuring fetal head size and for determining the number of fetuses—tasks that are normally too dangerous to do with x rays. Ultrasound has also been useful for locating such soft organs as the liver and kidneys.

As you can see from the photo, the resolution of ultrasound images is not as good as that of x rays, but it is improving. Another disadvantage of ultrasonic imaging is that almost 100% of the sound is reflected when the sound hits an air- or gas-filled region; thus, no image can be obtained from organs beyond the gas. For this reason, ultrasonic imaging is not feasible for the lungs and the intestines, both of which usually contain a lot of gas.

Another use of ultrasound is to heat sore joints and tissues to reduce pain and speed healing (diathermy). Most of the sound energy is converted into heat deep within the body, because ultrasound is quite penetrating. This is the main advantage of ultrasonic diathermy over local hot packs: The heat is actually produced inside the body, close to where it is needed. The heat from hot packs must penetrate the skin and other tissues.

Concepts

Sound waves
Condensations
Rarefactions
Pitch
Tone quality
Noise
White noise
Sound intensity
Loudness
Logarithmic response
Bel
Decibel
Doppler effect
Audiogram
Audiometer
Ultrasound
Ultrasonic imaging
Ultrasonic diathermy

Discussion Questions

1. Why can't sound waves in air be polarized?
2. Describe how the molecules move as a sound wave passes through the air.
3. Why can't there be any sound in outer space?

4. Explain the difference between the intensity of a sound wave and the loudness of a sound.

5. Explain the difference between the frequency of a sound wave and the pitch of a sound.

6. If you ran toward a person who was playing a note on a trumpet, would you hear a lower pitch note, a higher pitch note, or the same pitch note as a stationary listener? If you then ran away from the trumpet player, what change in pitch would you hear?

7. How is the Doppler effect used to detect the flow of blood (or a stoppage in the flow)?

8. Explain why a violin and a piano sound very different even when they are both playing the same note.

9. What do you feel is the difference between music and noise?

10. Give a short description of how the body produces vocal sounds. What parts of the body are the most important for speech?

11. Describe the ear and the function of its various parts in hearing.

12. What is an audiogram? How does a normal audiogram look?

13. What kinds of hearing loss are most common in the elderly? In people who have had a long-term exposure to loud noises?

14. What is *ultrasound* and how is it different from ordinary sound?

15. Give three examples of how ultrasound is commonly used.

16. What are some of the advantages of ultrasonic imaging of internal organs over x-ray imaging? What are some of the disadvantages?

Problems

1. If thunder is heard 2 seconds after lightning is seen, how far away is the storm?

2. If you knock on a 2-centimeter-thick glass door, how long does it take for the sound to pass through the door?

3. If you scream at a rock cliff 220 meters away, how long does it take your echo to return?

4. A marching band spreads out over a football field (which is about 100 meters long). To a person in the end zone, how much of a time lag is there between the notes played by the nearby band members and those played by the members across the field?

5. If one person playing a piano makes a 65-decibel sound, how loud will the sound of 10 pianos be?

6. How loud in decibels is a sound that is 100,000 times more intense than threshold sound?

7. If a person has a 10-decibel hearing loss for all frequencies, that person is
a. seriously hard-of-hearing.
b. noticeably hard-of-hearing.
c. able to hear almost as well as the average person.
d. able to hear far better than the average person.

8. If a 2-watt stereo system can produce 60-decibel sound, how powerful a system should you buy to produce 80-decibel sound?

Experiments

1. Demonstrate that doubling the sound intensity only produces a small increase (3 decibels) in the loudness in the following way: Adjust the two speakers of a stereo set so that they are equally loud. Then turn one off and listen. Next turn both on to double the sound intensity. How much does the loudness seem to change? You can also do this experiment with two radios or TVs tuned to the same station.

2. You can quickly demonstrate the importance of your mouth, teeth, and tongue in the speech process by opening your mouth and trying to speak without moving them. Try it with a friend and see if you can make any sounds that your friend can understand.

CHAPTER SIXTEEN

Light and Vision

Educational Goals

The student's goals for Chapter 16 are to be able to:

1. Give the principal types of electromagnetic radiation in order of increasing frequency (decreasing wavelength).
2. Explain what polarized light is and how polarizers can be used to reduce glare and to bring out hidden detail in microscope specimens.
3. Determine the position of the image in a mirror using the law of reflection.
4. Calculate how a light ray will be bent (refracted) when going into a flat glass of known index of refraction.
5. Find the image formed by a single lens.
6. Describe the optics of the eye and how glasses are used to correct vision defects.
7. Explain the basic theory of color vision.

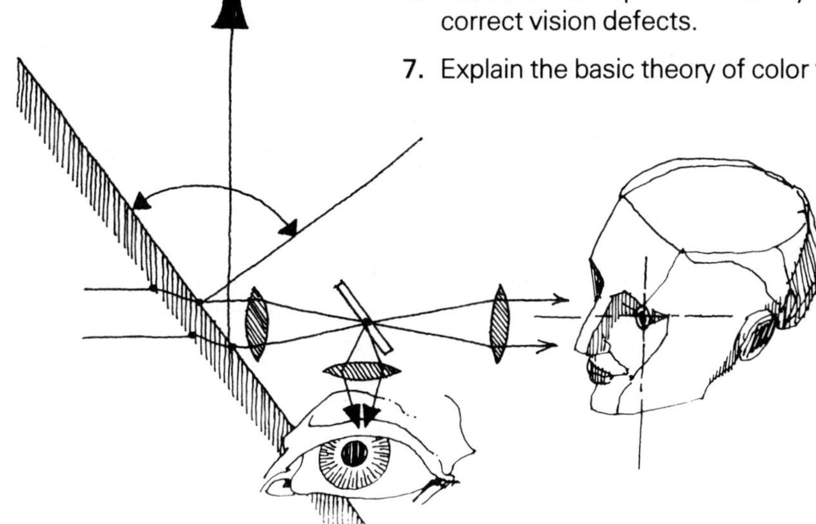

Our vision is our richest source of information about the world around us. In a sense it is too rich, because the brain can only handle a small fraction of the information received from the eyes, and it stores even less. This is one reason why people who have witnessed the same event seldom agree on exactly what happened. While their eyes saw the same images, their brains filed away different portions in their memories.

Because the body does not produce light and only produces images slowly by writing or drawing, visual communication was always rather slow until television became widespread in the 1950s. Television made the rapid creation and transmission of visual information possible, thus changing how people think. Quite likely, the advent of television in the twentieth century will mark an even bigger change in our culture than the invention of the printing press did in the fifteenth century. Whether the change will be for the better or worse remains to be seen.

The Electromagnetic Spectrum

Light is a transverse electromagnetic wave that travels at the incredible speed of 300,000 km/s in a vacuum. The speed in matter is always slower, less than 1% slower through air and 60% slower through a diamond. Incidentally, the speed of light is a universal speed limit; nothing can travel faster. (This is one of the key concepts of Einstein's theory of relativity.)

Since light can travel in a vacuum, where there is no matter, a likely question is, What is vibrating when a light wave travels through a vacuum if nothing is there to vibrate? The answer is complex, but it is essentially this: Although no matter is in a vacuum, that does not mean that *nothing* is there. Matter is only one aspect of the physical universe; energy and time are just as real as matter, and they can exist perfectly well in a vacuum. Something else that exists at every point in the universe—including inside a vacuum—is the influence of all the electric charges and magnets in the universe, which is called the electromagnetic field. Not only does the electromagnetic field exist in a vacuum, but it can also change and vibrate in a vacuum. These vibrations form light waves.

Light is not the only wave in the electromagnetic spectrum. Radio waves, infrared rays, ultraviolet waves, and x rays are others. The only difference between these waves is their frequency and wavelength. In fact, if you approached a radio station fast enough (at just under the speed of light), it would seem to send out x rays because of the Doppler effect. The entire electromagnetic spec-

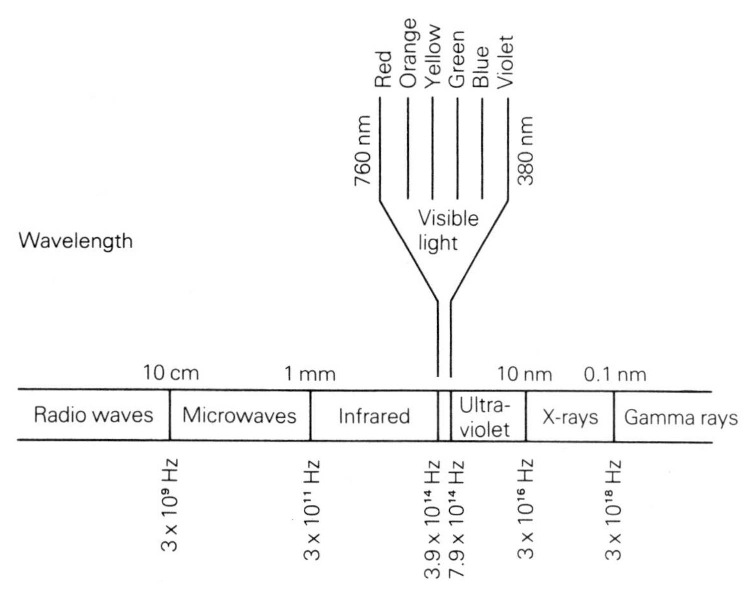

Figure 16-1 The electromagnetic spectrum.

trum is shown in Figure 16-1. The boundaries between the different regions are not precise. The zones blend smoothly from one to the next, just as the colors of the rainbow change smoothly from hue to hue.

Visible light is just one small slice of the electromagnetic spectrum, located between infrared rays on the low-frequency side and ultraviolet waves on the high-frequency side. Within the visible spectrum, the different frequencies correspond to different colors. The lowest visible frequency is deep red. As the frequency increases, the colors go to orange, yellow, green, blue, and then violet, the highest visible frequency.

Naturally, we must keep in mind the difference between light as a physical wave and the psychological sensations of color and brightness. This is especially important in the case of color, as we will see later.

The chart in Figure 16-1 gives both the frequency and the wavelength of electromagnetic radiation. In practice, the wavelength is much easier to measure, so electromagnetic radiation is usually described by its wavelength. Of course, the frequency can always be found from the relation $v = f\lambda$. The wavelength of visible light ranges from 760 nanometers for red light to 380 nanometers for ultraviolet light. These limits vary among people, but the variations among normal people are quite small.

Polarized Light

Light is a transverse wave—the vibrations of the electromagnetic field are perpendicular to the direction in which the light beam is moving. Thus, light beams can be either polarized, with the vibrations along one specific line, or unpolarized, with the vibrations in random directions perpendicular to the beam.

Certain materials can change unpolarized light into polarized light by blocking all vibrations that are not parallel to a certain direction. Two examples are Polaroid sunglasses and antiglare camera filters, used to cut down reflections on pictures taken through windows. We can see roughly how polarizers work by imagining a rope passing through a slot in a picket fence (Figure 16-2). If the slot was so narrow that the rope could only move up and down, the fence would act as a polarizer for waves on the rope. If an unpolarized wave was sent down the rope toward the fence, the fence would absorb the sideways vibrations, allowing only vertical vibrations to pass; that is, the wave would be polarized up and down. If the rope passed through a second fence, it would not change if the slot was vertical, as the slot in the first fence was. But if the slot was horizontal, the wave would not pass at all. This effect is easily seen by placing two Polaroid sunglass lenses over each other and rotating one of them. When the glasses are perpendicular to each other, very little light will pass. Of course, polarizing materials don't have tiny slots; they work by lining up the molecules in a way that blocks all but one polarization direction.

Polaroid sunglasses and antiglare filters work because the reflected glare from glass, water, and snow is highly polarized. The lenses of the sunglasses are turned so that their polarizing direction is perpendicular to the direction of the polarization of the glare. If you turn your head sideways, however, the glare will be able to pass through the glasses. Polaroid sunglasses only stop glare when your head is reasonably close to level and the water or snow that reflects the light is horizontal.

Polarized light can also bring out hidden detail in transparent objects. Many transparent materials change the direction of polarization of the light passing through them. (Just how they do this is complicated.) This effect is used in polarizing microscopes. The polarizing microscope illuminates the specimen with polarized light, and its eyepiece has a polarizer that can be rotated. In this way, two sections of an object that have no contrast under normal light can be clearly distinguished if their polarizing characteristics are different. Because polarizing microscopes make much of the internal structure of cells visible without the use of cell-killing stains, they are very useful for watching the processes of living cells.

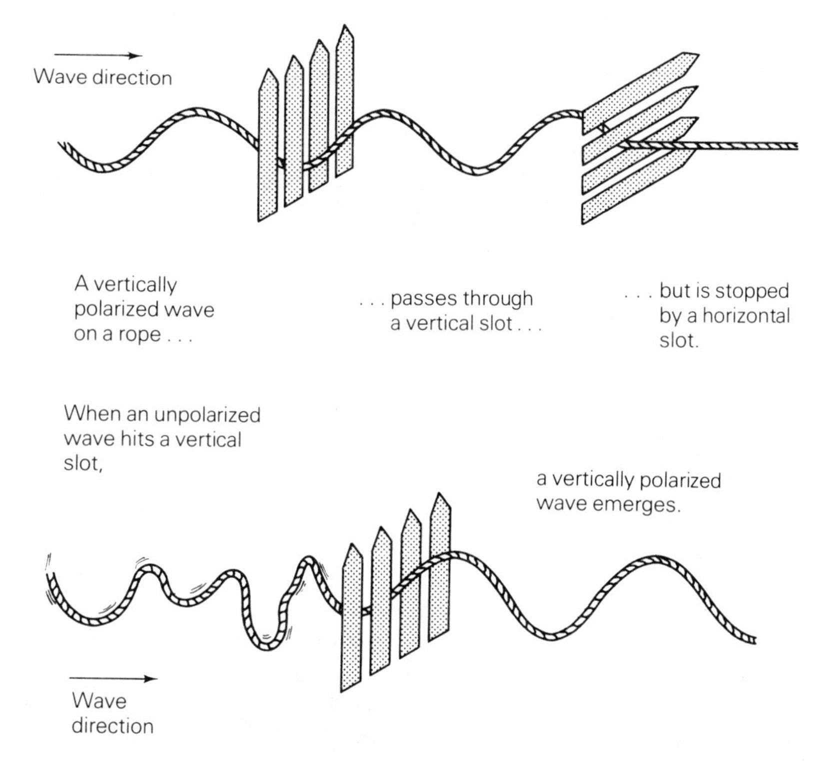

Figure 16-2

You can see the spectacular difference a polarizing microscope makes by looking at a bit of crumpled transparent plastic tape placed between two Polaroid sunglass lenses. Place the lenses perpendicularly to each other and then rotate one. This effect is surprisingly beautiful, and it is well worth hunting up two pairs of Polaroid glasses (or breaking a cheap pair in two) to see it.

Reflection

Light bounces off a mirror in the same way that a ball bounces off a sidewalk: The angle of the light beam with the surface is the same after reflecting as it was approaching.

Law of Reflection

The angle of incidence θ_i is equal to the angle of reflection θ_r (Figure 16-3).

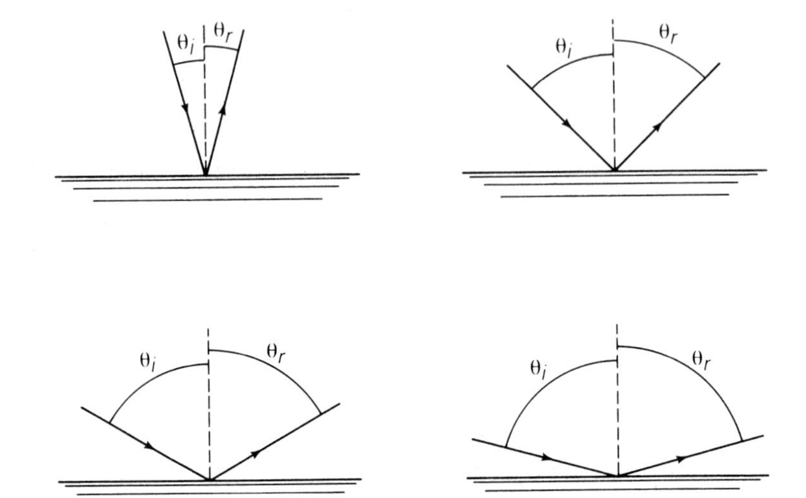

Figure 16-3 The law of reflection: θ_i is always equal to θ_r.

For technical reasons, the angles of incidence and reflection are normally measured from the line perpendicular to the surface instead of from the surface itself; however, the law is the same for either reference line.

Let's see how the law of reflection can be used to locate the image of an object behind a mirror (Figure 16-4). To do this, we simply follow the path of two rays as they go from the object,

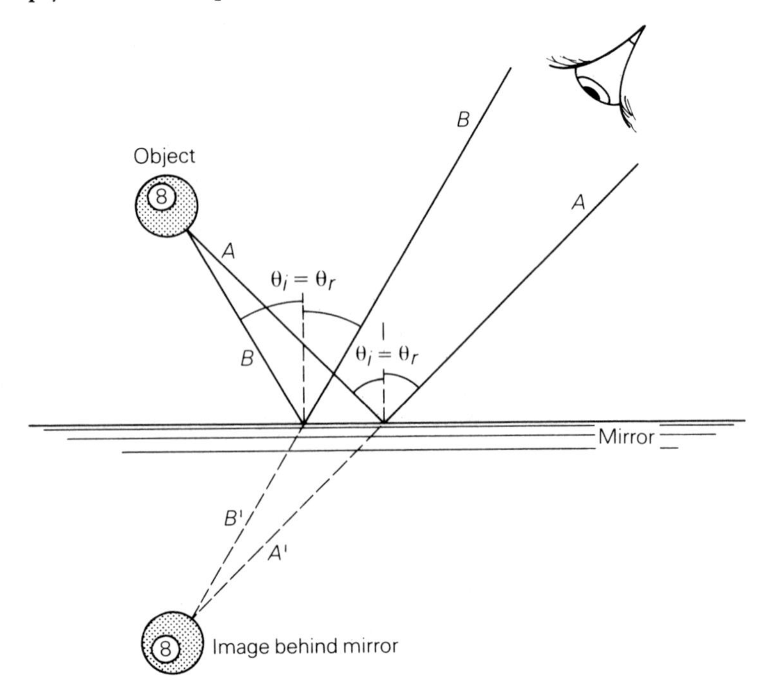

Figure 16-4

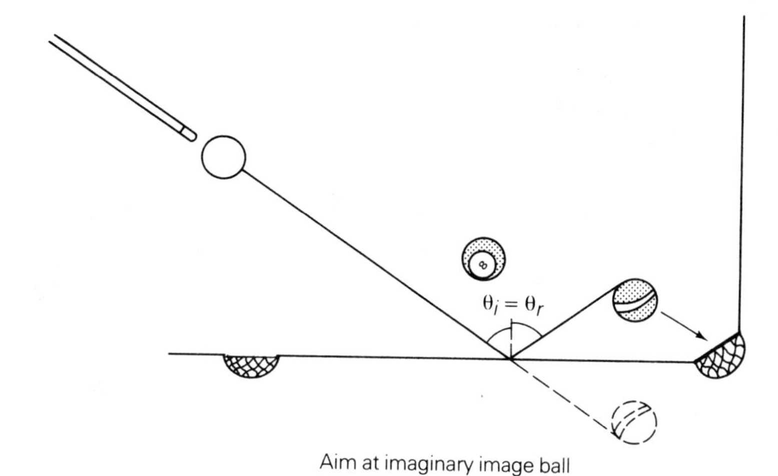

Figure 16-5 A practical application of the law of reflection.

bounce off the mirror, and go to our eyes. Rays A and B in the figure are two such rays. Each ray reflects off the mirror at an angle equal to the angle at which it struck the mirror, as shown in the figure. The eye, however, does not know that the rays were reflected, because the brain makes the reasonable assumption that the rays came along a straight line. Thus, the brain thinks that the rays came along the dotted lines behind the mirror, A' and B', that is, from an object located at the point where A' and B' cross. Therefore, we see an image at that point. Note that the image is located directly behind the mirror from the object and appears to be the same distance behind the mirror as the object is in front of the mirror.

The law of reflection also applies to billiard balls bouncing off the rails when the ball is not spinning. One of the ways to make a successful bank shot is to aim at an imaginary ball that is the same distance behind the rail as the ball that you want to hit is in front of the rail (Figure 16-5).

Refraction

From most angles, a spoon in a glass of water appears to be bent at the water's surface. Of course, the spoon isn't bent; the light rays are bent as they leave the water, a process known as *refraction.* Refraction occurs when light passes between two transparent materials through which light travels at different speeds (except when the light beam is exactly perpendicular to the surface between the materials).

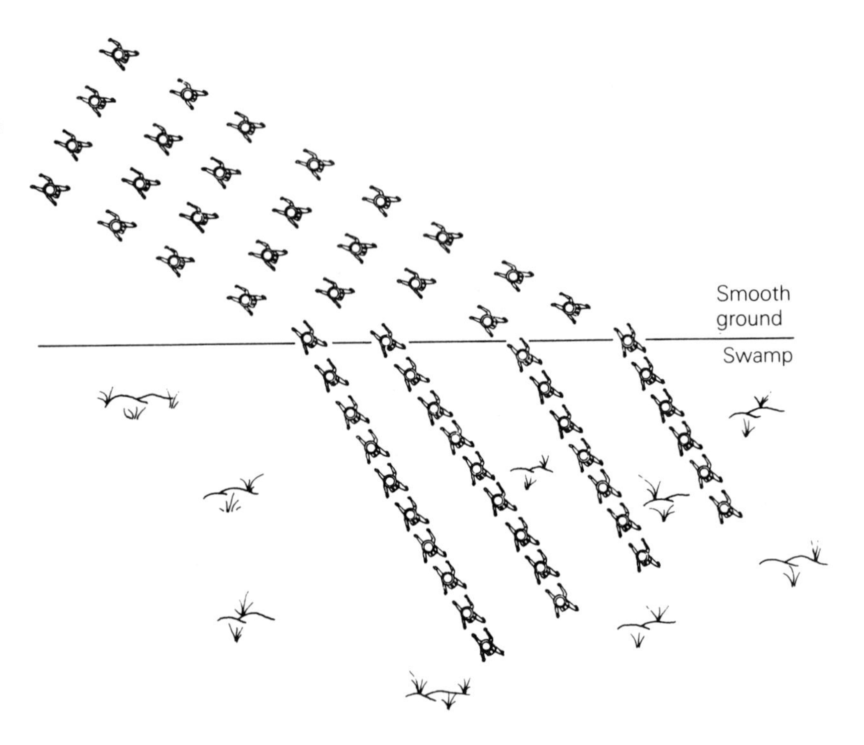

Figure 16-6 Refraction. In order to slow down *and* keep the ranks straight, the ranks must change the direction of their march.

The refraction of the light rays by the water severely distorts our view of the spoon.

Refraction occurs because the wave front (the line running along the crest of the wave) must have basically the same shape in both materials. If it starts out straight in the first material, it must also be straight in the new material. The only way for the wave front to stay straight and still change speed at the boundary is to change the direction of the wave. The situation is something like a squad of soldiers marching from smooth ground into a swamp. The soldiers want to keep their ranks straight, but if they hit the swamp at an angle, not all the soldiers in a given rank hit the swamp at the same time. To compensate for the slowness of the soldiers already in the swamp, the ranks must rotate the line of march around the soldiers in the swamp, which turns the line farther from the boundary (Figure 16-6). Naturally, the larger the speed change is, the more the direction of march will change.

Index of Refraction

The index of refraction of a material is the ratio of the speed of light in a vacuum to the speed of light in that material. The letter n is usually used for this index in formulas. Thus,

$$\text{index of refraction} = n = \frac{\text{speed of light in a vacuum}}{\text{speed of light in the material}}.$$

Since the speed of light in all materials is less than the speed of light in a vacuum, the index of refraction is always greater than 1. Note that the slower light travels in a material, the bigger the index of refraction.

The indexes of refraction for some typical optical materials are given in Table 16-1. The list ranges from air, whose index is just slightly greater than 1 (meaning that light travels almost as fast in air as it does in a vacuum) to diamond, whose index is 2.4, one of the highest known. This high index of refraction is why diamonds sparkle and shine with a brilliance that mere glass ($n = 1.5$) can never duplicate.

Table 16-1 Indexes of Refraction for Various Transparent Materials

Material	*Index of Refraction*
Vacuum	1.0000
Air	1.0003
Water	1.33
Eye	
Fluids	1.33
Cornea	1.38
Lens	1.40
Oil*	1.4
Glass*	1.5
Diamond	2.4

*Index varies with type.

Law of Refraction (Snell's Law)

When light enters a slower material (having a higher index of refraction), the light is bent closer toward a line perpendicular to the boundary, as we saw in the example of soldiers marching into a swamp. If it goes into a faster material (having a lower index of refraction), the light will be bent away from the perpendicular. The amount of bending depends on the ratio of the speeds in the two materials and therefore on the indexes of refraction. For small angles, that is, for light that hits the boundary nearly perpendicularly, the law of refraction states that the ratio of the angle of incidence to the angle of refraction is equal to the inverse of the ratio of the indexes of the two materials (Figure 16-7):

Law of Refraction (for Small Angles)

$$\frac{\theta_i}{\theta_r} = \frac{n_r}{n_i}$$

For angles up to 45°, this relation will not be off by more than about 10%. The smaller the angle is, the smaller the error. The exact relation for all angles is not much more complicated, but it does involve some trigonometric relations.

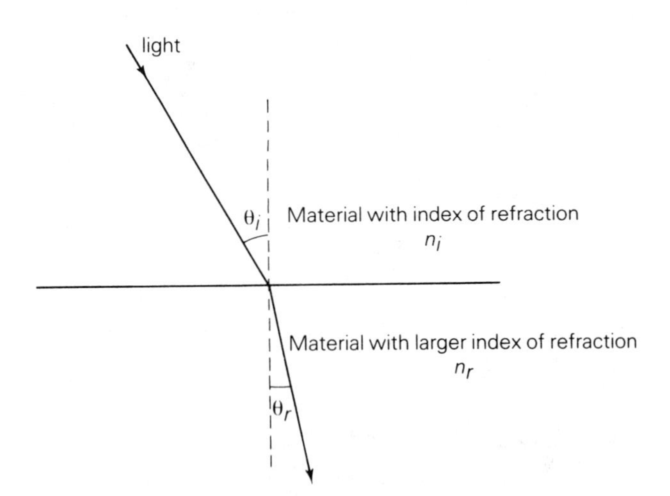

Figure 16-7

The location of a refracted image is found in much the same way that we found the location of an image behind a mirror: We trace the path of two rays as they go from the object to the eye. Consider a spoon in a glass of water as shown in Figure 16-8. We will trace the path of two rays A and B coming from the bowl of the spoon. Suppose ray A hits the surface of the water at an angle of incidence θ_{iA} of 15°. Water has an index of refraction of 1.33, so ray A would go into the air at an angle of

$$\theta_{rA} = \frac{n_i}{n_r}\,\theta_{iA}$$

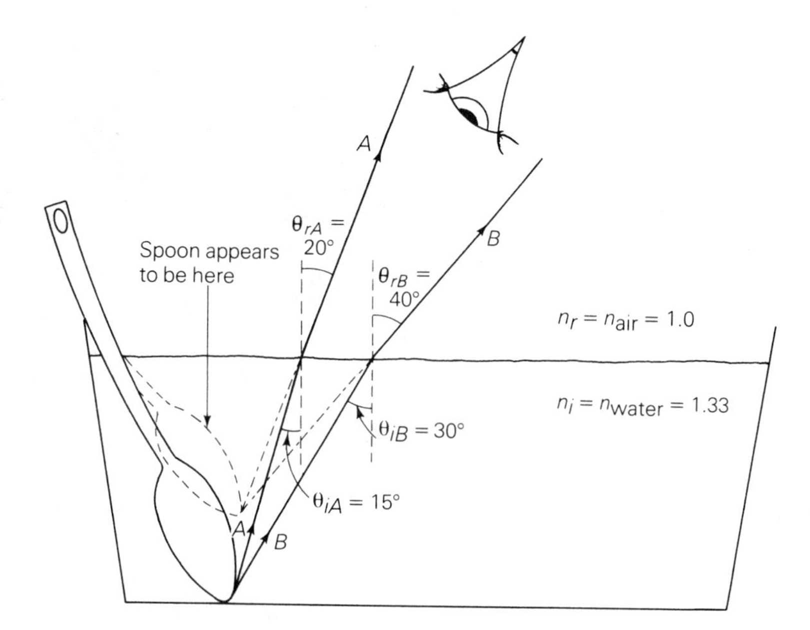

Figure 16-8 Refraction makes the spoon look bent at the surface of the water.

$$= \frac{1.33}{1.00} \times 15°$$

$$= 20°.$$

Similarly, ray *B*, which has an angle of incidence of 30°, would be refracted into the air at an angle of

$$\theta_{rB} = \frac{n_i}{n_r}\ \theta_{iB}$$

$$= 1.33 \times 30°$$

$$= 40°.$$

The eye and brain assume that the spoon bowl is located where the rays seem to come from: the spot where the dotted extensions of *A* and *B* cross. So, the image of the spoon bowl is located quite a bit above the true location, and the handle of the spoon seems to be bent at the water surface.

Because the speed of light in a material varies slightly with frequency, the index of refraction for a specific material varies somewhat with color. This effect is called *dispersion* (Figure 16-9). Since the index of refraction depends on the frequency and the amount of bending depends on the index of refraction, a refraction of a beam of white light—in which all frequencies are mixed

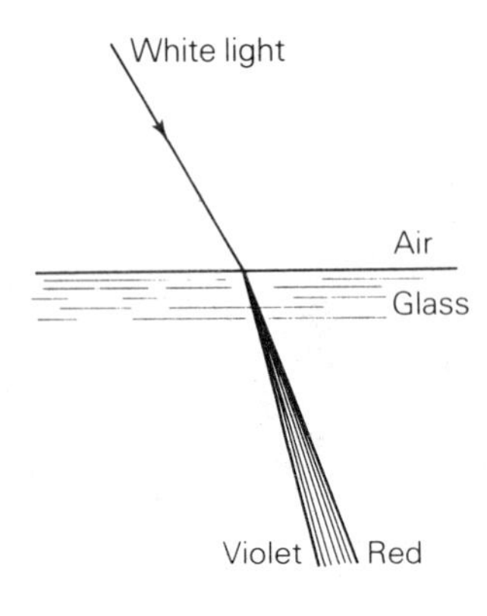

(a) Dispersion in glass. The amount depends on the type of glass.

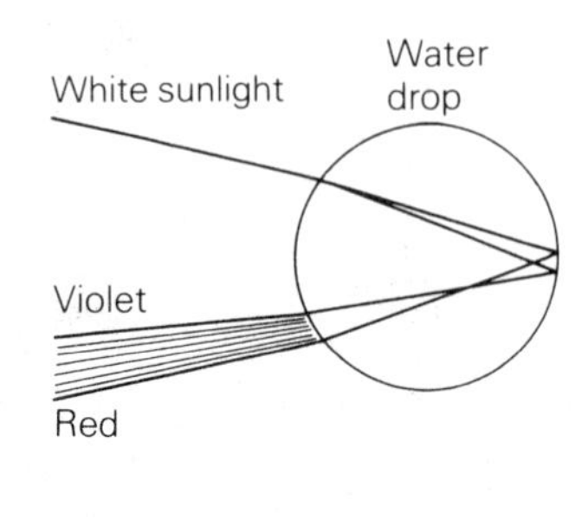

(b) Dispersion in a drop of water. This is what causes a rainbow.

Figure 16-9

equally—produces a full spectrum of colors. This effect causes the rainbow and the colored fringes on the edges of an image in a cheap telescope or microscope. High-quality optical instruments get rid of the colored fringes (or chromatic aberrations, as they are technically known) by carefully pairing lenses with different dispersion characteristics to make a compound lens with very little dispersion.

Lenses and Optical Instruments

Lenses are the principal components of most optical instruments. They are found in cameras, microscopes, telescopes, and hundreds of other things, including the human eye. They refract light so that images are formed. These images can be magnified or shrunk, or they can stay the same size, depending on our purposes and the arrangement of the lenses.

The image formation properties of a lens can be found by using the law of refraction to find the change in direction in the paths of rays through the lens. While this is a straightforward process, it is also a very tedious one. Figure 16-10 shows sketches of the process for two different lenses to establish one basic fact: Lenses that are thicker in the middle bring light beams together; that is, they converge the beam and are therefore known as *converging* lenses. Lenses that are thicker at the edges spread light beams apart; that is, they diverge the beam and are therefore known as *diverging* lenses (Figure 16-11).

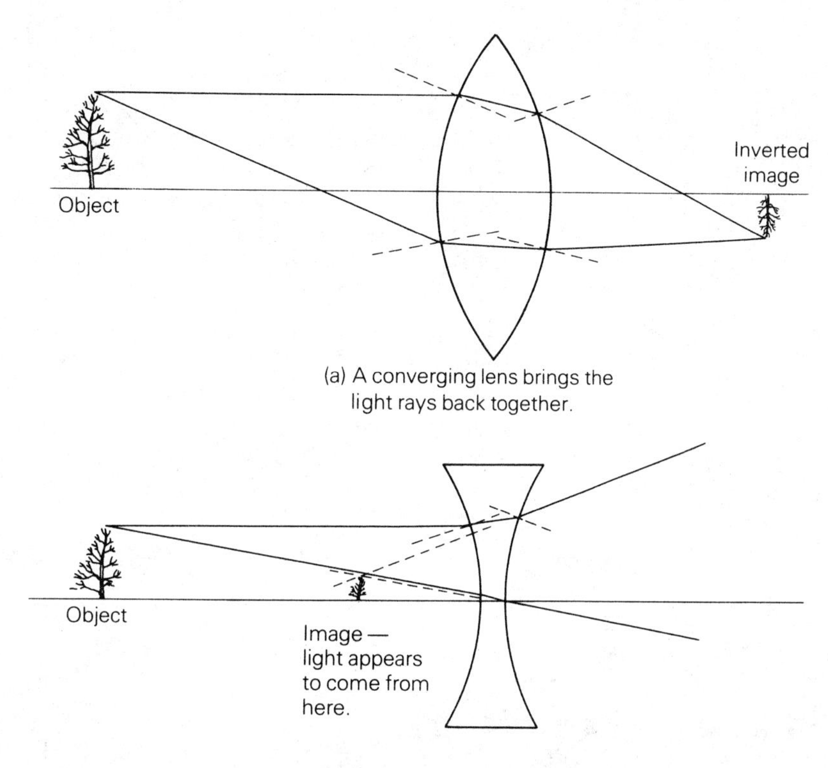

(a) A converging lens brings the light rays back together.

(b) A diverging lens spreads the light rays apart.

Figure 16-10

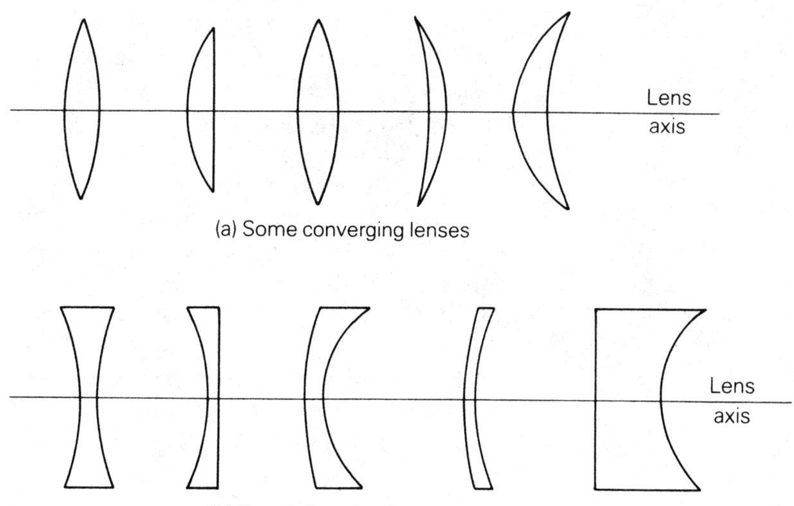

(a) Some converging lenses

(b) Some diverging lenses

Figure 16-11

Focal Length

One reason why the law of refraction is seldom used to locate images is because a far easier way can be used for most lenses. The first step is to locate what is called the *focal point* of the lens. By the way, this discussion will be much easier to follow if you use a magnifying glass or other converging lens and actually demonstrate the properties to yourself. The focal point of a converging lens is where a parallel beam of light is brought down to a small spot. It can be found by pointing the lens at a distant, bright object and then finding the point at which the inverted image of that object is on the other side of the lens (Figure 16-12). A distant object is used to insure that the light rays from the object are nearly parallel. The sun makes a fine distant, bright object, but its image is so bright that it will burn paper. So be careful if you use the sun as your object.

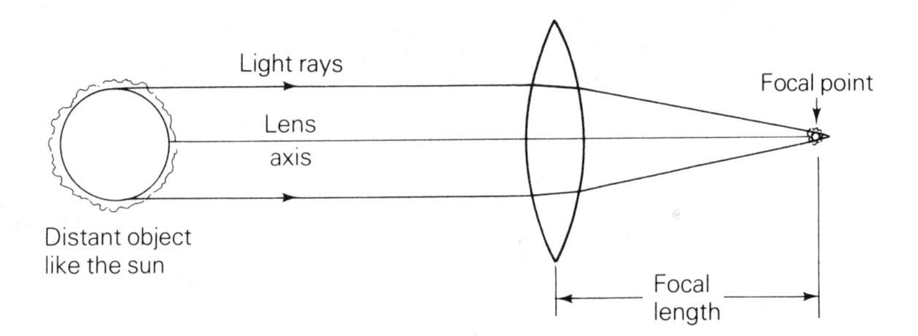

Figure 16-12 The focal point of a converging lens is where the image of a distant object is formed.

The distance from the focal point to the center of the lens is called the *focal length.* If you turn the lens over, the image position does not change; so a lens has two focal points, one on each side, of equal focal length. Generally speaking, the fatter and more curved a lens is, the shorter its focal length.

A diverging lens doesn't concentrate a parallel beam; it spreads it out. Thus, it doesn't bring light to one spot. However, if you look at the light leaving the lens, the rays *seem* to come from one point on the other side of the lens. This point is the focal point (Figure 16-13). (There is also a second focal point on the other side of the lens.) As with a converging lens, the distance from the focal point of a diverging lens to its center is the focal length. However, it is given as a negative number since the light is not brought to an actual spot but only *seems* to come from a single spot. The negative number also simplifies formulas used to calculate the image formation properties of diverging lenses. However, we won't go into the details of these formulas.

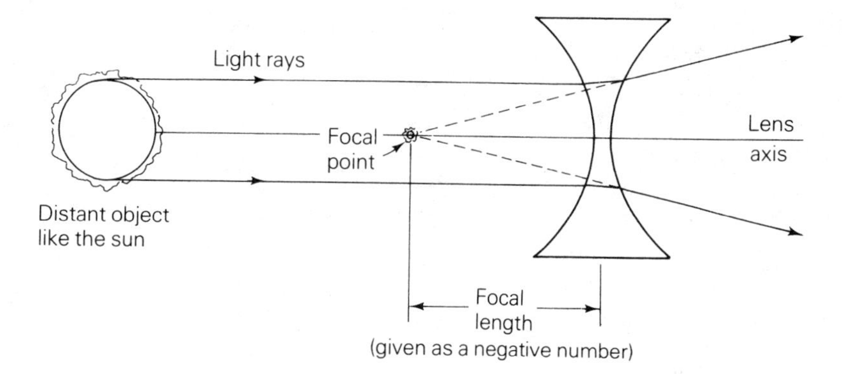

Figure 16-13 The focal point of a diverging lens is where the image of a distant object *seems* to be formed.

Image Formation

We can locate images by using the following property of the focal point: A light ray that is parallel to the axis of the lens will either go through the focal point or seem to come from the focal point after it goes through the lens. (The axis of the lens is the line going through the center of the lens and perpendicular to the lens.)

Also, any ray that passes through the center of the lens will not be bent significantly, because the refraction occurring on one side of the lens is exactly compensated by the refraction on the other side. This is the only other fact we need to locate images (Figure 16-14).

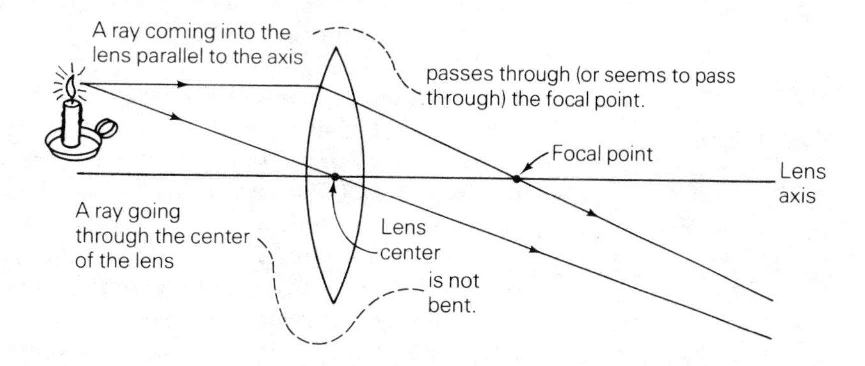

Figure 16-14 The two properties of lenses needed to locate images.

Let's try out these ideas by finding where the image is of a fly viewed under a magnifying glass. Again, if you have a magnifying glass handy, use it as we explain how it works. When a converging lens is used as a magnifying glass, the object—the fly in this case—is placed between the lens and its focal point. Now follow two rays

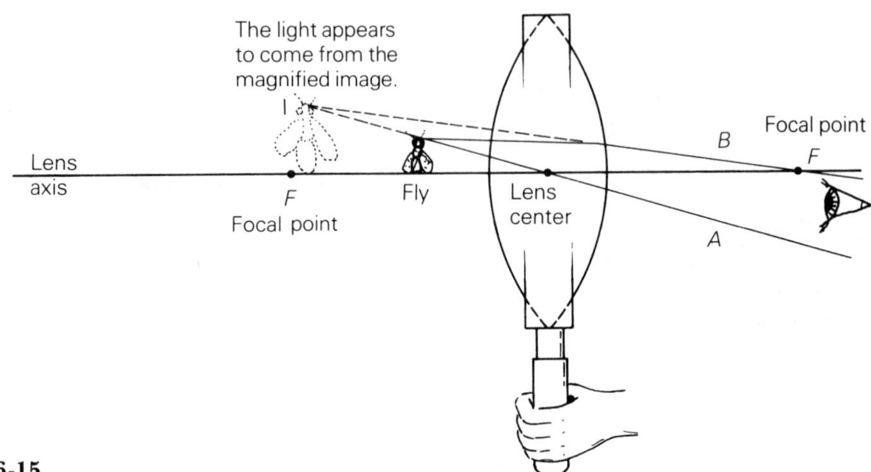

Figure 16-15

from the head of the fly, ray A going through the lens center and ray B going parallel to the lens axis (Figure 16-15). Ray A goes through the lens without bending because it goes through the center. Because ray B is parallel to the axis of the lens, it will go through the focal point F after it passes through the lens. The two rays will not come together to form an image on the right side of the lens, because they move apart after they pass through. To the eye, however, the two rays seem to come from point I on the left side of the lens, which is where the image of the fly's head is. The fly's tail happens to be on the axis of the lens, so its image will also lie on the axis directly below point I. Since the distance from I to the axis is greater than the actual length of the fly, the image of the fly is enlarged.

Next, let's see where the image is when the object, a house in this case, is beyond the focal point. Ray A again goes through the center of the lens without bending, and ray B, parallel to the axis of the lens, goes through the focal point F. In this instance, the two rays come together on the right side of the lens, forming an image of the house at point I. Note that the image of the house is upside down (Figure 16-16). When the lens is used in this way, do not look at the image through the lens; place a sheet of paper where the image is and it will be formed on the paper.

There is an important difference between the images in these last two cases. When we used the lens as a magnifying glass, there was no actual image of the fly behind the lens; the light only seemed to be coming from an image. In the second case, the light beams crossed, and an actual image was formed on the paper. Images of the first type are called *virtual images;* those of the second are called *real images.*

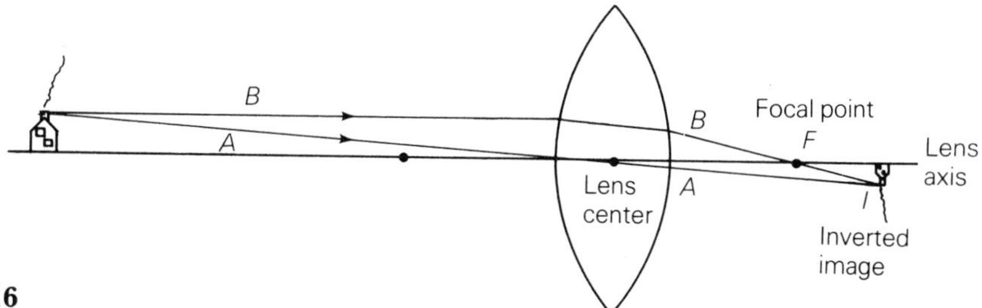

Figure 16-16

The Camera

To make any converging lens into a camera, we need only a box to keep out stray light, a shutter to open and close the lens, and a light-sensitive film to record the images (Figure 16-17).

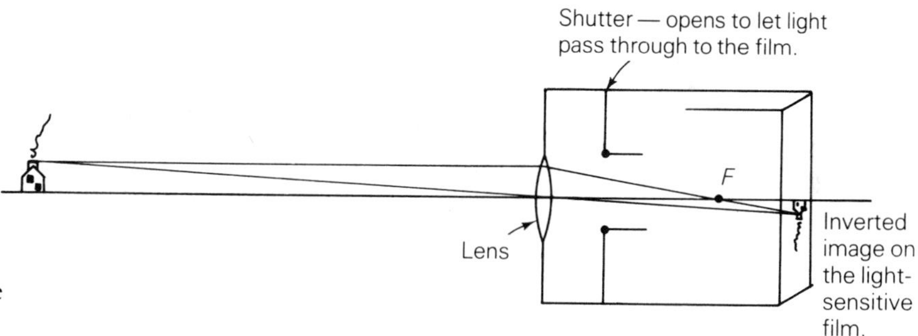

Figure 16-17 A simple camera.

Of course, you can add different lenses to get different image sizes and hundreds of accessories to make picture taking easier; however, the simple camera described above will take very good pictures. In fact, some of the finest photographs ever taken were made with such simple cameras.

The Projector

The projector is very similar to a camera except that the object and image are switched and a strong light is added to illuminate the object. The object, a picture on a piece of film, is placed just beyond the focal point of the lens at the point where the image would be if the lens were used in a camera. The light passes through the lens, forming an image on the screen where the object would be if the lens were used in a camera. A projector inverts the image, just as a

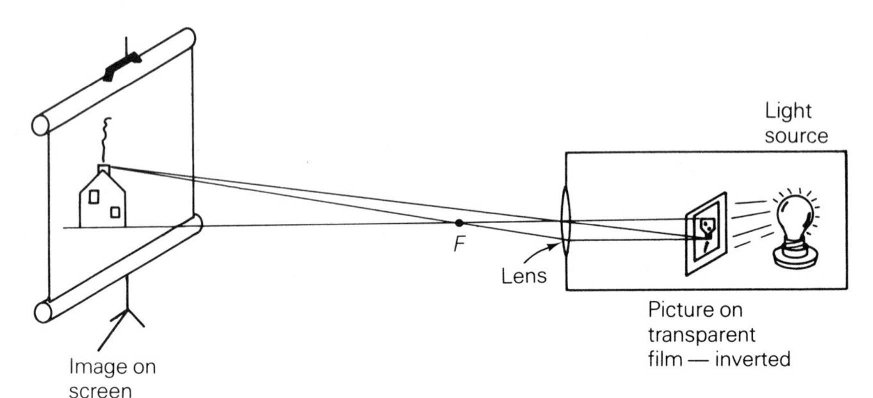

Figure 16-18 A simple slide projector.

camera does. This is why slides are put into a projector upside down (Figure 16-18).

The Eye

Functionally, the eye is very similar to a camera. It has a converging lens to focus the light rays to form an image of the viewed object, a dark covering to keep out stray light, and a light-sensitive surface, the retina, to detect images (Figure 16-19).

The Eye as an Optical Instrument

The lens of the eye is a compound lens, consisting of the cornea and the crystalline lens behind it. (The word *crystalline* refers to the clearness of the lens. The crystalline lens is not actually crystalline in structure.) Fortunately, the two lenses are close enough together to be regarded as a single lens for most purposes.

The cornea provides most of the focusing power. As you can see in Figure 16-19, the front surface of the cornea is the most curved. In addition, the difference between the index of refraction of air and the index of the cornea is much greater than the difference between the indexes of the crystalline lens and the two fluids—the aqueous and vitreous humors—on either side of it.

The location of the image behind the lens depends on how far away the object is. If it is very far away, the image will be at the focal point. As the object moves closer to the lens, the image moves farther beyond the focal point and deeper into the eye. As a result,

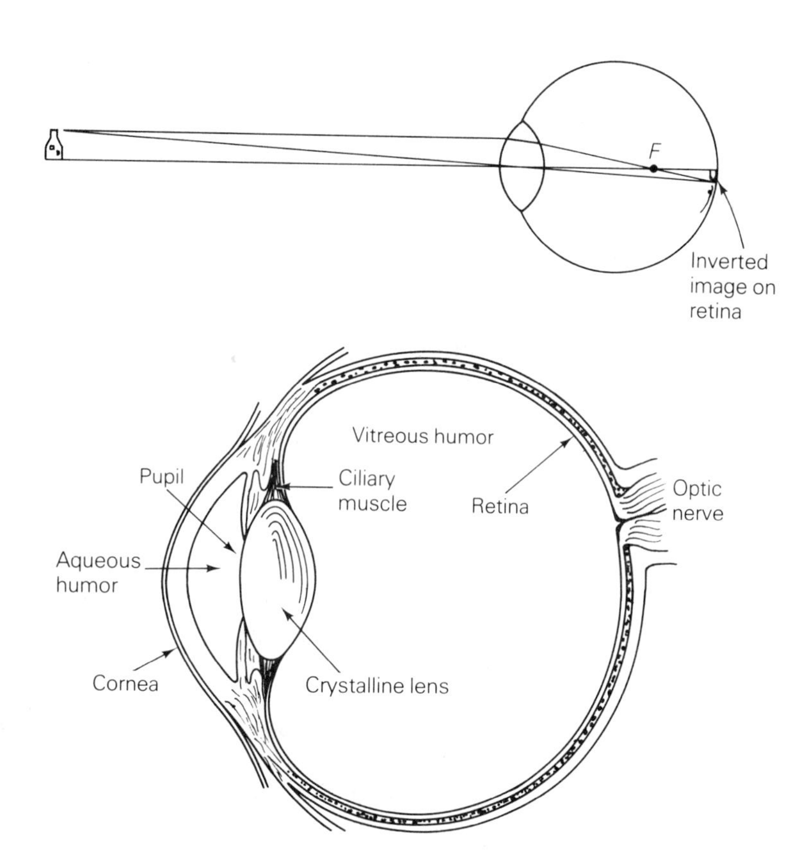

Figure 16-19 The eye.

the lens cannot form clear images of near and far objects simultaneously. A camera focuses by moving its lens until the object of interest forms a sharp image on the film. The eye accommodates in a different way: The focal length of the crystalline lens changes. The crystalline lens is held in place by the ciliary muscles, which pull on the edge of the lens. When more converging power is needed to view nearby objects, the ciliary muscles pull the crystalline lens into a more curved shape.

As people age, the ciliary muscles grow weaker and the lens becomes stiffer. Thus, older people can no longer make the lens as curved as they could when they were younger. A child can focus on objects as near as 10 centimeters, but the elderly often cannot focus on objects closer than 50 centimeters. Thus, people who need bifocals but refuse to wear them must read the newspaper at arm's length.

The iris, the ringlike muscle that gives the eye its color—blue, brown, or whatever—is located between the cornea and the crystalline lens. The opening in the iris, the pupil, regulates the amount of light reaching the retina. On a very bright day, the pupil shrinks to a very small hole, only letting in a small part of the light. When it is dark, the pupil expands to let in as much light as it can.

The opening and closing of the pupil has another effect on image quality. No lens is absolutely perfect; different sections of a lens have slightly different focal lengths. Therefore, the more of a lens in use, the fuzzier the image. On a bright day, when the pupil is narrowed and only a small part of the lens is in use, the image is sharper than when it is dark and the pupil is wide open. People with eye defects often squint to improve their vision for the same reason. In some cases, squinting improves image quality substantially. Of course, something is lost by squinting—less light gets in, so the image is not as bright.

Eye Defects

Judging by the number of people who wear glasses, eye defects are one of the most common human ailments. Fortunately, most lens imperfections are small, and either eyeglasses or contact lenses will restore adequate vision.

A lens with the wrong focal length for the size of the eyeball can cause either nearsightedness or farsightedness. The lens of the nearsighted (myopic) person has too short a focal length. The lens is too curved, and the light forms an image before it gets to the retina (except when the object is very near, in which case the image is formed on the retina). The cure for nearsightedness is to reduce the converging power of the lens by placing a diverging lens ahead of the eye. When the correct diverging lens is added, the image is formed on the retina and the person can see well at all distances (Figure 16-20).

The farsighted (hyperopic) person sees distant objects better than nearby objects because the focal length of the lens is too long. If the retina weren't there, the image would be formed behind the eyeball because the lens does not converge the light enough. The nearer the object is, the farther behind the retina the image would form. Farsightedness can be cured by adding a converging lens ahead of the eye to focus the light at the retina (Figure 16-21).

The third common lens defect is astigmatism, in which the lens is barrel shaped; that is, it curves more in one direction than in the other. Such a lens has no clearly defined focal length; horizontal lines might be brought to a sharp image at a different place than vertical lines. The standard test for astigmatism is based on this idea. As shown in Figure 16-22, line patterns are drawn in different directions. If any of the lines appear fuzzy while the others are

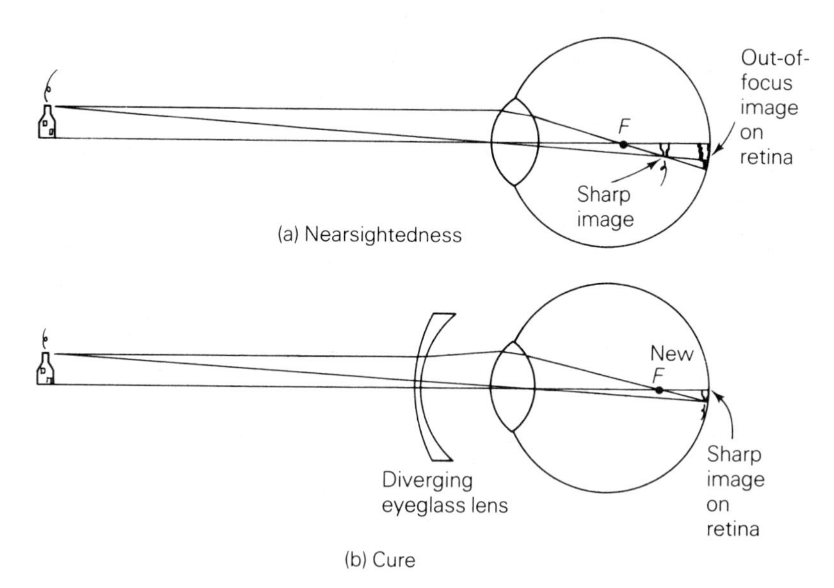

Figure 16-20 Nearsightedness (myopia) results from the lens of the eye having too short a focal length—the image is formed ahead of the retina. The cure is to add a diverging eyeglass lens that effectively increases the focal length.

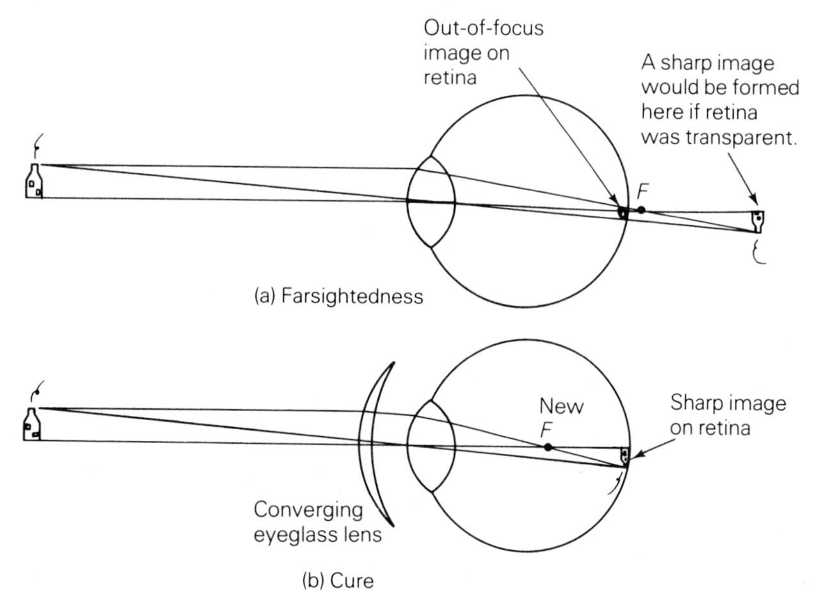

Figure 16-21 Farsightedness (hyperopia) results from the lens of the eye having too long a focal length—the image would be formed behind the eyeball if the retina were transparent. The cure is to add a converging eyeglass lens that effectively shortens the focal length.

clear, the person has astigmatism. The cure for astigmatism is to add a barrel-shaped eyeglass lens of the opposite curvature. Such a lens is called a cylindrical lens. Although cylindrical lenses need not look like cylinders, their optical characteristics are the same as those of a lens that is cylindrical.

The cylindrical lenses used to correct astigmatism must be properly aligned on the eye to work right. If the glasses are tipped by more than a few degrees, the person's vision will begin to blur.

Figure 16-22 Astigmatism test pattern. If some of the lines look fuzzy and light while others are clear and dark, you probably have astigmatism.

People with astigmatism should have frames that do not tilt easily, and the frames should be readjusted whenever they become loose.

Cataracts are a common eye defect in the elderly. When a cataract clouds the lens and ruins its focusing ability, the only known solution is to remove the clouded lens surgically. It can be replaced by surgically implanting a replacement lens, but these lenses are not suitable in all cases. If an implanted lens cannot be used, the only alternative is a strong converging eyeglass lens to replace the lost natural one. Of course, because the eyeglass lens has a fixed focal length, the patient will not be able to focus on objects at certain distances.

Stereoscopic Vision

Since the eyes are located several centimeters apart, the view of one eye is slightly different from the view of the other. This difference allows the brain to estimate the distance of an object accurately out to around 20 meters. One-eyed persons have a much harder time estimating distances because they must rely entirely on other clues, such as the apparent size of a distant object.

People seldom use both eyes equally. Most people are either right-eyed or left-eyed, just as they are right- or left-handed. Generally, a right-handed person is right-eyed, and conversely. You can easily find out which eye dominates by holding your finger out and lining it up with a distant object. Then cover up your left eye. If your finger and the object stay lined up, you are right-eyed. If the finger seems to jump to one side, you are left-eyed.

If one eye is significantly poorer than the other, or if the eyes are crossed so that they don't look in the same direction, the brain may ignore the information from one eye. This condition is called lazy eye, and it causes a loss of stereoscopic vision. Its development can often be stopped by correcting the defect, if the problem is caught early in life.

The Retina

The retina, forming the inside rear surface of the eye, converts light energy into nerve signals. It has two basic types of light detectors: the rods and the cones. The rods are very sensitive to light, but they are not color sensitive. The cones, on the other hand, are color sensitive but take more light to create a nerve signal than the rods do.

Because the rods and cones are not uniformly distributed in the retina, different sections of the retina have different sensitivities for color and amount of light. The central part, or fovea centralis, consists entirely of cones, and each cone has a fairly direct connection to the optic nerve. Thus, the central part has the greatest sensitivity to detail and color. When we concentrate on the details of an object, we look straight at it, putting the image of the object right on the fovea centralis.

As we go away from the fovea, the percentage of cones drops and the percentage of rods goes up. Thus, the farther from the fovea, the less the color sensitivity but the greater the light sensitivity. In addition, away from the fovea more than one detector will be connected to the same nerve, so fine details cannot be seen as clearly there. At the edge of the retina, there are no cones, only rods. This region is most sensitive to weak light, but it has no color vision. Since the greatest light sensitivity is near the edge of the retina, when you are looking for a weak light, you should not look directly at the point from which you think it will come.

Color Vision

The eye cannot sort out light frequencies nearly as well as the ear can sort out sound frequencies. For instance, the response of the ear to a particular note is unique: No other note or combination of notes will produce the same response. This is not true with light. Many combinations of frequencies will produce exactly the same sensation of color in the brain. Take the color orange, for example. One way to produce the sensation of orange is to shine pure, single-frequency orange light in the eye. But exactly the same response can be produced by shining a mixture of pure yellow light and pure red light, and there is no end to the number of other combinations that will also produce the same sensation of orange. Some colors, like purple and brown, do not even exist as pure, single frequencies of light.

The eye appears to sort out light frequencies in the following way: There are three different types of cones, each with a different pigment and a different response to different frequencies of light. We shall call the three types red cones, green cones, and blue cones. The red cones respond most strongly to pure red light, the green cones to green, and the blue cones to blue; but all three respond quite strongly to light that is anywhere close to their particular colors. The response curves for the three types of cones are shown in Figure 16-23.

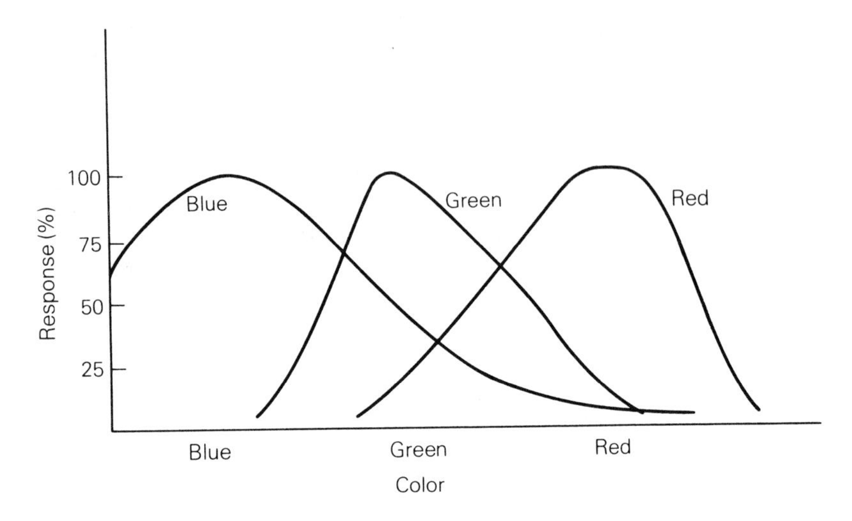

Figure 16-23 The response curves for the three types of cones in the eye.

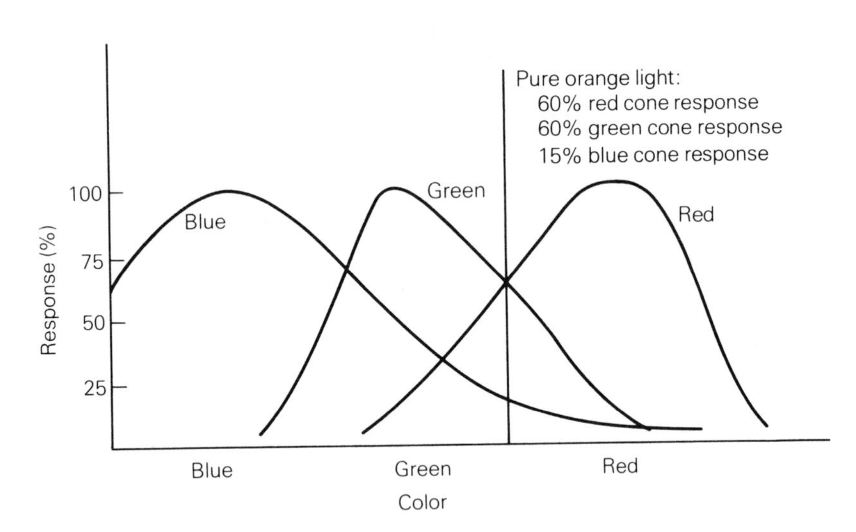

Figure 16-24 The response to pure orange light.

Let's look at two of the ways that the sensation of orange can be produced in the brain. The simplest is to shine pure orange light on the eye, which is halfway between red and green. If we draw a line on the response figure between red and green, we see that pure orange light excites the red cones and the green cones about equally and the blue cones only slightly (Figure 16-24). The brain thus responds to a strong red signal and an equally strong green signal that are mixed with a weak blue signal by producing the sensation of orange. As we have said, this sensation can be produced in various ways. If we shine equal beams of red and yellow light on the eye, we will have the situation shown in Figure 16-25. The red light will produce a strong red cone response, a weaker

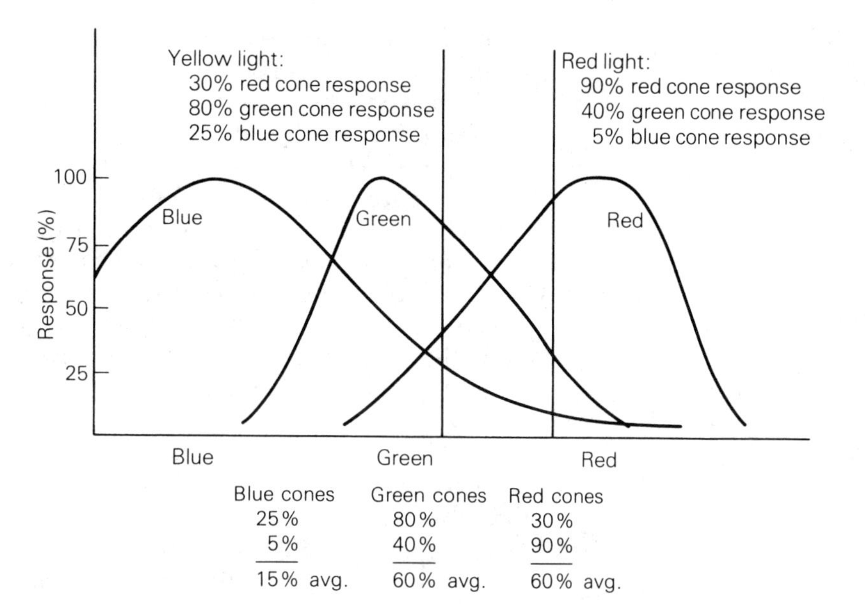

Figure 16-25 Average response to yellow and red light.

green cone response, and almost no blue cone response; the yellow light will produce a strong green cone response, a weak red cone response, and a still weaker blue cone response. Averaged together, the combination signal will produce about equal red and green cone responses mixed with a weak blue response—exactly the same response as to pure orange light. Consequently, the brain produces the same sensation as it does to pure orange light.

If you take three widely spaced pure colors, such as red, green, and blue, and control their relative proportions, you can get nearly every possible response from the three cones and thus produce the sensation of almost every color. All color photography, color television, and color printing processes use this principle. Examined closely, a color television picture consists of nothing but an alternating pattern of red, green, and blue dots or lines—no other colors. Yet when you look at the picture from a reasonable distance, the three colors produce the sensation of almost all the colors. Similarly, color films only have three different-colored dyes, and color illustrations in magazines are made from three colored inks (plus black).

When one (or more) of the different types of cones is missing, the person is said to be color-blind. However, the term *color-blind* is too strong to describe the common case where just one type of cone is missing. Such people are more properly called dichromats, meaning two-colored, while normal people should be called trichromats (three-colored). Dichromats do not see the world in black and white; they do have some color response, but their ability to distinguish colors is much more limited than a normal person's. For instance, a dichromat missing green cones would have a hard time telling red, orange, and green apart, because these colors would all produce a strong red cone response mixed with a weak blue cone response. However, this person would have no problem telling the difference between red and blue. Incidentally, only three types of dichromats have been observed, a fact that strongly supports the theory that there are three different types of cones. In fact, the likely color response patterns of the three different types of cones were determined by studying the color perception of dichromats. Only recently have the actual pigments in the cones been found and examined separately.

The Optic Nerve

Our perception of light and color does not occur solely in the rods and cones of the retina; the patterns must be analyzed and recorded by the nerves and brain. The analysis of the image starts in the eye itself. The signals from the rods and cones go to junctions in the nerves lying on the surface of the retina. Decisions seem to be made in these junctions about what to pass and what not to pass along the optic nerve to the brain. The brain ignores much of the information, while other bits are stored for life in the memory.

A good example of such differences in perception is our far greater sensitivity to moving objects than to stationary objects. This trait probably stems from the fact that moving enemies and predators threatened our ancestors more than stationary things, such as trees and rocks. This effect is very apparent in our peripheral (or side) vision. Look straight ahead and bring your hands around from the back of your head until they are just barely visible at the side of your head. Now wiggle your fingers. Most people see their fingers far more vividly when the fingers are moving than when they are stationary. Of course, this difference must occur in the nerves, because the image striking the retina is about the same whether or not the fingers are moving.

Concepts

Electromagnetic spectrum
Electromagnetic field
Polarized light
Reflection
Law of reflection
Refraction
Index of refraction
Law of refraction

Dispersion
Lens
Converging lens
Diverging lens
Focal point
Focal length
Real image
Virtual image
Nearsightedness (myopia)
Farsightedness (hyperopia)
Astigmatism
Stereoscopic vision
Color vision
Color blindness

Discussion Questions

1. What are the main types of electromagnetic radiation?
2. What is the principal difference between a radio wave and an x ray?
3. How is polarized light different from ordinary light?
4. How do Polaroid sunglasses reduce glare?
5. Polaroid sunglasses reduce glare from the road very effectively, but they don't stop glare from a window unless you turn your head sideways. Why?
6. Explain, using a ray diagram, why your image in a flat mirror appears to be the same distance behind the mirror that you are in front of it.
7. Explain why you cannot see clearly underwater without a face mask.
8. What is meant by the index of refraction of a substance?
9. Why does a diamond sparkle much more brilliantly than glass?
10. How do raindrops break up the white light of the sun into all the colors of the rainbow?
11. Draw a diagram of the eye and describe the functions of its principal parts.
12. Discuss the similarities between a camera and the eye. What are some of the important differences?
13. Explain why people tend to hold the newspaper farther from their eyes as they grow older.
14. Explain the causes of the three most common eye defects: nearsightedness (myopia), farsightedness (hyperopia), and astigmatism. What lens shape is used to cure each of these defects?
15. Why do colors seem much brighter on a sunny day than at twilight?
16. Explain why various combinations of three colors, such as red, green, and blue, can give the impression of almost any other color. Compare this to the case of hearing. Why can't three musical notes combine to give the impression of some other single note?

Problems

1. How long must a full-length mirror be for someone who is 180 centimeters tall? That is, what is the shortest mirror in which the person can see his or her whole body? Does it make any difference how far away from the mirror the person is?

2. Show by tracing rays for three or more different eye positions that the image of a stationary object in a mirror is in the same position behind the mirror, no matter where the eye is located.

3. When you stand in a 90° corner made of mirrors, you see three images of yourself, one on each wall and a third in the corner. Trace the rays to see how each image is formed. If you are right-handed, the two images on the wall will seem to be left-handed, but the image in the corner will be right handed. Explain.

4. What is the speed of light in a diamond?

5. A light ray strikes a glass plate at an angle of 27°. Inside the glass, the ray moves at an angle of 16° to a line perpendicular to the surface. What is the index of refraction of this glass?

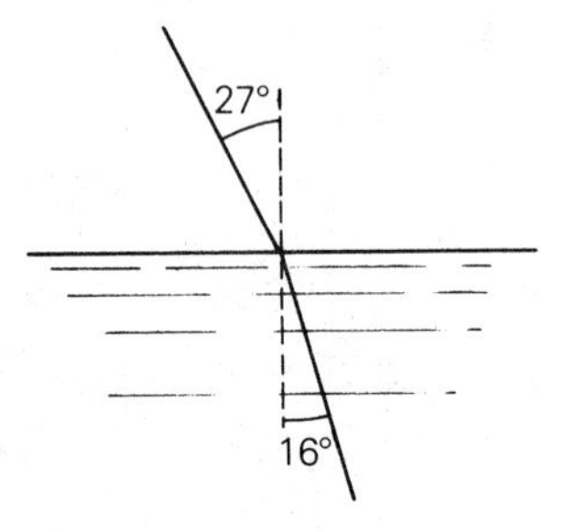

6. The sun appears to be at an angle of 21° from vertical to a person underwater. What is the true angle of the sun from vertical, as seen by someone at poolside?

7. A converging lens is used as a "burning glass" to start a fire by focusing sunlight on a small spot. If the sunlight is focused on a spot 7.5 centimeters from the lens, what is the focal length of the lens?

8. A converging lens has a focal length of 3 centimeters. By tracing rays, locate the image of an object 5 centimeters from the lens. If the object is 1 centimeter long, how long will the image be? Is the image upright or inverted?

9. A converging lens with a focal length of 4 centimeters is used as a magnifying glass to look at an object 3 centimeters from the lens that is 0.3 centimeter long. Where will the image be located and how long will it be? Is the image upright or inverted?

Experiments

1. Most of the important effects of polarized light can be demonstrated with two pairs of Polaroid sunglasses (or a cheap pair cut in half).
   a. Show that light can be polarized by placing one lens over another and rotating one until very little light passes through.
   b. Show how a polarizing microscope can show hidden detail by placing a piece of crumpled cellophane tape between two lenses. Look at a light through them while rotating one of the lenses.
   c. Show that reflected glare from a window is highly polarized by looking through one lens and rotating it until the glare is minimized.

2. Demonstrate that white light is made up of all colors of the rainbow by shining a flashlight on a piece of rhinestone jewelry in a dark room. You will be able to see the colors better if the walls are white (of if you tape a piece of white paper to the wall).

3. Shine a flashlight on a mirror. Measure the angles that the light makes as it hits and leaves the mirror. Show that they are the same.

4. Find the focal length of a converging lens by forming the image of a distant, bright light on a piece of paper. The focal length is the distance from the lens to the paper. (The sun is usually too bright to use, because its image will often set the paper on fire.) Now use the lens as a magnifying glass. Measure the distance from the lens to the object and also estimate the magnifying power of the lens; that is, how many times bigger is the image than the object? Next draw a ray diagram to locate the image and find its size. (Use the measured focal length and object locations to determine the location of the focal point and the object on the diagram.) How does the size of the image in the drawing compare with the size of the image that you observed?

5. Use a converging lens to form the image of a distant, bright object on a piece of white paper. Is the image erect or inverted? Bigger or smaller than the object? Next move close enough to the object so that it is only two or three focal lengths from the lens. How do you have to change the spacing of the lens and the paper to get a sharp image? How is the operation of a camera or an eye similar to this experiment?

6. Take a red, a green, and a blue light of about equal brightness and shine them on a white background. What color do you see? Next shut off one light at a time and look at the colors. Place an object between the lights and the background, and look at the various-colored shadows cast by the object. Can you explain the colors you see? Three flashlights covered with the three different colors of plastic will work fine. Three slide projectors showing the three different colors are ideal.

CHAPTER SEVENTEEN

X Rays and Radiation

Educational Goals

The student's goals for Chapter 17 are to be able to:

1. Describe the structure of the atom and explain how the atom produces light and x rays.
2. Determine the daughter nucleus of an alpha, beta, or gamma decay of a known radioisotope.
3. Calculate the strength of a radioactive source of known half-life as a function of the age of the source.
4. Give the three principal ways of detecting x rays and other ionizing radiations.
5. Explain the effects of ionizing and nonionizing radiations on the body.
6. Describe the construction of an x-ray tube.
7. Give the proper safety precautions for the use of radiation and radioisotopes.

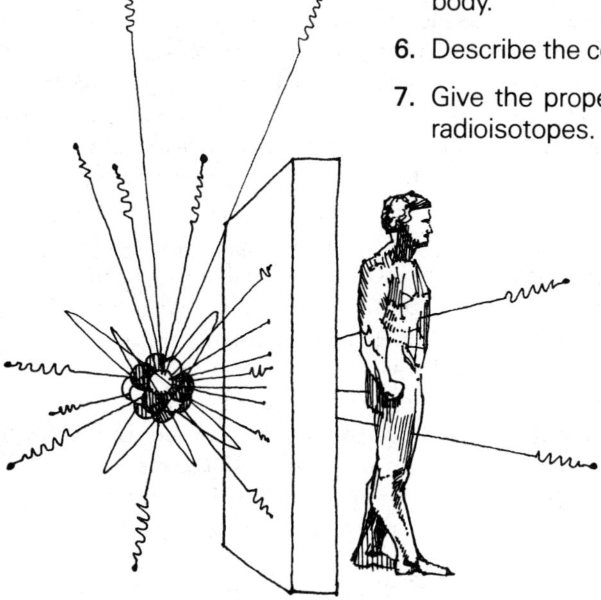

Almost every medical treatment has some undesirable side effects. In most cases they are mild, and normal caution is enough to prevent complications. However, in some cases the side effects are serious, and the treatment plan must be carefully balanced between the good and bad effects. X rays and other radiations definitely fall in this latter class. In addition, a serious mistake in the use of radiation is as likely to injure the technician giving the treatment as it is the patient. Thus, nearly everyone near the treatment area benefits from careful radiation safety practice.

The dangers of radiation are real, but they have been overblown by many newspaper and TV stories. Because of this sensationalism, some people erroneously believe that radiation inevitably means cancer and that cancer inevitably brings death. In actuality, the dangers of most x-ray procedures are no worse than the dangers of driving to the hospital to get the treatment. Like driving, working with radiation calls for constant caution and attention to detail. Under these circumstances, the benefits of radiation far outweigh the risks.

Structure of the Atom and Its Nucleus

The atom and its nucleus can store potential energy. Some of this energy will be electrical, and some will be from the other forces holding the nucleus together. When an atom releases this potential energy, it is usually carried away as radiation. Sometimes the radiation is electromagnetic—light, x rays, and so on—and at other times it consists of particles such as electrons and neutrons.

As discussed in our study of electricity, an atom consists of light negative electrons that orbit a heavy positive nucleus. The nucleus consists of protons, which are positively charged, and neutrons, which have no electric charge. Both neutrons and protons have roughly the same mass, which is about 1,800 times that of an electron. The electrons are held in their orbits by the electrical attraction between them and the nucleus. The nucleus is held together by a strong but poorly understood force, which we shall call the nuclear force. Electric forces are also a factor in the nuclear structure. The positive protons repel each other electrically, which tends to break apart the nucleus. For the light elements—those having few protons and neutrons—the nuclear forces are stronger than the repulsive electric forces of the protons, so the nucleus can be stable. For the heavy elements—those with more than 83 protons—the repulsive forces among the protons are larger than the attractive nuclear forces, making the nuclei unstable. That is, heavy atoms are radioactive if they exist at all. They eventually break up by emitting radiation. Sometimes they simply fly apart, an event known as nuclear fission.

Up to now, we have pictured the atom as sort of a miniature solar system with the light electrons orbiting the heavy nucleus in much the same way that the planets orbit the sun. This is a helpful model, but it ignores the possibility that small chunks of matter, such as subatomic particles, might not behave like large chunks of matter. This possibility turns out to be fact: All matter actually has wave characteristics. The wavelength of matter waves is very short, about the size of an atom. Thus, we don't notice the wave characteristics of the large pieces of matter that we ordinarily handle. But in small bits—such as electrons, protons, and neutrons—the wave nature becomes important, and it is a significant factor in atomic structure.

The wave character of atoms is very complex, but the following simplified description will give you a rough understanding. Because electrons have a wave character, they can only circle in orbits that have circumferences exactly equal to some whole number of electron wavelengths. In this way the electron wave stays in phase with itself as it orbits the nucleus. Each time the wave orbits the nucleus, it interferes constructively with itself to build up a stable pattern. Orbits of other circumferences are impossible because the wave would go out of phase with itself; as a result, it would eventually interfere destructively with itself and destroy itself (Figure 17-1).

Each allowed orbit can only hold a certain maximum number of electrons, because no two electrons can be in exactly the same state at the same time—a principle called the Pauli exclusion principle. This principle also largely explains why two objects can't be in the same place at the same time.

The fact that each orbit has a limited electron capacity is responsible for the periodic nature of the table of elements. All elements with filled orbits (or shells, as they are often called), such as helium, neon, argon, krypton, and xenon, behave about the same chemically. They are all extremely difficult to combine with other elements because of their filled electron orbits—they can neither gain nor lose electrons easily. The elements with one electron outside a filled shell, such as lithium, sodium, potassium, rubidium, and cesium, are all extremely reactive metals. They are metals because the lone outside electron is easily detached from the atom to conduct electricity, and they are reactive because this electron is easily transferred to other atoms to form new molecules. Likewise, all elements with the same number of electrons outside a closed shell tend to be quite similar chemically.

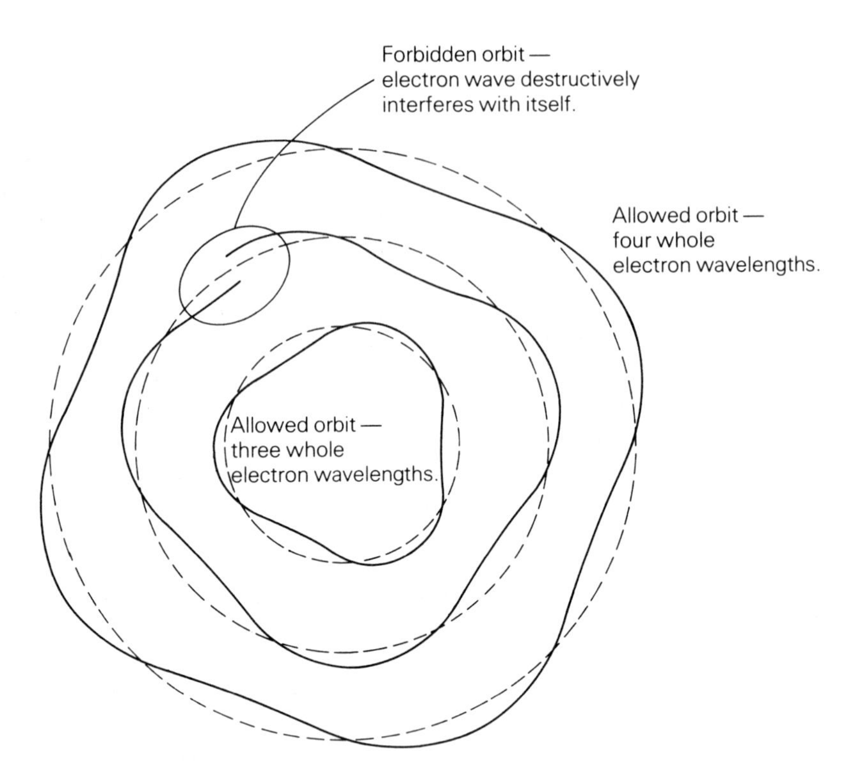

Figure 17-1 Some allowed and forbidden electron orbits.

Electrons can be bumped into higher orbits by several things —including collisions with other atoms, electric sparks, light waves, and chemical effects. An atom with an electron in a high orbit has a high potential energy, which is released when the electron falls back to its original orbit. This energy usually leaves the atom as electromagnetic radiation, such as light, ultraviolet rays, and x rays. Because the electron can only jump between allowed orbits, only certain energies of light rays can be emitted by a particular element. Since the frequency of light rays is directly proportional to their energy, a particular element can only emit certain frequencies. The particular frequency patterns are different for the different elements. These frequency patterns are thus the "fingerprints" of the elements, and chemists commonly analyze the light spectrum of unknown substances to determine what elements are present.

The farther the electron falls between orbits, the more energy it releases and the more energetic the electromagnetic radiation (Figure 17-2). If the two orbits are very close, the radiation will be in the form of microwaves or infrared radiation. For slightly larger jumps, the radiation will be visible light from the red end of the spectrum for low energies and from the blue end for high energies. Still larger jumps give off ultraviolet radiation and the longest jumps give off x rays.

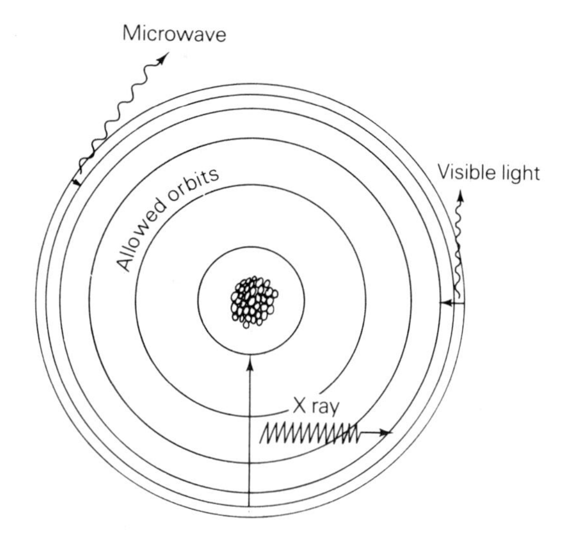

Figure 17-2 Electromagnetic radiation (such as microwaves, light, and x rays) is emitted when electrons drop to lower allowed orbits. The further the drop is, the more energetic the radiation.

The Nucleus

The nucleus is extremely small. The ratio of its size to that of an atom is about the same as the ratio of the size of a cell to the human body. Yet almost all of the mass of the atom is concentrated in this tiny nucleus. Thus, matter in the nucleus is trillions of times more concentrated than matter as we know it. Given its enormous density and very small size, it isn't surprising that we don't know exactly how the nucleus works and what holds it together.

The number of protons in the nucleus is called the *atomic number* of the atom. In the neutral atom, the number of electrons is equal to the number of protons, so the atomic number is also equal to the number of electrons. The chemical properties of an atom depend mainly on the number of electrons. All atoms with the same atomic number behave almost identically chemically; that is, they are atoms of the same element.

The number of neutrons in an atom is usually about the same to 50% larger than the number of protons. The total number of neutrons and protons is called the *nucleon number* of an atom (the word *nucleon* is the family name for neutrons and protons).

The nucleus also has a wave and a shell structure, but the behavior of a particular nucleus depends not only on the atomic number but also on the number of neutrons. The nuclei of all atoms of the same element may not behave at all alike if the number of neutrons differs, even though the chemical properties of the atoms are the same.

Atoms of the same element with different numbers of neutrons are called *isotopes*. If a particular isotope does not emit radiation, it is said to be a stable isotope; if it does, it is called a radioactive isotope. The chemical symbol is not enough to describe a particular isotope, since it only gives the atomic number. An isotope is symbolized by the chemical symbol of the element plus the nucleon number in the upper left corner. (So that you won't need to look up the atomic number in a chart of the elements, we will include it in the lower left corner. In actual practice, the atomic number is seldom added.) For instance, the isotope of sodium that has 22 nucleons is given the symbol

$$^{22}_{11}\text{Na},$$

meaning that the isotope has 11 protons and a total of 22 nucleons. Since the number of neutrons is just the difference between the nucleon number and the atomic number, $^{22}_{11}$Na has $(22 - 11) = 11$ neutrons. This isotope is often called sodium-22.

Some radioisotopes occur in nature, but most are made by bombarding a stable isotope with neutrons, protons, or other particles. Neutron bombardment is usually the most economical, because nuclear reactors create vast numbers of relatively cheap neutrons. For example, the radioisotope cobalt-60 ($^{60}_{27}$Co) is produced by simply putting ordinary, stable cobalt in a nuclear reactor. Ordinary cobalt is cobalt-59 ($^{59}_{27}$Co), so when a neutron sticks to cobalt-59, it forms cobalt-60. The reaction is written as

$$^{1}_{0}n + ^{59}_{27}\text{Co} \rightarrow ^{60}_{27}\text{Co}.$$

The n stands for the neutron, the 1 above the n denotes that the neutron is a single nucleon, and the 0 denotes that the neutron has no electric charge. (That is, the neutron has an atomic number of zero. Using the same scheme, a proton would be $^{1}_{1}p$.) Note that both the upper and lower numbers in the equation balance: The sums of the upper numbers on both sides of the arrow are equal, as are the sums of the lower numbers.

Proton bombardment is not used often unless there is no known way to make the isotope with neutrons, because proton beams cost more than neutron beams. A typical proton reaction is the one used to make cobalt-56 ($^{56}_{27}$Co). Stable iron-56 ($^{56}_{26}$Fe), the main isotope in normal iron, is bombarded with high-energy protons. The proton sticks to the iron and knocks out a neutron. Let's review the equation carefully to see how this reaction makes cobalt-56. We have a proton and iron-56 on the left side of the equation and a neutron plus an unknown nucleus on the right:

$$^{1}_{1}p + ^{56}_{26}\text{Fe} \rightarrow ^{1}_{0}n + ^{X}_{Y}?.$$

The lower numbers must balance on both sides of the equation, so Y, the atomic number of the unknown isotope, must be 27. Looking at the chart of the elements in Appendix II, we find that element number 27 is cobalt, so ? is Co. Thus, the equation is now

$$^{1}_{1}p + ^{56}_{26}\text{Fe} \rightarrow ^{1}_{0}n + ^{X}_{27}\text{Co}.$$

Since the upper numbers must also balance, X must be 56; therefore, the isotope produced by this reaction must be cobalt-56 ($^{56}_{27}\text{Co}$). The complete reaction equation is

$$^{1}_{1}p + ^{56}_{26}\text{Fe} \rightarrow ^{1}_{0}n + ^{56}_{27}\text{Co}.$$

Nuclear Radiations

Radioactive nuclei release their energy in a variety of ways. The four most common ways are alpha radiation, beta radiation, gamma radiation, and nuclear fission. The terms *alpha, beta,* and *gamma* are arbitrary names given to the radiations before their natures were known. However, the names have stuck, just as the name *x ray* has.

Gamma Radiation

Gamma radiation is a form of electromagnetic radiation, just as x rays or light rays are. The only difference is that gamma rays originate when nucleons drop into lower nuclear shells. Since much more potential energy is released when nucleons change shells than when electrons change shells, gamma rays are generally much more energetic than x rays. Since electromagnetic radiation has neither mass nor electric charge, the atomic number and nucleon number of the nucleus remain the same after a gamma ray is emitted. However, the nucleus does drop to a lower energy state.

Gamma rays are even more penetrating than x rays. This great penetrating power makes gamma rays especially useful for irradiating tumors deep inside the body. Gamma rays are too penetrating for taking x-ray pictures, but they are often used to look for cracks in heavy steel castings.

Beta Radiation

In beta decay, an electron is born within the nucleus and then promptly ejected. Thus, beta rays are simply very energetic electrons. However, beta-ray electrons are not always ordinary negative electrons. Some are positrons, which are exactly like ordinary electrons except that they have a positive charge. Positrons are perhaps best thought of as antimatter electrons; when a positron meets an ordinary electron, they immediately annihilate each other in a sort of miniature atomic explosion, changing into nothing but electromagnetic energy.

The ordinary negative electron is symbolized as $_{-1}^{0}e$. The zero means that its mass is so small compared with that of a nucleon that it can be ignored when balancing reaction equations. The -1 means that the electron has a charge of -1. The positron would be $_{1}^{0}e$, since its charge is positive. Since the nucleus gives off electric charge during a beta decay, the atomic number must also change, but the number of nucleons is left unchanged. For instance, the beta decay of sodium-22 ($_{11}^{22}Na$) by positron emission yields

$$_{11}^{22}\mathrm{Na} \rightarrow {}_{1}^{0}e + {}_{Y}^{X}?.$$

To balance the equation, the atomic number Y must be 10, which is the atomic number of neon (Ne). The nucleon number X is still 22, so when sodium-22 emits a positron, the nucleus neon-22 remains. The reaction equation is

$$_{11}^{22}\mathrm{Na} \rightarrow {}_{1}^{0}e + {}_{10}^{22}\mathrm{Ne}.$$

The nuclei that are left after a decay are called daughter nuclei, and the original nucleus is called the parent isotope.

Beta rays do not penetrate very far, but they can penetrate through dead skin layers to the living skin below. Since all of a beta particle's energy is deposited in the narrow layer where it stops, the damage in that layer is very severe. Thus, beta rays are very useful for delivering an intense dose of radiation to a confined tumor without harming the surrounding tissue. Unfortunately, getting the beta rays to the desired region is usually difficult.

Alpha Radiation

The alpha particle consists of two protons and two neutrons stuck together. A radioactive nucleus emits two protons and two neutrons rather than, say, two neutrons and one proton because that particular combination is held together by the nuclear forces more tightly than the other combinations.

The symbol for the alpha particle is $_{2}^{4}\alpha$, but two neutrons and two protons also happen to be the nucleus of the ordinary helium atom, helium-4 ($_{2}^{4}He$). Thus, the symbol $_{2}^{4}He$ can also be used for the alpha particle. In fact, most of the world's helium comes from the decay of alpha-emitting radioactive elements.

Alpha emission is caused by the repulsive electric forces between the protons. By spitting out two protons along with two neutrons, the net electrical potential energy of the nucleus is reduced. Alpha emission only occurs in nuclei that have many protons—the heavy elements with high atomic numbers. Since two protons leave with the alpha particle, the daughter nucleus will have an atomic number that is two smaller than that of the parent and a nucleon number that is four smaller. Radium-226 ($^{226}_{88}Ra$) is a typical alpha emitter. The reaction is

$$^{226}_{88}Ra \rightarrow {}^{4}_{2}\alpha + {}^{222}_{86}Rn.$$

Thus, the daughter is radon-222, which also happens to be radioactive.

Although they do intense damage for the short distance that they do travel, alpha rays are not very penetrating. They only go a few centimeters in air and are easily stopped by the dead layers of the skin. Therefore, they present no hazard to the body as long as they stay outside (except when the source is within a few centimeters of the cornea of the eye). Inside the body, where no air or dead skin can stop alpha particles, alpha emitters are extremely hazardous; even microgram quantities can be lethal. For these reasons, alpha particles are rarely used in medicine.

Nuclear Fission

The mutual electrical repulsion of the protons will sometimes cause a heavy nucleus to break into two large fragments plus a spray of neutrons. This process is called *nuclear fission.* Since nuclear fission releases a great deal of energy, it is used in nuclear reactors and atomic bombs (Figure 17-3).

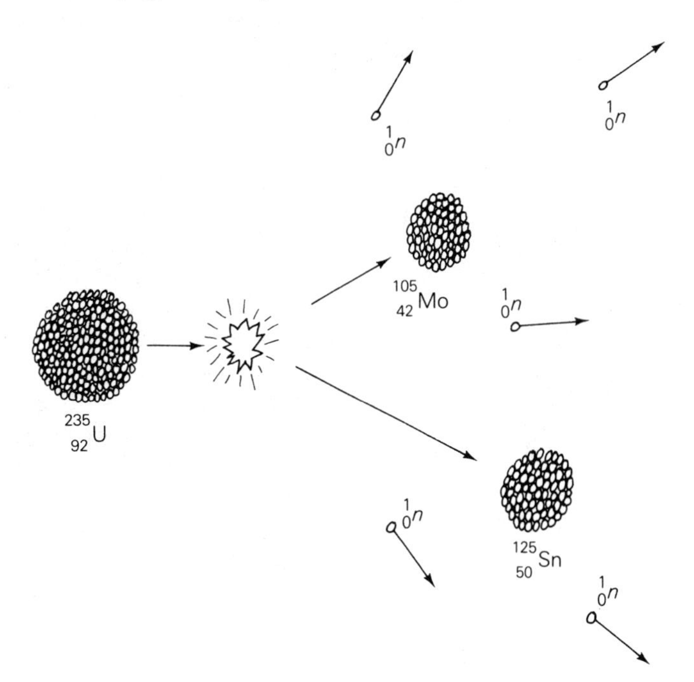

Figure 17-3 The fission of uranium-235 into molybdenum-105, tin-125, and five neutrons.

The flood of neutrons released in nuclear reactors has made the production of radioisotopes relatively cheap. Most of the commonly used radioisotopes are made by neutron irradiation in reactors.

Neutrons are very penetrating, and several meters of earth, concrete, or water are required to shield against them. Their penetrating power makes them very useful for the treatment of deep tumors.

Half-Lives of Radioactive Nuclei

The lifetime of a radioactive nucleus is not a fixed, predictable quantity. A given nucleus might live for only a microsecond, while another absolutely identical nucleus might live for days or centuries. Nor do nuclei show any sign of aging; they seem to stay exactly the same from birth to the moment of decay.

Since a nucleus doesn't age, the chances of a nucleus decaying during a particular moment are constant, which causes what is called an exponential decay of the radioactive material. To illustrate, suppose we had 360 players in a dice game, and each player had a pair of dice. They all throw their dice at the same time, and those that throw snake-eyes (a pair of ones) leave the game. The chances of throwing snake-eyes is 1 out of 36, because there are $6 \times 6 = 36$ possible ways for a pair of dice to land but only 1 way to get snake-eyes. In the first throw, about 10 players will leave the game, since there are 360 players and each one has a 1 in 36 chance of throwing snake-eyes. The number who will leave isn't necessarily 10—it could be any number between zero and 360, but usually around 10 will leave. As the game goes on, more and more players will leave. Because there are fewer players each time, the chances of someone in the group throwing snake-eyes go down. They remain 1 in 36 per person, but there are fewer players at each throw. Graphs of the number of players left after a certain number of throws are shown in Figure 17-4 for three games.

No two games are the same, yet the graphs have very similar shapes. The only noticeable difference is that the number of throws necessary to get the last player out varies considerably from game to game. In one game only 185 throws were needed, while another took 242 throws. How can we describe the length of the game using an average? The life of the longest-lived individual varies too much to be of value. Something that is fairly constant in all three graphs is the number of throws needed to get one-half of the players to leave the game; 27 in the first game, 25 in the second, and 23 in the third.

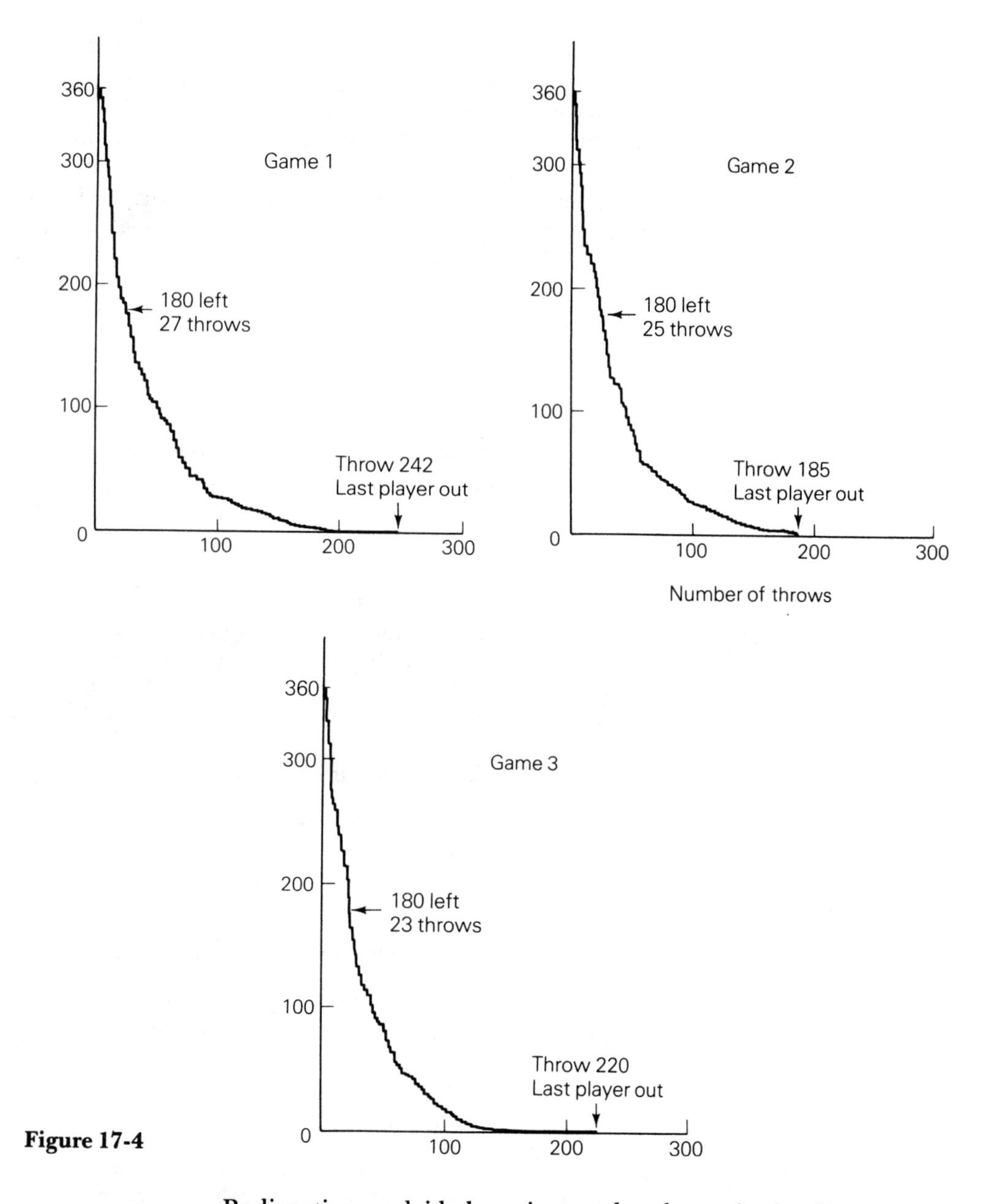

Figure 17-4

Radioactive nuclei behave just as the players in the dice game; the chances of any individual nucleus decaying in a certain time stay constant. The decay curve, the number left versus time, has the same basic shape as the curve in the dice game, which is called an *exponential* decay curve. Again, the time needed for every last nuclei to decay is very unpredictable, but the time for half of all the nuclei to decay is nearly constant as long as the number of remaining nuclei is fairly large. (This relation is always true for the radioactive sources used in medical applications.) The time it takes for one-half of the radioactive nuclei to decay is called the half-life.

Since the nuclei that remain after one half-life are identical in all respects to newly born nuclei, one-half of them will decay in the next half-life, leaving one-fourth of the original number. In the third half-life, one-half of that one-fourth will decay, leaving one-eighth of the original nuclei. This decay will proceed until so few nuclei are left that the laws of probability don't apply. This relationship is expressed in the following formula:

$$\text{amount remaining} = \text{original amount} \times \left(\frac{1}{2}\right)^n,$$

where n is the number of half-lives that have passed. Tables are available that give the value of $(1/2)^n$ when n is not a whole number, but in most cases you can make a satisfactory estimate by taking the nearest whole number.

EXAMPLE A certain cobalt-60 therapy irradiation machine took 2.5 minutes to give the proper amount of radiation for a tumor when the cobalt-60 source was new. How much time would be needed to deliver the same amount of radiation 10.5 years later? The half life of cobalt-60 is 5.25 years.

Answer The number of half-lives that have passed is

$$n = \frac{10.5 \text{ years}}{5.25 \text{ years}} = 2,$$

so

$$\begin{aligned} \text{amount remaining} &= \text{original amount} \times \left(\frac{1}{2}\right)^n \\ &= \text{original amount} \times \left(\frac{1}{2}\right)^2 \\ &= \text{original amount} \times \frac{1}{4}. \end{aligned}$$

Since only one-fourth of the source remains, the treatments would have to be 4 times longer than they were originally; 4×2.5 minutes $= 10$ minutes.

The half-lives of the various radioisotopes cover an enormous range, from less than a picosecond to billions of years. Whenever there is any chance of radioactive material spreading, as there always is when the material is injected into a patient, the short half-life isotopes are the safest. When a short-lived isotope is spilled, it can be safely handled by simply keeping people away until it decays to a safe level. When long-lived isotopes are spilled, every last speck must be picked up and accounted for—a very expensive procedure that involves ripping up floors, destroying contaminated furniture, and cleaning the entire area a few square centimeters at a time.

EXAMPLE Copper-64 and sodium-22 are two commonly used positron sources. Copper-64 has a half-life of 12.8 hours, while sodium-22 has a half-life of 2.6 years. If copper-64 was spilled in a room, how long would the room have to be sealed up for the source to decay to less than 1% of the original amount? If sodium-22 was spilled, how long would the room have to be sealed up?

Answer We want less than 1/100 of the source to remain. The two values of $(\frac{1}{2})^n$ that are closest to 1/100 are $(\frac{1}{2})^6 = 1/64$ and $(\frac{1}{2})^7 = 1/128$. To be safe, we will say that we must wait at least seven half-lives for the room to be safe. If copper-64 was used, this would take 7×12.8 hours $= 89.6$ hours $= 3.7$ days. This would be an inconvenience, but it would be less inconvenient than decontaminating the room. If sodium-22 was used for the same purpose, we would have to wait for 7×2.6 years $= 18.2$ years. Since this would be far too long to wait, the spilled radioactive material would have to be cleaned up at great expense.

I must point out that waiting seven half-lives is *not* a general rule for how long one must wait for an area to be safe. Hundreds of half-lives might be needed before handling a large source of a dangerous isotope was safe, but two or three half-lives might be enough for small amounts of relatively safe isotopes. The situation depends on how large the source is and how dangerous the isotope is.

Radiation Detection

As radiation passes through matter, it often strikes electrons and knocks them away from the atoms, thus ionizing the atoms. The ions can be detected in a number of ways, including photographically and electrically.

Photographic Film

Photographic film is made of specially treated silver compounds that react to light or other radiations. Exposed silver stays behind on the film, while unexposed silver is washed away by developing chemicals. Incidentally, used developing chemicals are a valuable source of silver.

Because photographic film is both sensitive to radiation and reasonably inexpensive, it is widely used for detecting and recording radiation. For instance, the vast majority of medical x-rays are made with a photographic film that is only slightly different from snapshot film.

Geiger Counters and Ionization Detectors

Ionization detectors and Geiger counters are reasonably inexpensive devices that gather ions produced by radiation. Generally, the counting tubes are filled with a gas so that the ions are free to move. The ions are gathered on highly charged electrodes, which produce a current. This current is then sent through amplifiers that are often sensitive enough to detect the passage of a single high-energy particle.

Because x rays and gamma rays don't produce many ions as they pass through the air, they cannot be detected very efficiently by Geiger counters and ionization detectors. Thus, these detectors cannot be used in any normal type of imaging device, such as a fluoroscope.

In medicine, these detectors are most commonly used as area survey meters, which measure the amount of radiation in an area and are used to locate spilled radiation.

Fluorescent Materials

The ions created by radiation don't stay ionized for long in most materials. The electrons quickly rejoin the ions and drop into their lowest possible orbits. In some materials, this activity will create visible light. Such substances are called *fluorescent materials* or *scintillators.*

Fluorescent screens have been used to watch x-rays of the motion of internal organs since the early days of radiology. While such fluoroscopic examinations have been extremely helpful in diagnosing internal illnesses, they require fairly large x-ray doses to produce a visible image. In addition to giving the patient a large radiation dose, the radiologist, who must sit close to the fluoroscope screen, also gets hit by scattered radiation. The amount received from one fluoroscopic examination isn't much, but the cumulative effect of several exams every day for several years can be hazardous to the radiologist.

In recent years, image intensifiers—very sensitive television cameras—have been made for watching fluoroscope screens. Be-

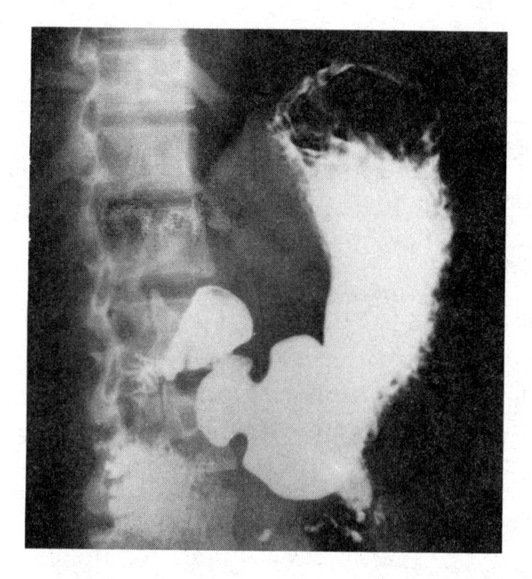

A fluoroscope of the stomach.

cause they are much more sensitive than the human eye, the patient is given a much smaller x-ray dose. Also, since the final image is on a TV screen, the radiologist can watch from beyond the radiation area. An additional advantage of image intensifiers is that the entire exam may be stored on magnetic recording tape, allowing it to be replayed for another person or to be viewed a second time for details that may have been missed.

Very low levels of radiation can be measured with a scintillation counter, which detects the light from fluorescent material with a photomultiplier tube (Figure 17-5). A photomultiplier tube is a very sensitive photoelectric tube that can detect the light from an electron dropping into a lower orbit in a single atom. A scintillation

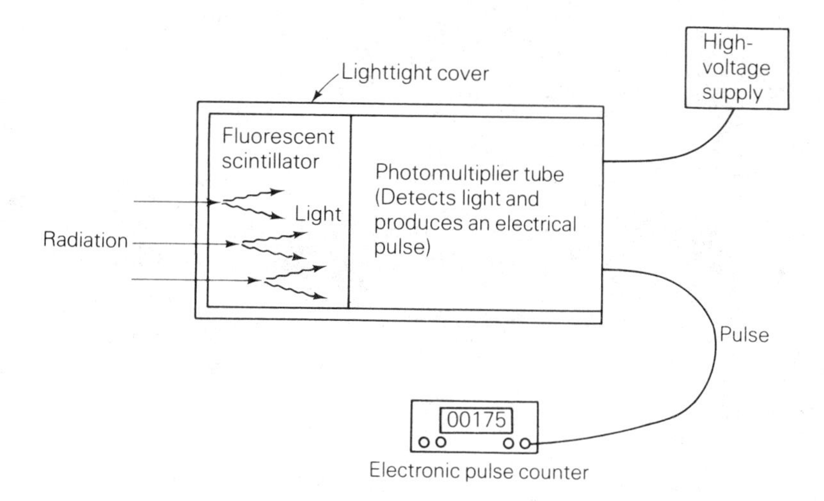

Figure 17-5 A scintillation counter.

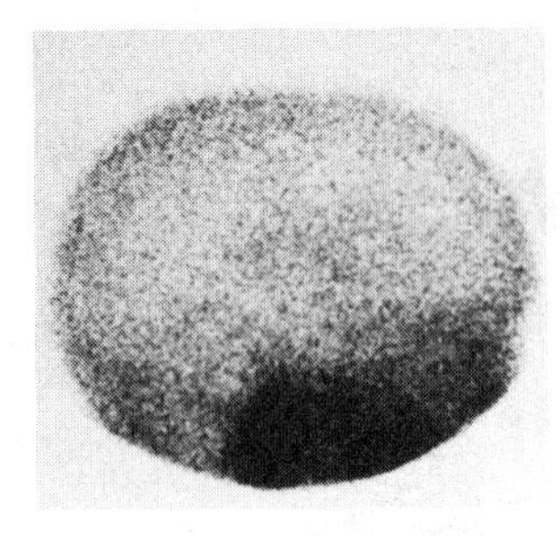

An Anger camera scan of the brain. Note the dark area, which indicates the presence of a tumor.

counter is often used to trace the motion of radioisotopes through a patient because the patient need only take in very small amounts of radioactive material.

When radioisotopes are used to locate and picture an organ, the radioactive material is located with a scanning camera. The first scanning camera was a single scintillation counter, which was passed back and forth over the body in a regular pattern until the entire organ was scanned. This required the patient to be immobilized for about an hour, which also made it impossible to watch the radioisotope as it moved through the patient to the organ.

The modern Anger scanning camera overcomes these problems by scanning the entire organ at once. The Anger camera contains a large scintillator that is placed behind a collimating screen. A collimating screen is a metal radiation shield in which holes are positioned so that only radiation coming directly from the section under the camera can enter. The scintillator is monitored by several photomultiplier tubes, which are hooked to a computer that can locate the position of the radiation by comparing the sizes of the signals received by the various photomultiplier tubes. A picture of the entire organ can be taken in only a few minutes.

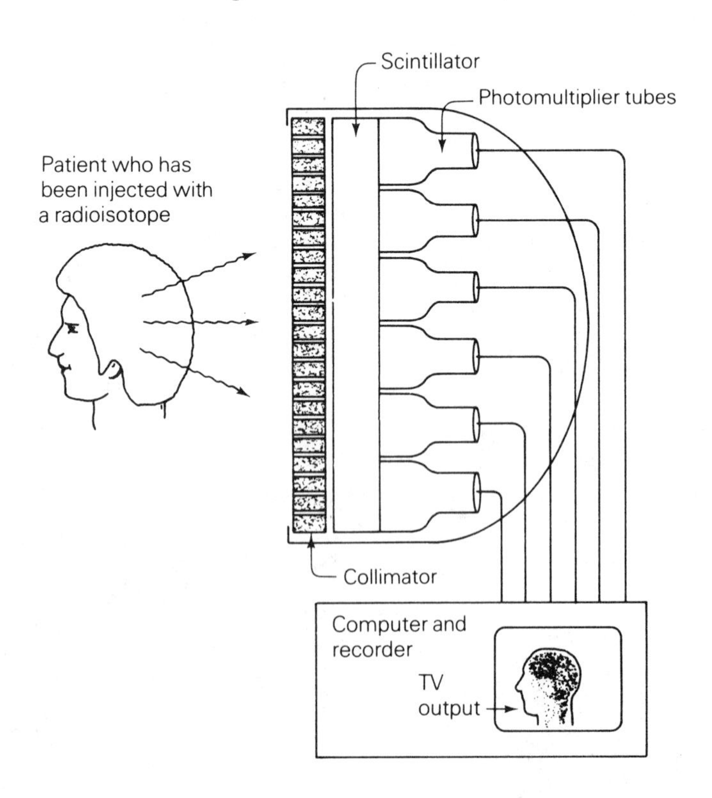

Figure 17-6 An Anger scanning camera.

Effect of Radiation on Tissue

The effects of high doses of radiation on tissue have been studied through animal experiments and investigations of accidents with humans. We know in detail the damage caused by moderate to high doses of radiation. The precise effects of very low doses, however, are poorly understood. Nor can we expect any sudden breakthroughs in understanding low-level radiation, because a very large test population is necessary to determine the effects, since they only show up in a small fraction of the population. For instance, suppose that a certain level of x rays was suspected of producing a tumor in 1 out of 10,000 rats. In order to get statistically reliable data, about a million rats treated with x rays would be needed, as well as a million more control rats that were not x-rayed. At about 10¢ per rat per day, it would cost $200,000 per day to take care of the rats. Since the rats would have to be kept for most of their lives, this experiment would cost about $200 million.

Until better ways are found to investigate low-radiation doses, our best estimates of their effects will be based on projections of high-dose data. These projections may or may not be right, but they will be conservative, because the damage is unlikely to be worse than the projections and it could be less. We simply don't know if low-radiation doses are beneficial or harmful; if they are harmful, we don't know how bad they are. As a result, it is probably safest to be conservative and not expose people to any more radiation than necessary.

The body can rapidly detect only two kinds of radiation: infrared rays and visible light. All other radiations will seriously injure the body long before any obvious signs appear. Since we have no built-in protection for these other types of radiation, we must always use radiation-emitting apparatus or material with extra care.

All forms of radiation alter tissue by releasing energy directly in the tissue, but the effects vary depending on the amount of radiation energy and how it is released. First, we will consider the effects of the various electromagnetic radiations in order of increasing radiation energy.

Microwaves and Radio Waves

Microwaves and radio waves are the lowest-energy electromagnetic radiations. Their energy is too low to knock electrons away from atoms, so they are classed as nonionizing radiation. These radiations do make the atoms vibrate more rapidly; that is, they heat the material through which they pass.

Since most nonmetals are semitransparent to radio waves and microwaves, these waves penetrate deeply into most substances, including tissue. This is why microwave ovens cook food so fast without burning—they produce heat uniformly throughout the food instead of dumping it onto the surface as an ordinary oven does.

The diathermy machine, used by physical therapists to relieve muscle and joint soreness, is basically a microwave transmitter. Since the microwaves penetrate deeply beneath the surface of the skin, a diathermy machine heats joints and muscles more effectively than external hot packs do. Such treatment is often very helpful in relieving pain and soreness.

The main dangerous effect of microwaves is overheating the tissue—killing it by cooking it. The eyes are particularly sensitive; overexposure of the eyes to microwaves will cause lens cataracts and blindness. As is the case with all types of radiation, the effects of low-level microwave exposures are not well understood; however, they don't seem to be too severe. As long as the exposure doesn't overheat the tissue, there doesn't seem to be observable permanent damage. However, when using microwaves you must keep in mind that the deep tissues of the body are not very temperature sensitive. They can easily be cooked long before the person notices that something is wrong. Of course, people with pacemakers should be especially careful around microwaves and other types of electrical interference that can disrupt the pacemaker's sensitive electronics.

Infrared Radiation and Visible Light

Infrared radiation and visible light are more energetic than microwaves, but they still don't have enough energy to ionize atoms. Thus, they can also damage tissue by overheating it. Neither infrared radiation or visible light is very penetrating unless the material happens to be transparent. Therefore, with the exception of the eye, they can only heat the upper layers of the skin.

Since the skin has sensitive infrared detectors and the eyes are very sensitive to visible light, there is very little danger of overexposure to either type of radiation. About the only common way that either type can hurt a person is if the person looks directly at the sun. The lens of the eye will concentrate the sunlight on a small spot on the retina, rapidly overheating and destroying that spot.

Ultraviolet Radiation

Ultraviolet (UV) radiation lies above visible light and below x rays in frequency and energy. Because it usually originates from electrons making moderately long jumps between atomic orbits, ultraviolet radiation has enough energy to knock outer electrons away from atoms and to ionize the atoms. UV radiation also has enough energy to break up the complex molecules necessary for life. Thus, ultraviolet radiation is quite damaging to exposed tissue. In fact, ultraviolet light is often used in sterilizers to kill bacteria. Ultraviolet light is also used in the operating room for sterilization. One drawback of ultraviolet sterilization is that the UV light will not penetrate into deep cracks or pores; thus, it is only effective on smooth-surfaced objects.

Ultraviolet light is not very penetrating, but it will penetrate to the living layers of the skin and cause reddening, peeling, and, in severe cases, burning and blistering—the classic symptoms of sunburn. Unfortunately, much of the damage that the sun does to the skin never heals completely. Exposure to the sun kills many of the skin cells, causing the skin to wrinkle prematurely; worse yet, it can trigger skin cancer. Every health professional should educate the public about the dangers of overexposure to the sun and convince people that a deep suntan is a very hazardous skin condition—not a mark of health. Of course, one should never look directly at an ultraviolet light, since the eyes are easily damaged by overexposure to UV.

Aside from some therapeutic uses, the only benefit of ultraviolet light is that it helps to synthesize vitamin D. However, the normal diet has so much vitamin D that the extra amount produced by the sun is of no benefit.

Ultraviolet light passes through water and water vapor to some extent—enough so that a light cloud cover provides little protection from the sun. Ultraviolet light will not pass through ordinary glass, although it will go through pure quartz glass, which is far too expensive to use in windows. Thus, when there is a window between you and the sun, there is little danger of skin damage or sunburn.

Most surfaces that reflect ordinary light also reflect ultraviolet radiation. Water and snow sports are particularly hazardous because both water and snow reflect ultraviolet radiation very efficiently. Thus, when you are on the water or snow, you must protect yourself from the reflected radiation as well as from the direct rays of the sun.

The atmosphere has a high, narrow layer that contains a lot of ozone (O_3), a gas composed of oxygen molecules that have three oxygen atoms. (The normal oxygen molecule, O_2, consists of two oxygen atoms.) Ozone absorbs ultraviolet radiation, so the ozone layer protects us, other animals, and the plants from the sun's radiation. There is a fear that fluorocarbons—which are used in refrigerators and air conditioners, and which were previously used in spray cans—will deplete the ozone layer and drastically increase the amount of ultraviolet radiation reaching the ground. The effect of such an increase is hard to predict, but it could be severe. Certainly the incidence of skin cancer would increase.

X Rays and Gamma Rays

Both x rays and gamma rays are more energetic than ultraviolet rays; thus, they are far more penetrating and can damage the deepest organs in the body. Both types of rays destroy tissue in the same way as ultraviolet rays—by ionizing atoms. Although all cells are damaged by x rays and gamma rays, the cells in the process of division (mitosis) seem to be much more sensitive than the others. This is the idea behind radiation therapy of tumors—radiation destroys the rapidly dividing tumor cells far more rapidly than the surrounding normal tissue cells. The cells of a fetus, and to a lesser extent those of an infant, also divide rapidly, so neither should be exposed to unnecessary radiation. Pregnant women should not work with radiation. Female x-ray technicians and radiation workers should stop working around radiation whenever they become pregnant. Some conservative radiation authorities urge female radiation workers to plan their families so that the fetus is not exposed to radiation from the moment of conception.

X rays and gamma rays cannot be focused by any known lens material, nor do they reflect from any surface. However, when they collide with electrons, they are randomly scattered in all directions (Figure 17-7). Because this scattered radiation is also hazardous, care must be taken to minimize one's exposure to it.

X-ray and gamma-ray sources are hard to shield because the rays are so penetrating. Although any material can be used for shielding, the denser the material is, the more efficiently it will stop these radiations. Concrete, water, and dirt are good shielding materials if the radiation source is not portable. Lead is used for most routine shielding because it is reasonably dense and not too expensive. When the shield must be as compact as possible, the superdense metal tungsten is used.

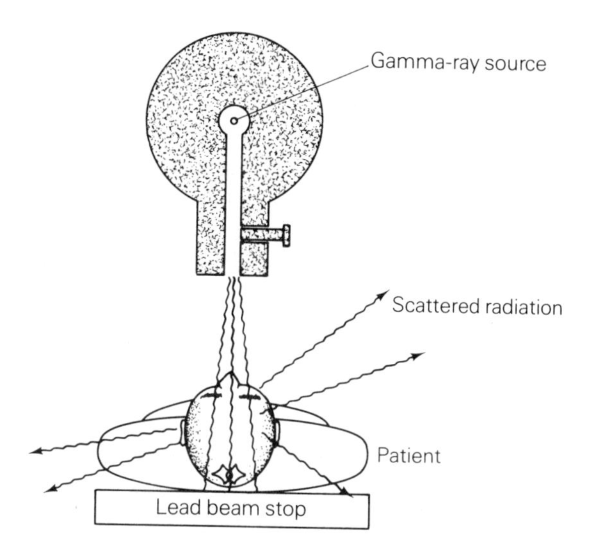

Figure 17-7 X rays and gamma rays are scattered in all directions when they pass through a patient.

Alpha and Beta Rays

Alpha and beta rays are particles, and they interact more strongly with electrons than electromagnetic radiation does. Because they interact so strongly, they don't penetrate very far. Consequently, damage is very concentrated along their path. Alpha-ray and beta-ray damage patterns are like the damage a heavy truck would do if it ran wild in a parking lot; electromagnetic radiation damage is more like that of an occasional rifle bullet shot into the parking lot. Both are damaging, but alpha rays are much more concentrated. Alpha and beta rays damage tissue by knocking electrons from atoms and ionizing them; in addition, they are heavy enough to knock atoms around and break up molecules.

Alpha rays can't penetrate through more than a few centimeters of air, nor can they penetrate the dead layers of the skin, so they are seldom used in medicine. Beta rays will penetrate about a millimeter of tissue, so they are used when concentrated radiation in a thin layer of tissue is needed. Both alpha- and beta-ray sources are very easy to shield; a metal can of normal thickness will suffice. However, many alpha- and beta-ray sources also give off gamma rays, in which case the shielding must protect against the most penetrating radiation, the gamma rays.

Radiation Damage and Dosage

The biological damage caused by radiation depends on the type of tissue and the type of radiation. The exact biological effects of a given type of radiation cannot be measured precisely, but a fairly good estimate can be made by simply assuming that the amount of damage is proportional to the energy deposited in the tissue. The unit of radiation dosage is therefore the amount of energy deposited per kilogram of tissue. In SI units, this unit is the joule per kilogram (J/kg), which is given the name **gray** (Gy) after a pioneer worker in the field of radiation effects on tissue.

Two older, non-SI units are still in common use: the rad and the roentgen (R). The rad is equal to 0.01 J/kg; that is,

$$1 \text{ rad} = 0.01 \text{ Gy} = 10 \text{ mGy}.$$

The roentgen is based on the amount of ionization radiation produced in air, which isn't closely related to tissue damage. The name has persisted long after the unit was actually used because 1 roentgen is very nearly equivalent to 1 rad. Thus, people still use the term *roentgen* to describe a radiation dose that is actually measured in rads. Soon most medical institutions will convert from rads to grays. The transition will demand extra caution to avoid errors caused by unfamiliarity with the units.

The short-term effects of large radiation doses have been studied in detail; the results for whole-body exposures are given in Table 17-1. Remember that the three common radiation dose units—the gray, rad, and roentgen—give the amount of radiation *per kilogram* of exposed tissue, not the total amount of energy deposited. Thus, the same *total* amount of energy would be needed to give a 1-kilogram organ a dose of 1 gray as would be needed to give a 100-kilogram person a dose of 0.01 gray.

During common medical procedures, the body receives from a milligray for soft tissue x-rays to several grays for tumor therapy. The doses of some representative procedures are listed in Table 17-2.

The long-term effects of radiation are hard to investigate, since a person must be observed throughout his or her lifetime. Even when this is done, it is hard to isolate radiation as the sole cause of a particular problem, because it is hard to find people who have never been exposed to any hazard except radiation. Much of our information on the long-term effects of radiation comes from the histories of early radiation workers, radiologists, and some of their patients who were frequently exposed to overdoses of radiation.

Large doses of radiation are definitely carcinogenic. Leukemia and malignant tumors are common among people who have been overexposed to radiation. However, as is the case with most carcinogens, there is no way to predict who will become diseased and who will not from overexposure. The exposure levels required to increase the risk of cancer noticeably are quite high, around 5 grays

Table 17-1 Short-Term Effects of Whole-Body Exposure to Radiation

Dose		*Probable Effect*
Gy	*rad*	
0–0.25	0–25	No apparent injury.
0.25–0.5	25–50	Blood cells show changes; no serious injury.
0.5–1.0	50–100	Changes in blood cells; some injury, but no permanent disability.
1–2	100–200	Injury, vomiting in about half the cases during first day after exposure. Possible disability, but most people recover within a few weeks.
2–4	200–400	Injury and disability certain. Vomiting in most cases during the first day. Hair loss in two weeks. Death possible. Four gray (400 rads) will be fatal within a few months to half the cases.
4–6	400–600	Death is highly likely. Above 6 grays (600 rads) there is little chance of survival.

Table 17-2 Doses of Some Common Radiological Procedures

Procedure	*Dose*	
Chest x ray	1 mGy	100 mrad
Lower back x ray	20 mGy	2 rad
Gastrointestinal fluoroscope exam		
Without image intensifier	500 mGy	50 rad
With image intensifier	200 mGy	20 rad
Iodine-131 thyroid scan	750 mGy	75 rad
Tumor therapy	60 Gy*	6,000 rad*

*These large doses are intended to be lethal to the malignant tumor tissue. The radiation is confined to the tumor as closely as possible to minimize damage to the surrounding tissue. However, complete confinement of the radiation is rarely possible; some normal tissue is nearly always injured.

(500 rads). Whether very small doses cause cancer has not been determined, because the number of such tumors may be too small to detect among all the other tumors. Since this question has not been answered, radiation safety standards are based on the conservative assumption that unnecessary small doses should be avoided.

Genetic damage is another long-term effect of radiation. Radiation can alter the genes, causing mutations that might show up in the next generation or several generations later. Although large radiation doses undoubtedly produce genetic damage, the hazards of low-level radiation are still relatively unknown. Again, the only safe practice is to be extra careful with radiation. Exposing the gonads of any person of reproductive age to radiation should be kept at an absolute minimum. Those people whose gonads are irradiated should be urged to not conceive children for several months.

It is unlikely that there are any serious unknown genetic effects of radiation, but the possibility cannot be ruled out until humans have used radiation for several centuries. In order to prevent irreversible damage to the human race from this very remote possibility, the exposure levels for the general population are about 10 times lower than those for radiation workers. In this way the consequences of any unsuspected genetic damage will be confined to a small part of the population. Since the standards for radiation workers are low enough to avoid any serious risk from known genetic effects, the standards for the general population should prevent any unforeseen hazards.

X-Ray Radiography

There are no lenses that focus x rays; x rays travel in straight lines until they are absorbed or scattered. Consequently, the images on x-ray radiographs are just the shadows of the internal organs. However, the shadows are not pure black and white, because many organs partially absorb the x rays and so create shadows that are various shades of gray.

X-Ray Tubes

X rays are usually produced by slamming high-energy electrons into a metal block. As the electrons quickly slow down in the block, they emit x rays and also knock electrons out of their orbits. As the electrons drop back to their proper orbits, the atoms emit more x

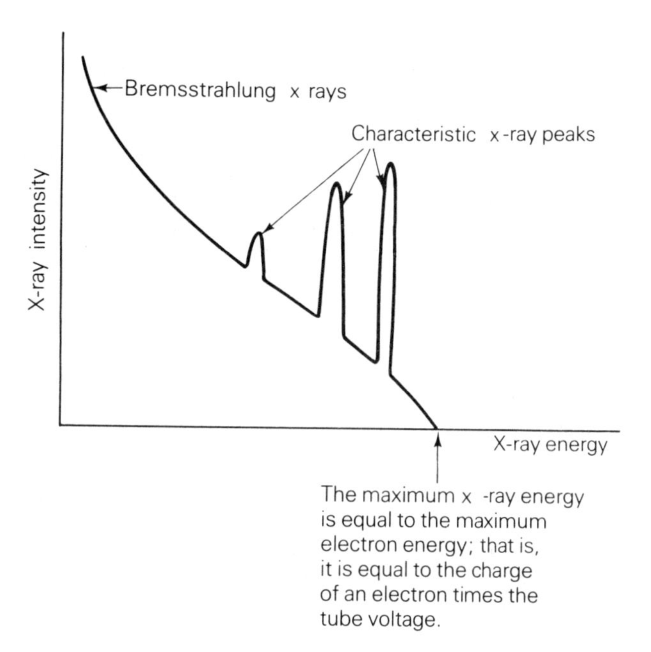

Figure 17-8 A typical x-ray tube spectrum.

rays. The x rays produced by the electrons slowing down are called *bremsstrahlung,* which means stopping radiation in German; those produced by the electrons dropping into lower atomic orbits are called *characteristic x rays.* An energy spectrum from a typical x-ray tube is shown in Figure 17-8. The bremsstrahlung part of the spectrum looks about the same for most target metals, with the maximum x-ray energy equal to the energy of the electrons. The energy of the characteristic peaks depends on the target material. This fact explains the origin of the term: The location of the peaks is a *characteristic* of the atomic shell structure of the target material, and it can be used to identify the material.

The electrons start in the x-ray tube from a hot wire filament, called the cathode. The electrons are "boiled off" by the vibrating atoms in a process known as thermionic emission. Other electrodes near the cathode focus the electrons into a tight beam, which is then attracted toward the metal target, called the anode. The anode is at a high positive voltage with respect to the cathode, so the electrons are accelerated to a very high energy as they travel to the anode. Since the electrons produce a great deal of heat as well as x rays when they slam into the anode, the anode is made from a high-melting-point metal, such as tungsten, and it is cooled by water, oil, or air. The entire assembly is enclosed in a glass tube from which the air has been pumped so that the electrons have a free path (Figure 17-9).

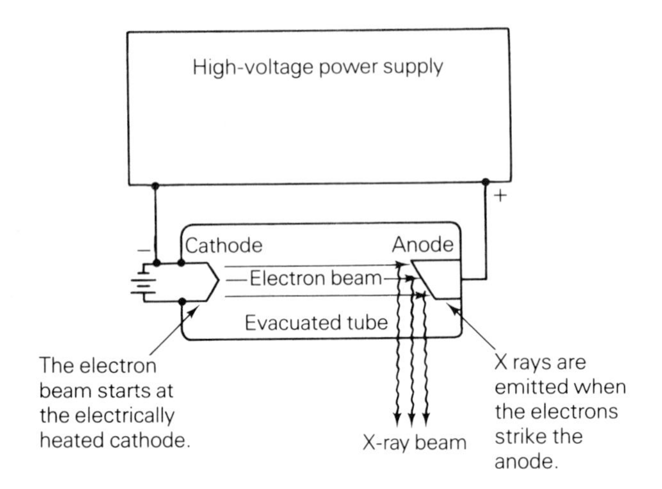

Figure 17-9 An x-ray tube and power supply.

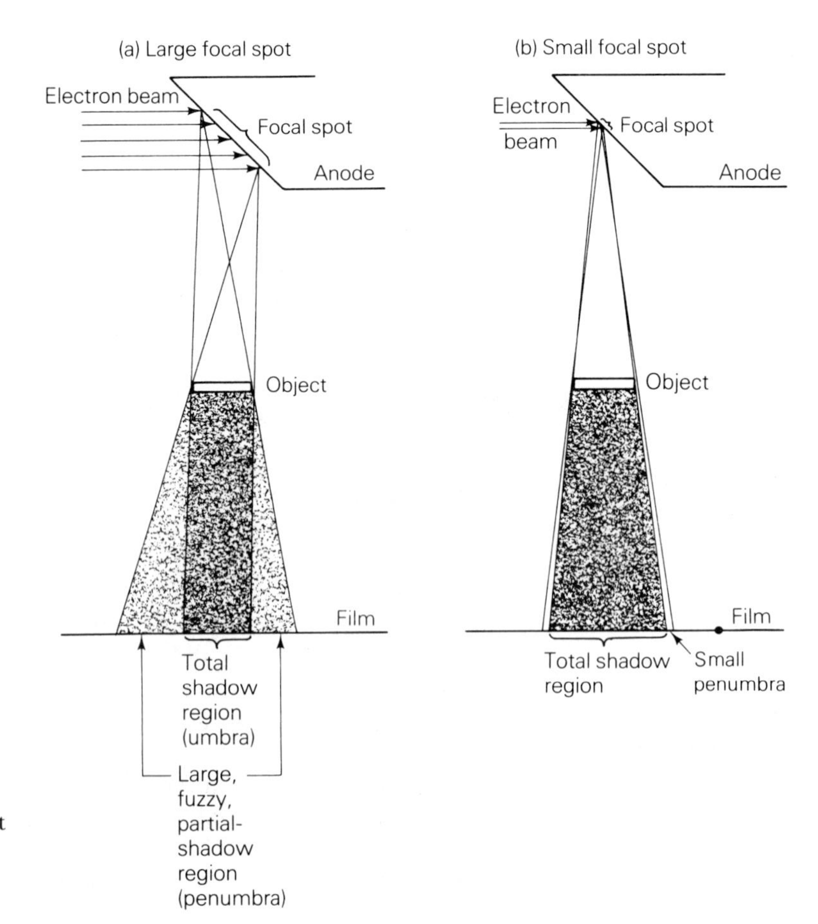

Figure 17-10 An x-ray tube with a small focal spot is needed to make sharp pictures.

Since x-rays are shadow pictures, the spot on the anode, called the *focal spot,* should be as small as possible. The smaller it is, the sharper the pictures of the organs. The effect of the size of the focal spot on image sharpness is shown in Figure 17-10.

To produce sharp pictures in diagnostic x-ray machines, the electron beam is concentrated on a very small spot on the anode. This focal spot will quickly overheat if the anode is not cooled between exposures. Many different schemes have been used to do this. One is to rotate the anode rapidly so the electrons don't always strike the same spot. However, anode overheating is still a problem in most diagnostic x-ray machines.

X-ray machines used for radiation therapy usually have a much bigger focal spot than diagnostic machines. Consequently, cooling the anode properly is not a problem. Of course, x-ray therapy machines will not produce sharp images either, but this is not important since the images from therapy machines are only used to help line up a patient for treatment.

X-Ray Machine Controls

X-ray machines have three basic controls: the voltage control, the electron current control, and an exposure timer. The voltage controls the energy of the electrons, so it also controls the energy of the x-ray beam. Of course, the more energetic the x rays are, the more penetrating they will be; thus, the voltage control also controls the penetrating power of the x rays. The current control regulates the number of electrons striking the anode, so it also controls the number of x rays in the beam; in other words, it controls the x-ray beam intensity. The current control has no effect whatever on the penetrating power. The timer controls the exposure time.

The same exposure can be obtained with either a high current and a short exposure time or a low current and a long time. The two conflicting factors that influence an x-ray technologist's choice are that a patient's wiggling will ruin the picture during a long exposure and that a high current will overheat the tube during a short exposure. Since patient wiggling is almost impossible to stop, most exposures are made with as short an exposure time as possible without damaging the tube.

Some of the factors that influence an x-ray technologist's choice of voltage and current are given in the section on x-ray technology in the last chapter.

Filters

Many of the bremsstrahlung x rays have a very low energy (see Figure 17-8). Thus, they have very little penetrating power and almost never get to the film. Consequently, they don't add any information to the radiograph; their sole contribution is to add to the patient's radiation exposure. In order to reduce this exposure, filters are often placed in the x-ray beam to cut down these low-energy x rays. X-ray filters are usually made from thin metal foils.

Collimators

Most x-ray machines are equipped with a collimator—a shield with a hole that confines the beam to a specific region. Many machines have adjustable collimator windows so that the beam can be confined to the part of the patient being radiographed.

Film Screens

X rays blacken ordinary photographic film, but not very rapidly. In order to improve the sensitivity of the film and cut down the patient's radiation exposure, a screen is usually added to the film. The screen consists of a layer of fluorescent material similar to the

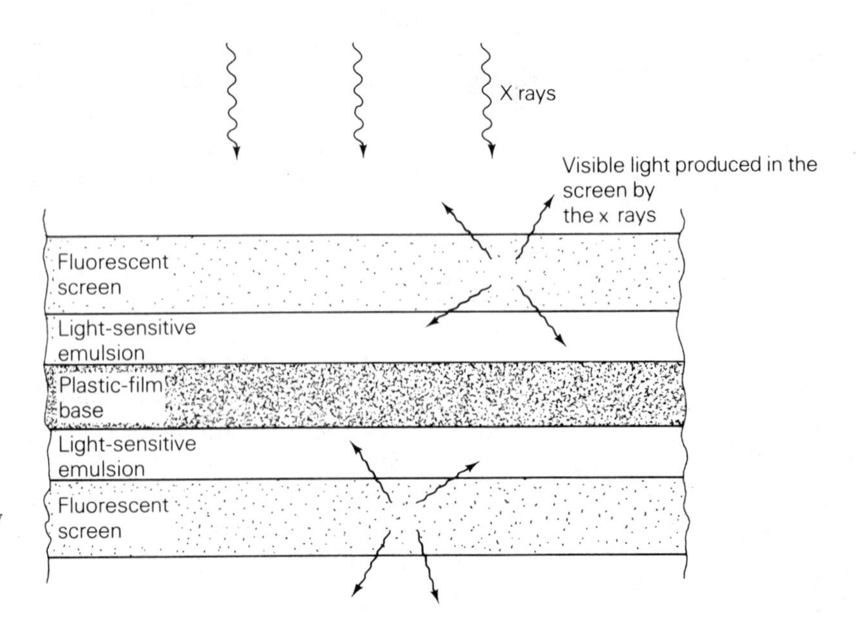

Figure 17-11 Fluorescent screens make x-ray film much more sensitive, thus cutting the patient's exposure.

material used in fluoroscope screens. When the x rays strike the screen, the screen gives off visible light that blackens the film faster than the x rays alone (Figure 17-11). This visible light is responsible for more than 90% of the film blackening, with direct x-ray exposure contributing the rest. Thus, screens play a very important part in keeping patient exposures to a minimum.

There doesn't seem to be any one perfect screen material. The screens that are most sensitive to x rays lack resolution—that is, they yield fuzzy images—while those with high resolution require large x-ray exposures.

Grids

As the x rays pass through the body, some are completely absorbed; however, many are merely scattered out of the original beam. These x rays do not carry any useful information, but they can still fog the film. This fogging reduces the contrast between the various organs and makes the film harder to interpret. In order to stop the scattered x rays and yet allow the direct beam through, a grid is often used. A grid consists of hundreds of lead strips with narrow

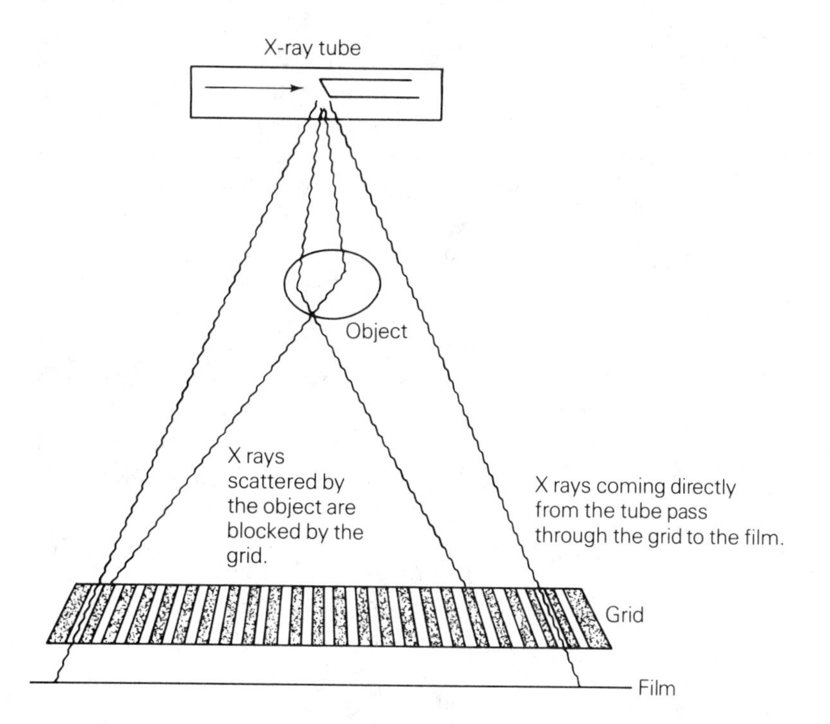

Figure 17-12 Grids cut down film fogging by scattered x rays by admitting only those x rays that come directly from the tube.

slots in between that point back toward the focal spot of the x-ray tube. The direct rays from the tube pass through the slots, but scattered x rays coming in at an angle hit the lead strips and are absorbed before getting to the film (Figure 17-12). Of course, the lead strips leave a shadow on the film and some information is lost; however, this effect can be minimized by moving the grid slightly during the exposure, which allows every part of the film to be exposed. Since the lead grid absorbs some x rays, a higher-intensity x-ray beam is needed to expose the film properly. As a result, the patient must be exposed to more radiation, but the added clarity and detail are often worth the small additional exposure. One note of caution on the use of a grid: Because lining up several hundred slots takes a lot of work, grids are expensive. They are also easily bent, so treat them with care.

Tomography

Imagine that you were standing on one side of a picket fence trying to see what was on the other side. You would only be able to see through the slots; the areas blocked by the boards would remain unseen. However, when you drive by the same fence, you can see everything behind the fence because *all* of the scene behind it is visible *part* of the time. The image isn't as sharp, because the fence appears as a blur that degrades the optical quality of the scene behind. You can see the same effect by looking through your outstretched fingers and waving your hand back and forth rapidly.

A similar problem occurs in x-raying a deep organ of the body: The shadows of other organs can block the view. The solution is similar to driving by a picket fence. The x-ray tube and film are moved so that only the particular organ appears to be stationary. Because the other organs appear to be in motion, their images don't appear on the film. This process is called *tomography.*

There are various methods of moving the film and x-ray tube, but you can understand the basic principle of tomography if you imagine that the tube is placed on one side of a wheel, the film on the other, and the organ to be x-rayed at the center (Figure 17-13). During the exposure, the wheel is rotated about a quarter turn.

The quality of the tomographic image isn't as good as a stationary x-ray would be if the other organs were removed, but if you could remove the other organs, you wouldn't need x-rays in the first place.

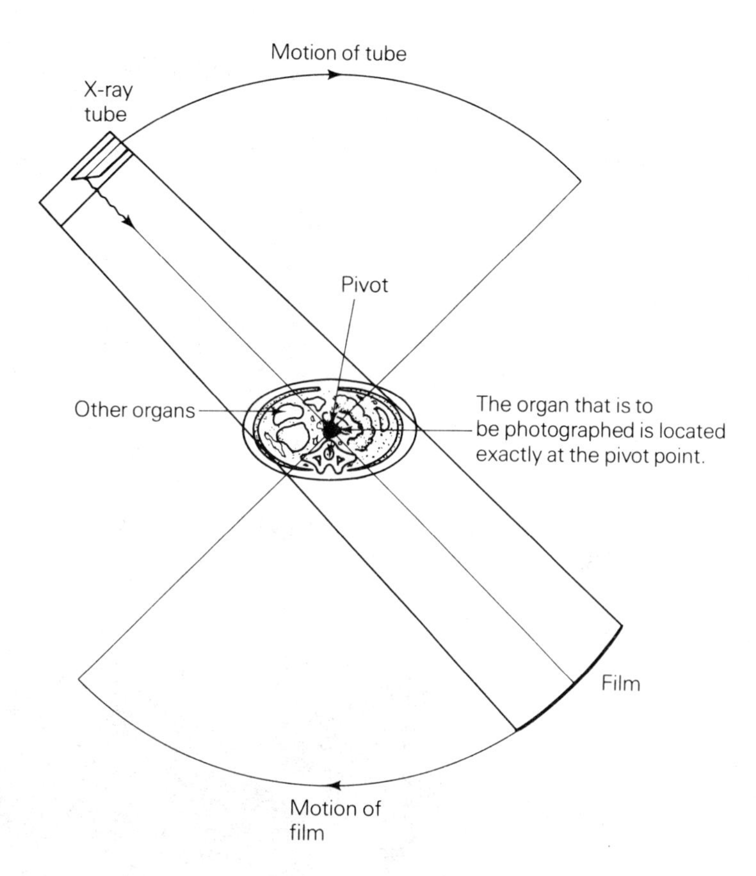

Figure 17-13 Tomography. By moving the x-ray tube and the film together around a pivot point while the exposure is made, only the organ at the pivot shows up clearly on the picture.

Computerized Axial Tomography (CAT)

Suppose you had a drill that could bore through a patient without serious harm. If you drilled the patient in hundreds of places and kept the drill shavings from each hole separate, you could eventually figure out where every organ was located (although you would probably need a computer to help reconstruct the image of the patient). No one shaving would tell you much, but with the information from hundreds of holes, you could accurately locate all the organs of the body.

Computerized axial tomography (CAT) uses the same idea. A pencil-thin beam of x rays is sent through the body to a detector, which records the percentage of x rays that go through the body completely. One beam alone doesn't provide much information. But the beam is eventually sent in almost every direction along the slice of the body being imaged. This information is then analyzed

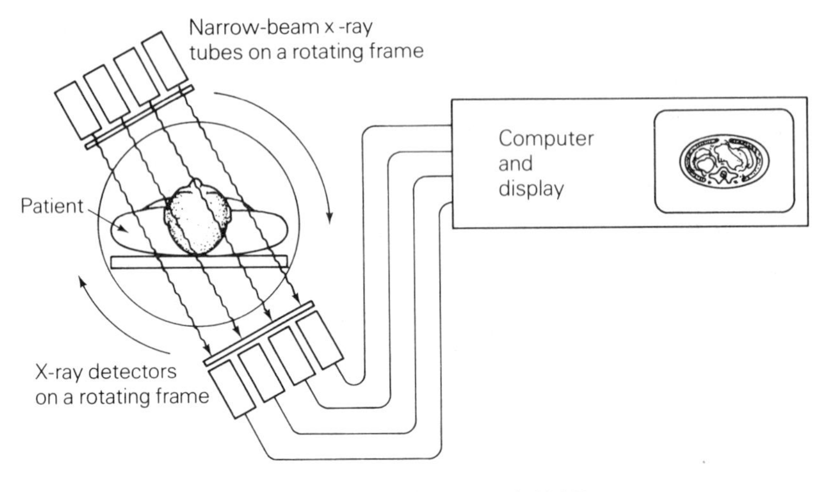

(a) A computerized axial tomograph (CAT)

Figure 17-14

(b) A CAT image of the body

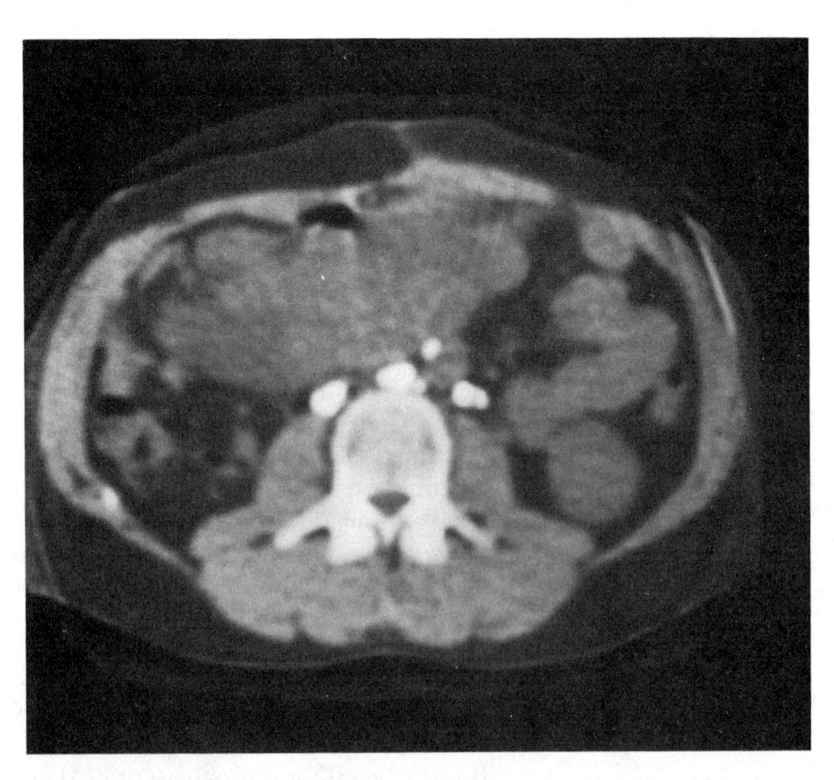

A CAT scan of a patient with suspected lymphoma. The image of the lower abdomen shows the spine muscles around the spine, and parts of the bowel (on the right). The white dots above the spine are lymph nodes that were dyed during lymphangiography.

by a computer, which reconstructs the image of that slice of the body (Figure 17-14).

Note that CAT images are completely different from ordinary x-rays or tomographs, in that the CAT image is a slice of the body parallel to the beam direction, not a shadow perpendicular to the beam direction. Thus, CAT images of organs can be made in directions that are impossible with ordinary methods. For instance, the image of the abdomen shown in Figure 17-14 could not have been made with an ordinary x ray. If it weren't for the dangers of radiation exposure, a CAT scanner could construct a three-dimensional image of the entire body.

Radiation Therapy

Because malignant tumors are more susceptible to radiation damage than healthy normal tissue, radiation therapy has proven to be a valuable supplement to surgery in the treatment of cancer. Ordinary x rays are used for shallow tumors, and the more penetrating gamma rays are usually used for deeper growths. The gamma rays can be produced by radioisotopes such as cobalt-60, or they can be artificially produced with "supervoltage" electron accelerators, which slam very high energy electrons into a metal target. Accel-

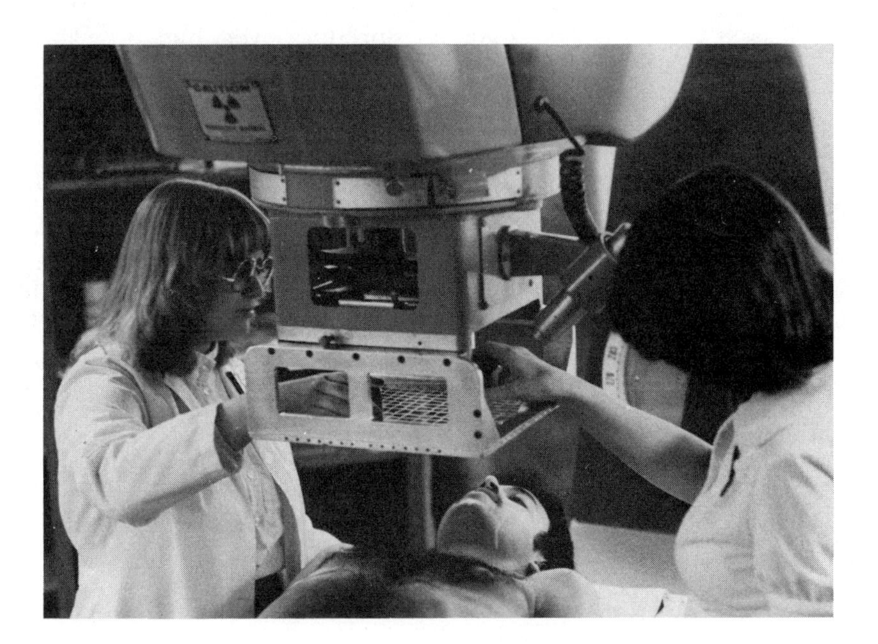

Cobalt-60 radiation therapy is very effective in the treatment of many types of cancer.

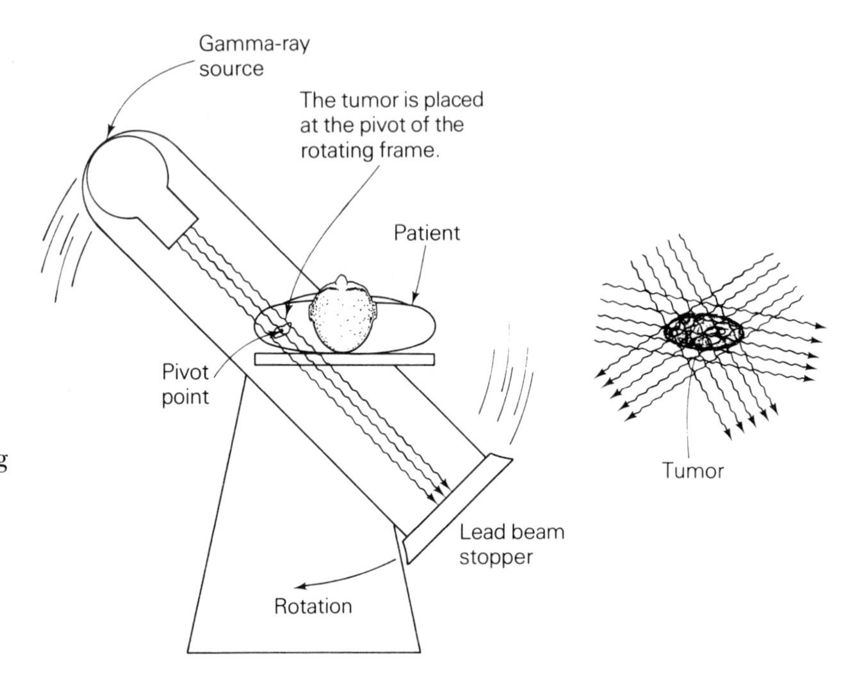

Figure 17-15 By sending the gamma rays at the tumor from many directions, the radiation dose at the tumor can be many times higher than the dose at the surrounding tissue.

erators have the advantage of controlling the gamma-ray energy, but cobalt-60 sources have the advantage of not needing much complicated equipment or maintenance—just shields with a beam shutter for the gamma rays. Isotope sources can also be surgically implanted directly at the tumor site—a method that can sometimes produce a very high dose at the tumor without exposing the surrounding tissue much.

Many radiation therapy machines can be rotated so that the radiation can be sent from different directions. This allows the dose at the tumor to be several times higher than the dose at the surrounding healthy tissue (Figure 17-15).

The operator of a cobalt-60 therapy unit should always remember that the source cannot really be turned off. The off position merely closes a metal shutter to confine the radiation from the source, which is still as potent as ever. These shutters are very well designed (Figure 17-16). They shut automatically when the electricity fails, there are several safety interlocks to prevent accidental opening, and they normally shut with no problem. However, on rare occasion, they don't shut when they should. Despite the very small chance of shutter failure, the careful operator should always double-check the shutter with a Geiger counter or some other means.

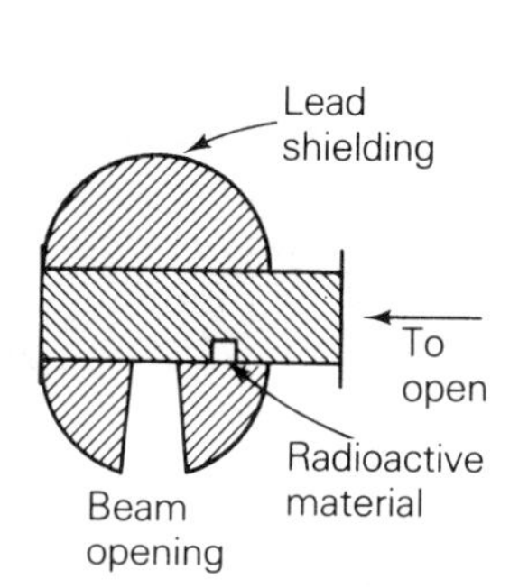

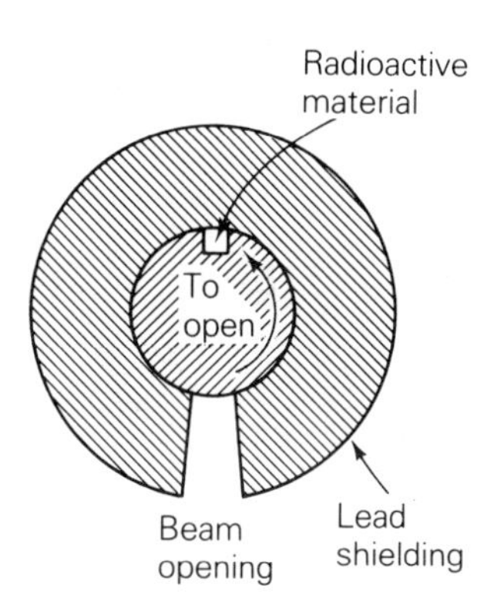

Figure 17-16 Two shutter designs for radioisotope therapy machines.

Radioisotopes

The relatively inexpensive radioisotopes made in nuclear reactors have opened up many new possibilities for the detection and treatment of diseases. New methods are continually being developed as a result of improvements in cameras and radiopharmaceuticals.

Radiopharmacists are always looking for new ways to tag drugs with radioisotopes so that the isotope will go to the desired organ. Tumors and other pathological tissue often absorb radiopharmaceuticals at a different rate than healthy tissue. If the isotope emits gamma rays, which escape the body easily, the tissues that have absorbed the isotope will show up as hot regions on a scanning camera. Liver tumors can be located in this way. Ultrafine gold powder injected into the bloodstream will be absorbed mainly by healthy liver tissue, but not liver tumors. When radioactive gold-198 is mixed with ordinary gold, healthy liver will show up as a hot region on the scanning image, while a tumor will show much lower activity. Thyroid abnormalities are found in the same way by using radioactive iodine-131. Abnormalities in the brain, lungs, kidneys, and heart are also diagnosed with radioisotopes.

Radioisotopes can be used for therapy when the isotope can be put into a pharmaceutical that is concentrated by the organ to be irradiated. Finding such radiochemicals is one of the major challenges of isotopic radiotherapy. If a radiopharmaceutical was found that was absorbed only by malignant tumors, radiation tumor therapy would become a far more effective cancer treatment.

A typical therapeutic use of radioisotopes is the reduction of thyroid activity with iodine-131. The iodine is quickly concentrated in the overactive thyroid, and the radiation from the iodine-131 destroys some of the tissue. This reduces the thyroid's activity to a normal level if the proper amount of iodine was administered. Since the thyroid concentrates the iodine rapidly, the other parts of the body receive only a minor radiation exposure. Since there is no incision, as there would be with surgery, the healing process is rapid and the chances of infection are very small.

Radiation Safety

The enormous benefits of radiation and radioisotopes are not without risks. But after all, drugs, electricity, automobiles, surgery, and airplanes have risks, too. By pointing out all the possible hazards of radiation, we may have given the impression that radiation is hard to use and control. This is not the case at all. With care, radiation is no more dangerous than electricity or strong chemicals. But as in the case of electricity, we have no built-in senses to warn us of radiation hazards and must use instruments to detect danger. How radiation safety practices affect patient care are discussed in the section on nursing in the last chapter.

Allowable Dose Levels

The allowable whole-body radiation dose for radiation workers is 50 milligrays per year (5 rads/year) with a maximum of 30 milligrays (3 rads) in a 90-day period. This is equivalent to 1 milligray per week (100 mrad/week). (The allowable doses for the general population are generally one-tenth of the corresponding doses for radiation workers.) Since young people are more sensitive to genetic damage than adults, 18 is the minimum age for a radiation worker.

Because the hands and forearms are not as sensitive to radiation as the trunk of the body, the hands may be safely exposed to higher levels. The allowable dose to the hands is 15 mGy/week (1.5 rad/week), 15 times higher than the allowed body exposure.

These levels are reasonably conservative. They are only 50 times higher than the small amount of natural radiation received from cosmic rays and natural radioisotopes, and they are far below the exposure needed to definitely increase the chances of tumors. If a person was exposed to these levels every day over a 40-year career, the total accumulated dose would only be 2 grays (200 rads)—which usually doesn't cause permanent injury even when received in a single dose. Nevertheless, the allowable doses should be treated as maximums.

Furthermore, every reasonable precaution should be taken to keep the exposure *as small as possible*. The handling techniques of a radiation worker who regularly exceeds 10% of the allowable dose should be investigated to see if the exposure can be reduced. Nine times out of ten the dose can be cut down. A task that involves a large radiation exposure should not be assigned to one individual but rotated among several people.

EXAMPLE While handling a certain radioactive source, a worker will be in an area where the radiation level is 5mGy/hr (500 mrad/hr). What is the maximum length of time that can be spent in that area?

Answer This question has no definite answer. The answer depends on the urgency of the task and whether the worker will be exposed to other radiation later on.

The dose is found by multiplying the radiation level and the exposure time:

$$\text{dose} = \text{radiation level} \times \text{exposure time},$$

so the time would be

$$\text{exposure time} = \frac{\text{dose}}{\text{radiation level}}.$$

If the worker had to do this task once daily (five exposures per week) and the total exposure was to be less than 10% of the allowable dose, the time to do the task once would have to be less than

$$\begin{aligned}\text{time} &= \frac{0.10 \times 0.001 \text{ Gy/week}}{5 \text{ exposures/week} \times 0.005 \text{ Gy/hr}} \\ &= 0.004 \text{ hr} \\ &= 14.4 \text{ s}.\end{aligned}$$

If the worker only had to do the task once per week and had no other significant radiation exposure, one performance of the task could take 5 times longer, 72 seconds.

Suppose the task was essential, but it couldn't be done in 72 seconds and the radiation could not be cut in any reasonable way. If it had to be done once a week and the worker was allowed to receive the maximum dose, the exposure time would be:

$$\begin{aligned}\text{exposure time} &= \frac{0.001 \text{ Gy}}{0.005 \text{ Gy/hr}} \\ &= 0.2 \text{ hr} \\ &= 12 \text{ min}.\end{aligned}$$

However, the worker could not be exposed to any further radiation for that week. Certainly, the task should be rotated among the radiation workers so that the exposure would not be concentrated on one person.

If the task was an essential emergency task, the worker is allowed to receive a full quarter-year dose, 30 milligray, in one exposure. The working time would be:

$$\text{exposure time} = \frac{0.030 \text{ Gy}}{0.005 \text{ Gy/hr}}$$

$$= 6 \text{ hr.}$$

That worker could not be exposed to any more radiation for 90 days and then only to small amounts (less than 20 milligray total) for the rest of a year. Only under an extreme emergency would someone be allowed to be exposed to such a large dose.

Radioisotope Safety

The SI unit for measuring the strength of a radioactive source is the **becquerel** (Bq), named in honor of the discoverer of radioactivity. A source of 1 becquerel averages a decay of 1 nucleus every second. Thus, a kilobecquerel source averages a decay rate of 1,000 nuclei per second, a megabecquerel source would have an average of 1 million decays per second, and so on. The word *average* is used because radioactive decay is a completely random process, and so it cannot be predicted exactly.

An older non-SI unit of source strength is the **curie** (Ci). Originally, 1 curie was defined as the equivalent of the activity in 1 gram of radium, but it was later redefined as a source averaging 37 billion disintegrations per second. Thus,

$$1 \text{ Ci} = 3.7 \times 10^{10} \text{ Bq.}$$

For a particular radioisotope, the average decay rate is directly proportional to the number of atoms present. Thus, the strength of a source is directly proportional to the number of atoms, which means that the source strength decays at the same rate as the number of atoms. That is, after one half-life, one-half of the source remains *and* the source has one-half its original strength. After two half-lives, one-fourth of the source remains and it has one-fourth its original strength, and so on.

The radiation exposure produced by a particular source varies among isotopes. A rule of thumb for most gamma-emitting isotopes is that a 1-gigabecquerel source will have a radiation level of around 200 μGy/hr at a distance of 1 meter. Based on an allowable weekly exposure of 1 milligray, such a source could be handled for a few hours per week at a distance of 1 meter. Sources in the low megabecquerel range would have radiation levels of around 2 μGy/hr at 1 meter, so at that distance such sources could be handled indefinitely without exceeding the allowable doses.

This rule of thumb expressed in terms of curies and rads states that a 1-curie gamma emitter has a radiation level of about 1 rad/hr at a distance of 1 meter. Since the allowed dose is 100 millirads a week, a 1-curie source can only be handled for a few minutes; sources in the millicurie range must be handled in minutes; and sources in the microcurie range can be handled for long periods of time if necessary. Of course, the handling time should always be kept at a minimum, no matter how small the source is.

The radiation level drops off very rapidly as you move away from a source. If you moved twice a given distance from a source, you would receive only one-fourth as much radiation. This is because your apparent height h and width w, as seen from the source, are both cut in half by moving twice as far away. Thus, your new apparent area is $(\frac{1}{2}h \times \frac{1}{2}w) = \frac{1}{4}hw$. In other words, the radiation level is inversely proportional to the *square* of the distance from the source:

$$\text{radiation level} = \text{radiation level at 1 meter} \times \frac{1}{r^2},$$

where r is the distance from the source in meters. This relation tells us two important things about handling radioactive material. First, when you handle a source directly with your fingers, the radiation levels are very high. Second, distance is one of the best (and cheapest) ways to reduce radiation from a source. When you are 10 meters from a source, the radiation level is only 1% of the level at 1 meter.

EXAMPLE Roughly what is the radiation level 1 centimeter, 1 meter, and 20 meters away from a 5-gigabecquerel gamma ray source?

Answer Using the rule of thumb that a 1-gigabecquerel source has a radiation level at a meter of 200 μGy/hr, a 5-gigabecquerel source would have a radiation level of 1 mGy/hr at 1 meter. Thus, at 1 centimeter,

$$\begin{aligned}\text{radiation level} &= \text{radiation level at 1 meter} \times \frac{1}{r^2}\\ &= 0.001\ \text{Gy/hr} \times \frac{1}{(0.01)^2}\\ &= 10\ \text{Gy/hr.}\end{aligned}$$

At 20 meters,

$$\begin{aligned} \text{radiation level} &= 0.001 \text{ Gy/hr} \times \frac{1}{(20)^2} \\ &= 0.0000025 \text{ Gy} \\ &= 2.5\ \mu\text{Gy}. \end{aligned}$$

Since the allowed weekly exposure is 1 milligray, this source could be handled for an hour at a distance of a meter. At 1 centimeter, the radiation level is so high that the source could be handled for less than a second; only an extreme emergency would justify the risk. At 20 meters, the radiation level is low enough that a worker could safely work with the source for a 40-hour week.

The basic ideas behind radiochemical safety are (1) don't spill any radioactive materials and (2) lay out the work area as if a spill were certain. The entire work area should be covered with absorbent paper having a waterproof backing to catch any spilled material. Before you leave the work area, you must account for all the radioactive material and be sure that none has been spread to your gloves, clothes, or shoes.

If a spill should occur, try to confine it to a small region to minimize cleanup problems. Since by backing away from the source, you can almost always cut the radiation to a harmless level, there is seldom any reason to rush when working with radioactivity. If you are having difficulty, just step away from the source and think out the problem without hurrying.

Film Badges and Survey Meters

X-ray technicians and radiation workers always wear a small badge that monitors their radiation exposure. The badge consists of photographic film and selective filters that absorb some types of radiation while letting others pass. The amount and type of radiation are determined from the blackening of the developed film. These badges are always worn somewhere on the trunk of the body. When radioactive material is handled with the hands, separate wrist and ring badges are also worn.

Film badges have one drawback: They don't indicate the amount of radiation exposure until *after the badge is developed,* which often occurs a month after the exposure. To measure the radiation levels *before* exposure, survey meters are used. Survey meters are hand-held Geiger counters or ionization counters that measure the radiation level in an area, usually in μGy/hr or mrad/hr. Survey meters are also used to check for spilled radioactivity, and they should always be used to check the work area and the worker after handling radiochemicals. Careful radiation workers never leave their survey meter very far away, using it to check whenever they have the slightest suspicion that radiation is present. In areas where there is a high chance of exposure to radiation, workers wear pocket survey meters. These are always turned on, and they sound a warning whenever the radiation exceeds a preset level. Some even record the worker's total exposure, beeping whenever the worker gets an additional milligray and sounding an alarm when the maximum allowable exposure is reached.

Radiation Shielding

Since the rays from alpha and beta sources don't penetrate far, they can be easily shielded by metal containers with walls thicker than 1 millimeter.

X rays and gamma rays are more difficult to shield. Nothing can stop them completely; at best, the radiation level can be reduced. Lead is the most common shielding material for x rays and gamma rays, but almost any material will work if enough is used. The rule of thumb is that 2 millimeters of lead are needed to cut down the radiation intensity of ordinary x rays by a factor of 10. Thus, 4 millimeters are needed to cut x rays down by a factor of 100, 6 millimeters to cut them down by 1,000, and so on. For the more penetrating gamma rays, 10 centimeters of lead are needed to cut down the radiation by a factor of 10. This figure also applies to radiations from supervoltage x-ray machines. These rules of thumb are safe because the amount that actually leaks through is always less than the calculated amount.

EXAMPLE A certain gamma-ray source has a radiation level of 0.5 Gy/hr on contact. How thick should the lead shield be if the radiation level on the outside is to be no more than 500 μGy/hr?

Answer We want to reduce the intensity of the radiation by a factor of

$$\frac{0.5\ \text{Gy/hr}}{0.000500\ \text{Gy/hr}} = 1{,}000.$$

Every 10 centimeters of lead cuts down the intensity by a factor of 10, so three 10-centimeter thicknesses will be needed to cut the intensity by a factor of 1,000. Thus, the shield must be at least 30 centimeters thick.

Ultraviolet Radiation Safety

Although the skin would be better off if it was never exposed to the sun, no one is going to give up outdoor activities. Nor is there any need to give them up if reasonable care is used to cut down exposure to the sun's ultraviolet radiation. The main precaution is: *Don't sunbathe.* Wear loose, long-sleeved clothes if you are going to be out in the sun for a long time. When you have a choice between the sun and the shade, pick the shade.

Whenever your skin must be exposed to the sun for any length of time, use a sunscreen lotion. However, it must be a good lotion; many suntan lotions are almost worthless. Pick one that says something like "stops almost all of the ultraviolet light" or "little or no tanning." Some of the best sunscreens contain PABA lotion. If you like outdoor activities, pick a good sunscreen lotion and use it. If you notice any signs of burning, look for a better sunscreen or stay away from the sun for a while.

If you follow these rules, you won't get burned, your skin won't look like a prune when you are 50, and your chances of getting skin cancer will be far smaller than the chances of people who carelessly expose their skin to the sun.

Microwave Safety

The two most common sources of microwaves in hospitals are diathermy machines and microwave ovens. Both are safe when used properly.

The operator of a diathermy machine should move away from the machine when giving treatments. Of course, the operator must be careful not to apply too large a microwave current to the patient. There is only one other precaution: When the microwaves are being applied near the patient's eyes, extra shielding must be placed between the eyes and the microwave electrode.

Microwave ovens do not leak much radiation when they are properly maintained. They all have interlocks that turn off the microwaves as soon as the door is opened, so there is little chance of serious exposure when the oven is in good condition. However, it is a good idea to turn the oven off before opening the door, and not staring through the door is advisable. If the oven is in good condition, no special precautions are needed. However, when the door shows the slightest sign of being loose or when the seals seem worn or loose, the oven should not be used until repairs have been made. Of course, people with pacemakers must avoid all microwave sources.

Concepts

- Atomic number
- Nucleon
- Nucleon number
- Isotopes
- X ray
- Gamma ray
- Beta ray
- Alpha ray
- Nuclear fission
- Half-life
- Exponential decay
- Geiger counter
- Scintillation counter
- Fluoroscope
- Anger camera
- Radiation dose
- Radiation exposure
- X-ray tube
- Bremsstrahlung
- Characteristic x rays
- Anode focal spot
- X-ray filters
- X-ray machine collimators
- X-ray grids
- Film screens
- Tomography
- Computerized axial tomography (CAT)
- Film badge
- Survey meter
- Radiation shielding
- Sunscreens

Discussion Questions

1. Explain how microwaves, infrared rays, visible light, ultraviolet rays, and x rays are emitted by atoms.
2. What are isotopes?
3. What does the symbol $^{29}_{14}Si$ mean?
4. Radioactive nuclei usually decay by emitting alpha rays, beta rays, or gamma rays. Explain what each type of radiation is and what, if anything, they are used for in diagnosis or therapy.
5. What does the term *half-life* mean?
6. If half of a radioactive isotope decays in three days, will it be gone in six days?
7. Describe at least three ways of detecting x rays and other ionizing radiations.
8. Why are screens added to most diagnostic x-ray films?
9. How are grids used to improve the quality of x rays?
10. Why are most cobalt-60 and supervoltage therapy machines mounted on rotating frames?
11. Describe the hazards and the safety precautions for the following types of radiation:

a. microwaves
b. ultraviolet radiation
c. x rays
d. gamma rays
e. beta rays
f. alpha rays

12. What precautions should be taken when working or playing in the sun?

Problems

1. What isotope would you start with if you wanted to make gold-198 by neutron irradiation in a nuclear reactor? That is, what is the isotope represented by the blank in the equation below?

$$^{1}_{0}n + \_\_\_\_ \rightarrow {}^{198}_{79}\text{Au}.$$

2. Gold-198 ($^{198}_{79}$Au) decays by giving off a negative beta particle ($_{-1}^{0}e$). What is the daughter isotope of this reaction?

3. Uranium-238 ($^{238}_{92}$U) decays by emitting an alpha particle. What is the daughter nucleus of that decay?

4. If the decay of cobalt-58 ($^{58}_{27}$Co) has the daughter nucleus iron-58 ($^{58}_{26}$Fe), does cobalt-58 decay by alpha, beta, or gamma decay?

5. Cobalt-60 ($^{60}_{27}$Co) decays by emitting a negative beta particle and two gamma rays. What is the daughter nucleus of this reaction?

6. If carbon-11 ($^{11}_{6}$C) is produced by bombarding boron-11 ($^{11}_{5}$B) with protons, what other particle is also produced by the bombardment? That is, fill in the blank in the equation below.

$$^{1}_{1}p + {}^{11}_{5}\text{B} \rightarrow {}^{11}_{6}\text{C} + \_\_\_\_ .$$

7. Radioactive carbon-14 ($^{14}_{6}$C), which has a half-life of 5,700 years, is constantly produced by the cosmic rays that strike the earth. Thus, all living things contain a small percentage of carbon-14, and the fraction of the carbon that is carbon-14 is the same in all living things. However, as soon as a living thing dies, the fraction of carbon-14 starts to decline as the carbon-14 decays. For instance, a mummy that had been dead for 5,700 years would have only half as much carbon-14 as a living person. This idea can be used to date ancient objects. The only condition is that the object must have either been alive or been made from something that was alive, as paper, cloth, wooden furniture, and many other things are. For example, if the charcoal from an ancient campfire has only one-eighth as much carbon-14 as new charcoal, when were the trees cut down to make that campfire?

8. Iodine-131 ($^{131}_{53}I$) has a half-life of just slightly over eight days. If you had an iodine-131 source that had a strength of 370,000 becquerels (10 microcuries), how many disintegrations per second would be occurring in that source 32 days later?

9. A technetium-99 ($^{99}_{43}Tc$) source had a strength of 1 megabecquerel (27 microcuries) when it was new. If it now has a strength of 250 kilobecquerels (7 microcuries), how old is the source? Technetium-99 has a half-life of six hours.

10. Roughly how long would you have to wait for a cesium-131 ($^{131}_{55}Cs$) source to decay to 10% of its present strength? Cesium-131 has a 9.7-day half-life.

11. A radiation worker wants to measure the half-life of an unknown radioactive source. She makes the following observations of the source strength:

Time	*Source Strength*
0 hr	85 GBq
1 hr	50 GBq
2 hr	30 GBq
4 hr	11 GBq
6 hr	4 GBq

Plot these results on a graph with source strength on the vertical axis and the time on the horizontal axis. Estimate the half-life of the source.

12. A radioactive source preparation room has a radiation level of 100 μGy/hr (10 mrad/hr). How many hours per week could a radiation worker spend in that room? Should a member of the general public be allowed to work in the room?

13. A radioactive cobalt-60 ($^{60}_{27}Co$) source produces a dose of 30 mGy/hr (3 rad/hr) at a distance of 1 meter. The source is fenced in so that no one can get closer than 30 meters. What is the dose rate at the fence? How many hours per week could someone work near the fence? Note: The last question cannot be answered completely by calculation; some judgment about general radiation safety practices must be used, too.)

14. Approximately what would the radiation level be 5 meters away from a 20-gigabecquerel gamma ray source? Would you classify this source as one that requires just routine care?

Experiments

1. Take 50 to 100 pennies and pretend that they are radioactive atoms that decay whenever they land tails. Count the pennies and then dump them on a table. Remove the tails. Count the heads, shake them up,

dump them on the table again, and pull off the tails. Repeat the process until all the pennies are gone or you get tired of dumping the pennies. Plot your results on a graph, with the number remaining on the vertical axis and the number of dumpings on the horizontal axis. What was the "half-life" in number of dumpings? Was the number remaining after four dumpings about the same as you would predict from the half-life?

2. Close one eye and then look through your fingers. Part of your view is naturally blocked by your fingers. Now wave your hand back and forth rapidly. You can now see everything behind your fingers. Why? Aren't your fingers always blocking part of your view? How does this experiment compare with the use of tomography for x-raying internal organs?

3. The effect of different focal spot sizes on the quality of x-ray pictures can be investigated by examining the effect of the light source size on the sharpness of ordinary shadows. Cut a 3-centimeter-diameter hole in a piece of cardboard and place the cardboard in front of an ordinary incandescent light bulb. Use this light to cast shadows of your hand on the wall. In particular, look at the sharpness of the edge of the shadows. Next, replace the cardboard with another having a 1-centimeter-diameter hole. Keep the position of everything else unchanged. Look at the sharpness of the shadow to see the improvement in image quality.

CHAPTER EIGHTEEN

Physics in the Health Professions

Educational Goals

The student's goal for Chapter 18 is to be able to give at least three examples of how physics is applied in his or her planned profession.

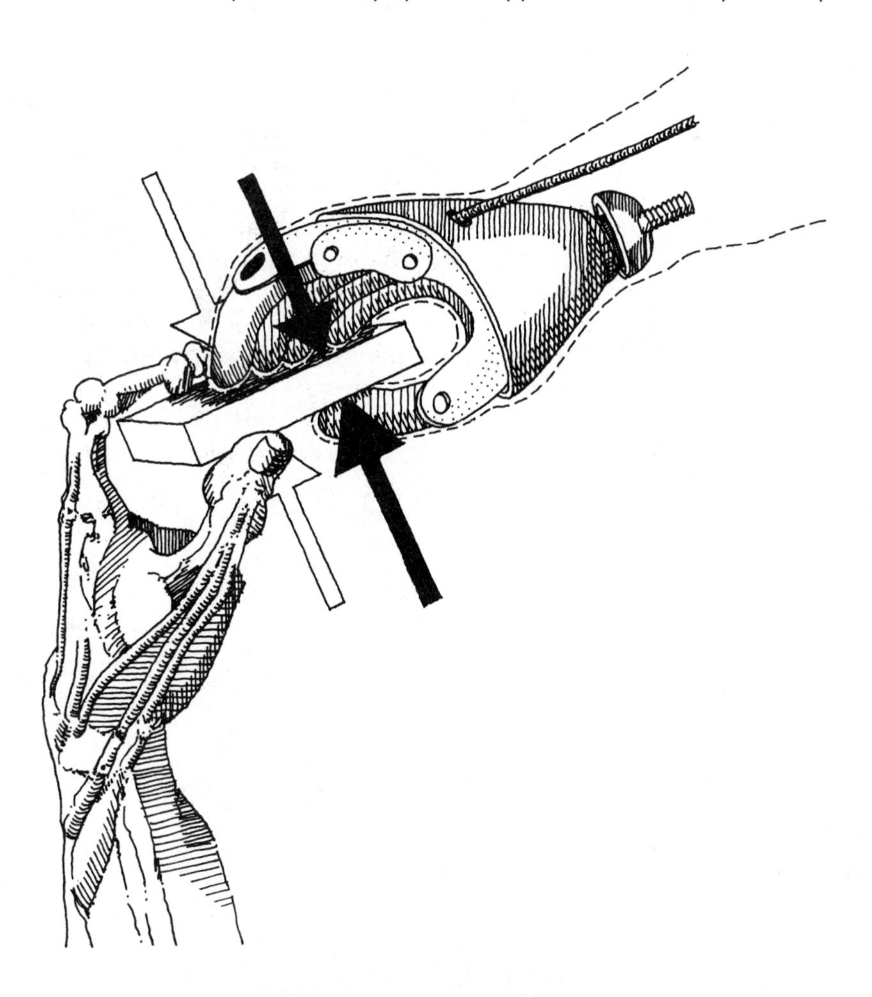

When you use physics in your future studies and in your profession, its ideas will usually be one link in a long chain of thought. While all the links—including that of physics—are important, finding a solution to a problem usually involves so many different thoughts that the role of any one field does not stand out.

The examples in this chapter were chosen to help you understand the role of physics in the health professions. All are real situations, but they are not typical in that they all draw on physics in a clear and obvious way. But typical or not, these examples will give you a good idea about how the material in this course fits into your work in the health professions.

Physical Therapy

A patient whose legs are too weak to support his or her full body weight can exercise on a tilt table. The force on the legs is adjusted by changing the angle of the table; the steeper the angle of the table is, the greater the load on the legs. The force on a patient's legs while exercising is generally prescribed as a certain percentage of the patient's weight. Thus, the question is, What angle should the table be at to put a given percentage of the patient's weight on his or her legs? When the table is flat ($\theta = 0°$), the load on the patient's legs will be zero. When the table is straight up ($\theta = 90°$), the load will be the patient's full weight. The percentage supported by the legs at intermediate angles cannot be found as quickly because the three forces on the patient—the patient's weight, the reaction force of the table, and the force of the footboard—are vectors and thus involve vector addition (Figure 18-1). For instance, when the table is tilted halfway up ($\theta = 45°$), the load on the patient's legs is not 50%—it is more than 70%.

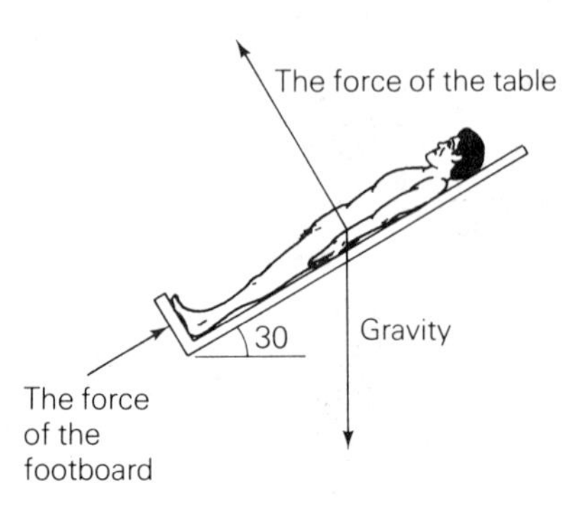

Figure 18-1 The three forces on a patient lying on a smooth tilt table.

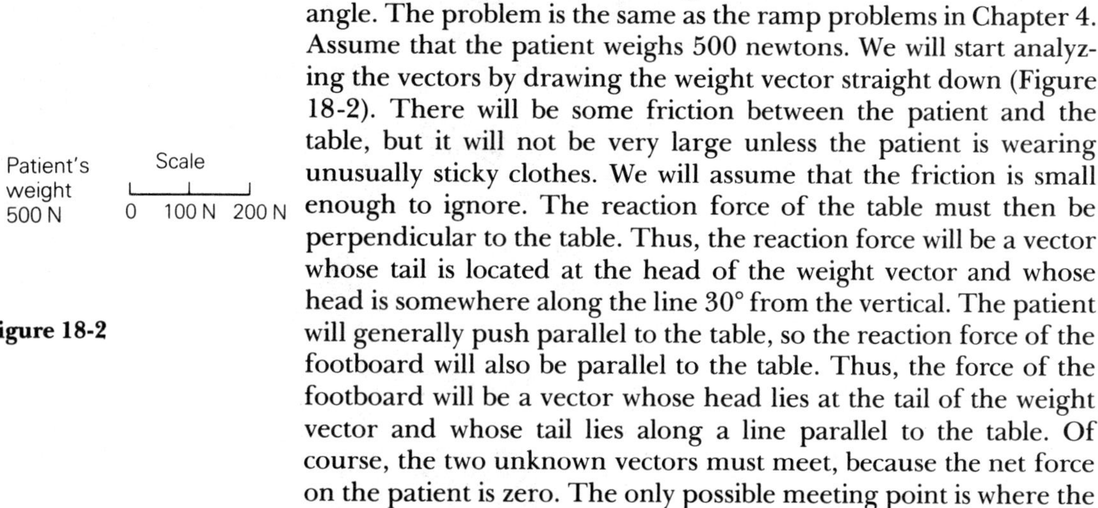

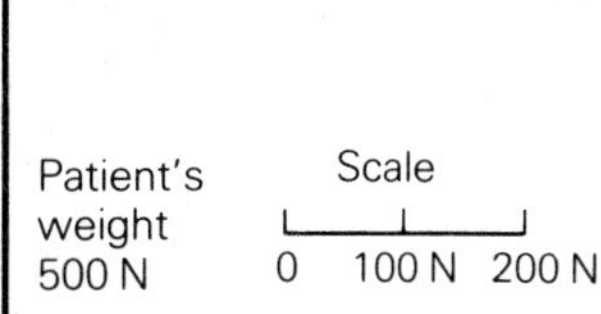

Figure 18-2

Let's calculate the percentage when the table is tilted at a 30° angle. The problem is the same as the ramp problems in Chapter 4. Assume that the patient weighs 500 newtons. We will start analyzing the vectors by drawing the weight vector straight down (Figure 18-2). There will be some friction between the patient and the table, but it will not be very large unless the patient is wearing unusually sticky clothes. We will assume that the friction is small enough to ignore. The reaction force of the table must then be perpendicular to the table. Thus, the reaction force will be a vector whose tail is located at the head of the weight vector and whose head is somewhere along the line 30° from the vertical. The patient will generally push parallel to the table, so the reaction force of the footboard will also be parallel to the table. Thus, the force of the footboard will be a vector whose head lies at the tail of the weight vector and whose tail lies along a line parallel to the table. Of course, the two unknown vectors must meet, because the net force on the patient is zero. The only possible meeting point is where the lines cross at point X. Measuring the footboard force, we find that it is 250 newtons, 50% of the patient's weight (Figure 18-3). By the third law of motion, this is equal in size to the force exerted by the patient.

The percentage of the patient's weight carried by the legs at a given table angle will be the same for all patients, regardless of their weights. Only the size of the vector triangle will change; the angles and proportions stay the same. Therefore, the ratio of the weight to the footboard force stays the same. The percentages for other angles can be worked out as above, and some are listed in Table 18-1.

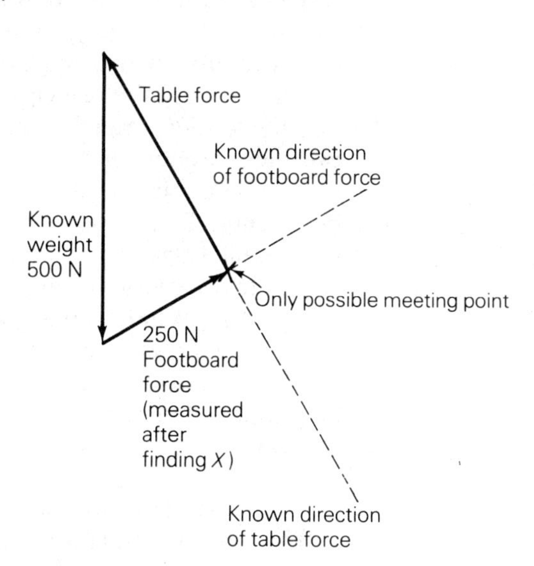

Figure 18-3

Table 18-1 Percentage of Body Weight Supported by the Legs as a Function of Tilt Table Angle

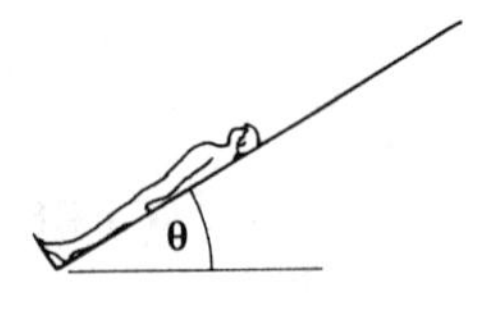

Angle θ	*Percentage*
0°	0
10°	17
20°	34
30°	50
40°	64
50°	77
60°	87
70°	94
80°	98
90°	100

Occupational Therapy

An occupational therapist's patients often have very poor strength and endurance. The guiding principle for the motions of these patients is: Save energy. A very useful first step in their rehabilitation is to give them a minicourse in the physics and physiology of energy by using the principles discussed in Chapter 8.

The patients need to know that gravity and friction are their two biggest enemies in saving energy, and they must learn that a muscle burns up energy whenever it is activated, whether it is moving or not. They also need to know that an object can be moved horizontally with very little energy when friction is small, but it always takes energy to lift something—including their arms, legs, and bodies. When patients understand what energy is, they will quickly see the value in conserving their energy.

The following are some standard energy-saving ideas that will help disabled persons conserve energy and improve their efficiency of motion:

1. Always sit down instead of standing; standing requires constant tension in several muscles.

2. Try to arrange tasks to avoid long reaches; again, long reaches require large, unnecessary muscle tensions.
3. Arrange work areas so that objects can be slid horizontally instead of lifted vertically. Unless the surface is extremely sticky, sliding takes much less energy than lifting.
4. When moving things, including the body, reduce the friction as much as possible. Use smooth counter tops, and move things on carts instead of carrying them. Use a wheelchair to move around part of the time instead of walking. Don't carpet floors on which wheelchairs will be used.
5. Use clamps to hold objects instead of holding them with the hands and arms.

Most tools and tasks are designed for normal people who have no shortage of energy. By looking at a task with an eye to saving energy, the therapist and the patient can usually cut the energy requirements so that the patient can handle the job again.

Respiration Therapy

The normal airflow through the tubes in the lungs is a smooth, low-speed, laminar flow that takes very little energy. However, when the tubes becomes partially obstructed, the flow often becomes turbulent. As we found out in Chapter 10, it takes more pressure to maintain a turbulent flow than an equivalent laminar flow, so more energy will be needed to move the air into and out of the lungs. The situation is compounded because the body will need more air to produce the extra energy. Thus, the person whose lungs are severely constricted must labor under a double burden that the patient's weakened lungs and heart may not be able to carry.

One way to reduce the energy used in breathing is to use a mixture of helium and oxygen instead of air, which is mainly nitrogen and oxygen. The nitrogen in air is basically inert—it goes in and out of the lungs unchanged. Helium is even more inert than nitrogen, so it doesn't have any therapeutic value; however, it is far lighter than nitrogen. Thus, the much lighter helium-oxygen mixture will flow past the obstructions more rapidly than normal air without breaking into turbulent flow. When turbulent flow does occur, breathing the lighter mixture will still use less energy. As a result, the patient can breathe with far less effort and devote more energy to other things.

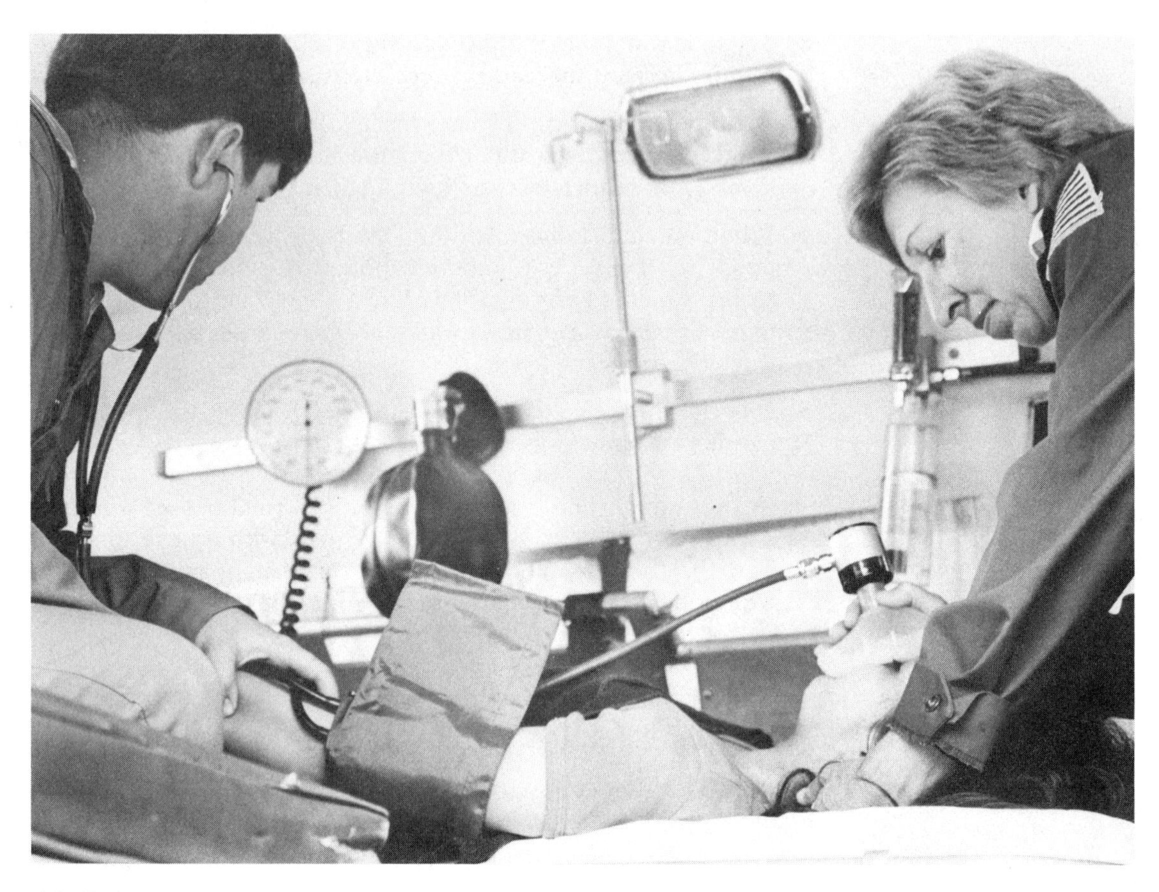

A helium-oxygen mixture can deliver more oxygen to the lungs than normal air can when the patient's lungs are obstructed.

Since helium is completely inert, mixtures of helium and oxygen are as safe to use as air. Mixtures of 80% helium–20% oxygen and 70% helium–30% oxygen are commonly available. The 80-20 mixture behaves about the same as normal air except for its faster flow. The 70-30 mixture is used with patients who need a slightly higher oxygen concentration. Since helium is so safe, only a few precautions are needed:

1. Since helium flows much faster than air, it also leaks faster. The system must be tight and the patient's face mask must fit tightly.
2. Helium is inert, so it cannot support life by itself. Pure helium can never be used—it must be mixed with oxygen.
3. Since the flow properties of helium-oxygen mixtures are different from those of oxygen, a flowmeter that is calibrated for oxygen will not give the correct reading.

4. Since the light helium-oxygen mixture will not retain aerosol sprays very well, medications cannot be effectively administered by aerosols while the patient is breathing a helium-oxygen mixture.

5. The speed of sound is higher in helium-oxygen than in air. Thus, sounds with the same wavelength must have a higher frequency; as $v = f\lambda$ for all waves. Since the size of the body stays the same, the wavelengths of the sounds produced by a person always stay about the same; as a result, the pitch of the voice goes up when breathing helium-oxygen. Consequently, patients who are breathing helium-oxygen talk in a high-pitched voice like that of Donald Duck. Patients must be told to expect this. It is not at all harmful, and it will go away as soon as the therapy is stopped.

X-Ray Technology

The x-ray technician's prime responsibility is to take the best possible radiographs with the smallest possible radiation exposure to the patient—and to him or herself. There are many factors involved in making a good radiograph: patient positioning, collimation, shielding, film and screen choices, grids and other scatter reduction methods, and film development. However, the most important is the x-ray machine itself—in particular, the settings of the voltage, current, and time controls, which we studied in Chapter 17.

The voltage control sets the energy of the electrons striking the target in the x-ray tube. Thus, it also indirectly controls the energy of the x rays. The voltage doesn't affect the number of electrons hitting the target, but it does have an effect on the x-ray beam intensity, because the harder the electrons strike the target, the more x rays they produce.

The main effect of increasing the tube voltage—and thus increasing the energy of the x rays—is to increase the penetrating power of the x rays. When the tube is run at a low voltage, the x rays don't penetrate dense tissue such as bone—they can only penetrate soft tissue. Thus, a radiograph made at a low tube voltage shows bones as completely white. No interior detail can be seen, and little is revealed about the bones except for their silhouettes. However, the soft tissue that isn't behind a bone shows considerable detail. When the tube voltage is raised, the x rays penetrate deeper tissues. More detail is visible in bony regions, but less soft tissue is visible. If the tube voltage is too high, the x rays are too penetrating and produce almost totally blackened film, which shows no detail at all. A further disadvantage of too high a voltage is the high amount of scatter from the increased x-ray energy. Low-voltage radiographs have few scatter problems, while high-voltage exposures require grids and other tricks to cut down the scatter. Of course, the thickness of the patient also plays a part in choosing the right voltage. The thicker the patient is, the greater the penetrating power needed; thus, the greater the voltage must be.

The current control regulates the number of electrons striking the target. The electron current is usually measured in milliamperes. At a fixed voltage, the number of x rays is proportional to the current. Thus, the current is one of the two major factors in controlling the exposure. The other is time. At a fixed voltage, the exposure depends on the total number of x rays that hit the film; when the product of the current and the time is the same, the exposure will be the same.

Nursing

The nurse often has the responsibility of telling a patient what to expect during and after a medical or surgical procedure. Such a talk can effectively relieve a patient's anxiety, because a patient's fears of what will or might happen are often far worse than the truth.

This is especially true of any procedure involving radiation, because most people know very little about radiation and often have misinformation based on scare stories. The explanation of radiation in Chapter 17 gives the nurse a basis for an honest discussion of radiation with a patient.

In addition to normal x rays, there are four basic types of treatment involving radiation: (1) diagnostic procedures, such as brain scans and liver scans, in which the patient either swallows or is injected with a radioisotope; (2) therapeutic treatments, in which a radioisotope is either swallowed or injected; (3) therapeutic application of a sealed radioactive source; and (4) therapy with radiation from a cobalt-60 source or a supervoltage accelerator.

Diagnostic Radioisotope Procedures

The details of the precautions vary with the procedure and the amount and half-life of the isotope, and the nurse must find out the details for each procedure. However, in most cases the amount

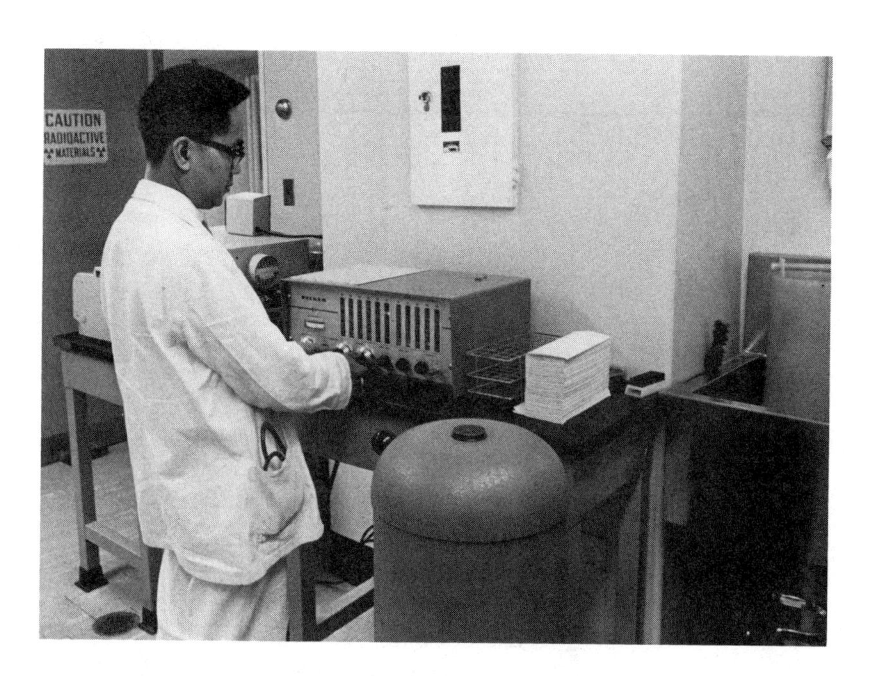

Radioactive materials are prepared under carefully controlled conditions.

of radioactive material used is not very large, and the following results are typical.

The patient will not experience any pain due to the radioisotope during or after the test. The only likely pain is the minor pain of a hypodermic needle if the isotope is injected. The patient can be told that the radiation is very similar to that of an ordinary x ray and that the radioisotope is unlikely to cause any sensation whatever.

The patient should be told that the scanning camera may be large and even scary looking, and it might make strange clicking sounds, but in most cases the camera won't touch the patient. Only the restraints used to keep the patient from moving might cause any discomfort.

The posttreatment precautions, which depend on the amount and half-life of the isotope, range from almost none at all to a few days of isolation for the patient and the patient's wastes. If isolation is necessary, the patient should be trained before the test to store bodily wastes and to provide as much care as the patient can for him or herself. The patient should be told that distance is the best way to reduce the exposure to others and he or she should stay a few meters away from people to avoid exposing them to radiation.

The amount of direct nursing care also depends on the amount of radioactivity present, but in most diagnostic cases the nurse should work rapidly and keep the amount of time spent near the patient to a minimum. Because the exposure from across the room will be almost nothing, a nurse shouldn't hesitate to cheer up the patient from that distance. However, don't spend a lot of time close to the patient.

Therapy with Unsealed Radioisotopes

Much radiation is required to cause changes in tissue, so the amount of radioisotope that is given during most therapeutic procedures is fairly large. Thus, the patient must stay in one area to avoid spreading the radioactive material; body wastes may have to be collected and saved; and care must be taken with bedding and other objects used by the patient. Until the isotope decays or is eliminated, nurses and visitors should stay at least several meters from the patient, and pregnant women and children should not be allowed to come near. When close contact is necessary, the nurse should wear a film badge, the radiation level near the patient should be measured, and the nurse should make a rough estimate of the exposure that he or she will receive to insure that safe limits are not exceeded. Close patient care should be rotated among several nurses to reduce exposure to any individual.

The patient should be told about likely reactions to the treatment. The reactions vary, but skin reddening, nausea, loss of appetite, and some discomfort are common. By knowing what to expect, the patient will not think that the normal reactions are signs of something going wrong.

Therapy with Sealed Radioisotopes

The precautions for patients exposed to sealed radioisotopes are the same as those for unsealed radioisotopes except that the body wastes and bedclothes need not be collected, because the radioactive material is sealed within a tight container. Thus, patient care is quite a bit easier. The main precaution is to see that the distance between the patient and others is sufficient to prevent unnecessary exposure.

Again, the patient should be told of likely reactions and informed that he or she will no longer be radioactive when the applicators are removed, at which time the patient can resume normal contact with friends and relatives.

External Radiation Therapy

The patient should be told what to expect during irradiation, since the machines that will be used are large and often spin around in a way that could frighten the unprepared patient. The patient should know that the actual radiation will not produce any pain or even any sensation, but that reactions to the radiation will show up shortly after the treatment, including skin reddening, possible nausea, and other effects.

Patients should know that they will not be made radioactive by the irradiation and that as soon as they leave the therapy room, they can resume normal contact with people. Nurses can treat the patient without any hazard to themselves, and no special radiation precautions are needed.

Athletics

One of a coach's most important jobs is to show athletes how to take advantage of the laws of motion. The laws of motion, treated in Chapters 3 through 8, cannot be broken, but they can be used in endless ways to improve athletic performance.

Consider jumping to catch a rebound in basketball. Should the player try to raise his or her hands as high as possible before the jump, as the legs push upward, or after being airborne? This is an important question in coaching basketball players, but it can't be answered with a few quick words; we must look into the laws of motion.

The first question to answer is, What do we want the basketball player to achieve? Of course, the answer is to take control of the ball; and the player who jumps the highest will have the best chance of doing that. So our next question is, Which of the above three ways allows the player to jump the highest?

The most important factor is the thrust of the legs during the jump. The harder and longer the legs can push upward, the further the player's center of gravity will rise. If the player's hands remain in the same position during the jump, either up or down, they can't have any effect on the thrust of the legs, and the center of gravity will move up by the same amount. However, when the player's hands are up to start with, the center of gravity is a few centimeters higher. Thus, the center of gravity and the hands will end up a few centimeters higher. So raising the hands before jumping is better than raising them after leaving the ground.

However, the remaining possibility, raising the hands as the legs push upward, is the best way to jump. As the arm muscles raise the arms, they create an equal but opposite reaction force on the rest of the body. This force is transmitted down through the spine to the legs, where the force of the leg muscles is *added* to it. Thus, the total force of the legs on the ground is greater by this amount.

As a result, the energy that the player produces is greater and the player jumps higher.

The same idea also applies to sprinting. The first steps are the most efficient, because the acceleration is greatest then. If the arms are moved forward during the first stride, the total force of the leg on the ground (or starting block) is greater; thus, the forward acceleration is greater.

Industrial Hygiene

It is impractical if not impossible to eliminate all industrial hazards. No one knows how to forge metal without making noise or to smelt ores without heat. These and many other jobs can never be risk free until robots do all our work. The industrial hygienist's job is to see that the dangers and discomforts of the work place are minimized. Almost every step taken to protect workers from the hazards of industry is based on physical principles. A good example is how workers are protected from the superhigh temperatures used to melt steel, which applies the heat transfer principles discussed in Chapter 11.

The first step is to determine how the heat is transferred from the block to the worker. Of the three possible processes—conduction, convection, and radiation—only radiation is very important. Air doesn't conduct very well, and convection carries the heat away from the worker, because the air rises along the red-hot block. Thus, the radiation of infrared and visible light carries the heat to the worker. Better ventilation, which would help against conduction and convection, won't help at all against radiation.

To stop the heat flow, a shield is needed to block the radiation. Almost any nontransparent shield will work. However, most materials will absorb the radiation and heat up the shield, which will reradiate part of the heat toward the worker. The best shield would *reflect* the heat back toward the hot block instead of absorbing it. Reflectors also reduce the amount of heat needed to keep the block hot, thus saving energy.

A cheap material that makes an excellent reflector is aluminum foil. Since the reflection takes place in the first few atomic layers, the thickness doesn't matter; it is the shiny surface that counts. Even under rather severe conditions, aluminum stays shiny for a long time. When it does get torn or dirty, a fresh layer can be added easily. When viewing windows are needed, special glass can be used that is nearly opaque to infrared rays but semitransparent to visible light.

Where shields are impractical, the worker's clothing must act as a shield. For mild exposures, loose-fitting clothing, a hat covered with aluminum foil, and foil covering any hot spots are often enough. For extreme exposures, special clothing that is aluminized on the outside and insulated on the inside is used. The head covering has an infrared opaque window to see through. Since the primary protection is in the reflective properties of the aluminum coating, the surface must be kept clean and undamaged.

Of course, for lengthy exposures, some means of removing the heat from the worker must be added. Many simple air and water cooling systems work well. With such suits, people can work in areas that are as hot as 500°C (dull red heat) for fairly long times, and they can work for short times above 800°C (bright red heat).

Discussion Question

1. Write an essay on the applications of physics in your present or your planned profession.

SI Units

Physical Quantity	*Unit*	*SI Abbreviation*	*Relation to Other Units*
Length	meter (metre)	m	
Mass	kilogram	kg	
Time	second	s	
Acceleration	meter per second squared	m/s^2	m/s^2
Activity (of a radioactive source)	becquerel	Bq	disintegrations/s
Area	square meter	m^2	m^2
Density	kilogram per cubic meter	kg/m^3	kg/m^3
Electric charge	coulomb	C	
Electric current	ampere	A	C/s
Electric potential difference	volt	V	J/C
Electric resistance	ohm	Ω	V/A
Energy	joule	J	N·m
Force	newton	N	$kg \cdot m/s^2$
Frequency	hertz	Hz	cycles/s
Power	watt	W	J/s
Pressure	pascal	Pa	N/m^2
Radiation dose	gray	Gy	J/kg
Temperature	kelvin	K	
Torque	newton-meter	N·m	N·m
Velocity	meter per second	m/s	m/s
Viscosity	pascal-second	Pa·s	Pa·s
Volume	cubic meter	m^3	m^3
Work	joule	J	N·m

Periodic Table of the Elements

Atomic Number — 11; Name — Sodium; Symbol — Na

Period	Group IA	IIA	IIIB	IVB	VB	VIB	VIIB	VIII	VIII	VIII	IB	IIB	IIIA	IVA	VA	VIA	VIIA	Noble Gases
1	1 Hydrogen **H**																	2 Helium **He**
2	3 Lithium **Li**	4 Beryllium **Be**											5 Boron **B**	6 Carbon **C**	7 Nitrogen **N**	8 Oxygen **O**	9 Fluorine **F**	10 Neon **Ne**
3	11 Sodium **Na**	12 Magnesium **Mg**											13 Aluminum **Al**	14 Silicon **Si**	15 Phosphorus **P**	16 Sulfur **S**	17 Chlorine **Cl**	18 Argon **Ar**
4	19 Potassium **K**	20 Calcium **Ca**	21 Scandium **Sc**	22 Titanium **Ti**	23 Vanadium **V**	24 Chromium **Cr**	25 Manganese **Mn**	26 Iron **Fe**	27 Cobalt **Co**	28 Nickel **Ni**	29 Copper **Cu**	30 Zinc **Zn**	31 Gallium **Ga**	32 Germanium **Ge**	33 Arsenic **As**	34 Selenium **Se**	35 Bromine **Br**	36 Krypton **Kr**
5	37 Rubidium **Rb**	38 Strontium **Sr**	39 Yttrium **Y**	40 Zirconium **Zr**	41 Niobium **Nb**	42 Molybdenum **Mo**	43 Technetium **Tc**	44 Ruthenium **Ru**	45 Rhodium **Rh**	46 Palladium **Pd**	47 Silver **Ag**	48 Cadmium **Cd**	49 Indium **In**	50 Tin **Sn**	51 Antimony **Sb**	52 Tellurium **Te**	53 Iodine **I**	54 Xenon **Xe**
6	55 Cesium **Cs**	56 Barium **Ba**	* 57 Lanthanum **La**	72 Hafnium **Hf**	73 Tantalum **Ta**	74 Wolfram **W** (Tungsten)	75 Rhenium **Re**	76 Osmium **Os**	77 Iridium **Ir**	78 Platinum **Pt**	79 Gold **Au**	80 Mercury **Hg**	81 Thallium **Tl**	82 Lead **Pb**	83 Bismuth **Bi**	84 Polonium **Po**	85 Astatine **At**	86 Radon **Rn**
7	87 Francium **Fr**	88 Radium **Ra**	** 89 Actinium **Ac**	104 Kurcha-tovium **Ku**	105 Hahnium **Ha**													

Series														
* Lanthanide Series 6	58 Cerium **Ce**	59 Prase-odymium **Pr**	60 Neodymium **Nd**	61 Promethium **Pm**	62 Samarium **Sm**	63 Europium **Eu**	64 Gadolinium **Gd**	65 Terbium **Tb**	66 Dysprosium **Dy**	67 Holmium **Ho**	68 Erbium **Er**	69 Thulium **Tm**	70 Ytterbium **Yb**	71 Lutetium **Lu**
** Actinide Series 7	90 Thorium **Th**	91 Protactinium **Pa**	92 Uranium **U**	93 Neptunium **Np**	94 Plutonium **Pu**	95 Americium **Am**	96 Curium **Cm**	97 Berkelium **Bk**	98 Californium **Cf**	99 Einsteinium **Es**	100 Fermium **Fm**	101 Mendelevium **Md**	102 Nobelium **No**	103 Lawrencium **Lr**

Answers to Selected Questions and Problems

Chapter 2 Discussion Questions

2. Although good arguments can be made for either choice, most technical people felt that getting rid of conversion factors would make their calculations faster and more accurate. After all, prefixes can be used to create a unit with a convenient size, but conversion factors cannot be eliminated after a measurement system is chosen. No one knows of a way to have both advantages.

Chapter 2 Problems

1. 63 gallons, 1 quart, 3 ounces, or 63.27 gallons (there are 231 cubic inches in a gallon); 0.30 m^3

2.
a. 0.27 m
b. 2.5 kg
c. 86,400 s
d. 300 s
e. 0.00005 kg
f. 0.1 m^3

3.
a. 2 millimeters = 2 mm
b. 50 kilograms = 50 kg
c. 70 megawatts = 70 MW
d. 10 nanoseconds = 10 ns
e. 10 micro–cubic meters = 10 μm^3 (10 milliliters in a non-SI unit)

4. 0.003 kg (3 g in a non-SI unit)

5. The errors would vary somewhat depending on the quality of the instruments. However, the errors would generally be as follows:
a. ± 1 mm
b. ± 0.1°C
c. ± 2°C
d. ± 3 kg
e. ± 1 kg (± 0.1 kg for a really good beam-balance scale)
f. ± 3 s
g. ± 0.1 s (± 0.01 s for a precision stopwatch)
h. ± 1 m

6.
a. 2
b. 3
c. 8
d. 1
e. 3
f. 5
g. 5
h. 3

7.
a. 0.9%
b. 14%
c. 0.24%
d. 30%

8.
a. 63
b. 300
c. 33
d. 1.6
e. 0.087

9.
a. 8.3×10^4
b. 2.7×10^{-5}
c. 3.4×10^{-3}
d. 3.7×10^{10}
e. 3.291×10^3
f. 7×10^{-1}

10.
a. 1.0×10^3
b. 1.3×10^2
c. 3.9×10^{-2}
d. 9.5×10^{-5}

Chapter 3 Discussion Questions

3. b.

4. Many pitchers say that they release the ball at the very end of their throw, when their hand is level with their shoulder, but this cannot possibly be. The hand is moving downward at this point; if the ball was released then, it too would go down, in accordance with Newton's first law of motion. The ball must have been released somewhat before then, while the pitcher's hand was moving toward the plate. This illustrates a problem all coaches must be aware of: What an athlete feels has happened and what actually happens are often two different things.

5. Since an object in uniform motion in a straight line remains that way unless acted upon by an outside force, a car's tires on a straight section need only supply a small steering force and a small force to overcome friction. To turn a corner, the tires must supply a large force to change the direction of the car's motion, even if the speed doesn't change. This doesn't mean that it is safe to drive fast on a slippery, straight road; it just says that corners are more dangerous than straight sections.

6. baby crawling, ~0.3 m/s; walking, ~1 to 2 m/s; running, ~10 m/s

Chapter 3 Problems

2. *Vectors*
   Weight of a person
   Force of a magnet on a nail
   Location of a distant city
   Location of a nearby table
   Velocity of a bullet
   Nonvectors (scalars)
   Mass of a person
   Time of day
   Date of a certain battle
   Volume of a bottle

3. 0.44 N

4. 306 kg

5. 19.6 N

6. 58,800 N

7. contact force A = 49 N; contact force B = 245 N; contact force C = 1,225 N

8. Even though the ball may be going around the circle at a constant speed, the *direction* is constantly changing. This requires a force, just as a change in speed does. Of course, the force is supplied by the string, and the direction of the force is along the string toward the center of the circle. After the string breaks, there is no horizontal force on the ball, so the ball continues on in a straight line toward the right.

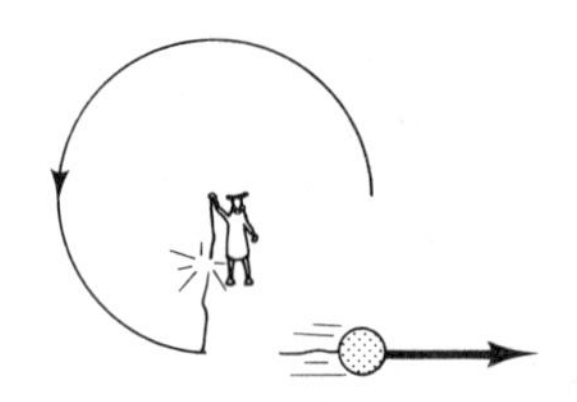

9. 20 m/s

10. 300 kg; 3 m

11. rope A, 49 N; rope B, 98 N; rope C, 157 N; rope D, 186 N

12. greatest possible sum, 8 N; smallest possible sum, 2 N

Chapter 4 Discussion Questions

2. the reaction force of the wall

3. Once the jumper is airborne, all the forces are internal to the jumper's body and cannot change the basic path of the jump. (However, this doesn't mean the motions are unimportant—they play an important role in positioning the jumper for a proper landing.)

Chapter 4 Problems

1. 27 N in the direction opposite to the force you apply

2. 18 m in the forward direction

4. yes; no

5. 58 m

6. 5 km approximately toward the southwest (29° west of south, or 209° on a compass)

7. 220 N straight ahead

8. 10 km/hr straight to the east

9. 175 N at 0.5 m; 275 N at 1.0 m; 325 N at 1.5 m; 350 N at 2.0 m

10. 5,000 N

11. The three forces acting on the body—its weight, the force of the rope, and the reaction force of the ground—must add to zero. If you do the calculation for various angles θ, you will see that the bigger θ is, the bigger the force of the rope will be. Assume that the reaction force is parallel to the body.

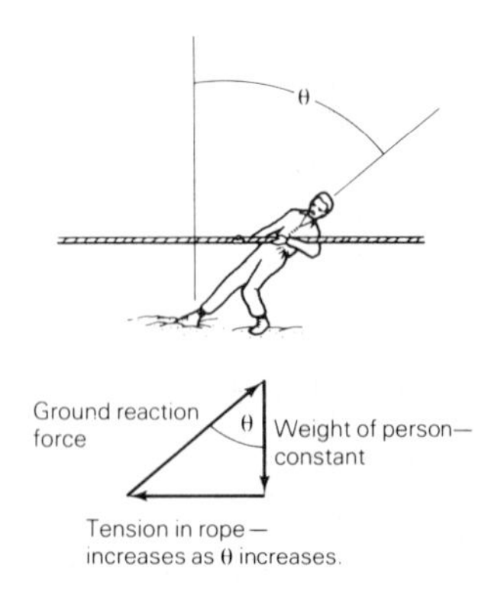

12. left section, 1,140 N; right section, 1,120 N (Note that both tensions are more than twice the woman's weight.)

13. We start by drawing the 1,000-N weight straight down. The sum of the forces must be zero, so we know that the *head* of the 100-N forward force vector must be at the tail of the weight vector. Thus, its *tail* must be somewhere on a circle of 100-N radius centered at the tail of the weight vector. The ramp reaction vector runs from the head of the weight vector to join the 100-N applied force vector on the circle. The largest possible angle occurs for the meeting point shown. That angle is equal to the maximum steepness of the ramp—6°. Note that this isn't a very steep ramp, rising only 10 cm for every meter of length. Wheelchair ramps should never be very steep, particularly if any of the patients are weak.

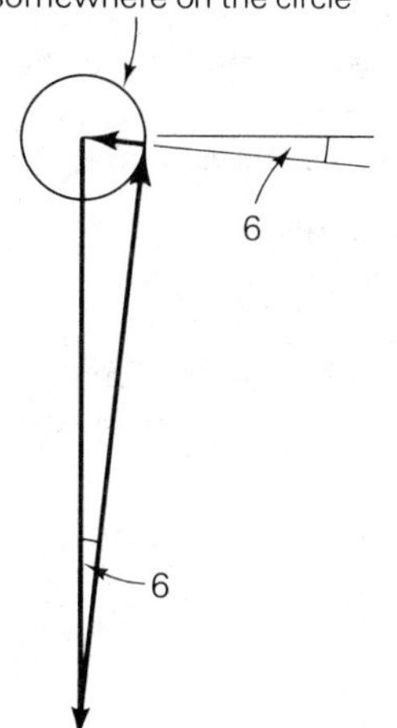

Chapter 5 Discussion Questions

5. The primary reason the arm produces the greatest pulling force when the lower arm is perpendicular to the upper arm is that the effective lever arm of the biceps muscle is greatest at that angle.

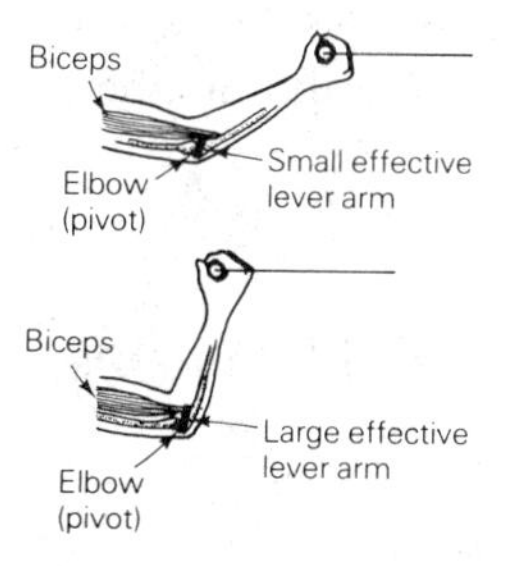

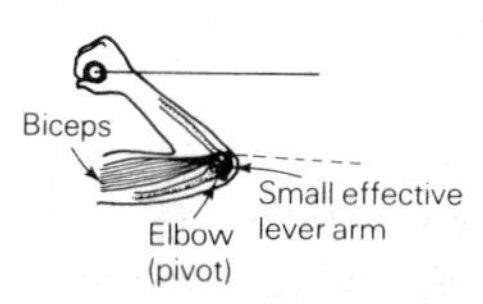

8. One reason is that platform shoes raise the CG of the body, which reduces stability. Reducing the stability increases the chances of taking a misstep. But the biggest danger is when a misstep actually occurs. The platform shoe lengthens the lever arm to the ankle joint, greatly increasing the torque at the ankle. Of course, this increases the risk of ankle sprains and fractures considerably.

10. Even though the *net* force is the same, the *torque* needed to pick up a long pole by one end is much greater than the torque needed to pick it up at its center.

Chapter 5 Problems

1. 2.0 m

2. 196 N

3. 1 m from the end carried by the two women

4. tension in the biceps, 980 N; contact force at the elbow, 882 N

5. 690 N

6. force at the hands, 515 N; force at the toes, 220 N; 70% of the athlete's weight

7. 49 N; 199 N

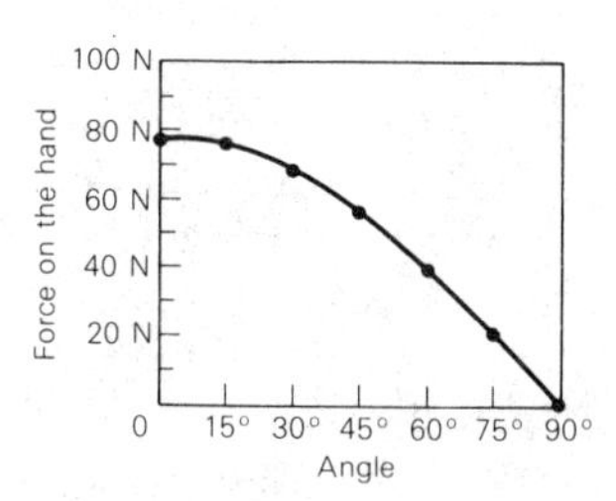

8. Note that the graph is not a straight line; the largest reductions occur for angles greater than 45°.

9. level carry, 79 N straight down; high carry, 60 N forward and down at an 18° angle

10. 100 N

11. 50 N

Chapter 6 Discussion Questions

3. If the binding doesn't release, the ski acts as a very long lever that puts a large torque on the bones of the lower leg.

9. In bowling, the leading foot must slide easily. Thus, that shoe should have a small coefficient of friction. In many dance movements the feet must slide on the floor, so dancing shoes usually have a moderately low coefficient of friction. Ice skates have an extremely low coefficient of friction.

10. It would be most difficult. Since the coefficient of friction between rock and most clothing is far less than 1.0, the belayer could be dragged along the rock by a force far smaller than the weight of a fallen climber. A safety rope isn't very safe unless the belayer is actually secure in a crack, behind a rock or tree, or tied down securely.

11. We will assume that the coefficient of friction with the hands is roughly the same everywhere on the object. (The answer is only slightly changed if it isn't.) From our work with levers, we know that the downward force of the rod will be smaller on the hand that is farther from the CG. Therefore, the frictional force will also be smaller on the hand farther from the CG. So as the hands start to move together, the object will slide on the hand farther from the CG until that hand becomes the hand closer to the CG. Then the object will slide on the other hand until the other hand gets closer to the CG. This process continues with the CG always staying somewhere in between the two hands until the hands are together and the CG is directly above them.

Chapter 6 Problems

1. 7.5 N

2. 3 N; 2.3 N

3. The girl will win unless she lets go of the rope, because the frictional force of her shoes on the concrete is much greater than the frictional force of the boy's shoes on the ice.

4. 39°

5. 0.49 m, or 49 cm

Chapter 7 Discussion Questions

1. There is very little force on the water drops, so when the hand stops or changes direction rapidly, the drops of water continue to move in a straight line away from the hands.

7. The main advantage of a light racket, bat, or club is that it is easy to accelerate. It will be easy to start moving and to change its motion. The biggest disadvantage is that a light racket (or bat or club) will not have much momentum to give to the ball. However, a heavy one will be very difficult to start moving, and it will be difficult to change its direction; but when it does strike the ball, the momentum transfer will be much larger. Therefore, choosing the best mass must be a compromise between handling ease and momentum transfer. Of course, the choice will be strongly influenced by the individual's condition; the right choice at the start of a season might be wrong by midseason, when the person is in better condition.

Chapter 7 Problems

1. $1 per month per month = $\$1/\text{month}^2$ = $12 per month per year

2. 3 m/s^2

3. 300 m/s

4. −3.2 m/s^2 (the minus sign indicates that the car was decelerating)

5. 2 m/s; 0 m/s; −2 m/s (the minus sign in the last answer means that the ship is backing up)

6. 49 m/s^2 forward and upward at a 55° angle

7. 3.4 m/s^2 northeast at a 36° angle from north
8. 130 N
9. 0.70 m/s^2
10. 7,500 N compared with the car's weight of 14,700 N
11. 0.9 s
12. 5 s; 50 s
13. 3,870 N (about 2,700 times the weight of the ball)
14. 2.5 m/s
15. 19.6 m/s; 29.4 m/s
16. The centripetal force, mv^2/r, cannot exceed the force of friction, μmg. Therefore, v must be less than $\sqrt{\mu rg} = 3.1$ m/s.

Chapter 8 Discussion Questions

4. When the gymnast stands on the platform, he or she has potential energy. During the fall to the trampoline, the potential energy is changed into kinetic energy. As the gymnast strikes the trampoline, his or her kinetic energy is stored as potential energy in the springs of the trampoline. Then the potential energy of the springs is retransformed into kinetic energy as the trampoline throws the gymnast upward. As the gymnast travels upward, his or her kinetic energy is gradually transformed into potential energy. When the gymnast stops on the platform, the energy is again all potential energy.
5. A vehicle traveling at 80 km/hr has 4 times more kinetic energy than it has at 40 km/hr. Therefore, there will be roughly 4 times more damage in a collision at 80 km/hr than at 40 km/hr.
8. As we have calculated in the examples, the amount of mechanical energy in a few food calories is quite large. Thus, stupendous amounts of exercise must be done to burn up a few calories of food energy. A few minutes of exercise a day can do wonders for the body, but exercise alone will not take off a significant amount of fat.

Chapter 8 Problems

1. 147 kJ, or 35 food calories (about equal to the amount of energy in a small piece of candy)
2. 740 J
3. 490 N
4. 34 m/s
5. at 50 km/hr on dry concrete, 10 m; at 100 km/hr on slush, 200 m (20 times farther)
6. 31 m/s, or 110 km/hr
7. 370
8. 24 billion joules (20 billion joules in potential energy, 4 billion joules in kinetic energy); $1,200
9. about 30,000 times
10. 35 days; 33¼ days
11. 97 W
12. 0.23 horsepower (175 W)
13. 950 W
14. 680 s
15. 50 kW

Chapter 9 Discussion Questions

2. In a gas, the molecules are very far apart. When a gas is compressed, the molecules are forced closer together, but the size of the individual molecules is not changed. In liquids and solids, the molecules are already in close contact, with no empty space between them. Thus, the molecules must actually be compressed, which is very hard to do.
5. A large pressure can be created with a small force by applying the force to a very small area. Some common, but important, examples are the pressure under the point of a needle, under the edge of a knife, and at the points of saw teeth.
6. Wheels have a much smaller area in contact with the ground than bulldozer or tank tracks do. Therefore, the

pressure under a wheeled vehicle is much greater than that under a tracked vehicle of equal mass. Of course, the greater the pressure is under a vehicle, the farther it sinks into the mud. Thus, tracked vehicles can travel on much softer mud than wheeled vehicles.

11. The flow spurts only when the cuff pressure is in between the systolic and diastolic pressure; thus, the sounds of the spurts, the Korotkoff sounds, only occur during this period. When the cuff pressure is above the systolic pressure, the flow is cut off completely; when the cuff pressure is less than the diastolic pressure, the flow is uninterrupted.

Chapter 9 Problems

1. 92 kg; 900 N

2. 41 kg

3. 1,240 kg/m^3, or 1.24 gm/cm^3; 1.24

4. 1,500 kg/m^3, or 1.5 gm/cm^3

5. 12

6. 744 kg/m^3, or 0.744 gm/cm^3

7. If we assume the average car is about 2 meters wide and 1½ meters tall, the car will push aside 2 m × 1½ m × 80,000 m = 240,000 m^3 of air in an hour. This volume of dry air has a mass of 312,000 kg = 312 tonnes. It is easy to understand why so much of a car's energy goes to overcoming air friction.

8. 0.51 kg

9. 0.15 kg, or 150 g

10. 1,600 Pa; 2,300 N. Note that the total force on the bottom can be found either by multiplying the pressure at the bottom by the area of the bottom or by simply calculating the weight of the water in the tub. Try both ways to check your work.

11. 20,000 Pa; 2,600 Pa (almost 8 times smaller)

13. 56,000 kg, or 56 tonnes

14. It floats but just barely; 99% of its volume would be underwater.

15. 19 cm; 1.6 m

16. If we assume that the person is about 180 cm tall and 30 cm wide, that person would be supported by about 1,350 nails when he or she lies on the bed of nails. If the person has a mass of 60 kg, he or she would weigh 590 N, so an average nail would exert a force of 0.4 N on the person. The pressure would thus be

$$P = \frac{F}{A} = \frac{0.4\text{ N}}{0.002\text{ m} \times 0.002\text{ m}}$$

$$= 100{,}000\text{ Pa.}$$

Pain doesn't start until the pressure is about 10 times higher.

17. 320 N

18. 620 N, which is equivalent to having a person stand on each lens

19. 950 tonnes; 980 tonnes

20. 12,000 kg/m^3, or 12 gm/cm^3. This is far less than the density of gold, so the jewelry is clearly not pure gold.

21. 92%; 89%

22. 27%

23. 11%; normal

24. about 13 kg

Chapter 10 Problems

1. 200 kW

2. 8.3 W in a trained athlete; 14 W in an untrained person. This is why all exercise programs must build up gradually. The heart might not be able to stand the strain of a sudden increase in activity.

3. 0.0013 m^3/s, or 1.3 l/s. Of course, as the water level in the tub drops, the flow rate will slow down.

4. 0.039 m^3/s, or 39 l/s

5. 52 cm; yes

6. The stream of water will go up to the level of the water in the tank; that is, $h = H$. The problem is done by finding the pressure at the bottom and then the speed of the water as it leaves the pipe. The height is then found by applying the fact that the kinetic energy of the water leaving the pipe is conserved and changed to potential energy at the top of its path.

7. 1 ml/s

8. 2.44 (that is, the flow will increase by 144%)

9. right

Chapter 11 Discussion Questions

6. Aluminum is a much better heat conductor than wood, so the body heat flows into the aluminum bat more rapidly. However, this only holds true on days when the bat is colder than the body. If the temperature of the bat is above 37°C, the aluminum bat will feel hotter.

7. When it is dry, goose down has millions of tiny air pockets, which stop convective air currents. Its white color also reflects heat back toward the body, and it is a poor conductor of heat. However, when down gets wet, the down fibers mat together, letting convective air currents flow and the body heat escape. For this reason many winter hikers prefer other types of jacket insulation, since they feel that it is particularly important to stay warm if you get wet.

8. The clouds reflect the radiation of the earth back down to the earth, thus cutting the heat loss. Just the opposite occurs during the day; the clouds reflect the sunlight back into space, thus reducing the amount of heat energy reaching the earth.

9. Wetting the finger increases the cooling effect of the wind by increasing the amount of evaporative cooling. This in turn cools the side of the finger facing the wind. Holding up a dry finger will also work, but it will not be as sensitive.

Chapter 11 Problems

1. −183°C

2. 873 K

3. around 295 K

4. very cold, most likely

6. 440 kPa

7. 160 kPa

8. 19%

9. 0.3 kg, or, in terms of volume, 0.3 l. This assumes that all of the sweat is evaporated. On a humid day, much of the sweat will simply drip off, reducing the efficiency of the evaporative cooling.

Chapter 12 Discussion Questions

2. In a small animal, the nutrients can move directly into the cells by diffusion. However, diffusion is very inefficient for moving things long distances. Therefore, in large animals the nutrients must be brought close to the cells by a pumped flow through the various tubes in the animal.

6. The high salt concentration in the seawater draws water from the cells by osmosis. This would reduce the water level in the cells even further, and the sailor would quickly die of dehydration.

10. Osmosis just involves the diffusion of molecules through the membrane. No outside energy is needed, and the net flow is from the high-concentration to the low-concentration side. In active membrane transport, the molecules are actually sorted out on one side and actively pumped to the other. This, of course, takes energy.

Chapter 13 Discussion Questions

1. The attraction between the electrons and protons holds the electrons in their orbits.

3. Synthetic fabrics are usually far better insulators than natural fibers, such as cotton.

5. Long, thin extension cords have a high resistance. Hooking them together compounds the problem because the resistances are added in series, and there will be a large voltage drop in the cords. The appliance will not get the full voltage, which it may need to run properly. This can damage the appliance, and possibly start it on fire. If the current is too large, it may start a fire in the extension cords.

13. Since a transformer has no direct electrical connection between the input and output coils, it can be used to isolate a circuit or appliance from the ground. This means that isolation transformers stop currents from flowing through the body to such things as water pipes and wet floors when the circuit is accidentally touched.

14. *Advantages*
    Portability
    Convenience
    Safety

    Disadvantages
    Very high energy cost
    Relatively short shelf life
    Difficulty in determining how much energy is left

15. First, disconnect the victim from the electricity without risking electrocuting yourself. Then apply cardiopulmonary resuscitation.

20. These items bring the electricity directly beneath the skin. Since the high resistance of the skin is bypassed, a relatively small voltage can cause a fatal shock.

Chapter 13 Problems

1. to the left, toward the positively charged object
2. straight down.
3. 480 C
4. 1,320 J
5. $1.08
6. 2,200 W
7. at least 50 A
8. 16
9. 97,000 J. The cost of energy from the battery would be $51 per megajoule, 2,600 times more expensive than energy from the wall socket.
10. 6.4 A; 17 Ω
11. 2.7 A
12. 75 W
13. a. 1 A
    b. 0.31 A
    c. 0.39 A
    d. 1.3 A
    e. 0.29 A
    f. 0.55 A
14. 235 Ω

Chapter 14 Discussion Questions

2. Musical instruments tend to produce sounds that have wavelengths similar in size to the instrument. Thus, a trumpet would produce shorter wavelengths than the larger tuba. Of course, shorter wavelengths mean higher frequency and thus a higher pitch.

3. As long as the water is much deeper than the amplitude of the wave, the water does not move horizontally. Of course, this is easily proved by watching something floating in the water. It will just bob up and down as the wave passes.

6. We know light waves are transverse because they can be polarized with Polaroid sunglasses, for instance. Longitudinal waves cannot be polarized.

Chapter 14 Problems

1. 4 s
2. 800 Hz
3. wavelength, 6 m; amplitude, 1 m
4. 15 m (Incidentally, such high waves are close to the record height for waves on the ocean.)
5. 0.66 m; 2.87 m
6. 423 m

Chapter 15 Discussion Questions

1. Sound waves in air are longitudinal waves, which cannot be polarized.

3. There is no air in outer space; thus, there cannot be any sound because there cannot be any vibrations of air molecules.

6. You would hear a higher pitch than a stationary listener as you ran toward the trumpet player and a lower pitch as you ran away.

14. The principal difference between ultrasound and ordinary sound is that ultrasound is higher in frequency, beyond the range of the human ear.

16. The main advantage of ultrasonic imaging over x-ray imaging is that it is much safer. There are no known hazards of low-intensity ultrasonic waves. In addition, some soft tissue structures show up better on ultrasound images. The disadvantages are that the image quality is generally poorer, and many organs cannot be satisfactorily imaged with ultrasound.

Chapter 15 Problems

1. 688 m
2. 4 μs
3. 1.3 s
4. 0.3 s (enough to cause a noticeable distortion in the music for all but the slowest songs)
5. 75 dB
6. 50 dB (still a quiet sound—something like a quiet conversation)
7. c
8. 200 W

Chapter 16 Discussion Questions

1. radio waves, microwaves, infrared rays, visible light, ultraviolet light, x rays, and gamma rays

2. An x ray has a much higher frequency.

3. The electromagnetic field vibrates in only one direction in polarized light. In ordinary unpolarized light, the electromagnetic field vibrates in all directions perpendicular to the direction in which the light is moving.

5. The glare from a road and the glare from a window are both polarized, but not in the same direction. Thus, Polaroid sunglasses cannot stop both types of glare at the same time. If you hold your head level, the glasses will stop the glare from horizontal surfaces but not from vertical surfaces.

7. Since the refractive index of water is much larger than that of air, the cornea loses almost all of its converging power underwater. Thus, the eye cannot form a sharp image on the retina when it is in contact with water.

9. Diamond has a much higher index of refraction than glass; thus, a diamond bends the light more and therefore sparkles more.

15. Color is only sensed by the cones in the retina, which are much less light sensitive than the rods. Thus, at twilight most of the signal going to the brain comes from the rods, so very little information about color is sent.

Chapter 16 Problems

1. 90 cm (however, it must be hung at exactly the right height on the wall); no

3. The key to this problem is to realize that the image in the corner is a double reflection—the rays bounce twice, once on each mirror, before returning to you.

4. 125,000,000 m/s
5. 1.7
6. 28°
7. 7.5 cm
8. 7.5 cm from the lens on the other side; 1.5 cm; inverted
9. 12 cm from the lens on the same side as the object; 1.2 cm; upright

Chapter 17 Discussion Questions

2. Isotopes are atoms of the same element that have different numbers of neutrons.

3. $^{29}_{14}Si$ is the isotope of silicon with 15 neutrons. Of course, this isotope has 14 protons, as do all silicon isotopes.

5. The half-life is the time it takes for one-half of a radioactive isotope to decay.

6. No. In six days three-quarters of the isotope will have decayed.

8. Screens make the film more sensitive to x rays, thus cutting the radiation exposure to the patient. Screens work by converting the x rays into visible light, which blackens the film more effectively than x rays do.

9. Grids reduce the amount of scattered x rays hitting the film, thus reducing the amount of fogging.

10. Rotating the gamma-ray beam lets it strike the patient from many directions. Thus, the dose to the healthy tissue can be held to a fraction of the dose to the tumor.

Chapter 17 Problems

1. $^{197}_{79}Au$ (gold-197, the only naturally occurring stable isotope of gold)

2. $^{198}_{80}Hg$ (mercury-198)

3. $^{234}_{90}Th$ (thorium-234)

4. beta decay (Specifically, it emits a position [$^{0}_{1}e$].)

5. $^{60}_{28}Ni$ (nickel-60) (The two gamma rays have neither charge nor mass.)

6. a neutron ($^{1}_{0}n$)

7. about 17,000 years ago

8. 23,000 (That is, the source would have a strength of 23,000 Bq [630 nCi].)

9. 12 hr

10. between three and four half-lives, around 35 days

11. approximately 80 min

12. Not more than 10. If possible, the exposure should be held below 1 hour per week. No one but a qualified radiation worker should work in the room.

13. 33 μGy/hr (3.3 mrad/hr). If we want to keep the dose below the allowed dose of 1 mGy (100 mrad) per week, the worker should not work by the fence for more than 30 hours per week. However, there would have to be some very good reason for exposing one worker to so much radiation. Three hours a week would be better, since that would keep the worker's dose below 10% of the allowed dose.

14. 160 μGy/hr. No radioactive source should get "just routine care"; however, this source is so large that it should be handled only with the greatest care and only with very good shielding.

Index

Units of measurement are listed in boldface type.